T0226342

Lecture Notes in Computer Science   10765

Commenced Publication in 1973
Founding and Former Series Editors:
Gerhard Goos, Juris Hartmanis, and Jan van Leeuwen

Editorial Board

David Hutchison
  Lancaster University, Lancaster, UK
Takeo Kanade
  Carnegie Mellon University, Pittsburgh, PA, USA
Josef Kittler
  University of Surrey, Guildford, UK
Jon M. Kleinberg
  Cornell University, Ithaca, NY, USA
Friedemann Mattern
  ETH Zurich, Zurich, Switzerland
John C. Mitchell
  Stanford University, Stanford, CA, USA
Moni Naor
  Weizmann Institute of Science, Rehovot, Israel
C. Pandu Rangan
  Indian Institute of Technology Madras, Chennai, India
Bernhard Steffen
  TU Dortmund University, Dortmund, Germany
Demetri Terzopoulos
  University of California, Los Angeles, CA, USA
Doug Tygar
  University of California, Berkeley, CA, USA
Gerhard Weikum
  Max Planck Institute for Informatics, Saarbrücken, Germany

More information about this series at http://www.springer.com/series/7407

Ljiljana Brankovic · Joe Ryan
William F. Smyth (Eds.)

Combinatorial Algorithms

28th International Workshop, IWOCA 2017
Newcastle, NSW, Australia, July 17–21, 2017
Revised Selected Papers

 Springer

Editors
Ljiljana Brankovic ⓘ
University of Newcastle Australia
Callaghan, NSW
Australia

Joe Ryan ⓘ
University of Newcastle Australia
Callaghan, NSW
Australia

William F. Smyth
McMaster University
Hamilton, ON
Canada

ISSN 0302-9743   ISSN 1611-3349   (electronic)
Lecture Notes in Computer Science
ISBN 978-3-319-78824-1   ISBN 978-3-319-78825-8   (eBook)
https://doi.org/10.1007/978-3-319-78825-8

Library of Congress Control Number: 2018939453

LNCS Sublibrary: SL1 – Theoretical Computer Science and General Issues

Printed on acid-free paper

This Springer imprint is published by the registered company Springer International Publishing AG
part of Springer Nature
The registered company address is: Gewerbestrasse 11, 6330 Cham, Switzerland

Preface

The 28th International Workshop on Combinational Algorithms (IWOCA 2017) was held during July 17–21, 2017, in Newcastle, Australia. This meeting was dedicated to the memory of Emeritus Professor Mirka Miller, who sadly passed away in January 2016. Mirka was a founding member of AWOCA, as IWOCA was then known, and attended many gatherings of this workshop since its initiation in 1988. Mirka was present at AWOCA 2006 at which it was decided to change the name to IWOCA to better reflect the growing international participation and interest. Mirka was a member of the IWOCA Steering Committee from its beginning until her death.

Obituaries have appeared in the Australian Mathematical Society Gazette (see http://www.austms.org.au/Publ/Gazette/2016/July16/ObitMiller.pdf) as well as the *AKCE International Journal of Graphs and Combinatorics* and the *Bulletin of The Institute of Combinatorics and Its Applications*. In 2017 a special issue of the *Australasian Journal of Combinatorics* was dedicated to Mirka. The introduction is available at https://ajc.maths.uq.edu.au/pdf/69/ajc v69 p292.pdf.

For IWOCA 2017, calls for papers were distributed widely around the world and participants came from Austria, Canada, China, Czech Republic, France, Germany, India, Italy, Japan, Singapore, Spain, UK, as well as Australia. The invited speakers were Martin Baca (Kosice, Slovakia), Henning Fernau (Trier, Germany), Costas Iliopolous (UK), and Diane Donovan and Jennifer Seberry both of Australia. All five invited papers are included in these LNCS proceedings. There were 55 contributed submissions. Each contributed submission was reviewed by at least two, though generally three Program Committee members. The committee decided to accept 30 papers.

We thank the sponsors of IWOCA 2017: CARMA, School of Electrical Engineering and Computing, EATCS, and Springer. We acknowledge and thank all members of the Program Committee and the Organizing Committee for their commitment to IWOCA and for their excellent and timely work. Last but not least, we thank EasyChair that we used for organizing the conference, as well the informal proceedings and these LNCS proceedings, which made our job so much easier.

July 2017

Ljiljana Brankovic
Joe Ryan
William F. Smyth

Organization

Program Committee

Donald Adjeroh	West Virginia University, USA
Hideo Bannai	Kyushu University, Japan
Cristina Bazgan	LAMSADE, Universite Paris-Dauphine, France
Ljiljana Brankovic	University of Newcastle, Australia
Pino Caballero-Gil	DEIOC, University of La Laguna, Spain
Charles Colbourn	Arizona State University, USA
Tatjana Davidovic	Mathematical Institute of Serbian Academy of Sciences and Arts, Serbia
Vlad Estivill-Castro	Griffith University, Australia
Michael Fellows	University of Bergen, Norway
Andrea Semanicova-Fenovcikova	Technical University Kosice, Slovakia
Gabriele Fici	Università di Palermo, Italy
Dalibor Froncek	University of Minnesota, USA
Serge Gaspers	University of New South Wales and Data61, CSIRO, Australia
Pinar Heggernes	University of Bergen, Norway
Seok-Hee Hong	University of Sydney, Australia
Peter Horak	University of Washington, USA
Jesper Jansson	The Hong Kong Polytechnic University, Hong Kong, SAR China
Ralf Klasing	CNRS and University of Bordeaux, France
Christian Komusiewicz	Philipps-Universität Marburg, Germany
Jan Kratochvil	Charles University, Czech Republic
Thierry Lecroq	University of Rouen, France
Zsuzsanna Liptak	University of Verona, Italy
Petra Mutzel	University of Dortmund, Germany
Kunsoo Park	Seoul National University, South Korea
Vangelis Paschos	LAMSADE, University Paris-Dauphine, France
Christophe Paul	CNRS - LIRMM, France
Solon Pissis	King's College London, UK
Simon Puglisi	University of Helsinki, Finland
Frances Rosamond	University of Bergen, Norway
Gordon Royle	University of Western Australia, Australia
Frank Ruskey	University of Victoria, Canada
Joe Ryan	The University of Newcastle, Australia
Oliver Schaudt	University of Cologne, Germany
Jamie Simpson	Curtin University, Australia

Michiel Smid	Carleton University, Canada	
William F. Smyth	McMaster University, Canada	
Sue Whitesides	University of Victoria, Canada	
Christos Zaroliagis	Computer Technology Institute and University of Patras, Greece	

Additional Reviewers

Ahn, Hee-Kap
Bampas, Evangelos
Basavaraju, Manu
Belazzougui, Djamal
Biniaz, Ahmad
Bredereck, Robert
Bökler, Fritz
Castiglione, Giuseppa
Charalampopoulos, Panagiotis
Chen, Jiehua
Cicalese, Ferdinando
De Oliveira Oliveira, Mateus
Droschinsky, Andre

French, Tim
Fujita, Takahiro
Gargano, Luisa
Grabowski, Szymon
Hsieh, Sun-Yuan
Jaffke, Lars
Kontogiannis, Spyros
Kuo, Jyhmin
Lee, Chia-Wei
Lee, Edward J.
Lefebvre, Arnaud
Macgillivray, Gary
Mihalak, Matus
Myrvold, Wendy
Patrignani, Maurizio

Prezza, Nicola
Rahman, Md. Saidur
Rechnitzer, Andrew
Rutter, Ignaz
Ryan, Patrick
Rümmele, Stefan
Saffidine, Abdallah
Salson, Mikaël
Schittekat, Patrick
Shur, Arseny
Soltys, Michael
Toubaline, Sonia
Umboh, Seeun William
Wasa, Kunihiro

Contents

Heuristics

Mixed Integer Programming

Polynomial Algorithms

Privacy

String Algorithms

Invited Papers

Entire H-irregularity Strength of Plane Graphs

Martin Bača[1(✉)], Nurdin Hinding[2], Aisha Javed[3],
and Andrea Semaničová-Feňovčíková[1]

[1] Department of Applied Mathematics and Informatics,
Technical University, Košice, Slovakia
martin.baca@tuke.sk
[2] Faculty of Mathematics and Natural Sciences,
Hasanuddin University, Makassar, Indonesia
[3] Abdus Salam School of Mathematical Sciences, GC University, Lahore, Pakistan

Abstract. We investigate an entire H-irregularity strength of plane graphs as a modification of the well-known total and entire face irregularity strengths. Estimations on this new graph characteristic are obtained and determined the precise values for graphs from two families of plane graphs to demonstrate that the obtained bounds are tight.

Keywords: Irregularity strength · Entire face irregularity strength
Entire H-irregularity strength

1 Introduction

We consider finite undirected graphs without loops and multiple edges. Denote by $V(G)$ and $E(G)$ the set of vertices and the set of edges of a graph G, respectively. By a *labeling* we mean any mapping that maps a set of graph elements to a set of numbers (usually positive integers), called *labels*. If the domain is the vertex set and the edge set then we call the labeling *total labeling*.

For a given total k-labeling $\phi : V(G) \cup E(G) \to \{1, 2, \ldots, k\}$ the associated total vertex-weight of a vertex x is $wt_\phi(x) = \phi(x) + \sum_{xy \in E(G)} \phi(xy)$ and the associated total edge-weight of an edge xy is $wt_\phi(xy) = \phi(x) + \phi(xy) + \phi(y)$. In [8] a total k-labeling ϕ is defined to be an *edge irregular total k-labeling* of a graph G if for every two different edges xy and $x'y'$ of G there is $wt_\phi(xy) \neq wt_\phi(x'y')$ and to be a *vertex irregular total k-labeling* of G if for every two distinct vertices x and y of G there is $wt_\phi(x) \neq wt_\phi(y)$.

The minimum k for which the graph G has an edge irregular total k-labeling is called the *total edge irregularity strength* of the graph G and is denoted by tes(G). Analogously, the *total vertex irregularity strength* of G, denoted by tvs(G), is the minimum k for which there exists a vertex irregular total k-labeling of G.

Ivančo and Jendrol' [13] posed a conjecture that for arbitrary graph G different from K_5 and maximum degree $\Delta(G)$, tes$(G) = \max\left\{ \lceil (|E(G)| + 2)/3 \rceil, \right.$

© Springer International Publishing AG, part of Springer Nature 2018
L. Brankovic et al. (Eds.): IWOCA 2017, LNCS 10765, pp. 3–12, 2018.
https://doi.org/10.1007/978-3-319-78825-8_1

$\lceil (\Delta(G) + 1)/2 \rceil$ }. This conjecture has been verified for complete graphs and complete bipartite graphs in [14,15], for the categorical product of two cycles in [3], for generalized Petersen graphs in [12], for generalized prisms in [10], for corona product of a path with certain graphs in [17] and for large dense graphs with $(|E(G)| + 2)/3 \leq (\Delta(G) + 1)/2$ in [11].

The bounds for the total vertex irregularity strength of a graph G with minimum degree $\delta(G)$ are given in [8] as follows

$$\left\lceil \frac{|V(G)| + \delta(G)}{\Delta(G) + 1} \right\rceil \leq \text{tvs}(G) \leq |V(G)| + \Delta(G) - 2\delta(G) + 1. \qquad (1)$$

Przybyło in [18] proved that $\text{tvs}(G) < 32|V(G)|/\delta(G) + 8$ in general and $\text{tvs}(G) < 8|V(G)|/r + 3$ for r-regular graphs. This was then improved by Anholcer et al. in [4] that $\text{tvs}(G) \leq 3\lceil |V(G)|/\delta(G) \rceil + 1 \leq 3|V(G)|/\delta(G) + 4$.

Recently Majerski and Przybyło in [16] based on a random ordering of the vertices proved that if $\delta(G) \geq (|V(G)|)^{0.5} \ln |V(G)|$, then $\text{tvs}(G) \leq (2 + o(1))|V(G)|/\delta(G) + 4$. The exact values for the total vertex irregularity strength for circulant graphs and unicyclic graphs are determined in [1,2,5], respectively.

Motivated by total irregularity strengths and a recent paper on entire colouring of plane graphs [19], Bača et al. in [6] studied irregular labelings of plane graphs with restrictions placed on the weights of faces. For a 2-connected plane graph $G = (V, E, F)$ with the face set $F(G)$, they defined a labeling $\varphi : V(G) \cup E(G) \cup F(G) \rightarrow \{1, 2, \ldots, k\}$ to be an *entire k-labeling*. The *weight of a face f* under an entire k-labeling φ, $w_\varphi(f)$, is the sum of labels carried by that face and all the edges and vertices surrounding it. An entire k-labeling φ is defined to be a *face irregular entire k-labeling* of the plane graph G if for every two different faces f and g of G there is $w_\varphi(f) \neq w_\varphi(g)$. The *entire face irregularity strength*, denoted by efs(G), of a plane graph G is the smallest integer k such that G has a face irregular entire k-labeling.

Bača et al. in [6] proved that for every 2-connected plane graph G with n_i i-sided faces, $i \geq 3$, efs$(G) \geq \lceil (2a + n_3 + n_4 + \cdots + n_b)/(2b + 1) \rceil$, where $a = \min\{i : n_i \neq 0\}$ and $b = \max\{i : n_i \neq 0\}$. In [7] there is described a face irregular entire 2-labeling of octahedron and it proves that this lower bound is tight. In the case if a 2-connected plane graph contains only one largest face, $n_b = 1$, and $c = \max\{i : n_i \neq 0, i < b\}$, then in [6] there is proved that efs$(G) \geq \lceil (2a + |F(G)| - 1)/(2c + 1) \rceil$. In [9] are estimated the lower and upper bounds of the entire face irregularity strength for the disjoint union of multiple copies of a plane graph and proved the sharpness of the lower bound in two cases.

An *edge-covering* of G is a family of subgraphs $H_1, H_2, \ldots, H_t$ such that each edge of $E(G)$ belongs to at least one of the subgraphs H_i, $i = 1, 2, \ldots, t$. Then it is said that G admits an $(H_1, H_2, \ldots, H_t)$-*(edge) covering*. If every subgraph H_i is isomorphic to a given graph H, then the graph G admits an H-*covering*. Note, that in this case all subgraphs of G isomorphic to H must be in the H-covering.

Let G be a plane graph admitting H-covering. For the subgraph $H \subseteq G$ under the entire k-labeling φ, we define the associated H-weight as

$$w_\varphi(H) = \sum_{v \in V(H)} \varphi(v) + \sum_{e \in E(H)} \varphi(e) + \sum_{f \in F(H)} \varphi(f).$$

An entire k-labeling φ is called an H-*irregular entire k-labeling* of the plane graph G if for every two different subgraphs H' and H'' isomorphic to H there is $w_\varphi(H') \neq w_\varphi(H'')$. The *entire H-irregularity strength* of a plane graph G, denoted $\mathrm{Ehs}(G, H)$, is the smallest integer k such that G has an H-irregular entire k-labeling. If H is isomorphic to the cycle C_n, then the C_n-irregular entire k-labeling is isomorphic to the face irregular entire k-labeling of a plane graph G with only n-sided faces and thus the entire C_n-irregularity strength of the plane graph G is equivalent to the entire face irregularity strength, that is $\mathrm{Ehs}(G, C_n) = \mathrm{efs}(G)$.

The main aim of the presented paper is to obtain estimations on the parameter $\mathrm{Ehs}(G, H)$ and determine the precise values of the entire H-irregularity strength for two families of plane graphs namely, ladders and fan graphs to demonstrate that the obtained bounds are tight.

2   Lower Bounds for the Entire H-irregularity Strength

Next theorem provides a lower bound on the parameter $\mathrm{Ehs}(G, H)$.

Theorem 1. *Let $G = (V, E, F)$ be a 2-connected plane graph admitting an H-covering given by t subgraphs isomorphic to H. Then*

$$\mathrm{Ehs}(G, H) \geq \left\lceil 1 + \frac{t - 1}{|V(H)| + |E(H)| + |F(H)|} \right\rceil.$$

Proof. Let φ be an H-irregular entire k-labeling of a 2-connected plane graph $G = (V, E, F)$ admitting an H-covering given by t subgraphs isomorphic to H with $\mathrm{Ehs}(G, H) = k$. The smallest weight of a subgraph H under the entire k-labeling is at least $|V(H)| + |E(H)| + |F(H)|$ and the largest H-weight admits the value at most $(|V(H)| + |E(H)| + |F(H)|)k$. It means that for every subgraph H we have

$$|V(H)| + |E(H)| + |F(H)| \leq w_\varphi(H) \leq (|V(H)| + |E(H)| + |F(H)|)k. \quad (2)$$

Since H-covering of G is given by t subgraphs thus from (2) it follows that

$$|V(H)| + |E(H)| + |F(H)| + t - 1 \leq (|V(H)| + |E(H)| + |F(H)|)k$$

and

$$k \geq \left\lceil 1 + \frac{t - 1}{|V(H)| + |E(H)| + |F(H)|} \right\rceil.$$

If H is isomorphic to the cycle C_n and G is a plane graph with only n-sided faces then from Theorem 1 follows the lower bound on the entire face irregularity strength given in [6].

Corollary 1. *Let $G = (V, E, F)$ be a plane graph having exactly t n-sided faces. Then*
$$\text{Ehs}(G, C_n) = \text{efs}(G) \geq \left\lceil \frac{2n + t}{2n + 1} \right\rceil.$$

Next theorem gives the exact value of the entire L_m-irregularity strength for ladders L_n, $2 \leq m \leq n$ and it proves that the lower bound in Theorem 1 is tight.

Theorem 2. *Let $L_n \cong P_n \square P_2$, $n \geq 2$, be a ladder admitting L_m covering, where m is a positive integer, $2 \leq m \leq n$. Then*
$$\text{Ehs}(L_n, L_m) = \left\lceil \frac{5m+n-3}{6m-3} \right\rceil.$$

Proof. Let $L_n \cong P_n \square P_2$, $n \geq 2$, be a ladder with the vertex set $V(L_n) = \{u_i, v_i : i = 1, 2, \ldots, n\}$ and the edge set $E(L_n) = \{u_i u_{i+1}, v_i v_{i+1} : i = 1, 2, \ldots, n - 1\} \cup \{u_i v_i : i = 1, 2, \ldots, n\}$. The ladder L_n contains $(n - 1)$ 4-sided faces and one external $2n$-sided face. Denote by f_i the 4-sided face surrounded by vertices $u_i, u_{i+1}, v_i, v_{i+1}$ and edges $u_i u_{i+1}, v_i v_{i+1}, u_i v_i, u_{i+1} v_{i+1}$, for $i = 1, 2, \ldots, n - 1$, and denote by f_{ext} the external $2n$-sided face. Clearly, for every m, $2 \leq m \leq n$, the ladder L_n admits a L_m-covering with exactly $n - m + 1$ subgraphs. According to Theorem 1 we have that
$$\text{Ehs}(L_n, L_m) \geq \left\lceil \frac{5m+n-3}{6m-3} \right\rceil.$$

Put $k = \lceil (5m + n - 3)/(6m - 3) \rceil$. To show that k is an upper bound for the entire L_m-irregularity strength of L_n we describe an entire k-labeling $\varphi_m : V(L_n) \cup E(L_n) \cup F(L_n) \to \{1, 2, \ldots, k\}$, $m = 2, 3, \ldots, n$, as follows:

$$\varphi_m(u_i) = \left\lceil \frac{i+3m-2}{6m-3} \right\rceil \qquad \text{for } i = 1, 2, \ldots, n,$$

$$\varphi_m(v_i) = \left\lceil \frac{i+5m-3}{6m-3} \right\rceil \qquad \text{for } i = 1, 2, \ldots, n,$$

$$\varphi_m(u_i u_{i+1}) = \left\lceil \frac{i+2m-1}{6m-3} \right\rceil \qquad \text{for } i = 1, 2, \ldots, n - 1,$$

$$\varphi_m(v_i v_{i+1}) = \left\lceil \frac{i+4m-2}{6m-3} \right\rceil \qquad \text{for } i = 1, 2, \ldots, n - 1,$$

$$\varphi_m(u_i v_i) = \left\lceil \frac{i+m-1}{6m-3} \right\rceil \qquad \text{for } i = 1, 2, \ldots, n$$

$$\varphi_m(f_i) = \left\lceil \frac{i}{6m-3} \right\rceil \qquad \text{for } i = 1, 2, \ldots, n - 1,$$

$$\varphi_m(f_{ext}) = 1.$$

We can see that all vertex, edge and face labels are at most k. Every ladder L_m^j, $j = 1, 2, \ldots, n - m + 1$, in L_n has the vertex set $V(L_m^j) = \{u_{j+i}, v_{j+i} :$

$i = 0, 1, \ldots, m - 1\}$, the edge set $E(L_m^j) = \{u_{j+i}v_{j+i} : i = 0, 1, \ldots, m - 1\} \cup \{u_{j+i}u_{j+i+1}, v_{j+i}v_{j+i+1} : i = 0, 1, \ldots, m - 2\}$ and the face set $F(L_m^j) = \{f_{j+i} : i = 0, 1, \ldots, m - 2\}$. One can see that every edge of L_n belongs to at least one ladder L_m^j if $m = 2, 3, \ldots, n$.

For the L_m-weight of the ladder L_m^j, $j = 1, 2, \ldots, n - m + 1$, under the total labeling φ_m, $m = 2, 3, \ldots, n$, we get

$$w_{\varphi_m}(L_m^j) = \sum_{v \in V(L_m^j)} \varphi_m(v) + \sum_{e \in E(L_m^j)} \varphi_m(e) + \sum_{f \in F(L_m^j)} \varphi_m(f).$$

Since vertex labels, edge labels and face labels form non-decreasing sequences thus it is enough to prove that $w_{\varphi_m}(L_m^j) < w_{\varphi_m}(L_m^{j+1})$, $j = 1, 2, \ldots, n - m$. In fact, for the difference of weights of subgraphs L_m^{j+1} and L_m^j for $j = 1, 2, \ldots, n - m$ we get

$$
\begin{aligned}
w_{\varphi_m}(L_m^{j+1}) - w_{\varphi_m}(L_m^j) = & \varphi_m(u_{j+m}) + \varphi_m(u_{j+m-1}u_{j+m}) + \varphi_m(u_{j+m}v_{j+m}) \\
& + \varphi_m(v_{j+m}) + \varphi_m(v_{j+m-1}v_{j+m}) + \varphi_m(f_{j+m-1}) \\
& - \varphi_m(u_j) - \varphi_m(u_ju_{j+1}) - \varphi_m(u_jv_j) \\
& - \varphi_m(v_j) - \varphi_m(v_jv_{j+1}) - \varphi_m(f_j) \\
= & \left\lceil \tfrac{j+4m-2}{6m-3} \right\rceil + \left\lceil \tfrac{j+3m-2}{6m-3} \right\rceil + \left\lceil \tfrac{j+2m-1}{6m-3} \right\rceil + \left\lceil \tfrac{j+6m-3}{6m-3} \right\rceil \\
& + \left\lceil \tfrac{j+5m-3}{6m-3} \right\rceil + \left\lceil \tfrac{j+m-1}{6m-3} \right\rceil - \left\lceil \tfrac{j+3m-2}{6m-3} \right\rceil - \left\lceil \tfrac{j+2m-1}{6m-3} \right\rceil \\
& - \left\lceil \tfrac{j+m-1}{6m-3} \right\rceil - \left\lceil \tfrac{j+5m-3}{6m-3} \right\rceil - \left\lceil \tfrac{j+4m-2}{6m-3} \right\rceil - \left\lceil \tfrac{j}{6m-3} \right\rceil \\
= & \left\lceil \tfrac{j+6m-3}{6m-3} \right\rceil - \left\lceil \tfrac{j}{6m-3} \right\rceil = 1.
\end{aligned}
$$

Thus $w_{\varphi_m}(L_m^j) < w_{\varphi_m}(L_m^{j+1})$, for $m = 2, 3, \ldots, n$ and $j = 1, 2, \ldots, n - m$, and the labeling φ_m is a desired L_m-irregular entire k-labeling.

Let G be a plane graph admitting H-covering. By the symbol $\mathbb{H}_r^S = (H_1^S, H_2^S, \ldots, H_r^S)$, we denote the set of all subgraphs of G isomorphic to H such that the graph S, $S \ncong H$, is their maximum common subgraph. Thus $V(S) \subset V(H_i^S)$, $E(S) \subset E(H_i^S)$ and $F(S) \subset F(H_i^S)$ for every $i = 1, 2, \ldots, r$. Next theorem provides another lower bound on the entire H-irregularity strength.

Theorem 3. *Let $G = (V, E, F)$ be a 2-connected plane graph admitting an H-covering. Let S_i, $i = 1, 2, \ldots, s$, be all subgraphs of G such that S_i is a maximum common subgraph of r_i, $r_i \geq 2$, subgraphs of G isomorphic to H. Then*

$$
\mathrm{Ehs}(G, H) \geq \max \left\{ \left\lceil 1 + \tfrac{r_1 - 1}{|V(H/S_1)| + |E(H/S_1)| + |F(H/S_1)|} \right\rceil, \right.
$$

$$
\left. \ldots, \left\lceil 1 + \tfrac{r_s - 1}{|V(H/S_s)| + |E(H/S_s)| + |F(H/S_s)|} \right\rceil \right\}.
$$

Proof. Let G be a plane graph admitting an H-covering. Let $\mathbb{H}_{r_i}^{S_i}$, $i = 1, 2, \ldots, s$, be the set of all subgraphs $H_1^{S_i}, H_2^{S_i}, \ldots, H_{r_i}^{S_i}$, where each of them is isomorphic to H, and S_i is their maximum common subgraph. Let ψ be an H-irregular entire k-labeling of G. Clearly, the $H_j^{S_i}$-weights

$$w_\psi(H_j^{S_i}) = \sum_{v \in V(S_i)} \psi(v) + \sum_{e \in E(S_i)} \psi(e) + \sum_{f \in F(S_i)} \psi(f) + \sum_{v \in V(H_j^{S_i}/S_i)} \psi(v)$$

$$+ \sum_{e \in E(H_j^{S_i}/S_i)} \psi(e) + \sum_{f \in F(H_j^{S_i}/S_i)} \psi(f),$$

for $j = 1, 2, \ldots, r_i$, are all distinct and each of them contains the value

$$\sum_{v \in V(S_i)} \psi(v) + \sum_{e \in E(S_i)} \psi(e) + \sum_{f \in E(S_i)} \psi(f).$$

The largest among these $H_j^{S_i}$-weights cannot be less than

$$\sum_{v \in V(S_i)} \psi(v) + \sum_{e \in E(S_i)} \psi(e) + \sum_{f \in E(S_i)} \psi(f)$$
$$+ |V(H/S_i)| + |E(H/S_i)| + |F(H/S_i)| + r_i - 1.$$

This weight is the sum of at most $|V(H/S_i)| + |E(H/S_i)| + |F(H/S_i)|$ labels. So at least one label is at least $\lceil 1 + (r_i - 1)/(|V(H/S_i)| + |E(H/S_i)| + |F(H/S_i)|) \rceil$, for $i = 1, 2, \ldots, s$. Thus for the entire H-irregularity strength of a plane graph G we get the desired result.

The lower bound from Theorem 3 is tight for fan graphs.

Theorem 4. *Let F_n, $n \geq 2$, be a fan on $n + 1$ vertices admitting F_m-covering, where m is a positive integer, $2 \leq m \leq n$. Then*

$$\mathrm{Ehs}(F_n, F_m) = \left\lceil \frac{3m+n-2}{4m-2} \right\rceil.$$

Proof. Let F_n, $n \geq 2$, be a *fan graph* obtained by joining all vertices of a path P_n to a further vertex, called the centre. Thus F_n contains $n + 1$ vertices, say, $u_1, u_2, \ldots, u_n, w$, $2n - 1$ edges, say, $u_i w$, $i = 1, 2, \ldots, n$, and $u_i u_{i+1}$, $i = 1, 2, \ldots, n - 1$, $(n - 1)$ 3-sided faces, say, f_i, $i = 1, 2, \ldots, n - 1$, and finally the external $(n + 1)$-sided face, say, f_{ext}. The fan graph F_n, $n \geq 2$, admits F_m-covering with exactly $n - m + 1$ fan graphs F_m, $2 \leq m \leq n$. So every fan graph F_m^j, $j = 1, 2, \ldots, n - m + 1$, in F_n has the vertex set $V(F_m^j) = \{u_{j+i}, w : i = 0, 1, \ldots, m - 1\}$, the edge set $E(F_m^j) = \{u_{j+i} w : i = 0, 1, \ldots, m - 1\} \cup \{u_{j+i} u_{j+i+1}, : i = 0, 1, \ldots, m - 2\}$ and the face set $F(F_m^j) = \{f_{j+i} : i = 0, 1, \ldots, m - 2\}$. Evidently every edge of F_n belongs to at least one fan graph F_m^j if $m = 2, 3, \ldots, n$.

Since every fan graph F_m^j contains the vertex w as the maximum common subgraph it follows that $V(S_1) = w$, $r_1 = n - m + 1$ and from Theorem 3 we have

$$\text{Ehs}(F_n, F_m) \geq \left\lceil 1 + \frac{r_1 - 1}{|V(H/S_1)| + |E(H/S_1)| + |F(H/S_1)|} \right\rceil = \left\lceil 1 + \frac{n-m}{4m-2} \right\rceil = \left\lceil \frac{3m+n-2}{4m-2} \right\rceil.$$

Let $k = \lceil (3m + n - 2)/(4m - 2) \rceil$. To show that k is an upper bound for entire F_m-irregularity strength of F_n it suffices to prove the existence of an optimal entire k-labeling $\psi_m : V(F_n) \cup E(F_n) \cup F(F_n) \to \{1, 2, \ldots, k\}$. For $m = 2, 3, \ldots, n$, we construct the function ψ_m in the following way:

$$\psi_m(u_i) = \left\lceil \frac{i+2m-1}{4m-2} \right\rceil \qquad \text{for } i = 1, 2, \ldots, n,$$

$$\psi_m(u_i u_{i+1}) = \left\lceil \frac{i+3m-1}{4m-2} \right\rceil \qquad \text{for } i = 1, 2, \ldots, n-1,$$

$$\psi_m(u_i w) = \left\lceil \frac{i+m-1}{4m-2} \right\rceil \qquad \text{for } i = 1, 2, \ldots, n$$

$$\psi_m(f_i) = \left\lceil \frac{i}{4m-2} \right\rceil \qquad \text{for } i = 1, 2, \ldots, n-1,$$

$$\psi_m(f_{ext}) = 1.$$

We can see that all vertex, edge and face labels are at most k. For the F_m-weight of the fan graph F_m^j, $j = 1, 2, \ldots, n - m + 1$, under the total labeling ψ_m, $m = 2, 3, \ldots, n$, we get

$$w_{\psi_m}(F_m^j) = \sum_{v \in V(F_m^j)} \psi_m(v) + \sum_{e \in E(F_m^j)} \psi_m(e) + \sum_{f \in F(F_m^j)} \psi_m(f).$$

Since vertex labels, edge labels and face labels form non-decreasing sequences it follows that it is enough to consider the difference of weights of subgraphs F_m^{j+1} and F_m^j for $j = 1, 2, \ldots, n - m$. Observe that for every $m = 2, 3, \ldots, n$ and $j = 1, 2, \ldots, n - m$ we get

$$w_{\psi_m}(F_m^{j+1}) - w_{\psi_m}(F_m^j) = \psi_m(u_{j+m-1} u_{j+m}) + \psi_m(u_{j+m}) + \psi_m(u_{j+m} w)$$
$$+ \psi_m(f_{j+m-1}) - \psi_m(u_j) - \psi_m(u_j u_{j+1}) - \psi_m(u_j w)$$
$$- \psi_m(f_j)$$

$$= \left\lceil \frac{j+4m-2}{4m-2} \right\rceil + \left\lceil \frac{j+3m-1}{4m-2} \right\rceil + \left\lceil \frac{j+2m-1}{4m-2} \right\rceil + \left\lceil \frac{j+m-1}{4m-2} \right\rceil$$
$$- \left\lceil \frac{j+2m-1}{4m-2} \right\rceil - \left\lceil \frac{j+3m-1}{4m-2} \right\rceil - \left\lceil \frac{j+m-1}{4m-2} \right\rceil - \left\lceil \frac{j}{4m-2} \right\rceil$$

$$= \left\lceil \frac{j+4m-2}{4m-2} \right\rceil - \left\lceil \frac{j}{4m-2} \right\rceil = 1.$$

In fact, the labeling ψ_m has been chosen in such a way that for every $m = 2, 3, \ldots, n$ and $j = 1, 2, \ldots, n - m$ $w_{\psi_m}(F_m^j) < w_{\psi_m}(F_m^{j+1})$. So the labeling ψ_m has the required properties of F_m-irregular entire k-labeling. This concludes the proof.

3   Upper Bound for the Entire H-irregularity Strength

Next theorem gives an upper bound of the parameter $\mathrm{Ehs}(G, H)$ and shows that this graph invariant is always finite.

Theorem 5. *Let $G = (V, E, F)$ be a 2-connected plane graph admitting an H-covering given by t subgraphs isomorphic to H. Then*

$$\mathrm{Ehs}(G, H) \leq 2^{|F(G)|-2}.$$

Proof. Let G be a 2-connected plane graph admitting H covering given by subgraphs $H_1, H_2, \ldots, H_t$. Let us denote the internal faces of G arbitrarily by the symbols $f_1, f_2, \ldots, f_{|F(G)|-1}$. We define a total $2^{|F(G)|-2}$-labeling φ of G in the following way.

$$\begin{aligned}
\varphi(v) &= 1, & &\text{for } v \in V(G), \\
\varphi(e) &= 1, & &\text{for } e \in E(G), \\
\varphi(f_i) &= 2^{i-1}, & &\text{for } i = 1, 2, \ldots, |F(G)| - 1, \\
\varphi(f_{ext}) &= 1.
\end{aligned}$$

Let us define the labeling θ such that

$$\theta_{i,j} = \begin{cases} 1, & \text{if } f_i \in F(H_j), \\ 0, & \text{if } f_i \notin F(H_j), \end{cases}$$

where $i = 1, 2, \ldots, |F(G)| - 1$, $j = 1, 2, \ldots, t$.

The H-weights are the sums of all vertex labels, edge labels and face labels of vertices, edges and faces in the given subgraph. Thus, for $j = 1, 2, \ldots, t$ we have

$$\begin{aligned}
wt_\varphi(H_j) &= \sum_{v \in V(H_j)} \varphi(v) + \sum_{e \in E(H_j)} \varphi(e) + \sum_{f \in F(H_j)} \varphi(f) \\
&= \sum_{v \in V(H_j)} 1 + \sum_{e \in E(H_j)} 1 + \sum_{f_i \in F(H_j)} 2^{i-1} \\
&= |V(H_j)| + |E(H_j)| + \sum_{i=1}^{|F(G)|-1} \theta_{i,j} 2^{i-1}.
\end{aligned} \tag{3}$$

As we have

$$\begin{aligned}
|V(H_j)| &= |V(H)| \\
|E(H_j)| &= |E(H)|
\end{aligned}$$

for every $j = 1, 2, \ldots, t$, for proving that the H-weights are all distinct it is enough to show that the sums $\sum_{i=1}^{|F(G)|-1} \theta_{i,j} 2^{i-1}$ are distinct for every $j = 1, 2, \ldots, t$. However, this is evident if we consider that the ordered $(|F(G)| - 1)$-tuple $(\theta_{|F(G)|-1,j}\theta_{|F(G)|-2,j} \ldots \theta_{2,j}\theta_{1,j})$ corresponds to binary code representation of the sum (3). As different subgraphs isomorphic to H can not have the same face sets we immediately get that the $(|F(G)| - 1)$-tuples are different for different subgraphs.

4   Conclusion

In the paper, we estimated the lower and the upper bounds of the entire H-irregularity strength of plane graphs and determined the precise values of this parameter for ladders and fan graphs. These two cases proved the sharpness of the lower bounds of the entire H-irregularity strength.

Acknowledgement. The research for this article was supported by APVV-15-0116 and by VEGA 1/0233/18.

References

1. Ahmad, A., Bača, M.: On vertex irregular total labelings. Ars Comb. **112**, 129–139 (2013)
2. Ahmad, A., Bača, M., Bashir, Y.: Total vertex irregularity strength of certain classes of unicyclic graphs. Bull. Math. Soc. Sci. Math. Roumanie **57**(2), 147–152 (2014)
3. Ahmad, A., Bača, M., Siddiqui, M.K.: On edge irregular total labeling of categorical product of two cycles. Theory Comput. Syst. **54**(1), 1–12 (2014)
4. Anholcer, M., Kalkowski, M., Przybyło, J.: A new upper bound for the total vertex irregularity strength of graphs. Discret. Math. **309**(21), 6316–6317 (2009)
5. Anholcer, M., Palmer, C.: Irregular labellings of circulant graphs. Discret. Math. **312**, 3461–3466 (2012)
6. Bača, M., Jendrol', S., Kathiresan, K., Muthugurupackiam, K.: On the face irregularity strength. Appl. Math. Inf. Sci. **9**(1), 263–267 (2015)
7. Bača, M., Jendrol', S., Kathiresan, K., Muthugurupackiam, K., Semaničová-Feňovčíková, A.: A survey of irregularity strength. Electron. Notes. Discret. Math. **48**, 19–26 (2015)
8. Bača, M., Jendrol', S., Miller, M., Ryan, J.: On irregular total labellings. Discret. Math. **307**, 1378–1388 (2007)
9. Bača, M., Lascsáková, M., Naseem, M., Semaničová-Feňovčíková, A.: On entire face irregularity strength of disjoint union of plane graphs. Appl. Math. Comput. **307**, 232–238 (2017)
10. Bača, M., Siddiqui, M.K.: Total edge irregularity strength of generalized prism. Appl. Math. Comput. **235**, 168–173 (2014)
11. Brandt, S., Miškuf, J., Rautenbach, D.: On a conjecture about edge irregular total labellings. J. Graph Theory **57**, 333–343 (2008)
12. Haque, K.M.M.: Irregular total labellings of generalized Petersen graphs. Theory Comput. Syst. **50**, 537–544 (2012)
13. Ivančo, J., Jendrol', S.: Total edge irregularity strength of trees. Discuss. Math. Graph Theory **26**, 449–456 (2006)
14. Jendrol', S., Miškuf, J., Soták, R.: Total edge irregularity strength of complete and complete bipartite graphs. Electron. Notes Discret. Math. **28**, 281–285 (2007)
15. Jendrol', S., Miškuf, J., Soták, R.: Total edge irregularity strength of complete graphs and complete bipartite graphs. Discret. Math. **310**, 400–407 (2010)
16. Majerski, P., Przybyło, J.: Total vertex irregularity strength of dense graphs. J. Graph Theory **76**(1), 34–41 (2014)

17. Nurdin, Salman, A.N.M., Baskoro, E.T.: The total edge-irregular strengths of the corona product of paths with some graphs. J. Combin. Math. Combin. Comput. **65**, 163–175 (2008)
18. Przybyło, J.: Linear bound on the irregularity strength and the total vertex irregularity strength of graphs. SIAM J. Discret. Math. **23**, 511–516 (2009)
19. Wang, W., Zhu, X.: Entire colouring of plane graphs. J. Comb. Theory Ser. B **101**, 490–501 (2011)

Combinatorial Questions: How Can Graph Labelling Help?

Diane Donovan$^{(\boxtimes)}$ and Thomas A. McCourt$^{(\boxtimes)}$

School of Mathematics and Physics, The University of Queensland,
Brisbane 4072, Australia
dmd@maths.uq.edu.au, tom.a.mccourt@gmail.com

Abstract. We highlight some connections between graph labelling, combinatorial design theory and information theory. We survey results on the construction and enumeration of Skolem labellings and related structures. This includes discussion of two constructions of low density parity check codes from Skolem labellings. We raise several pertinent questions and suggestions for future research directions.

In Honour of Emeritus Professor Mirka Miller
9/05/1949 to 2/01/2016
An esteemed colleague and a good friend.

1 Introduction

The purpose of this article is to highlight some connections between combinatorial objects and exploit these connections in applied problems. In particular this article will discuss connections between Skolem sequences, Skolem labelled graphs, additive permutations, Steiner triple systems and partial triple systems. The later sections of the article focus on methods for constructing parity check matrices for low density parity check codes from Skolem labellings for paths. We present a progression of ideas starting with how to use Skolem sequences and labellings to construct Steiner triple systems and partial triple systems leading to two different methods for constructing low density parity check codes. The main purpose is to raise and partially address the question

'Can we harness ideas from graph labellings to solve problems in combinatorics?'

To aid the reader each combinatorial structure considered is introduced as needed. We begin with a discussion of Skolem sequences and Skolem labelled graphs.

© Springer International Publishing AG, part of Springer Nature 2018
L. Brankovic et al. (Eds.): IWOCA 2017, LNCS 10765, pp. 13–23, 2018.
https://doi.org/10.1007/978-3-319-78825-8_2

2 What are Skolem Sequences and Skolem Labelled Graphs, and How are the Two Connected?

A *Skolem Sequence* of order n, denoted SS(n), is a sequence of $2n$ positive integers, $[s_1, s_2, \ldots, s_{2n}]$, that satisfies the property

$$\forall k \in \{1, \ldots, n\}, \exists i, j \in \{1, \ldots, 2n\}, \text{ such that } s_i = s_j = k \text{ and } |j - i| = k.$$

For example consider the Skolem sequence, SS(5), $[4, 1, 1, 5, 4, 2, 3, 2, 5, 3]$, of length $2n = 2 \times 5 = 10$. Here $s_1 = s_5 = 4$, $s_2 = s_3 = 1$, $s_4 = s_9 = 5$, $s_6 = s_8 = 2$, and $s_7 = s_{10} = 3$. It is easy to check that for each $k \in \{1, 2, 3, 4, 5\}$ there exists precisely two terms in the sequence s_i, s_j such that $s_i = s_j = k$ and $|j - i| = k$.

Skolem sequences were first introduced by Skolem in [23] where he presented arguments to verify that a Skolem sequence SS(n) exists if and only if $n \equiv 0$ or $1 \pmod 4$. For more details see [5].

Mendelsohn and Shalaby [19] introduced the concept of a Skolem labelled graph in 1991. Let u and v be distinct vertices in a connected graph G, then dist(u, v) denotes the number of edges in a shortest path between u and v in G. An undirected graph $G = (V(G), E(G))$, where $|V(G)| = 2n$ for some positive integer n, admits a *Skolem labelling* if there exists an onto function $f : V(G) \to \{d, d+1, d+2, \ldots, d+n-1\}$, for d an integer greater than 0, such that

(a) $\forall i \in \{0, \ldots, n-1\}$, $\exists u, v \in V(G)$, such that $f(u) = f(v) = d + i$ and dist$(u, v) = d + i$;

(b) for $G' = (V(G), E(G'))$ where $E(G') \subset E(G)$, f applied to G' violates (a).

If G admits a Skolem labelling, then G alongside its labelling is said to be a *Skolem labelled graph*. It is also worth noting that the definition of a Skolem labelling and a Skolem-graceful labelling (see [2,11]) are different.

Mendelsohn and Shalaby [19] established a number of results, including results on Skolem labelling of paths and cycles. In particular, Theorem 1 of [19] states that the existence of Skolem sequence, SS(n), of order n implies the existence of a Skolem labelling of the path P_{2n} of length $2n - 1$. The following example, where $d = 1$, illustrates the construction. The Skolem sequence SS(5) given above provides a Skolem labelling of P_{10}, with vertex set $V(P_{10}) = \{s_1, \ldots, s_{10}\}$, through the mapping $f : V(P_{10}) \to \{1, \ldots, 10\}$ where $f(s_1) = f(s_5) = 4$, $f(s_2) = f(s_3) = 1$, $f(s_4) = f(s_9) = 5$, $f(s_6) = f(s_8) = 2$ and $f(s_7) = f(s_{10}) = 3$, as displayed in Fig. 1.

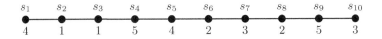

Fig. 1. A Skolem labelling for the graph P_{10}.

The remainder of this article we will be primarily concerned with Skolem labelled paths P_{2n} where $n \equiv 0$ or $1 \pmod 4$ and $d = 1$. For further details on

Skolem labellings, Skolem-graceful labellings and other graph labellings, see the 2013 survey by Gallian [11] and the 2006 survey [2] by Baca et al.

A Skolem labelling of P_{2n} ($d = 1$) defines a set of triples; that is, for each $k \in \{1, 2, \ldots, n\}$ there is precisely one triple $\{k, u, v\}$, $u, v \in V(P_{2n})$, such that $f(u) = f(v) = k$ and $\mathrm{dist}(u, v) = k$. Label the vertices of P_{2n} by $\{n + 1, n + 2, \ldots, 3n\}$ so that consecutively labelled vertices are joined by an edge. Then the set of triples $\{\{k, u, v\} \mid 1 \le k \le n\}$ (where $u + k = v$) partitions $\{1, \ldots, n\} \cup \{n + 1, \ldots, 3n\}$. This is important property of Skolem labellings will be used throughout this article. Figure 2 illustrates this partition for the Skolem labelling given in Fig. 1.

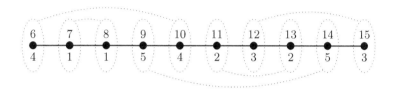

$$\{1, \ldots, 15\} = \{6, 4, 10\} \cup \{7, 1, 8\} \cup \{9, 5, 14\} \cup \{11, 2, 13\} \cup \{12, 3, 15\}$$

Fig. 2. Using a SS(5) to partition $\{1, \ldots, 5\} \cup \{6, \ldots, 15\}$.

2.1 How Many Distinct Skolem Labellings of P_{2n} Exist?

The above decomposition of the labelling into triples was exploited by Abrham in [1] to prove a lower bound on the number of distinct Skolem sequences SS(n), and hence the number of distinct Skolem labellings of P_{2n}, is $2^{\lfloor n/3 \rfloor}$. Later, in Bennett et al. [3] used similar counting arguments to extend the results to various generalizations of Skolem sequences. More recently Donovan and Grannell [6] focused on developing methods for partitioning the set $\{1, \ldots, n\} \cup \{n + 1, \ldots, 3n\}$ into triples and obtained improved lower bounds for the number of distinct Skolem labellings of P_{2n}. Donovan and Grannell began by highlighting a connection to additive permutations.

An *additive permutation* of order t, AP(t), on the set of integers $X = \{-t, \ldots, t\}$, is a permutation $\pi : X \to X$ such that the mapping $\sigma : X \to X$, $\sigma(x) = x + \pi(x)$, $x \in X$, is also a permutation. A permutation may be visualised as a $3 \times (2t + 1)$ array, where the first row contains $-t, \ldots, t$ in the given order, the second row contains the image of the first row under the permutation π and the third row is the sum of the first two rows; the permutation is additive if this third row is a permutation of $\{-t, \ldots, t\}$. For example, Fig. 3 documents an additive permutation π for $t = 3$.

The connection to Skolem labellings is through the triples $\{\{x, \pi(x), x + \pi(x)\} \mid x \in X\}$ of an additive permutation. However, these triples do not partition some underlying set and so in some sense they need to be 'spread out'.

x	-3	-2	-1	0	1	2	3
$\pi(x)$	0	3	-1	2	-2	1	-3
$x + \pi(x)$	-3	1	-2	2	-1	3	0

Fig. 3. An additive permutation of order 3.

To achieve this Donovan and Grannell [6] embedded the triples of an AP(t) into a Skolem labelling of P_{14t+8}, where $t \equiv 0, 3 \pmod 4$. Let $V(P_{14t+8}) = \{-(7t + 3), \ldots, 7t + 4\}$ and define a function $F : V(P_{14t+8}) \to \{1, \ldots, 7t + 4\}$ using the following algorithm.

1. Under the assumption that there exists a Skolem labelling g of P_{2t+2}, where $V(P_{2t+2}) = \{-t, \ldots, t + 1\}$ and $g(0) = g(t + 1) = t + 1$, set $F(x) = g(x)$, for all $x \in V(P_{2t+2}) \setminus \{0, t + 1\}$
2. Under the assumption there exists an additive permutation AP(t), for each $x \in \{-t, \ldots, t\}$ and hence each triple $\{x, \pi(x), x + \pi(x)\}$ in the AP(t) choose either

$$F(-(x + \pi(x) + 6t + 3)) = F(-(\pi(x) + 4t + 2)) = x + 2t + 1,$$
$$F(-(x + 2t + 1)) = F(\pi(x) + 4t + 2) = x + \pi(x) + 6t + 3,$$
$$F(x + 2t + 1) = F(x + \pi(x) + 6t + 3) = \pi(x) + 4t + 2,$$

or

$$F(\pi(x) + 4t + 2) = F(x + \pi(x) + 6t + 3) = x + 2t + 1,$$
$$F(-(\pi(x) + 4t + 2)) = F(x + 2t + 1) = x + \pi(x) + 6t + 3,$$
$$F(x + 2t + 1) = F(x + \pi(x) + 6t + 3) = \pi(x) + 4t + 2.$$

3. $F(0) = F(7t + 4) = 7t + 4$.

It was verified by Donovan and Grannell that for each $k \in \{1, \ldots, 7t + 4\}$ there are precisely two vertices of P_{14t+8} that are labelled with k and these vertices are distance k apart.

Thus we see that each distinct triple $\{x, \pi(x), x + \pi(x)\}$, for $x \in X$, of the additive permutation AP(t), $t \equiv 0, 3 \pmod 4$, is used to label three triples $\{w, k, u\}$, where $w + k = u$, associated with the path P_{14t+8}. Hence the number of distinct Skolem labellings of P_{14t+8} is at least 2^{2t+1} times the number of additive permutations of order t. The number of additive permutations has been computed for $t = 1$ to 11, and appear in Sloane's sequence encyclopedia, Sequence A002047, [24]. They are reproduced in Table 1.

To count the numbers of additive permutations for large values of t we follow the work of Cavenagh and Wanless [4]. It is well know that the addition table for the cyclic group of order $2t + 1$ can be represented as a set of $(2t + 1)^2$ distinct triples of the form $C_{2t+1} = \{(i, j, i + j \pmod{2t + 1})) \mid 0 \leq i, j \leq 2t\}$. Cavenagh and Wanless [4] established lower bounds on the number of distinct transversal T in C_{2t+1}. A *transversal* T is a subset of $2t + 1$ triples, such that for

Table 1. The number of additive permutations of order $2t + 1$.

t	Number of AP(t)
3	2
5	6
7	28
9	244
11	2 544
13	35 600
15	659 632
17	15 106 128
19	425 802 176
21	14 409 526 080
23	577 386 122 880

distinct triples $(u, v, w), (x, y, z) \in T$, $u \neq x$, $v \neq y$ and $w \neq z$. That is, a subset of $2t + 1$ triples drawn from distinct rows, columns and entries of the addition table. Indexing the rows, columns and entries by the set $\{-t, \ldots, t\}$ instead of $\{0, \ldots, 2t\}$ and using only transversals where all triples satisfy $(i, j, i + j)$, the permutation $\pi(i) = j$, $-t \leq i \leq t$ is an additive permutation AP(t). By way of an example the additive permutation given in Fig. 3 corresponds to the transversal (indicated by underlined entries) in Fig. 4. (Note only those entries of the form $(i, j, i + j)$ have been displayed.)

	-3	-2	-1	0	1	2	3
-3				*-3*	-2	-1	0
-2			-3	-2	-1	0	*1*
-1		-3	*-2*	-1	0	1	2
0	-3	-2	-1	0	1	*2*	3
1	-2	*-1*	0	1	2	3	
2	-1	0	1	2	*3*		
3	*0*	1	2	3			

Fig. 4. A relabelled version of the addition table of the cyclic group of order 7 with transversal underlined.

Thus for large t counting the number of additive permutations of order t, is equivalent to counting the number of transversals in this restricted region of the addition table of the cyclic group of order $2t + 1$. By subdividing the addition table into subarrays of size 23, Cavenagh and Wanless showed that a lower bound for the number of transversal in addition table for the cyclic group of order $2t + 1$ is 3.246^{2t+1}, where 3.246 is the 23rd root of the number of additive permutations of order 23, 577 386 122 880. The transversals included in the counting arguments of Cavenagh and Wanless are all contained in the restricted region and hence correspond to additive permutations.

Consequently Donovan and Grannell [6] showed that for sufficiently large t, the number of distinct Skolem labellings of P_{14t+8}, where $t \equiv 0, 3 \pmod 4$, is at least

$$2^{2t+1} \times 3.246^{2t+1} = (6.492)^{2t+1} > 2^{\lfloor (7t+4)/3 \rfloor}.$$

Donovan and Grannell [6] also provided other lower bounds for sporadic orders.

It was conjecture by Vardi [26] that the number of transversals in the addition table for the cyclic group of order $2t + 1$ exceeds $c^{2t+1}(2t + 1)!$ for some constant $c \in (0, 1)$. Eberhard, Manners and Mrazović, published a recent arXiv paper [8] claiming to prove this conjecture. However, it is important to note that not all transversals are suitable for constructing additive permutations. But unfortunately it is also easy to see that the Cavenagh and Wanless construction does not produce all additive permutations, leading to the following open questions.

Question: Can the lower bound on the number of additive permutation of order t, given in [4], be improved?

Question: Can the lower bound on the number of distinct Skolem labellings of order n, given in [6] be improved?

2.2 What is the Connection Between Skolem Labellings, Steiner Triple Systems and Partial Triple Systems?

A *Steiner triple system* of order v, STS(v), is an ordered pair (V, B) where V is a set of cardinality v and B is a collection of triples chosen in such a way that each pair of distinct elements of V occurs in precisely one triple. Counting arguments show that the number of triples is $b = v(v - 1)/6$, implying that a STS(v) exists only if $v \equiv 1, 3 \pmod 6$. Sufficiency has also been established, see for instance [5]. A Steiner triple systems, STS(v), with $V = \{0, 1, \ldots, v - 1\}$ is said to be cyclic if it possesses the triple preserving automorphism $\alpha : V \rightarrow V$, where $\alpha(x) = x + 1 \pmod v$. It is known that cyclic STS(v) exist for all admissible values of v except $v = 9$ [5]. Skolem showed that cyclic STS(v), $v = 6n + 1$, can be constructed from Skolem sequences of order n and hence they can be constructed from Skolem labellings of P_{2n}.

We explain this construction using our Skolem labelling notation. Take the set of n triples associated with a Skolem labelling f of P_{2n}, with $V(P_{2n}) = \{1, \ldots, 2n\}$; that is, the set $T = \{(x, y, k) \mid k = f(x) = f(y), 1 \le k \le n\}$ and use it to construct a set of *starter triples* $D = \{\{0, n + x, n + y\} \mid (x, y, k) \in T\}$. Note that, working modulo $6n + 1$,

$$\cup_{1 \le k \le n} \{\pm(n + x), \pm(n + y), \pm(k = f(x) = f(y))\} = \{1, \ldots, 6n\}.$$

Hence, again working modulo $6n+1$, the set of triples $B = \{\{i, n+x+i, n+y+i\} \mid \{0, n + x, n + y\} \in D, 0 \le i \le 6n\}$ forms a cyclic STS($6n + 1$).

A common representation for a STS(v), (V, B), is an *incidence matrix*; that is, a $v \times b$ matrix $M = [m_{ij}]$, with the rows indexed by the points of V, the columns indexed by the triples of $B = \{B_j \mid 1 \leq j \leq b\}$, and

$$m_{ij} = \begin{cases} 1, & \text{if point } i \text{ occurs in triple } B_j, \\ 0, & \text{otherwise.} \end{cases}$$

If a STS(v) is cyclic, then the corresponding incidence matrix is said to be *quasi-cyclic* in that the columns can be partitioned into n subsets of size $6n + 1$ and each column, within the n subsets, is a cyclic shift of the previous column modulo $6n + 1$.

A *partial triple system* PTS(v), with $\lambda = 1$, is a pair (V, B) where V is a set of cardinality v and B is a collection of triples chosen in such a way that each pair of distinct elements of V occurs in at most one triple. As in the case of a STS(v), a partial triple system PTS(v) can be represented in terms of its incidence matrix.

It is the incidence matrices of STS(v) and PST(v) that will be used in the next section to construct parity check matrices for low density parity check codes.

3 How Can We Use Skolem Labellings to Construct Parity Check Matrices for Linear Codes?

A $[n, k]$ *linear block code* over GF(2) encodes binary blocks $\mathbf{u} = [u_1 \, u_2 \, \ldots \, u_k]$ to *binary codewords* $\mathbf{x} = [x_1 \, x_2 \, \ldots \, x_n]$ of length n, via a binary *generating matrix*, over GF(2), $G = [I_k|A^T]$, where A is a $(n-k) \times k$ matrix and $\mathbf{x} = \mathbf{u}G$. Decoding is via a binary *parity check matrix* $H = [A|I_{n-k}]$, where $H\mathbf{x}^T = \mathbf{0}$. The generating and parity check matrices satisfy $GH^T = HG^T = \mathbf{0}$. Further, the set of all codewords is given by the null space of the parity check matrix H. Thus the number of linearly independent rows of H determines the number of codewords. For more details see for instance [22].

For example, consider a $[6, 3]$ code, with generating matrix

$$G = \begin{pmatrix} 1\,0\,0\,1\,0\,1 \\ 0\,1\,0\,1\,1\,1 \\ 0\,0\,1\,1\,1\,0 \end{pmatrix}$$

and parity check matrix

$$H = \begin{pmatrix} 1\,1\,1\,1\,0\,0 \\ 0\,1\,1\,0\,1\,0 \\ 1\,1\,0\,0\,0\,1 \end{pmatrix}.$$

Here a message $\mathbf{u} = [u_1 u_2 u_3]$ is encoded to the codeword $\mathbf{x} = [x_1 x_2 x_3 x_4 x_5 x_6]$, where $x_1 = u_1, x_2 = u_2, x_3 = u_3$ and x_4, x_5, x_6 are check digits chosen such that $H\mathbf{x}^T = \mathbf{0}$. Thus the parity check equations are:

$$x_4 = x_1 + x_2 + x_3,$$
$$x_5 = x_2 + x_3,$$
$$x_6 = x_1 + x_2.$$

In this case $\mathbf{u} = [011]$ is encoded to the codeword $\mathbf{x} = [011001]$.

A linear block code over GF(2) with a sparse parity check matrix is called a *Low Density Parity Check* code (LDPC code). The goal is to construct LDPC codes with a high information rate R ($R = k/n$) and large error correction capacity. The code will correct $\lfloor \frac{1}{2}(d-1) \rfloor$ errors where d is the minimum Hamming distance of the code (the minimum weight of non-zero codewords).

Low density parity check (LDPC) codes were first introduced by Gallager in [9], and later popularised by McKay and Neal [16,17] in the 1990s. Their main attraction is that they can achieve information rates close to the Shannon bound. Initially, Gallager constructed such codes by pseudorandomly generating parity check matrices, then identifying 'good' LDPC codes through computer searches [9,10]. However, the random structure of the generated codes made them somewhat difficult to encode and it was often hard to determine the minimum distance.

More recently, researchers have taken a more systematic approach to their construction. Kou et al. [14] constructed LDPC codes using finite geometries, Tanner et al. [25] used various group-structures to construct LDPC codes, and Rosenthal and Vontobel constructed LDPC codes from Ramanujan graphs [21]. More recently, combinatorial designs, and in particular their incidence matrices, have been used to construct LDPC codes, see [13,18,27,28].

Vasic and Milenkovic [27] used the incidence matrix of cyclic STS(v), (V, B), as the parity check matrix for an LDPC code. The cyclic structure of the incidence matrix was used to obtain a lower bound for the size of the null space of the parity check matrix, and hence on the number of codewords. The structure of the Steiner triple system also provides a bound on the information rate of the code and the minimum distance of the code. The fact that every pair of points of V occurs in precisely one triple of B implies that the inner product of any two columns of the incidence matrix is at most one. This fact can be used to prove the minimum distance of such codes is at least four, a result that is often validated using a bipartite *Tanner graph*, see [28] for more details. If the underlying triples of a cyclic Steiner triple system do not contain a subsets of triples of the form $\{0, y, z\}, \{0, u, w\}, \{x, y, u\}, \{x, z, w\}$—that is, they are *anti-Pasch systems*—then the resulting LDPC code can be shown to have a minimum distance of at least 6. Anti-Pasch STS(v) exist if and only if $v \equiv 1$ or $3 \pmod 6$ and $v \notin \{7, 13\}$ [12,15]. However if we add the requirement that the systems are cyclic (which may be desirable for hardware implementation [27]) known orders are far fewer.

More recently Donovan et al. [7] provided an alternate construction for LDPC codes based on partial triple systems. They began by constructing a set of $2n-1$ starter triples $D_j = (0, j, 2j + 1)$, $0 \leq j \leq n - 1$ and $D_j = (0, j, 2(j - n))$, $n + 1 \leq j \leq 2n - 1$. Using these starter triples a PTS($6n$), (V, B), can be constructed where $B = \{B_j \mid 0 \leq j \leq 2n - 1, j \neq n\}$, and

$$B_j = \{a, j + a \,(\text{mod } 2n) + 2n, j' + a \,(\text{mod } 2n) + 4n\}$$
$$\text{where } (0, j, j') = D_j, \; 0 \leq a \leq 2n - 1.$$

The result is a PTS($6n$), (V, B) where $V = \{0, \ldots, 6n-1\}$ and $|B| = 2n(2n-1) = 4n^2 - 2n$. Notice that there is no starter triple of the form $(0, n, 0)$. This restriction is necessary to ensure that each pair of points of V occur in at most one triple of PTS($6n$). Indeed if $y, z \in V$ the number of triples, $\lambda_{y,z}$, containing the pair y, z is

$$\lambda_{y,z} = \begin{cases} 0, \text{ if } y, z \in \{0, \ldots, 2n - 1\}, \\ 0, \text{ if } y, z \in \{2n, \ldots, 4n - 1\}, \\ 0, \text{ if } y, z \in \{4n, \ldots, 6n - 1\}, \\ 0, \text{ if } z = y + 3n \text{ or } z = y + 4n \text{ and } y \in \{0, \ldots, 2n - 1\}, \\ 0, \text{ if } z = y + 3n \text{ and } y \in \{2n, \ldots, 3n - 1\}, \\ 0, \text{ if } z = y + n \text{ and } y \in \{3n, \ldots, 4n - 1\}, \\ 1, \text{ otherwise.} \end{cases}$$

These triples can now be used to construct a $6n \times (4n^2 - 2n)$ incidence matrix $M = [m_{xy}]$ that can be used as a parity check matrix for an LDPC code. The structure of the resulting PTS($6n$) can be used to verify the existence of an infinite family of LDPC codes of length $6n$, for n a positive integer. The parity check matrix, H, has rank $6n - 2$, consequently there are $6n - 2$ linearly independent rows. Thus, the number of codewords (the cardinality of the null space of H) equals the length of the codewords less the rank of H; that is, there are $(4n^2 - 2n) - (6n - 2) = 4n^2 - 8n + 2$ codewords. Hence it is possible to explicitly determine the rate of the code, namely $(4n^2 - 8n + 2)/(4n^2 - 2n)$, which is greater than 0.8 for codes of length greater than or equal to 240. In addition, these codes have minimum distance 6 when n is odd and minimum distance 4 when n is even.

3.1 What is the Connection to a Skolem Labelling?

Skolem was the first researcher to use Skolem labelling of P_{2n} (or equivalently SS(n)) for the construction of a cyclic STS($6n + 1$). Then Vasic and Milenkovic [27] drew the connection between incidence matrices of cyclic STS($6n + 1$) and parity check matrix for a LDPC code. Later Park et al. [20] specially drew the connection with Skolem's construction. By [6], and as discussed above, there are at least $(6.492)^{2t+1}$ Skolem labellings of P_{14t+8}, where $t \equiv 0, 3 \pmod 4$. Raising the following question.

> **Question:** If we take a random Skolem labelling of P_{2n}, what are the properties of the LDPC code constructed from the incidence matrix of the associated cyclic STS($6n + 1$) (via the construction given in Subsect. 2.2)?

In addition, recent work has shown that it is possible to take a Skolem labelling of P_{2n} and use it to replicate, in a more general setting, the construction by Donovan, Rao and Yazıcı. Starting with the Skolem labelling of P_{2n} given in Fig. 1 this construction 'doubles' the sequence to yield the starter triples $(0, 10, 7)$, $(0, 2, 3)$, $(0, 6, 8)$, $(0, 1, 5)$, $(0, 4, 9)$, $(0, 8, 4)$, $(0, 3, 1)$, $(0, 7, 6)$ and $(0, 9, 2)$.

As before these starter triples can be cyclically developed to generate a PTS($6n$) where every pair of points of V occurs in at most one triple. The corresponding incidence matrix can be used as a parity check matrix for a LDPC code, leading to our last question.

Question: If we take a random Skolem labelling of P_{2n} and use it to construct the incidence matrix of the associated $PTS(6n)$, what properties does the corresponding LDPC code exhibit?

4   Closing Remarks

In this article we have highlighted connections between graph labelling, combinatorial design theory and information theory. All areas in which Mirka Miller made significant contributions. The main focus of this article has been the development of Skolem labelling of paths P_{2n}, leading to a number of questions about their enumeration and internal structure. We have also demonstrated how Skolem labellings can be used to construct highly sought after low density parity check (LDPC) codes. While there is existing research documenting constructions for these LDPC codes, there has been little research into the properties of individual Skolem labellings and how these manifest in distinct LDPC codes. In particular:

Question: Can we identify classes of Skolem labellings of P_{2n} that lead to 'optimal' LDPC codes?

We hope that we have enthused the reader to explore some of these codes, as Mirka Miller enthused many researchers to explore the finer structure of graph labellings.

References

1. Abrham, J.: Exponential lower bounds for the number of Skolem and extremal Langford sequences. Ars Comb. **22**, 187–198 (1986)
2. Baca, B., Baskoro, E.T., Miller, M., Ryan, J., Simanjuntak, R., Sugeng, K.A.: Survey of edge antimagic labelings of graphs. J. Indones. Math. Soc. **12**, 113–130 (2006)
3. Bennett, G.K., Grannell, M.J., Griggs, T.S.: Exponential lower bounds for the numbers of Skolem-type sequences. Ars Comb. **73**, 101–106 (2004)
4. Cavenagh, N.J., Wanless, I.M.: On the number of transversals in Cayley tables of cyclic groups. Discret. Appl. Math. **158**(2), 136–146 (2010)
5. Colbourn, C.J., Dinitz, J.H. (eds.): Handbook of Combinatorial Designs. CRC Press, Boca Raton (2006)
6. Donovan, D., Grannell, M.J.: On the number of additive permutations and Skolem-type sequences. Ars Math. Contemp. **14**(2), 415–432 (2017)
7. Donovan, D., Rao, R., Yazıcı, E.S.: High rate LDPC codes from difference covering arrays. arXiv:1701.05686 (2017)
8. Eberhard, S., Manners, F., Mrazovic, R.: Additive triples of bijections, or the toroidal semiqueens problem. arXiv:1510.05987 (2015)
9. Gallager, R.G.: Low density parity check codes. IRE Trans. Inf. Theory **IT–8**, 21–28 (1962)
10. Gallager, R.G.: Low Density Parity Check Codes. MIT Press, Cambridge (1963)
11. Gallian, J.A.: A dynamic survey of graph labeling. Electron. J. Comb. **16**, #DS6 (2013)

12. Grannell, M.J., Griggs, T.S., Whitehead, C.A.: The resolution of the anti-Pasch conjecture. J. Comb. Des. **8**, 300–309 (2000)
13. Johnson, S.J., Weller, S.R.: Regular low-density parity-check codes from combinatorial designs. In: Proceedings 2001 IEEE Information Theory Workshop, Cairns, Australia, pp. 90–92, 27 September 2001
14. Kou, Y., Lin, S., Fossorier, M.P.C.: Low-density parity-check codes based on finite geometries: a rediscovery and new results. IEEE Trans. Inf. Theory **47**, 2711–2736 (2001)
15. Ling, A.C.H., Colbourn, C.J., Grannell, M.J., Griggs, T.S.: Construction techniques for anti-Pasch Steiner triple systems. J. London Math. Soc. **61**(2), 641–657 (2000)
16. MacKay, D.J.C., Neal, R.M.: Near Shannon limit performance of low density parity-check codes. Electron. Lett. **33**(6), 457–458 (1997)
17. MacKay, D.J.C.: Good error-correcting codes based on very sparse matrices. IEEE Trans. Inf. Theory **45**(2), 399–432 (1999)
18. Mahadevan, A., Morris, J.M.: On RCD SPC codes as LPDC codes based on arrays and their equivalence to some codes constructed from Euclidean geometries and partial BIBDs. Technical report no.: CSPL TR:2002–1, Communications and Signal Processing Laboratory, Computer Science and Electrical Engineering Department University, of Maryland, USA
19. Mendelsohn, E., Shalaby, N.: Skolem labelled graphs. Discret. Math. **97**(1–3), 301–317 (1991)
20. Park, H., Hong, S., No, J.S., Shin, D.J.: Construction of high-rate regular quasi-cyclic LDPC codes based on cyclic difference families. IEEE Trans. Commun. **61**(8), 3108–3113 (2013)
21. Rosenthal, J., Vontobel, P.O.: Construction of LDPC codes using Ramanujan graphs and ideas from Margulis. In: Proceedings of 2001 IEEE International Symposium Information Theory, Washington, DC, p. 4, June 2001
22. Shokrollahi, A.: LDPC codes: an introduction. In: Feng, K., Niederreiter, H., Xing, C. (eds.) Coding, Cryptography and Combinatorics, pp. 85–110. Birkhäuser, Basel (2004). https://doi.org/10.1007/978-3-0348-7865-4_5
23. Skolem, T.: On certain distributions of integers in pairs with given differences. Math. Scand. **5**, 57–68 (1957)
24. Sloane, N.J.A.: The on-line encyclopedia of integer sequences. https://oeis.org/
25. Tanner, R.M., Srkdhara, D., Fuja, T.: A class of group-structured LDPC codes. http://www.cse.ucsc.edu/tanner/pubs.html
26. Vardi, I.: Computational Recreations in Mathematics. Addison-Wesley, Boston (1991)
27. Vasic, B., Milenkovic, O.: Combinatorial constructions of low-density parity-check codes for iterative decoding. IEEE Trans. Inf. Theory **50**(6), 1156–1176 (2004)
28. Zhang, L., Huang, Q., Lin, S., Abdel-Ghaffar, A., Blake, I.F.: Quasi-cyclic LDPC codes: an algebraic construction, rank analysis, and codes on Latin squares. IEEE Trans. Commun. **58**(11), 3126–3139 (2010)

Extremal Kernelization:
A Commemorative Paper

Henning Fernau[⊠]

Fachbereich 4, Informatikwissenschaften, CIRT, Universität Trier, Trier, Germany
fernau@uni-trier.de

Abstract. We try to describe in this paper how methods from extremal combinatorics play an important role in the development of parameterized algorithms, also sketching further venues how this influence could be even increased in order to obtain quick $\mathcal{FPT}$ classification results. Conversely, we show how certain notions that have become of importance within parameterized algorithmics can be useful to keep in mind for combinatorialists. We hope that this account initiates fruitful future discussions between the different scientific sub-communities that usually comprise IWOCA and that make this event quite special.

1 Introduction

This paper is a rather personal account, covering several encounters with the late Mirka Miller. Actually, on her last travel to Europe, Mirka Miller and Joe Ryan visited me. Unfortunately, she did not feel very well, so she basically stayed in the hotel while Joe worked with my group in Trier. They then went to her beloved Western Bohemia, hoping that she would recover from the hitherto too stressful travel, as they still thought. As this did not have the desired effect, they went home to Australia, where the sad reason of her unwellness was soon found out; unfortunately, too late, as we know now.

I came to know Mirka and Joe back in 2002 when I took over a 2-year position as a Senior Lecturer at the University of Newcastle. I still remember these times quite well, also with a bit of melancholia, as somehow I worked together or became friends with most of the staff members, something that I still miss in Germany, where mostly the different groups do separate research.

For instance, in those days I learnt to know about graph labeling, most likely in the many talks that were delivered in Mirka's group. Together with Kiki Sugeng and Joe, we worked on sum labelings of particular graphs. This way, I also learnt a bit more about the differences between Mathematics and Theoretical Computer Science, the research area that I am usually working in. Namely, upon presenting the mentioned paper at the British Combinatorics Conference (BCC) in Durham in 2005, when I was working at the University of Hertfordshire, I confidently entered the room where my talk was scheduled with my laptop computer in my hand, ready to show the slides that I prepared. Only then I discovered that not all lecture halls enjoyed the feature of having projecting facilities. Well, so

L. Brankovic et al. (Eds.): IWOCA 2017, LNCS 10765, pp. 24–36, 2018.
https://doi.org/10.1007/978-3-319-78825-8_3

I turned to chalk and blackboard to deliver our talk. Even in those days, it was rather standard to present papers using laptops within Computer Science conferences. The topic of the mentioned paper, finally published in [25], is also quite fitting to a commemorative paper like this one: generalised friendship graphs.

In 2003, Mirka asked me to give a talk at an event that was still known as AWOGL in these days. Similarly as AWOCA has now turned into IWOCA, AWOGL is now known as IWOGL (International Workshop on Graph Labelling). Noticeably, the conference series A/I WOGL was initiated by Mirka herself. Already at AWOGL 2003, I tried to bring together the mostly combinatorialists' community of AWOGL with one of my other interests, which is algorithms and complexity. I still think that the possible connections between graph labeling and combinatorial algorithms are largely unexplored today.

But let me not deviate too much from the main scientific topic of this paper, which is reflected in the title of *extremal kernelization*. This title was inspired by the work Elena Prieto-Rodríguez did in 2002 and 2003 when I was in Newcastle, see [35], where Michael Fellows, to whom I actually owe the chance to come to Newcastle for the mentioned two years, asked her to look into problems where extremal combinatorial methods could be used to prove so-called *boundary lemmas* to show kernelization results. At about the same time, Christian Sloper wrote his PhD thesis with a similar direction [40]; he also visited Newcastle for a couple of months. But, as we will see in the following, it is also often useful for people working in Parameterized Complexity to make use of already known results from Combinatorics, off-the-shelf, as Mike would say. We will show several examples in this paper, including the use of relatively general combinatorial machinery. As we will see, knowing of these results can be very helpful for those working in Parameterized Complexity. Conversely, Parameterized Complexity has developed its own methods and notions in the meantime, which can be useful to know and apply for combinatorialists as well, as we will explain. Finally, taking it to the extremes, so to speak, extremal kernelization tries to explore the limits of its discipline. I will therefore finally mention several lower bound results that, to some extent, also started out when I was in Newcastle.

2   Parameterized Complexity and Kernelization

Let me briefly summarize the basic notions of this field that we need in the following. For more details, we can easily refer to the two textbooks that appeared in recent years [14, 17].

Parameterized Complexity was developed in the early 90s, basically by Rod Downey and Mike Fellows. The idea was to go beyond the barrier established in the 70s when the $\mathcal{P}$ vs. $\mathcal{NP}$-question was first made explicit, together with the intuition that problems in $\mathcal{P}$ (having polynomial-time algorithms) are the good guys, as opposed to $\mathcal{NP}$-hard problems that were usually thought of as *infeasible*. However, in practice computer scientists had to deal with them and developed many ways to overcome this barrier of infeasibility. Nowadays, solvers for (mixed) integer linear programs or for SAT-variants can deal with instances of practically relevant sizes. Also, (meta-)heuristics can help solving relatively large instances. However, it is rarely well understood why these approaches work that well. Parameterized Complexity can be seen as one way of trying to explain the success of them. Namely, the idea is to use more information than just the number of bits needed to encode the given problem instance in order to describe and estimate the behavior of algorithms. This secondary measurement is called the *parameter*. So, if n is the overall size of the instance and k the (size of the) parameter, then one aims at algorithms running in time $\mathcal{O}(f(k)p(n))$, where f is some arbitrary (computable) function and p is some polynomial. In other words, if the parameter k is fixed, then the running time of such an algorithm is polynomial. The class of (parameterized) problems that admit such algorithms is usually called $\mathcal{FPT}$, standing for *fixed-parameter tractable*. This notion clearly generalizes the class $\mathcal{P}$, ignoring the parameter. Apart from this new class of good guys, there are also (actually there is a whole hierarchy of classes of) bad guys, the smallest bad class being W[1]. The hypothesis $\mathcal{FPT} \neq W[1]$ is stronger yet similar in spirit to the famous hypothesis $\mathcal{P} \neq \mathcal{NP}$.

As with every good type of Complexity Theory, a notion of reducibility comes along that preserves the good guys and allows to show hardness results. The simplest form of such a reduction for Parameterized Complexity is a polynomial-time computable function that maps an instance I of a parameterized problem P, where I is of size n and has the parameter $\kappa(I)$ associated to it, to an instance I' of a possibly different parameterized problem P', where I' is then of size n' and has the parameter $\kappa'(I')$ associated to it, where $\kappa'(I') \leq f(\kappa(I))$ for some (computable) function f dependent on P and P' but independent of n. If $P' = P$, $\kappa' = \kappa$ and $n' = g(\kappa(I))$, then such a reduction is also known as a *kernelization*, and instance I' is called a *kernel*. It is one of the simple yet fascinating results of Parameterized Complexity that a parameterized problem belongs to $\mathcal{FPT}$ if and only if it admits a kernelization, or, in other words, if it is kernelizable. This makes kernelization a key notion in this theory. But this notion also has a very practical motivation: Kernelizations are often built from so-called reduction rules, which is a collection of very simple algorithms that help simplify a given instance. It is exactly this strategy that is the key of many successful (heuristic) algorithms to attack $\mathcal{NP}$-hard problems. So, kernelization can provide a mathematical analysis of this success.

Interestingly, observing the proof why kernelization characterizes $\mathcal{FPT}$, one sees that we can also require that $\kappa'(I') \leq \kappa(I)$, a variant called *proper* in [2]. Actually, at the first WorKer (Workshop on Kernels) in Bergen, back in 2009, there was quite a discussion what the correct definition of a kernel should be; in the end, this was not completely clear, though. Empirically, most reduction rules (comprising reductions) are proper.

3   Combinatorial Results for Kernelization

We are going to explain how (known) combinatorial results can be used for or interpreted as algorithmic results, i.e., as kernelizations.

3.1   Special Combinatorial Results

I am now coming back to some sort of Australian experience. Back in 2003, we were looking into MINIMUM DOMINATING SET from the parameterized perspective. Recall that a *dominating set* of an undirected graph $G = (V, E)$ is a vertex set D such that $N[D] = V$, where $N[x] = \{y \in V \mid xy \in E \vee x = y\}$ is the closed neighborhood of x and $N[D] = \bigcup_{x \in D} N[x]$. Of course, the straightforward question would lead to the following parameterized problem:

DOMINATING SET (DS)
Given: A graph $G = (V, E)$, a positive integer k
Parameter: $\kappa(G, k) = k$
Question: Is there a *dominating set* $D \subseteq V$ with $|D| \leq k$?
<u>Well-known</u>: DS is W[2]-complete, i.e., most likely not in $\mathcal{FPT}$.
Therefore, we considered the *dual parameterization* of this problem. We call the complement of a dominating set a *nonblocker set*. This concept is known under different names, e.g., enclaveless set, in [39].
NONBLOCKER SET (NB)
Given: A graph $G = (V, E)$, a positive integer k_d
Parameter: $\kappa(G, k_d) = k_d$
Question: Is there a *nonblocker set* $N \subseteq V$ with $|N| \geq k_d$?

We quickly found the following result in the good old book by Ore [34]:

Theorem 1. *If a graph $G = (V, E)$ has minimum degree at least one, then the size of its minimum dominating set is at most $(1/2) \cdot |V|$.*

Of course, this minimum degree condition should only rule out graphs with isolated vertices, where the claim is obviously wrong. I chose this more complicated looking condition, as there are ways to further generalize this, as we will see. From this theorem, we can immediately deduce:

Corollary 1. *If a graph $G = (V, E)$ has minimum degree at least one, then the size of its maximum nonblocker set is larger than $(1/2) \cdot |V|$.*

We can easily use these rules to obtain a kernel for NB, employing only the following reduction rule:

Reduction Rule 1. *Delete isolated vertices (without changing the parameter).*

Let us make this simple algorithm explicit:

1. Apply reduction rule.
2. If reduced graph has at least $2k_d$ vertices, YES.
3. Otherwise: We have a graph instance with less than $2k_d$ vertices.

Strictly speaking, instead of answering YES, our algorithm should output a trivial YES-instance, but this way of presenting kernelizations is now established. The kernel that we obtain looks a bit like cheating, as we simply output the original graph instance itself (without isolates). However, this allows us to conclude:

Corollary 2. *NB admits a kernel of size at most $2k_d$.*

Of course, we were not completely happy with this rather trivial result. We worked pretty hard to find new reduction rules and also to prove new boundary lemmas. Only then, we discovered two more (similar) theorems that would have saved us a lot of work had we known them before. Obviously, there was a lesson to be learnt for us: first study combinatorial results and check out how to use them. Blank [6] and McCuaig and Shepherd [32] have shown:

Theorem 2. *If a connected graph $G = (V, E)$ has minimum degree of at least two and is not one of seven exceptional graphs (each of them having at most seven vertices), then the size of its minimum dominating set is at most $2/5 \cdot |V|$.*

Reed showed a bit later:

Theorem 3 [36]. *If a graph $G = (V, E)$ has minimum degree of at least three, then the size of its minimum dominating set is at most $3/8 \cdot |V|$.*

In [16], we showed how to turn Theorem 2 into a kernelization result, which reads as follows.

Theorem 4. *There is an algorithm that provides a kernel of size upper-bounded by $5/3 \cdot k_d + 3$ for any NONBLOCKER SET-instance (G, k_d), where the problem size is measured in terms of the number of vertices.*

We developed reduction rules to get rid of vertices of degree two that have a neighbor of degree two, but we did not find reduction rules to cope with all vertices of degree two. This leads to two natural questions, one more directed towards algorithms people and one more towards combinatorialists.

Open Problem 1. *Devise reduction rules in order to turn Theorem 3 into a kernelization result.*

Open Problem 2. *Prove an upper bound $\alpha|V|$ on the size of a minimum dominating set of any graph $G = (V, E)$ of minimum degree at least two, with no two adjacent vertices of degree, such that $3/8 < \alpha < 2/5$.*

Remark 1. The reduction rules from [16], devised to exploit Theorem 2, destroy graph properties like planarity, as they are based on merging far-away vertices. It would be nice to avoid this, for instance, by having a new kernelization result that produces only kernels that are induced subgraphs of the original input [2].

3.2   General Combinatorial Results

Is it always necessary to recur to rather special combinatorial results in order to prove kernelization algorithms? Possibly surprisingly, sometimes way more general results from combinatorics also do the job, leading to an approach that I like to advertise as *quick kernels*, because they do not require any advanced understanding of the combinatorial algorithmic question at hand, but quickly prove that the parameterized problem under consideration is in $\mathcal{FPT}$. Recall that it is believed that not all parameterized problems belong to $\mathcal{FPT}$. Clearly, quick kernels are usually not really small, although sometimes they are not that bad at all, as our first concrete example will show. The two general methods I will describe are based on Ramsey Theory and on the Degree-Diameter Problem. Also other general combinatorial statements should be explored for quick kernels.

Ramsey Theory. Many will recall one or other form of *Ramsey's Theorem*:

Theorem 5. *There exists a least positive integer $R(b, r)$ for which every blue-red edge coloring of the complete graph on $R(b, r)$ vertices contains a blue clique on b vertices or a red clique on r vertices.*

It is well-known that $R(b, r) \leq \binom{b+r-2}{b-1}$. This gives an immediate quadratic kernel for INDEPENDENT SET ON TRIANGLE-FREE GRAPHS, parameterized by a lower-bound k on the independent set, even slightly better, as $R(3, r) \leq \frac{r^2}{\log(r)}$ due to Kim [30]. Namely, if the input graph G has at least $R(3, k)$ many vertices, we know, as G contains no complete graphs of three or more vertices, that G must contain an independent set of size at least k, so that we can answer YES immediately. If the order of G is smaller than $R(3, k)$, we already have the desired kernel bound. This example also shows that kernels derived from Ramsey arguments need not be that big. But of course, having used this heavy and general combinatorial machinery, a natural question always asks if one can do better, using different methodologies. Let us make this explicit in this case.

Open Problem 3. *Does* INDEPENDENT SET ON TRIANGLE-FREE GRAPHS, *parameterized by a lower-bound k on the independent set, admit a kernel where the order of the graph is linear in k?*

Another classical result based on Ramsey's Theorem is due to Erdős and Szekeres [19].[1]

Theorem 6. *For each $k \geq 1$, there exists an $n_k \geq 1$ such that every point set (in the plane) of size n_k contains a convex subset of k points.*

This result holds in two-dimensional and in higher-dimensional Euclidean spaces. The question if there is a polynomial-time algorithm for finding such a set in 2D can be decided in polynomial time, as shown in [8,18]. However, Giannopoulos, Knauer and Werner have shown in [27] that CONVEX SET IN 3D Euclidean spaces (and in higher dimensions) is $\mathcal{NP}$-complete. In addition, they observed that Theorem 6 can be (ab)used to obtain a kernel result, which is of exponential size. They also ask the following question:

Open Problem 4. *Does* CONVEX SET IN 3D, *parameterized by a lower-bound k on the convex point set, admit a kernel of size polynomial in k?*

---

[1] For related results, we refer to P. Valtr's paper in this proceedings.

The idea of using Ramsey-style arguments in kernelization has been used by different groups of authors before, most notably in papers co-authored by Lozin and Rautenbach. Let us give one concrete example.[2]

Lemma 1 [15]. *For any natural numbers t and k, there is a number $N(t,k)$ such that every bipartite graph with a matching of size at least $N(t,k)$ contains either a bi-clique $K_{t,t}$ or an induced matching M_k.*

Corollary 3. *For each fixed t, the k-INDUCED MATCHING problem is fixed-parameter tractable in the class of $K_{t,t}$-free bipartite graphs.*

Open Problem 5. *Does the k-INDUCED MATCHING on $K_{t,t}$-free bipartite graphs allow a smaller kernel, avoiding Ramsey arguments?*

Degree-Diameter Problem. Let G be a graph of maximum degree d and diameter k. Forming a breadth-first tree shows that G has order at most $1 + d\sum_{i=0}^{k-1}(d-1)^i$; this number is known as the *Moore bound*. The problem of finding the largest possible graph for a given maximum degree and diameter is known as the *degree-diameter problem*. The Moore bound sets limits on this, but finding optimal graphs is in general an open combinatorial problem. Actually, this was one of Mirka's favorite ones, so she is still record holder for constructing various hitherto largest graphs for given d, k, as can be seen in the Wikipedia entry on the degree-diameter problem. Her survey [33] is the most-cited among her papers according to Google Scholar. In fact, we started to work on one concrete instance of the degree-diameter problem on the last trip Mirka could do, together with Joe. Katrin Casel still has quite a pile of papers with drawings and sketches on this work.

In order to make use of the relation of maximum degree and diameter bounds, the Moore bound is the main tool. Let us see some examples.

DOMINATING SET WITH DEGREE BOUND
Given: A graph $G = (V, E)$ with maximum degree at most d
Parameter: positive integers d, k
Question: Is there a *dominating set* $D \subseteq V$ with $|D| \leq k$?

Lemma 2. *If G has diameter $> 3k$, then G has no dominating set of size k.*

This gives the following reduction: If in the instance (G, d, k) graph G has more than $1 + d\sum_{i=0}^{3k-1}(d-1)^i$ many vertices, then this is a NO-instance.

Theorem 7. DOMINATING SET WITH DEGREE BOUND *is in $\mathcal{FPT}$.*

With a very similar argument, we can also derive a quick kernel for NB: if a graph has a large diameter, then it has also a large nonblocker set; if a graph has a vertex of large degree, then it also has a large nonblocker set. So, both situations can be ruled out by easy reduction rules, which means that both diameter and

---

[2] Further examples can be found in the paper of V. Lozin in this proceedings.

degree are linearly bounded by the size of the nonblocker set, which immediately gives a kernel for NB. However, this example also shows that kernels based on this type of argument are often unnecessarily big.

More effectively, these techniques can be applied to alliance problems. These were shown to belong to $\mathcal{FPT}$ in [23] using a search-tree argument. Yet, these problems (again) link to Mirka in several ways. For instance, Mirka collaborated with several graph theorists in Spain, see [13, 28], and so do I, see [22, 38] although so far with different sets of Spanish co-authors. We even had different sets of co-authors from Tarragona, not the largest city in Spain!

We omit details here for reasons of space, but mention that we can obtain kernels of sizes bounded by functions like $(k!)!$ (yes, that big).

Open Problem 6. *Devise better kernelization algorithms for alliance problems.*

More Contributions of Mirka to Our Community. I like to emphasize two further contributions of Mirka at this place. First, she was very active in helping especially the mathematical community in Indonesia to flourish. For myself, this brought along my hitherto only Indonesian co-author [25]. This will surely make a difference for the development of Discrete Mathematics in that region. Secondly, she founded the *Electronic Journal of Graph Theory and Applications*, which is now led by Edy Tri Baskoro, a former Ph.D. student of Mirka's. In connection with the alliance problem we studied above, I would like to mention [24], a paper that appeared in Mirka's journal, as it discusses several graph parameters similar to alliances, to which similar $\mathcal{FPT}$-membership arguments apply.

4 $\mathcal{FPT}$ Ideas for Combinatorialists

Let us now consider two variants of domination problems that have been introduced in the literature (independently) before.

A *Roman domination* function of a graph $G = (V, E)$ is $R : V \to \{0, 1, 2\}$ with

$$\forall v \in V : R(v) = 0 \Rightarrow \exists x \in N(v) : R(x) = 2.$$

The historical background seems to be the Roman Empire in the times of Emporer Constantine's who wanted to arrange his main armies in the main regions of the empire in a way that guards all regions. Here, regions without any armies can only be guarded if some neighboring regions have more than one army put on it. This domination variant was first discussed in [37] and more intensively in [12] with γ_R as a graph parameter; the following parameterized problem was studied in [20]. Actually, there are now dozens of papers on this problem, so it is hard to list them all.

ROMAN DOMINATION (ROMAN)
Given: A graph $G = (V, E)$, a positive integer k

Parameter: k
Question: Is there a *Roman domination* function R such that

$$R(V) := \sum_{x \in V} R(x) \leq k?$$

In 2011, I met Joe and Mirka again in Budapest at Eurocomb, see http://www.renyi.hu/conferences/ec11/participants.html. I had a presentation of some extremal combinatorial results on a graph parameter called *differential* that was brought to my attention by Sergio Bermudo who spent large parts of his sabbatical with me in Trier. These results were later published in [3].

Let $G = (V, E)$ be a graph. For $D \subseteq V$, $B(D) := N(D) \setminus D$ and $C(D) :=$ $V \setminus (D \cup B(D))$. The *differential of D* is $\partial(D) = |B(D)| - |D|$ and the *differential of G*, written $\partial(G)$, is equal to $\max\{\partial(D) \mid D \subseteq V\}$. This graph parameter was introduced in [31]. Our main results in [3] were as follows.

Theorem 8. *For any connected graph G of order $n \geq 3$, $\partial(G) \geq n/5$.*

Theorem 9. *For any connected graph G of order n that has minimum degree two, $\partial(G) \geq \frac{3n}{11}$, apart from five exceptional graphs listed below.*

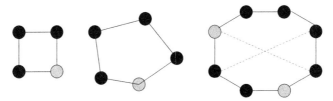

Red dotted edges may be present or not.

We were quite proud about our results, but pretty soon we discovered that they were already kind of known. Namely, as the main result of [5], we showed:

Theorem 10. *If G is a graph of order n, then $\gamma_R(G) = n - \partial(G)$.*

Now, both Theorems 8 and 9 follow from [9]. We simply forgot about the idea of looking into dual parameters! Our only consolation was that both Roman domination and differential were studied by similar sets of people. Hence, others did overlook this relation between γ_R and ∂, as well. So, most consequences that we list in [5] deal with improvements that were possible on earlier results on Roman domination by looking at the corresponding results for the differential of a graph, and vice versa.

Let us once more turn towards parameterized algorithms. As shown in [20], ROMAN DOMINATION is W[2]-complete, so most likely not in $\mathcal{FPT}$. Conversely, for the dual parameterization, coinciding with the standard parameterization for DIFFERENTIAL, or DIF for short (the natural parameterized decision problem associated to this graph parameter), we could prove in [4], using quite some intricate combinatorial arguments and quite a number of reduction rules:

Theorem 11. *There exists a kernelization algorithm for DIF that runs in linear time and that, given (G, k_d), provides a kernel (G', k_d) with $|V'| \leq 4k_d$, or that directly (and correctly) answers* YES.

Notice that this improves on the $5k_d$-vertex kernel that is rather immediate from Theorem 8. With a different approach we could even meet our improved bound of Theorem 9; see [1].

5   How Extreme Can Kernels Be?

Nowadays, it has become standard to base lower-bound results for kernels, in particular ruling out polynomial-size kernels, on the assumption that the polynomial-time hierarchy would not collapse to the third level; see [7,26]. However, in the world of approximation algorithms, where also a plentitude of different complexity assumptions are considered for inapproximability results, it is usually preferred if the assumption is that $\mathcal{P} \neq \mathcal{NP}$. Johnson [29] calls this *the gold standard* of inapproximability. Are there gold-standard lower-bound results for kernels? Indeed, there are, and they actually pre-date the lower-bound results that have nowadays become the standard ones.

I will report on these in the following, as they also relate to the times when I worked at The University of Newcastle. Back in 2003, I got the simple idea to rule out certain forms of kernel results, assuming that linear-size kernel exists both for the standard parameterization and the dual parameterization.

Theorem 12 [10]. *Let $\mathcal{P}$ be an $\mathcal{NP}$-hard parameterized problem. If $\mathcal{P}$ has an αk-size kernel and if its dual $\mathcal{P}_d$ has an $\alpha_d k_d$-size kernel $(\alpha, \alpha_d \geq 1)$, then $(\alpha - 1)(\alpha_d - 1) \geq 1$ unless $\mathcal{P} = \mathcal{NP}$.*

Notice that different results could be obtained depending on the chosen size function. However, there are only few problems where both primal and dual variants admit linear-size kernels. Most of them are restrictions of graph problems to planar graphs (or more generally, graphs of bounded genus). For instance, VERTEX COVER admits a $2k$-vertex kernel on general and on planar graphs, and its dual INDEPENDENT SET (only on planar graphs) allows a $4k_d$-vertex kernel due to the Four-Color-Theorem. This has the drawback that we can rule out kernels for VERTEX COVER smaller than $\frac{4}{3}k$ only if these reductions are fit for planar graphs, i.e., if they preserve planarity. More generally, such a lower-bound result holds true for four-colorable graphs, where the reduction rules should preserve four-colorability. Analogously, we can use Theorem 11 to conclude:

Corollary 4. *For any $\epsilon > 0$, there is no $(4/3 - \epsilon)k$ kernel for* PLANAR k-ROMAN DOMINATION *unless* $\mathcal{P} = \mathcal{NP}$.

Open Problem 7. *Design a linear-size kernel for* PLANAR k-ROMAN DOMINATION *that comes close to the lower-bound expressed in Corollary 4.*

It must have been in 2008 upon an excursion at a conference in China when Mike Fellows told me about another gold-standard lower-bound on kernel result, namely for ROOTED LONG PATH. This result (together with other similar ones) was published in [11] but (somehow surprisingly) did not find the popularity of the non-gold-standard approaches. In [21] we developed a somewhat more general framework in which one could show gold-standard lower bound results. For instance, this approach can be used for annotated versions of alliance problems.

Acknowledgements. This paper has profited a lot from discussions we had with many colleagues, in particular, with Faisal Abu-Khzam, Ljiljana Brankovic and Mike Fellows. Thanks to Frances Rosamond and Ulrike Stege for proofreading.

References

1. Abu-Khzam, F.N., Bazgan, C., Chopin, M., Fernau, H.: Data reductions and combinatorial bounds for improved approximation algorithms. J. Comput. Syst. Sci. **82**(3), 503–520 (2016)
2. Abu-Khzam, F.N., Fernau, H.: Kernels: annotated, proper and induced. In: Bodlaender, H.L., Langston, M.A. (eds.) IWPEC 2006. LNCS, vol. 4169, pp. 264–275. Springer, Heidelberg (2006). https://doi.org/10.1007/11847250_24
3. Bermudo, S., Fernau, H.: Lower bounds on the differential of a graph. Discret. Math. **312**, 3236–3250 (2012)
4. Bermudo, S., Fernau, H.: Combinatorics for smaller kernels: the differential of a graph. Theoret. Comput. Sci. **562**, 330–345 (2015)
5. Bermudo, S., Fernau, H., Sigarreta, J.M.: The differential and the Roman domination number of a graph. Appl. Anal. Discret. Math. **8**, 155–171 (2014)
6. Blank, M.: An estimate of the external stability number of a graph without suspended vertices. Prikl. Mat. i Programmirovanie Vyp. **10**, 3–11 (1973). (in Russian)
7. Bodlaender, H.L., Downey, R.G., Fellows, M.R., Hermelin, D.: On problems without polynomial kernels. J. Comput. Syst. Sci. **75**, 423–434 (2009)
8. Boyce, J.E., Dobkin, D.P., (Scot) Drysdale III, R.L., Guibas, L.J.: Finding extremal polygons. SIAM J. Comput. **14**(1), 134–147 (1985)
9. Chambers, E.W., Kinnersley, B., Prince, N., West, D.B.: Extremal problems for Roman domination. SIAM J. Discret. Math. **23**, 1575–1586 (2009)
10. Chen, J., Fernau, H., Kanj, I.A., Xia, G.: Parametric duality and kernelization: lower bounds and upper bounds on kernel size. SIAM J. Comput. **37**, 1077–1108 (2007)
11. Chen, Y., Flum, J., Müller, M.: Lower bounds for kernelizations and other preprocessing procedures. Theory Comput. Syst. **48**(4), 803–839 (2011)
12. Cockayne, E.J., Dreyer Jr., P.A., Hedetniemi, S.M., Hedetniemi, S.T.: Roman domination in graphs. Discret. Math. **278**, 11–22 (2004)
13. Conde, J., Miller, M., Miret, J.M., Saurav, K.: On the nonexistence of almost Moore digraphs of degree four and five. Math. Comput. Sci. **9**(2), 145–149 (2015)
14. Cygan, M., Fomin, F., Kowalik, Ł., Lokshtanov, D., Marx, D., Pilipczuk, M., Pilipczuk, M., Saurabh, S.: Parameterized Algorithms. Springer, Cham (2015). https://doi.org/10.1007/978-3-319-21275-3
15. Dabrowski, K.K., Demange, M., Lozin, V.V.: New results on maximum induced matchings in bipartite graphs and beyond. Theoret. Comput. Sci. **478**, 33–40 (2013)

16. Dehne, F., Fellows, M., Fernau, H., Prieto, E., Rosamond, F.: NONBLOCKER: parameterized algorithmics for MINIMUM DOMINATING SET. In: Wiedermann, J., Tel, G., Pokorný, J., Bieliková, M., Štuller, J. (eds.) SOFSEM 2006. LNCS, vol. 3831, pp. 237–245. Springer, Heidelberg (2006). https://doi.org/10.1007/11611257_21

17. Downey, R.G., Fellows, M.R.: Fundamentals of Parameterized Complexity. Texts in Computer Science. Springer, London (2013). https://doi.org/10.1007/978-1-4471-5559-1

18. Eppstein, D., Overmars, M.H., Rote, G., Woeginger, G.J.: Finding minimum area k-gons. Discret. Comput. Geom. **7**, 45–58 (1992)

19. Erdős, P., Szekeres, G.: A combinatorial problem in geometry. Compositio Mathematica **2**, 463–470 (1935)

20. Fernau, H.: Roman domination: a parameterized perspective. Int. J. Comput. Math. **85**, 25–38 (2008)

21. Fernau, H., Fluschnik, T., Hermelin, D., Krebs, A., Molter, H., Niedermeier, R.: Diminishable parameterized problems and strict polynomial kernelization. Technical report, Cornell University, arXiv:1611.03739 (2016)

22. Fernau, H., Rodríguez, J.A., Sigarreta, S.M.: Offensive r-alliances in graphs. Discret. Appl. Math. **157**, 177–182 (2009)

23. Fernau, H., Raible, D.: Alliances in graphs: a complexity-theoretic study. In: van Leeuwen, J., Italiano, G.F., van der Hoek, W., Meinel, C., Sack, H., Plášil, F., Bieliková, M. (eds.) SOFSEM 2007, Proceedings, vol. II, pp. 61–70. Institute of Computer Science ASCR, Prague (2007)

24. Fernau, H., Rodríguez-Velázquez, J.A.: A survey on alliances and related parameters in graphs. Electron. J. Graph Theory Appl. **2**(1), 70–86 (2014)

25. Fernau, H., Ryan, J.F., Sugeng, K.A.: A sum labelling for the generalised friendship graph. Discret. Math. **308**, 734–740 (2008)

26. Fortnow, L., Santhanam, R.: Infeasibility of instance compression and succinct PCPs for NP. In: Dwork, C. (ed.) ACM Symposium on Theory of Computing, STOC, pp. 133–142. ACM (2008)

27. Giannopoulos, P., Knauer, C., Werner, D.: On the computational complexity of Erdős-Szekeres and related problems in $\mathbb{R}^3$. In: Bodlaender, H.L., Italiano, G.F. (eds.) ESA 2013. LNCS, vol. 8125, pp. 541–552. Springer, Heidelberg (2013). https://doi.org/10.1007/978-3-642-40450-4_46

28. Gómez, J., Miller, M.: On the existence of radial Moore graphs for every radius and every degree. Eur. J. Comb. **47**, 15–22 (2015)

29. Johnson, D.S.: The NP-completeness column: the many limits on approximation. ACM Trans. Algorithms **2**, 473–489 (2006)

30. Kim, J.H.: The Ramsey number $R(3, t)$ has order of magnitude $t^2/\log t$. Random Struct. Algorithms **7**(3), 173–208 (1995)

31. Mashburn, J.L., Haynes, T.W., Hedetniemi, S.M., Hedetniemi, S.T., Slater, P.J.: Differentials in graphs. Utilitas Math. **69**, 43–54 (2006)

32. McCuaig, B., Shepherd, B.: Domination in graphs of minimum degree two. J. Graph Theory **13**, 749–762 (1989)

33. Miller, M., Širáň, J.: Moore graphs and beyond: a survey of the degree/diameter problem. Electron. J. Comb. **1000**, DS14 (2005)

34. Ore, O.: Theory of Graphs. Colloquium Publications, vol. XXXVIII. American Mathematical Society, Providence (1962)

35. Prieto, E.: Systematic kernelization in FPT algorithm design. Ph.D. thesis, The University of Newcastle, Australia (2005)

36. Reed, B.: Paths, stars, and the number three. Comb. Probab. Comput. **5**, 277–295 (1996)
37. ReVelle, C.S., Rosing, K.E.: Defendens imperium Romanum: a classical problem in military strategy. Am. Math. Mon. **107**, 585–594 (2000). http://www.jhu.edu/~jhumag/0497web/locate3.html
38. Sigarreta, J.M., Bermudo, S., Fernau, H.: On the complement graph and defensive k-alliances. Discret. Appl. Math. **157**, 1687–1695 (2009)
39. Slater, P.J.: Enclaveless sets and MK-systems. J. Res. Natl. Bur. Stand. **82**(3), 197–202 (1977)
40. Sloper, C.: Techniques in parameterized algorithm design. Ph.D. thesis, University of Bergen, Norway (2005)

Recent Advances of Palindromic Factorization

Mai Alzamel[1,2(✉)] and Costas S. Iliopoulos[1]

[1] Department of Informatics, King's College London, London, UK
{mai.alzamel,costas.iliopoulos}@kcl.ac.uk
[2] King Saud University, Riyadh, Kingdom of Saudi Arabia

Abstract. This paper provides an overview of six particular problems of palindromic factorization and recent algorithmic improvements in solving them.

1 Introduction

1.1 General Definitions

Let $S = S[1]S[2] \cdots S[n]$ be a *string* of *length* $|S| = n$ over an alphabet Σ. We consider the case of an integer alphabet; in this case each letter can be replaced by its rank so that the resulting string consists of integers in the range $\{1, \ldots, n\}$. For two positions i and j, where $1 \leq i \leq j \leq n$, in S, we denote the *factor* $S[i]S[i+1] \cdots S[j]$ of S by $S[i \mathinner{.\,.} j]$. We denote the reverse string of S by S^R, i.e. $S^R = S[n]S[n-1] \cdots S[1]$. The empty string (denoted by ε) is the unique string over Σ of length 0. A string S is said to be a *palindrome* if and only if $S = S^R$. If $S[i \mathinner{.\,.} j]$ is a palindrome, the number $\frac{i+j}{2}$ is called the center of $S[i \mathinner{.\,.} j]$. Let $S[i \mathinner{.\,.} j]$, where $1 \leq i \leq j \leq n$, be a palindromic factor in S. It is said to be a *maximal palindrome* if there is no longer palindrome in S with center $\frac{i+j}{2}$. Note that a maximal palindrome can be a factor of another palindrome.

Note that any single letter is a palindrome and, hence, every string can always be factorized into palindromes. However, not every string can be factorized into maximal palindromes; e.g. consider $S = \texttt{abaca}$ [2].

In this paper we present a survey of six novel algorithms of palindromic factorization. We start with maximal palindromic factorization presented by [2] in Sect. 2. Later, we explain palindromic factorization with gaps, maximal palindromic factorization with errors and maximal palindromic factorization with gaps and errors presented by [1] in Sects. 3, 4 and 5.

Finally, we show in Sect. 6 an efficient algorithm of palindromes in weighted strings presented by [3]

M. Alzamel—Fully supported by the Saudi Ministry of Higher Education and partially supported by the Onassis Foundation.
C.S. Iliopoulos—Partially supported by the Onassis Foundation.

L. Brankovic et al. (Eds.): IWOCA 2017, LNCS 10765, pp. 37–46, 2018.
https://doi.org/10.1007/978-3-319-78825-8_4

2   Maximal Palindromic Factorization

In this section we present an algorithm to compute the maximal palindromic factorization of a given string S presented by Alatabbi et al. [2]. They first present some notions required to present the algorithm. First of all, they use $\mathcal{MP}(S)$ to denote the set of center distinct maximal palindromes of S. They further extend this notation as follows. They use $\mathcal{MP}(S)[i]$, where $1 \leq i \leq n$ to denote the set of maximal palindromes with center i. Further, for the string S, they denote the set of all *prefix palindromes* (*suffix palindromes*) as $\mathcal{PP}(S)$ ($\mathcal{SP}(S)$).

Proposition 1. *The position i could be the center of at most two maximal palindromic factors, therefore; $\mathcal{MP}(S)[i]$ contains at most two elements, where $1 \leq i \leq n$, hence; there are at most $2n$ elements in $\mathcal{MP}(S)$.*

On the other hand, they use $\mathcal{MPL}(S)[i]$ to denote the set of the lengths of all maximal palindromes ending at position i,where $1 \leq i \leq n$ in S.

$$\mathcal{MPL}(s)[i] = \{2\ell - 1 \ |s[i - \ell + 1 \ldots i + \ell - 1] \in \mathcal{MP}(s)\}$$
$$\cup \{2\ell' \ |s[i - \ell' \ldots i + \ell' - 1] \in \mathcal{MP}(s)\} \quad (1)$$

where $1 \leq i \leq n$, with 2ℓ and $2\ell' + 1$ are the lengths of the odd and even palindromic factors respectively.

Proposition 2. *The set $\mathcal{MPL}(S)$ (Eq. 1) can be computed in linear time from the set $\mathcal{MP}(s)$.*

They define the list $\mathcal{U}(S)$ such that for each $1 \leq i \leq n$,
    $\mathcal{U}(S)[i]$ stores the position j such that $j + 1$ is the starting position of a maximal palindromic factors ending at i and j is the end of another maximal palindromic substring.
    Clearly, this can be easily computed once $\mathcal{MPL}(S)$ is computed.

$$\mathcal{U}[i][j] = i - \mathcal{MPL}(s)[i][j] \quad (2)$$

One can observe, from 1, that the sets $\mathcal{MPL}(S)$ and $\mathcal{U}(S)$ contain at most $2n$ elements. Given the list $\mathcal{U}(S)$ for a string S, they define a directed graph $\mathcal{G}_s = (\mathcal{V}, \mathcal{E})$ as follows. There are $\mathcal{V} = \{i \mid 1 \leq i \leq n\}$ and $\mathcal{E} = \{(i, j) \mid j \in \mathcal{U}(S)[i]\}$. Note that (i, j) is a directed edge where the direction is from i to j. The steps of the proposed algorithms are as follows.

MPF Algorithm: Maximal Palindromic Factorization Algorithm
Input: A String S of length n
Output: Maximal Palindromic Factorization of S

1. Compute the set of maximal palindromes $\mathcal{MP}(S)$ and identify the set of prefix palindromes $\mathcal{PP}(S)$.
2. Compute the list $\mathcal{MPL}(S)$.
3. Compute the list $\mathcal{U}(S)$.

4. Construct the graph $\mathcal{G}_s = (\mathcal{V}, \mathcal{E})$.
5. Do a breadth first search on $\mathcal{G}_s$ assuming the vertex n as the source.
6. Identify the shortest path $P \approx n \rightsquigarrow v$ such that v is the end position of a palindrome belonging to $\mathcal{PP}(S)$. Suppose $P \approx \langle n = p_k \ p_{k-1} \ \ldots \ p_2 \ p_1 = v \rangle$.
7. Return $S = S[1..p_1] \ S[p_1 + 1..p_2] \ \ldots \ S[p_{k-1} + 1..p_k]$.

Theorem 1. *Given a string S of length n, (Maximal Palindromic Factorization (MPF)) Algorithm correctly computes the maximal palindromic factorization of S in $O(n)$ time.*

3   Palindromic Factorization with Gaps

In this section we present an efficient solution to the palindromic factorization with gaps problem has been introduced by Adamczyk et al. [1].

It is based on several transformations of the algorithm for computing a palindromic factorization by Fici et al. [5]. For a string S of length n this algorithm works in $\mathcal{O}(n \log n)$ time. The algorithm consists of two steps:

1. Let P_j be the sorted list of starting positions of all palindromes ending at position j in S. This list may have size $\mathcal{O}(j)$. However, it follows from combinatorial properties of palindromes that the sequence of consecutive differences in P_j is non-increasing and contains at most $\mathcal{O}(\log j)$ distinct values. Let $P_{j,\Delta}$ be the maximal sublist of P_j containing elements whose predecessor in P_j is smaller by exactly Δ. Then there are $\mathcal{O}(\log j)$ such sublists in P_j. Hence, P_j can be represented by a set G_j of size $\mathcal{O}(\log j)$ which consists of triples of the form (i, Δ, k) that represent $P_{j,\Delta} = \{i, i + \Delta, \ldots, i + (k-1)\Delta\}$. The triples are sorted according to decreasing values of Δ and all starting positions in each triple are greater than in the previous one. Fici et al. show that G_j can be computed from G_{j-1} in $\mathcal{O}(\log j)$ time.
2. Let $PL[j]$ denote the number of palindromes in a palindromic factorization of $S[1..j]$. Fici et al. show that it can be computed via a dynamic programming approach, using all palindromes from G_j in $\mathcal{O}(\log j)$ time. Their algorithm works as follows. Let $PL_\Delta[j]$ be the minimum number of palindromes we can factorized $S[1..j]$ in, provided that we use a palindrome from $(i, \Delta, k) \in G_j$. Then $PL_\Delta[j]$ can be computed in constant time using $PL_\Delta[j - \Delta]$ based on the fact that if $(i, \Delta, k) \in G_j$ and $k \geq 2$, then $(i, \Delta, k-1) \in G_{j-\Delta}$. Exploiting this fact, $PL_\Delta[j]$ can be computed by only considering $PL_\Delta[j - \Delta]$ and the shortest palindrome in (i, Δ, k).
   Finally, $PL[j]$ can be computed from all such $PL_\Delta[j]$ values.

To solve the PALINDROMIC FACTORIZATION WITH GAPS problem, Adamczyk et al. [1] algorithm firstly modify each of the triples in G_j to reflect the length constraint (m). More precisely, due to the length constraint, in each G_j some triples will disappear completely, and at most one triple will get *trimmed* (i.e. the parameter k will be decreased).

The algorithm then computes an array $MG[1..n][0..g]$ such that $MG[j][q]$ is the minimum possible total length of gaps in a palindromic factorization of

$S[1 . . j]$, provided that there are at most q gaps. Simultaneously, Adamczyk et al. [1] algorithm computes an auxiliary array $MG'[1 . . n][0 . . g]$ such that $MG'[j][q]$ is the minimum possible total length of gaps up to position j provided that this position belongs to a gap: at most the q-th one.

The following formula for $j > 0$ and $q \geq 0$:

$$MG[j][q] = \min(MG'[j][q], \min_{\Delta}\{MG_{\Delta}[j][q]\})$$

where $MG_{\Delta}[j][q]$ is the partial minimum computed only using palindromes from $(i, \Delta, k) \in G_j$. The formula means: either there is a gap at position j, or using a palindrome ending at position j. Also $MG[0][q]$ is filled by zeros for any $q \geq 0$.

Adamczyk et al. [1] algorithm computes $MG_{\Delta}[j][q]$ for $(i, \Delta, k) \in G_j$ using the same approach as Fici et al. [5] used for PL_{Δ}, ignoring the triples that disappear due to the length constraint. If there is a triple that got trimmed, then the corresponding triple at position $j - \Delta$ (from which they reuse the values in the dynamic programming) must have got trimmed as well. More precisely, if the triple (i, Δ, k) is trimmed to (i, Δ, k') at position j, then at position $j - \Delta$ there is a triple $(i, \Delta, k-1)$ which is trimmed to $(i, \Delta, k'-1)$; that is, by the same number of palindromes. Consequently, to compute $MG_{\Delta}[j][q]$ from $MG_{\Delta}[j - \Delta][q]$, they need to include one additional palindrome (the shortest one in the triple) just as in Fici et al.'s approach.

Finally, for $j > 0$ and $q > 0$ they compute MG' using the following formula:

$$MG'[j][q] = \min(MG'[j - 1][q], MG[j - 1][q - 1]) + 1.$$

The first case corresponds to continuing the gap from position j, whereas the second to using a palindrome finishing at position $j-1$ or a gap finishing at position $j - 1$ (the latter will be suboptimal). Here the border cases are $MG'[j][0] = \infty$ for $j \geq 0$ and $MG'[0][q] = \infty$ for $q > 0$.

Thus Adamczyk et al. [1] arrive at the complete solution to the problem.

Theorem 2. *The Palindromic Factorization with Gaps problem can be solved in $\mathcal{O}(n \log n \cdot g)$ time and $\mathcal{O}(n \cdot g)$ space.*

4   Computing Maximal Palindromes with Errors

We show an algorithm presented by Adamczyk et al. [1] to compute maximal δ-palindromes under the edit distance within $O(n \cdot \delta)$. If u is a δ-palindrome under the edit distance, then there exists a palindrome v such that the minimal number of edit operations (insertion, deletion, substitution) required to transform u to v is at most δ. The following simple observation shows that it can restrict edit operations to deletions and substitutions only, which Adamczyk et al. [1] call in what follows the *restricted edit operations*. Intuitively, instead of inserting at position i a character to match the character at position $|u| - i + 1$, the character can be deleted at position $|u| - i + 1$.

Observation 3. *Let u be a δ-palindrome and v a palindrome such that the edit distance between u and v is minimal. Then there exists a palindrome v′ such that the number of restricted edit operations needed to transform u to v′ is equal to the edit distance between u and v.*

Definition 1. *A (LGPal-queries) is a maximal palindromes are computed using Gusfield's approach* [6].

Adamczyk et al. [1] extend a maximal δ-palindrome $S[i \mathinner{.\,.} j]$ to a maximal $(δ+1)$-palindrome in three ways; either ignore the letter $S[i-1]$ and then perform an LGPal-query, or ignore the letter $S[j+1]$ and then perform an LGPal-query, or ignore both and then perform the LGPal-query. More formally:

Definition 2. *Assume that $S[i \mathinner{.\,.} j]$ is a δ-palindrome. Then it says that each of the factors $S[i′ \mathinner{.\,.} j′]$ for:*

- $i′ = i - 1 - d$, $j′ = j + d$, where $d = LGPal(i - 2, j + 1)$
- $i′ = i - d$, $j′ = j + 1 + d$, where $d = LGPal(i - 1, j + 2)$
- $i′ = i - 1 - d$, $j′ = j + 1 + d$, where $d = LGPal(i - 2, j + 2)$

is an extension of $S[i \mathinner{.\,.} j]$. If the index $i′$ is smaller than 1 or the index $j′$ is greater than $|S|$, the corresponding extension is not possible. They also say that $S[i \mathinner{.\,.} j]$ can be extended to any of the three strings $S[i′ \mathinner{.\,.} j′]$.

Clearly, the extensions of a δ-palindrome are always $(δ + 1)$-palindromes.

To facilitate the case of δ-palindromes being prefixes or suffixes of the text, they also introduce the following *border-reductions* for $S[i \mathinner{.\,.} j]$ being a δ-palindrome:

- If $i = 1$, a border reduction leads to $S[1 \mathinner{.\,.} j - 1]$.
- If $j = n$, a border reduction leads to $S[i + 1 \mathinner{.\,.} n]$.

If any of the reductions is possible, they also say that $S[i \mathinner{.\,.} j]$ can be border-reduced to the corresponding strings. As previously, border-reductions of a δ-palindrome are always $(δ + 1)$-palindromes.

Lemma 1. *Given a maximal δ-palindrome $S[i′ \mathinner{.\,.} j′]$ with $δ > 0$, there exists a maximal $(δ - 1)$-palindrome $S[i \mathinner{.\,.} j]$ which can be extended or border-reduced to $S[i′ \mathinner{.\,.} j′]$.*

The combinatorial characterization of Lemma 1 yields an algorithm for generating all maximal d-palindromes, for all centers and subsequent $d = 0, \ldots, δ$.

Recall maximal 0-palindromes are computed using Gusfield's approach (LGPal-queries). For a given $d < δ$, they consider all the maximal d-palindromes and try to extend each of them in all three possible ways (and border-reduce, if possible). This way they obtain a number of $(d+1)$-palindromes amongst which, by Lemma 1, are all maximal $(d + 1)$-palindromes. To exclude the non-maximal ones, they group the $(d + 1)$-palindromes by their centers (in $\mathcal{O}(n)$ time via bucket sort) and retain only the longest one for each center.

They arrive at the following intermediate result.

Lemma 2. *Under the edit distance, all maximal δ-palindromes in a string of length n can be computed in $\mathcal{O}(n \cdot δ)$ time and $\mathcal{O}(n)$ space.*

5   Maximal Palindromic Factorization with Gaps and Errors

We show an algorithm presented by Adamczyk et al. [1] to solve maximal palin-dromic factorization with gaps and errors problem in $\mathcal{O}(n \cdot (g + \delta))$ time and $\mathcal{O}(n \cdot g)$ space.

Let $\mathcal{F}$ be a set of factors of the text $S[1 .. n]$. In this section they develop a general framework that allows to factorized S into factors from $\mathcal{F}$, allowing at most g gaps. They call such a factorization a $(g, \mathcal{F})$-factorization of S.

The goal is to find a $(g, \mathcal{F})$-factorization of S that minimizes the total length of gaps. The authors aim at the time complexity $\mathcal{O}((n + |\mathcal{F}|) \cdot g)$ and space complexity $\mathcal{O}(n \cdot g + |\mathcal{F}|)$.

In the proposed solution Adamczyk et al. [1] use dynamic programming to compute two arrays, similar to the ones used in Sect. 3:

$MG[1 .. n][0 .. g]$: $MG[j][q]$ is the minimum total length of gaps in a $(q, \mathcal{F})$-factorization of $S[1 .. j]$.

$MG'[1 .. n][0 .. g]$: $MG'[j][q]$ is the minimum total length of gaps in a $(q, \mathcal{F})$-factorization of $S[1 .. j]$ for which the position j belongs to a gap.

They use the following formulas, for $j > 0$ and $q > 0$:

$$MG[j][q] = \min(MG'[j][q], \min_{S[a .. j] \in \mathcal{F}} MG[a - 1][q])$$

$$MG'[j][q] = \min(MG[j - 1][q - 1], MG'[j - 1][q]) + 1$$

The border cases are exactly the same as in Sect. 3.

They apply this approach to maximal δ-palindromes in each of the considered metrics (see the classic result from [6] for the Hamming distance and Lemma 2 for the edit distance) to obtain the following result.

Theorem 4. *The* MAXIMAL δ-PALINDROMIC FACTORIZATION WITH GAPS *problem under the Hamming distance or the edit distance can be solved in* $\mathcal{O}(n \cdot (g + \delta))$ *time and* $\mathcal{O}(n \cdot g)$ *space.*

6   Maximal Palindromic Factorization of Weighted String

In this section, we show an algorithm to compute a smallest maximal z-palindromic factorization of a given weighted string X of length n for a given cumulative threshold $1/z \in (0, 1]$ has been presented by Alzamel et at [3]. Our algorithm follows the one of Alatabbi et al. for computing a smallest maximal palindromic factorization of standard strings [2] with some crucial modifications. Recall by $\mathcal{MP}(x)$, we denote the set of center-distinct maximal palindromes of string x We will use the below two facts related to palindromes:

Fact 5 ([6]). *Given a string x, $\mathcal{MP}(x)$ can be computed in time $\mathcal{O}(|x|)$.*

Fact 6 (Trivial). *Let $x[i..j]$ be a palindrome of string x with center c and let u, $|u| < j - i + 1$, be a factor of x with center c. Then u is also a palindrome.*

Note that for clarity we use upper case letters for weighted strings, e.g. X, and lower case letters, e.g. x, for standard strings.

We start with some definitions related to weighted strings:

Definition 3. *A weighted string X on an alphabet Σ is a finite sequence of n sets. Every $X[i]$, for all $0 \leq i < n$, is a set of ordered pairs $(s_j, \pi_i(s_j))$, where $s_j \in \Sigma$ and $\pi_i(s_j)$ is the probability of having letter s_j at position i. Formally, $X[i] = \{(s_j, \pi_i(s_j)) \mid s_j \neq s_l \text{ for } j \neq l, \text{ and } \Sigma \pi_i(s_j) = 1\}$. A letter s_j occurs at position i of X if and only if the occurrence probability of letter s_j at position $i, \pi_i(s_j)$, is greater than 0.*

Definition 4. *A string u of length m is a factor of a weighted string X if and only if it occurs at starting position i with cumulative probability $\prod_{j=0}^{m-1} \pi_{i+j}(u[j]) > 0$. Given a cumulative weight threshold $1/z \in (0, 1]$, we say factor u is z-valid, if it occurs at position i with cumulative probability $\prod_{j=0}^{m-1} \pi_{i+j}(u[j]) \geq 1/z$.*

Definition 5. *Given a cumulative weight threshold $1/z \in (0, 1]$, a weighted string X of length m is a z-palindrome if and only if there exists at least one z-valid factor u of X of length m which is a palindrome.*

If the weighted string $X[i..j]$ is a z-palindrome, we analogously define the number $\frac{i+j}{2}$ as the center of $X[i..j]$ in X and $\frac{j-i+1}{2}$ as the radius of $X[i..j]$.

Definition 6. *Let X be a weighted string of length n, $1/z \in (0, 1]$ a cumulative weight threshold, and $X[i..j]$, where $0 \leq i \leq j \leq n - 1$, a z-palindrome. Then $X[i..j]$ is a maximal z-palindrome if there is no other z-palindrome in X with center $\frac{i+j}{2}$ and larger radius.*

We proceed as follows: By $\mathcal{MP}(X, z)$, we denote the set of center-distinct maximal z-palindromes of our weighted string X. We present a z-palindrome with center c and radius r by (c, r). For each position of X we define the *heaviest letter* as the letter with the maximum probability (breaking ties arbitrarily). We consider the string obtained from X by choosing at each position the heaviest letter. We call this the *heavy string* of X.

We define a collection $\mathcal{Z}_X$ of $\lfloor z \rfloor$ *special-weighted strings* of X, denoted by $\mathcal{Z}_k$, $0 \leq k < \lfloor z \rfloor$. Each $\mathcal{Z}_k$ is of length n and it has the following properties. Each position j in $\mathcal{Z}_k$ contains at most one letter with positive probability and it corresponds to position j in X. If f is a z-valid factor occurring at position j of X, then f occurs at position j in some of the $\mathcal{Z}_k$'s. The combinatorial observation telling us that this is possible is due to Barton et al. [4]. For clarity of presentation we write $\mathcal{Z}_k$'s as standard strings.

Lemma 3 ([4]). *Given a weighted string X of length n and a cumulative weight threshold $1/z \in (0, 1]$, the $\lfloor z \rfloor$ special-weighted strings of X can be constructed in time and space $\mathcal{O}(nz)$.*

Fact 7. *Given a weighted string X of length n and a cumulative weight threshold $1/z \in (0, 1]$, we have that $\mathcal{MP}(X, z) \subseteq \mathcal{MP}(\mathcal{Z}_0, z) \cup \mathcal{MP}(\mathcal{Z}_1, z) \cup \ldots \cup \mathcal{MP}(\mathcal{Z}_{\lfloor z \rfloor - 1}, z)$.*

There are two steps for the correct computation of $\mathcal{MP}(X, z)$. First, we compute the set $\mathcal{A}_k$ of all maximal palindromes of the heavy string of $\mathcal{Z}_k$, for all $0 \le k < \lfloor z \rfloor$, using Fact 5. We then need to adjust the radius of each reported palindrome for $\mathcal{Z}_k$ to ensure that it is z-valid in X (the center should not change). To achieve this, we compute an array $\mathcal{R}_k$, for each $\mathcal{Z}_k$, such that $\mathcal{R}_k[2c]$ stores the radius of the longest factor at center c in $\mathcal{Z}_k$ which is a z-valid factor of X at center c, e.g. $\mathcal{R}_k[2c] = \frac{j-i+1}{2}$, $c = (i + j)/2$, if $\mathcal{Z}_k[i \ldots j]$ is a z-valid factor of X centered at c, and $\mathcal{Z}_k[i - 1 \ldots j + 1]$ is not a z-valid factor of X. By Fact 7, we cannot guarantee that all (c, r) in $\mathcal{MP}(\mathcal{Z}_k, z)$ are necessarily in $\mathcal{MP}(X, z)$. Hence, the second step is to compute $\mathcal{MP}(X, z)$ from $\mathcal{MP}(\mathcal{Z}_k, z)$ by taking the maximum radius per center and filtering out everything else.

Lemma 4. *Given a weighted string X of length n, a cumulative weight threshold $1/z \in (0, 1]$, and the special-weighted strings $\mathcal{Z}_X$ of X, each $\mathcal{R}_k$, $0 \le k < \lfloor z \rfloor$, can be computed in time $\mathcal{O}(n)$.*

After computing $\mathcal{A}_k$ and $\mathcal{R}_k$, we perform the following check for each palindrome $(c, r) \in \mathcal{A}_k$. If $r > \mathcal{R}_k[2c]$, the palindrome with radius r is not z-valid but the factor with radius $\mathcal{R}_k[2c]$ is z-valid and maximal (by definition) and palindromic (by Fact 6); if $r \le \mathcal{R}_k[2c]$, the palindrome with radius r_i must be z-valid and it is maximal. Therefore we set $(c, r) \in \mathcal{MP}(\mathcal{Z}_k, z)$, such that $r = \min\{r, \mathcal{R}_k[2c]\}$, $0 \le 2c \le 2n - 2$, and $r \ge 1/2$.

To go from $\mathcal{MP}(\mathcal{Z}_k, z)$ to $\mathcal{MP}(X, z)$ we need to take the maximum radius for each center. Therefore for each center $c/2$, $0 \le c \le 2n - 2$, we set $(c/2, r) \in \mathcal{MP}(X, z)$, such that $r = \max\{r_k | (c/2, r_k) \in \mathcal{MP}(\mathcal{Z}_k, z), 0 \le k < \lfloor z \rfloor\}$. We thus arrive at the first result of this article.

Theorem 8. *Given a weighted string X of length n and a cumulative weight threshold $1/z \in (0, 1]$, all maximal z-palindromes in X can be computed in time and space $\mathcal{O}(nz)$.*

After the computation of $\mathcal{MP}(X, z)$, we are in a position to apply the algorithm by Alatabbi et al. [2] to find the smallest maximal z-palindromic factorization. We define a list $\mathcal{F}$ such that $\mathcal{F}[i]$, $0 \le i \le n - 1$, stores the set of the lengths of all maximal z-palindromes ending at position i in X. We also define a list $\mathcal{U}$ such that $\mathcal{U}[i]$, $0 \le i \le n - 1$, stores the set of positions j, such that $j + 1$ is the starting position of a maximal z-palindrome in X and i is the ending position of this z-palindrome. Thus for a given $\mathcal{F}[i] = \{\ell_0, \ell_1, \ldots, \ell_q\}$, we have that $\mathcal{U}[i] = \{i - \ell_0, i - \ell_1, \ldots, i - \ell_q\}$. Note that $\mathcal{U}[i]$ can contain a "-1" element if there exists a maximal z-palindrome starting at position 0 and ending at position i. Note that the number of elements in $\mathcal{MP}(X, z)$ is at most $2n - 1$, and, hence, $\mathcal{F}$ and $\mathcal{U}$ can contain at most $2n - 2$ elements. The lists $\mathcal{F}$ and $\mathcal{U}$ can be computed trivially from $\mathcal{MP}(X, z)$. Finally, we define a directed graph $\mathcal{G}_X = (\mathcal{V}, \mathcal{E})$, where

$\mathcal{V} = \{i \mid -1 \leq i \leq n-1\}$ and $\mathcal{E} = \{(i,j) \mid j \in \mathcal{U}[i]\}$. Note that (i,j) is a directed edge from i to j. We do a breath first search on $\mathcal{G}_X$ assuming the vertex $n-1$ as the source and identify the shortest path from $n-1$ to -1, which gives the factorization. We formally present the above as Algorithm SMPF for computing a smallest maximal z-palindromic factorization and obtain the following result.

Theorem 9. *Given a weighted string X of length n and a cumulative weight threshold $1/z \in (0,1]$, Algorithm SMPF correctly solves the problem* SMALLEST MAXIMAL z-PALINDROMIC FACTORIZATION *in time and space $\mathcal{O}(nz)$.*

1 **Algorithm** SMPF$(X, n, 1/z)$

2   Construct the set $\mathcal{Z}_X$ of special-weighted strings of X;

3   **foreach** $\mathcal{Z}_k \in \mathcal{Z}_X$ **do**

4          $\mathcal{A}_k \leftarrow$ maximal palindromes of the heavy string of $\mathcal{Z}_k$;

5   Compute $\mathcal{R}_k$ for $\mathcal{Z}_k$;

6          $\mathcal{MP}(\mathcal{Z}_k, z) \leftarrow$ EMPTYLIST();

7   **foreach** $(c, r) \in \mathcal{A}_k$ **do**

8              $r \leftarrow \min(r, \mathcal{R}_k[2c])$;

9   **if** $r \geq \frac{1}{2}$ Insert (c, r) in $\mathcal{MP}(\mathcal{Z}_k, z)$;

10      $\mathcal{MP}(X, z) \leftarrow$ EMPTYLIST();

11   **foreach** $c \in [0, 2n-2]$ **do**

12          $r \leftarrow \max\{r_k \mid (c/2, r_k) \in \mathcal{MP}(\mathcal{Z}_k, z), 0 \leq k < \lceil z \rceil\}$;

13   Insert $(c/2, r)$ in $\mathcal{MP}(X, z)$;

14      $\mathcal{F} \leftarrow$ EMPTYLIST();

15      $\mathcal{U} \leftarrow$ EMPTYLIST();

16   **foreach** $(c, r) \in \mathcal{MP}(X, z)$ **do**

17          $j \leftarrow \lceil c + r \rceil$;

18   Insert $2r$ in $\mathcal{F}[j]$;

19   Insert $j - 2r$ in $\mathcal{U}[j]$;

20   Construct directed graph $\mathcal{G}_X = (\mathcal{V}, \mathcal{E})$, where $\mathcal{V} = \{i \mid -1 \leq i \leq n-1\}$, $\mathcal{E} = \{(i,j) \mid j \in \mathcal{U}[i]\}$ and (i,j) is a directed edge from i to j;

21   Breadth first search on $\mathcal{G}_X$ assuming the vertex $n-1$ as the source;

22   Identify the shortest path $P \approx \langle n-1 = p_\ell, p_{\ell-1}, \ldots, p_2, p_1, p_0 = -1 \rangle$;

23   **Return** $X[0 .. p_1], X[p_1 + 1 .. p_2], \ldots, X[p_{\ell-1} + 1 .. p_\ell]$;

7   Conclusion

In this paper we present a review of recent advances of palindromic factorization.

References

1. Adamczyk, M., Alzamel, M., Charalampopoulos, P., Iliopoulos, C.S., Radoszewski, J.: Palindromic decompositions with gaps and errors. arXiv preprint arXiv:1703.08931 (2017)
2. Alatabbi, A., Iliopoulos, C.S., Rahman, M.S.: Maximal palindromic factorization. In: Proceedings of Prague Stringology Conference 2013, pp. 70–77. Czech Technical University, Prague (2013)
3. Alzamel, M., Gao, J., Iliopoulos, C.S., Liu, C., Pissis, S.P.: Efficient computation of palindromes in sequences with uncertainties. In: Accepted at Mining Humanistic Data Workshop (2017)
4. Barton, C., Kociumaka, T., Liu, C., Pissis, S.P., Radoszewski, J.: Indexing weighted sequences: neat and efficient. CoRR, abs/1704.07625 (2017)
5. Fici, G., Gagie, T., Kärkkäinen, J., Kempa, D.: A subquadratic algorithm for minimum palindromic factorization. J. Discret. Algorithms **28**(C), 41–48 (2014)
6. Gusfield, D.: Algorithms on Strings, Trees, and Sequences: Computer Science and Computational Biology. Cambridge University Press, New York (1997)

A Construction for {0, 1, −1} Orthogonal Matrices Visualized

N. A. Balonin[1] and Jennifer Seberry[2(✉)]

[1] Saint Petersburg State University of Aerospace Instrumentation,
67, B. Morskaia Street, 190000 St. Petersburg, Russian Federation
`korbendfs@mail.ru`
[2] School of Computing and Information Technology, EIS,
University of Wollongong, Wollongong, NSW 2522, Australia
`jennifer_seberry@uow.edu.au`

Dedicated to the Unforgettable Mirka Miller

Abstract. *Propus* is a construction for orthogonal ±1 matrices, which is based on a variation of the Williamson array, called the *propus array*

$$\begin{bmatrix} A & B & B & D \\ B & D & -A & -B \\ B & -A & -D & B \\ D & -B & B & -A \end{bmatrix}.$$

This array showed how a picture made is easy to see the construction method. We have explored further how a picture is worth ten thousand words.

We give variations of the above array to allow for more general matrices than symmetric Williamson propus matrices. One such is the *Generalized Propus Array (GP)*.

Keywords: Hadamard matrices · *D*-optimal designs
Conference matrices · Propus construction · Williamson matrices
Visualization · 05B20

1 Introduction

Hadamard matrices arise in statistics, signal processing, masking, compression, combinatorics, error correction, coil winding, weaving, spectroscopy and other areas. They been studied extensively. Hadamard showed [14] the order of an Hadamard matrix must be 1, 2 or a multiple of 4. Many constructions for ±1 matrices and similar matrices such as Hadamard matrices, weighing matrices, conference matrices and *D*-optimal designs use skew and symmetric Hadamard matrices in their construction. For more details see Seberry and Yamada [30]. Different constructions are most useful in different cases. For example the Paley I

© Springer International Publishing AG, part of Springer Nature 2018
L. Brankovic et al. (Eds.): IWOCA 2017, LNCS 10765, pp. 47–57, 2018.
https://doi.org/10.1007/978-3-319-78825-8_5

construction for spectroscopy and the Sylvester construction for Walsh functions (discrete Fourier transforms) for signal processing.

An Hadamard matrix of order n is an $n \times n$ matrix with elements ± 1 such that $HH^\top = H^\top H = nI_n$, where I_n is the $n \times n$ identity matrix and $\top$ stands for transposition. A skew Hadamard matrix $H = I + S$ has $S^\top = -S$. For more details see the books and surveys of Seberry (Wallis) and others [30, 34] cited in the bibliography.

Propus is a construction method for symmetric orthogonal ± 1 matrices, using four matrices A, $B = C$, and D, where

$$AA^\top + 2BB^\top + DD^\top = \text{constant } I,$$

based on the array

$$\begin{bmatrix} A & B & B & D \\ B & D & -A & -B \\ B & -A & -D & B \\ D & -B & B & -A \end{bmatrix}.$$

It gives aesthetically pleasing visual images (pictures) when converted using MATLAB (we show some below).

We show how finding propus-Hadamard matrices using Williamson matrices and D-optimal designs can be easily seen through their pictures. These can be generalized to allow non-circulant and/or non-symmetric matrices with the same aim to give symmetric Hadamard matrices.

We illustrate two constructions to show the construction method (these are proved in [2])

- $q \equiv 1 \pmod 4$, a prime power, such matrices exist for order $t = \frac{1}{2}(q + 1)$, and thus propus-Hadamard matrices of order $2(q + 1)$ (this uses the Paley II construction);
- $t \equiv 3 \pmod 4$, a prime, such that D-optimal designs, constructed using two circulant matrices, one of which must be circulant and symmetric, exist of order $2t$, then such propus-Hadamard matrices exist for order $4t$.

We note that appropriate *Williamson type* matrices may also be used to give propus-Hadamard matrices but do not pursue this avenue in this paper. There is also the possibility that this propus construction may lead to some insight into the existence or non-existence of symmetric conference matrices for some orders. We refer the interested reader to mathscinet.ru/catalogue/ propus/.

1.1 Definitions and Basics

Two matrices X and Y of order n are said to be *amicable* if $XY^\top = YX^\top$.

A *D-optimal design* of order $2n$ is formed from two commuting or amicable (± 1) matrices, A and B, satisfying $AA^\top + BB^\top = (2n - 2)I + 2J$, J the matrix of all ones, written in the form

$$DC = \begin{bmatrix} A & B \\ B^\top & -A^\top \end{bmatrix} \text{ and } DA = \begin{bmatrix} A & B \\ B & -A \end{bmatrix}.$$

respectively. In Fig. 1 the structure is clear to see.

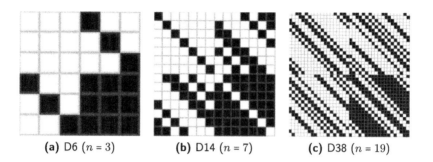

(a) D6 ($n = 3$)   **(b)** D14 ($n = 7$)   **(c)** D38 ($n = 19$)

Fig. 1. D-optimal designs for orders $2n$

Symmetric Hadamard matrices made using propus like matrices will be called *symmetric propus-Hadamard matrices.*

We define the following classes of propus like matrices. We note that there are slight variations in the matrices which allow variant arrays and non-circulant matrices to be used to give symmetric Hadamard matrices, All propus like matrices A, $B = C$, D are ± 1 matrices of order n satisfy the *additive property*

$$AA^\top + 2BB^\top + DD^\top = 4nI_n. \tag{1}$$

We make the definitions following [2]:

- *propus matrices*: four circulant symmetric ± 1 matrices, A, B, B, D of order n, satisfying the additive property (use P);
- *propus-type matrices*: four pairwise amicable ± 1 matrices, A, B, B, D of order n, $A^\top = A$, satisfying the additive property (use P);
- *generalized-propus matrices*: four pairwise commutative ± 1 matrices, A, B, B, D of order n, $A^\top = A$, which satisfy the additive property (use GP).

We use two types of arrays into which to plug the propus like matrices: the Propus array, P, or the generalized-propus array, GP. These can also be used with generalized matrices [33].

$$P = \begin{bmatrix} A & B & B & D \\ B & D & -A & -B \\ B & -A & -D & B \\ D & -B & B & -A \end{bmatrix} \text{ and } GP = \begin{bmatrix} A & BR & BR & DR \\ BR & D^\top R & -A & -B^\top R \\ BR & -A & -D^\top R & B^\top R \\ DR & -B^\top R & B^\top R & -A. \end{bmatrix}.$$

Symmetric Hadamard matrices made using propus like matrices will be called *symmetric propus-Hadamard matrices.*

2 Symmetric Propus-Hadamard Matrices

We first give the explicit statements of two well known theorem, Paley's Theorem [28], for the Legendre core Q, and Turyn's Theorem [31], in the form in which we will use them.

Theorem 1 [Paley's Legendre Core [28]]. *Let p be a prime power, either $\equiv 1 \pmod 4$ or $\equiv 3 \pmod 4$ then there exists a matrix, Q, of order p with zero diagonal and other elements ± 1 satisfying $QQ^\top = (q+1)I - J$, Q is or symmetric or skew-symmetric according as $p \equiv 1 \pmod 4$ (Paley I) or $p \equiv 3 \pmod 4$ (Paley II).*

Theorem 2 [Turyn's Theorem [31]]. *Let $q \equiv 1 \pmod 4$ be a prime power then there are two symmetric matrices, P and S of order $\frac{1}{2}(q+1)$, satisfying $PP^\top + SS^\top = qI$: P has zero diagonal and other elements ± 1 and S elements ± 1.*

2.1 Simple Propus-Hadamard Matrices of 12 and 20

2.2 $B = C = D$

There are only two starting Hadamard matrices, of orders 12 and 28, based on skew Paley core $B = C = D = Q + I$ (constructed using Legendre symbols). This special set is finite because $12 = 3^2 + 1^2 + 1^2 + 1^2$ and $28 = 5^2 + 1^2 + 1^2 + 1^2$ and these are the only orders for which a symmetric circulant A can exist with $B = C = D$. Figure 2 clearly shows the structure.

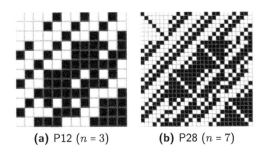

(a) P12 ($n = 3$)   **(b)** P28 ($n = 7$)

Fig. 2. Propus-Hadamard matrices using three back circulants $B = C = D$

There are two simple propus-Hadamard matrices of orders 12 and 20 based on symmetric Paley cores $A = J$, $B = C = J - 2I$, $D = J - 2I$ for $n = 3$, and $A = Q + I$, $B = C = J - 2I$, $D = Q - I$ (constructed using Legendre symbols) for $n = 5$. This second construction can be continued with back-circulant matrices $C = B$ which allows the symmetry property of A to be conserved.

Note how the slightly different construction of $P12$ in Figs. 2 and 3 can be easily seen.

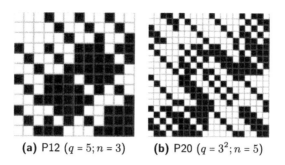

(a) P12 ($q = 5; n = 3$)   **(b)** P20 ($q = 3^2; n = 5$)

Fig. 3. Simple Propus-Hadamard matrices for orders 12 and 20

2.3 Order $4n$ from Williamson Matrices Using q a Prime Power

Lemma 1. *Let $q \equiv 1$ (mod 4), be a prime power, then propus matrices exist for orders $n = \frac{1}{2}(q + 1)$ which give symmetric propus-Hadamard matrices of order $2(q + 1)$.*

Proof. We note that for $q \equiv 1$ (mod 4), a prime power, Turyn (Theorem 2 [31]) gave Williamson matrices, $X + I$, $X - I$, Y, Y, which are circulant and symmetric for orders $n = \frac{1}{2}(q + 1)$. Then choosing

$$A = X + I, B = C = Y, D = X - I$$

gives the required propus-Hadamard matrices.   □

This gives propus-Hadamard matrices for 45 orders $4n$ where $n \leq 200$ [2]. Some of these cases arise when q is a prime power, however the Delsarte-Goethals-Seidel-Turyn construction means the required circulant matrices also exist for these prime powers (see Figs. 4 and 5).

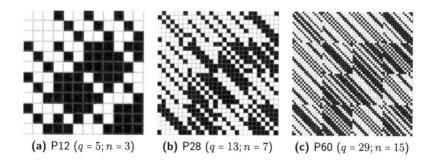

(a) P12 ($q = 5; n = 3$)   **(b)** P28 ($q = 13; n = 7$)   **(c)** P60 ($q = 29; n = 15$)

Fig. 4. Propus-Hadamard matrices for orders $4q$ for q prime, $q \equiv 1$ (mod 4)

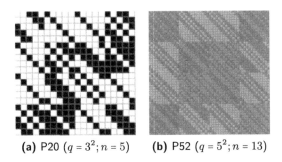

(a) P20 ($q = 3^2; n = 5$)   **(b)** P52 ($q = 5^2; n = 13$)

Fig. 5. Propus-Hadamard matrices for orders $4q$, q a prime power.

2.4 Propus-Hadamard Matrices from D-optimal Designs

Lemma 2. *Let $n \equiv 3 \pmod 4$, be a prime, such that D-optimal designs, constructed using two circulant matrices, one of which is symmetric, exist for order $2n$. Then propus-Hadamard matrices exist for order $4n$.*

Djoković and Kotsireas in [9, 23] give 43 D-optimal designs, constructed using two circulant matrices, for $n < 200$. We are interested in those cases where the D-optimal design is constructed from two circulant matrices one of which must be symmetric.

Suppose D-optimal designs for orders $n \equiv 3 \pmod 4$, a prime, are constructed using two circulant matrices, X and Y. Suppose X is symmetric. Let $Q + I$ be the Paley matrix of order n. Then choosing

$$A = X, \quad B = C = Q + I, \quad D = Y,$$

to put in the array GP gives the required propus-Hadamard matrices.

Hence we have propus-Hadamard matrices, constructed using D-optimal designs, for orders $4n$ where n is in $\{3, 7, 19, 31\}$. The results for $n = 19$ and 31 were given to us by Djoković.

We see clearly, looking first at $GP28$ in Fig. 6 where the D-optimal design is highlighted in purple, the construction method. Now the method will also be clear in $GP12$ and $GP76$.

2.5 The Propus Construction

We have shown [2] that if $X_1 = A$, $X_2 = B$, $X_3 = B$, $X_4 = D$ are pairwise amicable, symmetric Williamson type matrices of order $2n + 1$, where $X_2 = X_3 = B$, and satisfy the additive property, they can be used as in the appropriate array, G or GP, to form symmetric propus Hadamard matrix of order $(4(2n+1)$. For example from Paley's theorem (Corollary 1) for $p \equiv 3 \pmod 4$ we use the backcirculant or type 1, symmetric matrices QR and R instead of Q and I; whereas for $p \equiv 1 \pmod 4$ we use the symmetric Paley core Q.

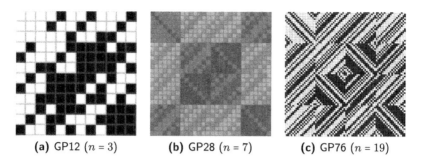

| (a) GP12 ($n = 3$) | (b) GP28 ($n = 7$) | (c) GP76 ($n = 19$) |

Fig. 6. Order $4n$ propus-Hadamard matrices constructed using D-optimal designs (Colour figure online)

Many powerful corollaries arose and new results were obtained by making suitable choices for X_1, X_2, X_3, X_4 in the arrays P and GP to ensure that the propus construction can be used to form symmetric Hadamard matrices of order $4(2n + 1)$.

From Turyn's result (Corollary 2) we set, for $p \equiv 1 \pmod 4$ $X_1 = P + I$, $X_2 = X_3 = S$ and $X_4 = P - I$.

Hence we have:

Corollary 1. *Let $q \equiv 1 \pmod 4$ be a prime power and $\frac{1}{2}(q + 1)$ be a prime power or the order of the core of a symmetric conference matrix (this happens for $q = 89$). Then there exist symmetric Williamson type matrices of order $q + 2$ and a symmetric propus-type Hadamard matrix of order $4(q + 2)$.*

3   Propus-Hadamard Matrices from Conference Matrices: Even Order Matrices

A powerful method to construct propus-Hadamard matrices for n even is using conference matrices.

Lemma 3. *Suppose M is a conference matrix of order $n \equiv 2 \pmod 4$. Then $MM^\top = M^\top M = (n - 1)I$, where I is the identity matrix and $M^\top = M$. Then using $A = M + I$, $B = C = M - I$, $D = M + I$ gives a propus-Hadamard matrix of order $4n$.*

We use the sixteen conference matrix orders of even order $n \leq 100$ from [1] to give propus-Hadamard matrices of orders $4n$. The conference matrices in Fig. 7 are made two circulant matrices A and B of order n where both A and B are symmetric.

Then using the matrices $A + I$, $B = C$ and $D = A - I$ in P gives the required construction.

The conference matrices in Fig. 8 are made from two circulant matrices A and B of order n where both A and B are symmetric. However here we use $A + I$, $BR = CR$ and $D = A - I$ in P to obtain the required construction.

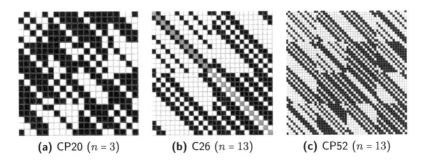

(a) CP20 $(n = 3)$ **(b)** C26 $(n = 13)$ **(c)** CP52 $(n = 13)$

Fig. 7. Conference matrices for orders $2n$ using two circulants: propus-Hadamard matrices for orders $4n$

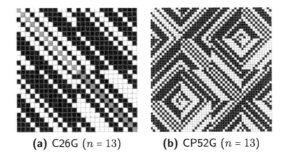

(a) C26G $(n = 13)$ **(b)** CP52G $(n = 13)$

Fig. 8. Conference matrices for orders $2n$ using two circulant and back-circulants: propus-Hadamard matrices for orders $4n$

There is another variant of this family which uses the symmetric Paley cores $A = Q + I$, $D = Q - I$ (constructed using Legendre symbols) and one circulant matrix of maximal determinant $B = C = Y$.

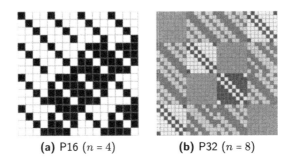

(a) P16 $(n = 4)$ **(b)** P32 $(n = 8)$

Fig. 9. Matrices P16 and P32

3.1   Propus-Hadamard Matrices for n Even

Fig. 9 gives visualizations (images/pictures) of propus-Hadamard matrices orders 16, 32. These have even n.

4   Conclusion and Future Work

Using the results of Lemma 1 and Corollary 1 and the symmetric propus-Hadamard matrices of Di Matteo et al. given in [5], we see that the unresolved cases for symmetric propus-Hadamard matrices for orders $4n$, $n < 200$ odd, are where $n \in$.

$$\{17, 23, 29, 33, 35, 47, 53, 65, 71, 73, 77, 93, 95, 97, 99,$$
$$101, 103, 107, 109, 113, 125, 131, 133, 137, 143, 149, 151, 153,$$
$$155, 161, 163, 165, 167, 171, 173, 179, 183, 185, 189, 191, 197.\}$$

There are many constructions and variations of the propus theme to be explored in future research. Visualizing the propus construction gives aesthetically pleasing examples of propus-Hadamard matrices. The visualization also makes the construction method clearer. There is the possibility that these visualizations may be used for quilting.

References

1. Balonin, N.A., Seberry, J.: A review and new symmetric conference matrices. Informatsionno-upravliaiushchie sistemy, **71**(4), 2–7 (2014)
2. Balonin, N.A., Seberry, J.: Two infinite families of symmetric Hadamard matrices. Australas. Comb. **69**(3), 349–357 (2017)
3. Baumert, L.D.: Cyclic Difference Sets. LNM, vol. 182. Springer, Heidelberg (1971). https://doi.org/10.1007/BFb0061260
4. Cohn, J.H.E.: A D-optimal design of order 102. Discret. Math. **1**(102), 61–65 (1992)
5. Di Matteo, O., Djoković, D., Kotsireas, I.S.: Symmetric hadamard matrices of order 116 and 172 exist. Spec. Matrices **3**, 227–234 (2015)
6. Djoković, D.Z.: On maximal $(1, -1)$-matrices of order $2n$, n odd. Radovi Matematicki **7**(2), 371–378 (1991)
7. Djoković, D.Z.: Some new D-optimal designs. Australas. J. Comb. **15**, 221–231 (1997)
8. Djoković, D.Z.: Cyclic $(v; r, s; \lambda)$ difference families with two base blocks and $v \leq 50$. Ann. Comb. **15**(2), 233–254 (2011)
9. Djoković, D.Z., Kotsireas, I.S.: New results on D-optimal matrices. J. Comb. Des. **20**, 278–289 (2012)
10. Djoković, D.Z., Kotsireas, I.S.: email communication from I. Kotsireas, 3 August 2014 1:13 pm
11. Fletcher, R.J., Gysin, M., Seberry, J.: Application of the discrete Fourier transform to the search for generalised Legendre pairs and Hadamard matrices. Australas. J. Comb. **23**, 75–86 (2001)

12. Fletcher, R.J., Koukouvinos, C., Seberry, J.: New skew-Hadamard matrices of order and new D-optimal designs of order $2 \cdot 59$. Discret. Math. **286**(3), 251–253 (2004)

13. Roderick, R.J., Seberry, J.: New D-optimal designs of order 110. Australas. J. Comb. **23**, 49–52 (2001)

14. Hadamard, J.: Résolution d'une question relative aux déterminants. Bull. des Sci. Math. **17**, 240–246 (1893)

15. Gysin, M.: New D-optimal designs via cyclotomy and generalised cyclotomy. Australas. J. Comb. **15**, 247–255 (1997)

16. Gysin, M.: Combinatorial Designs, Sequences and Cryptography, Ph.D. Thesis, University of Wollongong (1997)

17. Gysin, M., Seberry, J.: An experimental search and new combinatorial designs via a generalisation of cyclotomy. J. Combin. Math. Combin. Comput. **27**, 143–160 (1998)

18. Holzmann, W.H., Kharaghani, H.: A D-optimal design of order 150. Discret. Math. **190**(1), 265–269 (1998)

19. Kotsireas, I.S., Pardalos, P.M.: D-optimal matrices via quadratic integer optimization. J. Heuristics **19**(4), 617–627 (2013)

20. Koukouvinos, C., Kounias, S., Seberry, J.: Supplementary difference sets and optimal designs. Discret. Math. **88**(1), 49–58 (1991)

21. Koukouvinos, C., Seberry, J., Whiteman, A.L., Xia, M.: Optimal designs, supplementary difference sets and multipliers. J. Stat. Plan. Inference **62**(1), 81–90 (1997)

22. Georgiou, S., Koukouvinos, C., Seberry, J.: Hadamard matrices, orthogonal designs and construction algorithms. In: Wallis, W.D. (ed.) Designs 2002: Further Combinatorial and Constructive Design Theory, pp. 133–205. Kluwer Academic Publishers, Norwell (2002)

23. Geramita, A.V., Seberry, J., Designs, O.: Quadratic Forms and Hadamard Matrices. Marcel Dekker, New York-Basel (1979)

24. Hall Jr., M.: A survey of difference sets. Proc. Am. Math. Soc. **7**, 975–986 (1956)

25. Hall Jr., M.: Combinatorial Theory, 2nd edn. Wiley, Hoboken (1998)

26. Miyamoto, M.: A construction for Hadamard matrices. J. Comb. Theoy. Ser. A **57**, 86–108 (1991)

27. Mitrouli, M.: D-optimal designs embedded in Hadamard matrices and their effect on the pivot patterns. Linear Algebra App. **434**, 1751–1772 (2011)

28. Paley, R.E.A.C.: On orthogonal matrices. J. Math. Phys. **12**, 311–320 (1933)

29. Plackett, R.L., Burman, J.P.: The design of optimum multifactorial experiments. Biometrika **33**, 305–325 (1946)

30. Seberry, J., Yamada, M.: Hadamard matrices, sequences, and block designs. In: Dinitz, J.H., Stinson, D.R. (eds.) Contemporary Design Theory: A Collection of Surveys, pp. 431–560. Wiley, New York (1992)

31. Turyn, R.J.: An infinite class of Williamson matrices. J. Comb. Theory Ser. A. **12**, 319–321 (1972)

32. Sloane, N.J.A.: AT&T on-line encyclopedia of integer sequences. http://www.research.att.com/~njas/sequences/

33. Wallis, J.S.: Williamson matrices of even order. In: Holton, D.A. (ed.) Combinatorial Mathematics. Lecture Notes in Mathematics, vol. 403, pp. 132–142. Springer, Heidelberg (1974). https://doi.org/10.1007/BFb0057387

34. Wallis, W.D.: Room Squares. Combinatorics: Room Squares, Sum-Free Sets, Hadamard Matrices. LNM, vol. 292, pp. 30–121. Springer, Heidelberg (1972). https://doi.org/10.1007/BFb0069907

35. Whiteman, A.L.: A family of D-optimal designs. Ars Comb. **30**, 23–26 (1990)
36. Yamada, M.: On the Williamson type j matrices of order 4.29, 4.41, and 4.37. J. Comb. Theory, Ser. A, **27**, 378–381 (1979)
37. Wallis, J.S.: On the existence of Hadamard matrices. J. Comb. Theory (Ser. A), **21**, 186–195 (1976)
38. Craigen, R.: Signed groups, sequences and the asymptotic existence of Hadamard matrices. J. Comb. Theory (Ser. A), **71**, 241–254 (1995)
39. Ghaderpour, E., Kharaghani, H.: The asymptotic existence of orthogonal designs. Australas. J. Combin. **58**, 333–346 (2014)
40. de Launey, W., Kharaghani, H.: On the asymptotic existence of cocyclic Hadamard matrices. J. Comb. Theory (Ser. A), **116**(6), 1140–1153 (2009)

Approximation Algorithms and
Hardness

On the Maximum Crossing Number

Markus Chimani[1], Stefan Felsner[2], Stephen Kobourov[3], Torsten Ueckerdt[4], Pavel Valtr[5], and Alexander Wolff[6]($\boxtimes$)

[1] Universität Osnabrück, Osnabrück, Germany
markus.chimani@uni-osnabrueck.de
[2] Technische Universität Berlin, Berlin, Germany
felsner@math.tu-berlin.de
[3] University of Arizona, Tucson, USA
kobourov@cs.arizona.edu
[4] Karlsruhe Institute of Technology, Karlsruhe, Germany
torsten.ueckerdt@kit.edu
[5] Charles University, Prague, Czech Republic
valtr@kam.mff.cuni.cz
[6] Universität Würzburg, Würzburg, Germany
http://www1.informatik.uni-wuerzburg.de/wolff

Abstract. Research about crossings is typically about minimization. In this paper, we consider *maximizing* the number of crossings over all possible ways to draw a given graph in the plane. Alpert et al. [Electron. J. Combin., 2009] conjectured that any graph has a *convex* straight-line drawing, that is, a drawing with vertices in convex position, that maximizes the number of edge crossings. We disprove this conjecture by constructing a planar graph on twelve vertices that allows a non-convex drawing with more crossings than any convex one. Bald et al. [Proc. COCOON, 2016] showed that it is NP-hard to compute the maximum number of crossings of a geometric graph and that the weighted geometric case is NP-hard to approximate. We strengthen these results by showing hardness of approximation even for the unweighted geometric case and prove that the unweighted topological case is NP-hard.

1 Introduction

While traditionally in graph drawing one wants to minimize the number of edge crossings, we are interested in the opposite problem. Specifically, given a graph G, what is the maximum number of edge crossings possible, and what do embeddings[1] of G that attain this maximum look like? Such questions have first been asked as early as in the 19th century [3,23]. Perhaps due to the counterintuitive nature of the problem (as illustrated by the disproved conjecture below) and due to the lack of established tools and concepts, little is known about maximizing the number of crossings.

---

[1] We consider only embeddings where vertices are distinct points in the plane and edges are continuous curves containing no vertex points other than those of their end vertices.

© Springer International Publishing AG, part of Springer Nature 2018
L. Brankovic et al. (Eds.): IWOCA 2017, LNCS 10765, pp. 61–74, 2018.
https://doi.org/10.1007/978-3-319-78825-8_6

Besides the theoretical appeal of the problem, motivation for this problem can be found in analyzing the worst-case scenario when edge crossings are undesirable but the placement of vertices and edges cannot be controlled.

There are three natural variants of the crossing maximization problem in the plane. In the *topological* setting, edges can be drawn as curves, so that any pair of edges crosses at most once, and incident edges do not cross. In the straight-line variant (known for historical reasons as the *rectilinear* setting), edges must be drawn as straight-line segments. If we insist that the vertices are placed in convex position (e.g., on the boundary of a disk or a convex polygon) and the edges must be routed in the interior of their convex hull, the topological and rectilinear settings are equivalent, inducing the same number of crossings: the number only depends on the order of the vertices along the boundary of the disk. In this *convex setting*, a pair of edges crosses if and only if its endpoints alternate along the boundary of the convex hull.

The topological setting. The maximum crossing number was introduced by Ringel [20] in 1963 and independently by Grünbaum [11] in 1972.

Definition 1. ([21]). *The* maximum crossing number *of a graph G, max-cr(G), is the largest number of crossings in any topological drawing of G in which no three distinct edges cross in one point and every pair of edges has at most one point in common (a shared endpoint counts, touching points are forbidden).*

In particular, max-cr(G) is the maximum number of crossings in the topological setting. Note that only independent pairs of edges, that is those edge pairs with no common endpoint, can cross. The number of independent pairs of edges in a graph $G = (V, E)$ is given by $M(G) := \binom{|E|}{2} - \sum_{v \in V} \binom{\deg(v)}{2}$, a parameter introduced by Piazza *et al.* [18]. For every graph G, we have max-cr$(G) \leq M(G)$, and graphs for which equality holds are known as *thrackles* or *thrackable* [26]. Conway's Thrackle Conjecture [16] states that thrackles are precisely the pseudoforests (graphs in which every connected component has at most one cycle) in which there is no cycle of length four and at most one odd cycle. Equivalently, this famous conjecture states that max-cr$(G) = M(G)$ implies $|E(G)| \leq |V(G)|$ [26].

Another famous open problem is the Subgraph Problem posed by Ringeisen et al. [19]: Is it true that whenever H is a subgraph or induced subgraph of G, then we have max-cr$(H) \leq$ max-cr(G)?

Let us remark that allowing pairs of edges to only touch without properly crossing each other, would indeed change the problem. For example, the 4-cycle C_4 has two pairs of independent edges, and C_4 can be drawn with one pair crossing and the other pair touching, but C_4 is not thrackable; it is impossible to draw C_4 with both pairs crossing, i.e., max-cr(C_4) is 1 and not 2.

It is known that max-cr$(K_n) = \binom{n}{4}$ [20] and that every tree is thrackable, i.e., max-cr$(G) = M(G)$ whenever G is a tree [18]. We refer to Schaefer's survey [21] for further known results on the maximum crossing numbers of several graph classes.

The straight-line setting. The maximum rectilinear crossing number was introduced by Grünbaum [11]; see also [8].

Definition 2. *The* maximum rectilinear crossing number *of a graph G, $max\text{-}\overline{cr}(G)$, is the largest number of crossings in any straight-line drawing of G.*

For every graph G, we have $max\text{-}\overline{cr}(G) \leq max\text{-}cr(G) \leq M(G)$, where each inequality is strict for some graphs, while equality is possible for other graphs. For example, for the n-cycle C_n we have $max\text{-}\overline{cr}(C_n) = max\text{-}cr(C_n) = M(C_n) = n(n-3)/2$ for odd n [26], while $max\text{-}\overline{cr}(C_n) = M(C_n) - n/2 + 1$ and $max\text{-}cr(C_n) = M(C_n)$ for even n different than four [1,24]. For further rectilinear crossing numbers of specific graphs we again refer to Schaefer's survey [21].

For several graph classes, such as trees, the maximum (topological) crossing number $max\text{-}cr(G)$ is known exactly, while little is known about the maximum rectilinear crossing number $max\text{-}\overline{cr}(G)$. For planar graphs, Verbitsky [25] studied what he called the *obfuscation number*. He defined $obf(G) = max\text{-}\overline{cr}(G)$ and showed that $obf(G) < 3|V(G)|^2$. Note that this holds only for planar graphs. For maximally planar graphs, that is, triangulations, Kang et al. [14] give a $(56/39 - \varepsilon)$-approximation for computing $max\text{-}\overline{cr}(G)$.

The convex setting. It is easy to see that in the convex setting we may assume, without loss of generality, that all vertices are placed on a circle and edges are drawn as straight-line segments. In fact, if the vertices are in convex position and edges are routed in the interior of the convex hull of all vertices, then a pair of edges is crossing if and only if the vertices of the two edges alternate in the circular order along the convex hull.

Definition 3. *The* maximum convex crossing number *of a graph G, $max\text{-}cr^\circ(G)$, is the largest number of crossings in any drawing of G where the vertices lie on the boundary of a disk and the edges in the interior.*

From the definitions we now have that, for every graph G,

$$max\text{-}cr^\circ(G) \leq max\text{-}\overline{cr}(G) \leq max\text{-}cr(G) \leq M(G), \tag{1}$$

but this time it is not clear whether or not the first inequality can be strict. It is tempting (and rather intuitive) to say that in order to get many crossings in the rectilinear setting, all vertices should always be placed in convex position. In other words, this would mean that the maximum rectilinear crossing number and maximum convex crossing number always coincide. Indeed, this has been conjectured by Alpert et al. in 2009.

Conjecture 1. (Alpert et al. [1]). Any graph G has a drawing with vertices in convex position that has $max\text{-}\overline{cr}(G)$ crossings, that is, $max\text{-}\overline{cr}(G) = max\text{-}cr^\circ(G)$.

Our contribution. Our main result is that Conjecture 1 is false. We provide several counterexamples in Sect. 3. Before we get there, we discuss the four parameters in (1) and relations between them in more detail, and introduce some new problems in Sect. 2. Finally, in Sect. 4, we investigate the complexity and approximability of crossing maximization and show that the topological problem is NP-hard, while the rectilinear problem is even hard to approximate.

2   Preliminaries and Basic Observations

Here we discuss the chain of inequalities in (1) and extend it by several items. Recall that for a graph G, $M(G)$ denotes the number of independent pairs of edges in G. By (1) we have max-cr$^{\circ}(G) \leq M(G)$. We next show that this inequality is tight up to a factor of 3. The first part of the next lemma is due to Verbitsky [25].

Lemma 1. *For every graph G, we have $M(G)/3 \leq$ max-cr$^{\circ}(G)$. Moreover, if G has chromatic number at most 3, then $M(G)/2 \leq$ max-cr$^{\circ}(G)$.*

Proof. First, let G be any graph. We place the vertices of G on a circle in a circular order chosen uniformly at random from the set of all their circular orders. Then each pair of independent edges of G is crossing with probability $1/3$ and there must be an ordering witnessing max-cr$^{\circ}(G) \geq M(G)/3$.

Second, assume that G can be properly colored with at most three colors. In this case we place the vertices of G on a circle in such a way that the three color classes occupy three pairwise disjoint arcs. In each color class, we order the vertices randomly, choosing each linear order with the same probability. Doing this independently for each color class, each pair of independent edges is crossing with probability $1/2$. Hence, there must be an ordering witnessing max-cr$^{\circ}(G) \geq M(G)/2$. □

By Lemma 1 we can extend the chain of inequalities in (1) as follows: For every graph G, we have

$$M(G)/3 \leq \text{max-cr}^{\circ}(G) \leq \text{max-}\overline{\text{cr}}(G) \leq \text{max-cr}(G) \leq M(G). \tag{2}$$

The constant $1/3$ in the first inequality in (2) cannot be improved: Consider the six edges connecting a 4-tuple of vertices in a rectilinear drawing of the complete graph K_n. There is exactly one crossing among them if the four vertices are in convex position, and there is no crossing among them otherwise. It follows that the maximum rectilinear crossing number of K_n is attained if and only if the vertices are in convex position, and in this case there are $M(K_n)/3 = \binom{n}{4}$ crossings. Since Ringel [20] proved max-cr$(K_n) = \binom{n}{4}$, we get max-cr$^{\circ}(K_n) =$ max-$\overline{\text{cr}}(K_n) =$ max-cr$(K_n) = M(K_n)/3 = \binom{n}{4}$.

We now introduce another item in the chain of inequalities (2). We say that a rectilinear drawing of a graph G is *separated* if there is a line ℓ that intersects every edge of G. Clearly, this is only possible if G is bipartite and in this case the line ℓ separates the vertices of the two color classes of G.

Fig. 1. The smallest tree G that is not a caterpillar with a topological drawing with max-cr$(G) = M(G) = 9$ crossings (left), a 2-layer drawing with bcr$(G) = 1$ crossings (middle) and a 2-layer drawing with max-$\overline{\text{cr}}(G) = M(G) - \text{bcr}(G) = 8$ crossings (right).

Particularly nice are *separated convex drawings*, i.e., separated drawings with vertices in convex position; see Fig. 1 for an example. Drawing bipartite graphs in the separated convex model is equivalent to the 2-layer model where the vertices of the two color classes are required to be placed on two parallel lines. In this 2-layer model, the minimum number of crossings of a bipartite graph G has been studied under the name *bipartite crossing number*, denoted bcr(G).

Lemma 2. *For every bipartite graph G, the maximum number of crossings among all separated convex drawings of G is exactly $M(G) - \text{bcr}(G)$.*

Proof. Consider any separated convex drawing of any bipartite graph G. A pair of independent edges is crossing if and only if their endpoints alternate along the convex hull. So if $e_1 = u_1 v_1$ and $e_2 = u_2 v_2$ with u_1, u_2 being above the separating line ℓ and v_1, v_2 below, then e_1 and e_2 are crossing if in the circular order we see $u_1 - u_2 - v_1 - v_2$, and non-crossing if we see $u_1 - u_2 - v_2 - v_1$. In particular, reversing the order of all vertices below the separating line ℓ transforms crossings into non-crossings and vice versa. This shows that for a separated convex drawing with k crossings, reversing results in exactly $M(G) - k$ crossings, which concludes the proof. □

Applying Lemma 2 to the chain of inequalities (2) shows that for every bipartite graph G we have

$$M(G)/2 \leq M(G) - \text{bcr}(G) \leq \text{max-cr}^\circ(G) \leq \text{max-}\overline{\text{cr}}(G) \leq \text{max-cr}(G) \leq M(G). \tag{3}$$

It remains open whether the new inequality $M(G) - \text{bcr}(G) \leq \text{max-cr}^\circ(G)$ in (3) is attained with equality for every bipartite graph G. For example, for a tree G it is known, see e.g. [26], that max-cr$(G) = M(G)$, but it is not hard to see that max-$\overline{\text{cr}}(G) = M(G)$ if and only if G is a caterpillar[2]. (Hence max-$\overline{\text{cr}}(G) <$ max-cr(G) holds for every tree which is not a caterpillar.) Moreover, it is equally easy to see that a tree G has a crossing-free 2-layer drawing if and only if G is a caterpillar. Thus, for every tree G, we have that $M(G) - \text{bcr}(G) = M(G)$ if and only if max-$\overline{\text{cr}}(G) = M(G)$. We again refer to Fig. 1 for an illustration.

The expanded chain on inequalities (3), leads to two natural questions:

---

[2] A caterpillar is a tree in which all non-leaf vertices lie on a common path.

Problem 1. Does every bipartite graph G have a separated drawing with max-$\overline{\mathrm{cr}}(G)$ many crossings? Does every tree G have a separated convex drawing with max-$\overline{\mathrm{cr}}(G)$ crossings, i.e., is max-$\overline{\mathrm{cr}}(G) = M(G) - \mathrm{bcr}(G)$?

Let us mention that Garey and Johnson [9] have shown that bipartite crossing minimization is NP-hard. The problem remains NP-hard if the ordering of the vertices on one side is prescribed [7]. On trees, bipartite crossing minimization can be solved efficiently [22]. For the one-sided two-layer crossing minimization, Nagamochi [17] gave an 1.47-approximation algorithm, improving upon the well-known median heuristic, which yields a 3-approximation [7]. The weighted case, which we define formally in Sect. 4, admits a 3-approximation algorithm [5].

3   Counterexamples for Conjecture 1

In this section we present counterexamples for the convexity conjecture. After some preliminary work we provide a counterexample $H(4)$ on 37 vertices. To show that this graph is a counterexample, we need to analyze only two cases. The graph $H(2)$ with 19 vertices is smaller, but requires more work. In the full version of this paper [6] we prove that a planar subgraph of $H(2)$ with 12 vertices and 16 edges is a counterexample.

A set of vertices $X \subset V$ in a graph $G = (V, E)$ is a set of *twins* if all vertices of X have the same neighborhood in G (in particular X is an independent set). A *vertex split* of vertex v in G consists in adding a new vertex v' to G such that v' is a twin of v, that is, for any edge vu, there is an edge $v'u$, and these are all the edges at v'.

Lemma 3. *For any graph G there is a convex drawing of G maximizing the number of crossings among all convex drawings of G, such that each set of twins forms an interval of consecutive vertices along the convex hull of the drawing.*

Proof. Suppose $V_1, \ldots, V_s$ are the maximal sets of twins in G. Consider a convex drawing of G maximizing the number of crossings. It clearly suffices to show that for any set V_i we may move all the points of V_i next to one of the points of V_i without decreasing the number of crossings, since this procedure done iteratively s times, once for each of the sets $V_1, \ldots, V_s$, results in a desired convex drawing of G.

We call a crossing k-*rich* if there are k vertices of V_i among the four vertices of the edges forming the crossing. Since V_i is independent, k is 0, 1 or 2 for each crossing. If we move only vertices of V_i then 0-rich crossings remain in the drawing. If the vertices of V_i appear in consecutive order along the convex hull of the drawing then the number of 2-rich crossings is maximized due to the following argument. For any two vertices u, v of V_i and for any two neighbors x, y of V_i, the 4-cycle $uxvy$ is self-crossing which gives rise to a 2-rich crossing. Since every 2-rich crossing appears in a single 4-cycle and every 4-cycle can give rise to at most one crossing, the number of 2-rich crossings is indeed maximized whenever the vertices of V_i appear in consecutive order along the convex hull.

 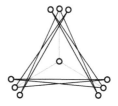

Fig. 2. Left: The graph $H(k)$. Each circle represents k independent vertices, each black line segment represents a bundle of k^2 edges, each gray line segment represents k edges. Right: A non-convex drawing of $H(k)$.

It remains to show that there is a vertex v in V_i such that we can move the other vertices next to v without decreasing the number of 1-rich crossings. Each 1-rich crossing involves exactly one vertex of V_i. The number of 1-rich crossings involving a given vertex of V_i is affected only by the position of that vertex and of the vertices of $V \setminus V_i$. Thus, if we choose v as the vertex involved in the largest number of 1-rich crossings and move all the other vertices of V_i next to v, every vertex will be involved in at least as many 1-rich crossings as it was before the vertices were moved. □

The construction of $H(k)$. For the construction of our example graphs $H(k)$, we start with a 9-cycle on vertices $v_0, \ldots, v_8$ with edges v_i, v_{i+1} where $i + 1$ is to be taken modulo 9. Add a 'central' vertex z adjacent to v_0, v_3, v_6. This graph on 10 vertices is the base graph H. The example graph $H(k)$ is obtained from H by applying k vertex splits to each of the nine cycle vertices v_i. The graph $H(k)$ thus consists of nine independent sets V_i of size k and the central vertex z. In total it has $9k + 1$ vertices and $9k^2 + 3k$ edges. Figure 2 (left) shows a schematic drawing of $H(k)$, where each black edge represents a "bundle" of k^2 edges of $H(k)$ and each gray edge represents a set of k edges. We will show that for $k \geq 4$ the drawing in Fig. 2 (right) has more crossings than any drawing with vertices in convex position.

From Lemma 3 we know that, in convex drawings of $H(k)$ with many crossings, the twin pairs of vertices can be assumed to be next to each other. Drawings of $H(k)$ of this kind are essentially determined by the corresponding drawings of H, in which each set of twins is represented just by one representative; see Fig. 2. This justifies that later on we only look at convex drawings of H with weighted crossings, and not of the full $H(k)$.

An independent set of edges of $H(k)$ is *weak* if the corresponding edges in the base graph H are not independent; it is *strong* otherwise. The next lemma shows that our drawing of $H(k)$ realizes as many crossings on weak pairs of independent edges as possible. This allows us to focus on strong pairs in the subsequent analysis.

Lemma 4. *The drawing of $H(k)$ on the right side of Fig. 2 maximizes the number of crossings on weak pairs of independent edges.*

Proof. Each edge v_i, v_{i+1} of H maps to a $K_{k,k}$ in $H(k)$. In the given drawing the $K_{k,k}$ is represented by a red edge. Since $V_i \cup V_{i+1}$ are in separated convex position the $K_{k,k}$ contributes $\binom{k}{2}^2$ crossing.

A pair of adjacent edges v_{i-1}, v_i and v_i, v_{i+1} in H maps to a $K_{k,2k}$ in $H(k)$. We know that max-$\overline{\mathrm{cr}}(K_{k,2k}) = \binom{k}{2}\binom{2k}{2}$ and this number of crossings is realized with separated convex position. In the drawing V_i and $V_{i-1} \cup V_{i+1}$ are in separated convex position.

A pair of adjacent edges v_i, z and v_i, v_{i+1} in H maps to a $K_{k,k+1}$ in $H(k)$. Now max-$\overline{\mathrm{cr}}(K_{k,k+1}) = \binom{k}{2}\binom{k+1}{2}$, and this number of crossings is realized with separated convex position of the vertices. In the drawing $V_i, V_{i+1} \cup \{z\}$ are in separated convex position. The case of adjacent edges v_i, z and v_{i-1}, v_i is identical. $\qquad\square$

The remaining crossings of the drawing of $H(k)$ correspond to crossings of two independent edges of H. These are either two red edges or a red and a green edge of H. Red edges represent a bundle of k^2 edges of $H(k)$ and green edges a bundle of k edges of $H(k)$. Hence a crossing of two red edges represents k^4 individual crossing pairs and a crossing of a red and a green edge represent k^3 individual crossing pairs. We devide by k^3 and speak about a crossing of two red edges as a crossing of weight k and of a red green crossing as a crossing of weight 1. In the given drawing of $H(k)$ every pair of red edges is crossing but every red edge has a unique independent green edge which is not crossed. Hence, the weight of the independent not crossing pairs of edges of H is 9. We summarize by saying that the given drawing has a weighted loss of 9.

The loss of convex drawings. We now study the weighted loss of convex drawings of H. In a convex drawing every red edge splits the 7 non-incident cycle vertices into those on one side and those on the other side. The *span* of a red edge is the number of vertices on the smaller side. Hence, the span of an edge is one of 0, 1, 2, 3.

Let us consider the case where the 9-cycle is drawn with zero loss, i.e., each red edge has span 3 and contributes a crossing with 6 other red edges. The cyclic order of the cycle vertices is $v_0, v_2, v_4, v_6, v_8, v_1, v_3, v_5, v_7$. Any two neighbors of z have the same distance in this cyclic order. Therefore, we may assume that z is in the short interval spanned by v_0 and v_6. Every edge of the 9-cycle is disjoint from at least one of the two green edges z, v_0 and z, v_6 and the edge v_7, v_8 is disjoint from both. This shows that the weighted loss of this drawing is at least 10.

A sequence of eight consecutive edges of span 3 forces the last edge to also have span 3. Hence, we have at least two red edges e and f of span at most 2. Each of these edges is disjoint from at least two independent red edges. Since the two edges may be disjoint they contribute a weighted loss of at least $3k$. For $k > 4$ this exceeds the weighted loss of the drawing of Fig. 2.

4 Complexity

Very recently, Bald et al. [2] showed, by reduction from MAXCUT, that it is NP-hard to compute the maximum rectilinear crossing number max-$\overline{\mathrm{cr}}(G)$ of a given graph G. Their reduction also shows that it is hard to approximate the weighted case better than ≈ 0.878 assuming the Unique Games Conjecture and better than $16/17$ assuming $\mathcal{P} \neq \mathcal{NP}$. In the convex case, one can "guess" the permutation; hence, this special case is in $\mathcal{NP}$. Bald et al. also stated that rectilinear crossing maximization is similar to rectilinear crossing minimization in the sense that the former "inherits" the membership in the class of the existential theory of the reals ($\exists \mathbb{R}$), and hence in PSPACE, from the latter. They also showed how to derandomize Verbitsky's approximation algorithm [25] for max-$\overline{\mathrm{cr}}$, turning the expected approximation ratio of $1/3$ into a deterministic one.

We now tighten the hardness results of Bald et al. by showing APX-hardness for the *unweighted* case. Recall that MAXCUT is NP-hard to approximate beyond a factor of $16/17$ [13]. Under the Unique Games Conjecture, MAXCUT is hard to approximate even beyond a factor of ≈ 0.878 [15]—the approximation ratio of the semidefinite programming approach of Goemans and Williamson [10] for MAXCUT. For a graph G, let max-cut(G) be the maximum number of edges crossing a cut, over all cuts of G.

Theorem 1. *Given a graph G, max-$\overline{\mathrm{cr}}(G)$ cannot be approximated better than* MAXCUT.

Proof. As Bald et al., we reduce from MAXCUT. In their reduction, they add a large-enough set I of independent edges to the given graph G. They argue that max-$\overline{\mathrm{cr}}(G+I)$ is maximized if the edges in I behave like a single edge with high weight that is crossed by as many edges of G as possible. Indeed, suppose for a contradiction that, in a drawing with the maximum number of crossings, an edge $e \in I$ crosses fewer edges than another edge e' in I. Then e can be drawn such that its endpoints are so close to the endpoints of e' that both edges cross the same edges—and each other. This would increase the number of crossings; a contradiction. W.l.o.g., we can make the "heavy edge" so long that its endpoints lie on the convex hull of the drawing. Then the heavy edge induces a cut of G. The cut is maximum as the heavy edge can be made arbitrarily heavy.

Instead of adding a set I of independent edges to G, we add a star S_t with $t = \binom{m}{2} + 1$ edges, where $m = |E(G)|$. Then, max-$\overline{\mathrm{cr}}(G) < t$. The advantage of the star is that all its edges are incident to the same vertex and, hence, cannot cross each other. Let $G' = G + S_t$ be the resulting graph. Exactly as for the set I above, we argue that all edges of S_t must be crossed by the same number of edges of G, and must in fact form a cut of G. Hence, we get

$$t \cdot \text{max-cut}(G) \leq \text{max-}\overline{\mathrm{cr}}(G') \leq t \cdot \text{max-cut}(G) + \text{max-}\overline{\mathrm{cr}}(G) < t \cdot (\text{max-cut}(G)+1).$$

This yields max-cut(G) = $\lfloor \text{max-}\overline{\mathrm{cr}}(G')/t \rfloor$. Hence, any α-approximation for maximum rectilinear crossing number yields an α-approximation for MAXCUT. $\square$

With the same argument, we also obtain hardness of approximation for max-cr°, which was only shown NP-hard by Bald et al. [2]. The reason is that in the convex setting, too, the "heavy obstacle" splits the vertex set into a "left" and a "right" side.

Corollary 1. *Given a graph G, max-cr°(G) cannot be approximated better than* MaxCut.

The *weighted* topological case is defined as follows. For a graph G with positive edge weights $w \colon E \to \mathbb{Q}_{>0}$ and a drawing D of G, let max-wt-cr$(D) = \sum_{e \text{ crosses } e'} w(e) \cdot w(e')$ be the weighted maximum crossing number of D, and let max-wt-cr(G) be the maximum weighted crossing number of G, that is, the maximum of wt-cr(D) over every drawing D of G. Let MaxWtCrNmb be the problem of computing the weighted maximum crossing number of a given graph.

Compared to the rectilinear and the convex case above, the difficulty of the topological case is that an obstacle (such as the heavy star above) does not necessarily separate the vertices into "left" and a "right" groups any more. Instead, our new obstacle separates the vertices into an "inner" group and an "outer" group, which allows us to reduce from a cut-based problem.

Our new starting point is the NP-hard problem 3MaxCut [27], which is the special case of MaxCut where the input graph is required to be 3-regular.

Theorem 2. *Given an edge-weighted graph G and a rational number $c > 0$, it is NP-complete to decide whether max-wt-cr$(G) \geq c$.*

Proof. Clearly, topological crossing maximization is in $\mathcal{NP}$ since we can guess a rotation system for the given graph and, for each edge, the ordered subset of the other edges that cross it. In polynomial time, we can then check whether (a) the weights of the crossings sum up to the given threshold c, and (b) the solution is feasible, simply by realizing the crossings via dummy vertices of degree 4 and testing for planarity of the so-modified graph.

To show NP-hardness, we reduce from 3MaxCut. Given an instance of 3MaxCut, that is, a 3-regular graph G and an integer $k > 0$, we construct an instance of topological crossing maximization, that is, a weighted graph G' and a rational number $c' > 0$ such that G has a cut crossed by at least k edges if and only if G' has a drawing with weighted crossing number at least c'. Let G' be the disjoint union of G with edges of weight 1 and a single triangle T with edges of (large) weight t. Let n be the number of vertices and m the number of edges of G. Due to the 3-regularity of G, we have $m = 3n/2$. We set $t = 9n^2/8$ and $c' = t(2m + k)$.

Let (V_1, V_2) be a solution of 3MaxCut, that is, a cut of G crossed by k edges. We need to show that this implies max-wt-cr$(G') \geq c'$. We construct a drawing D' of G' as in Fig. 3. For $i \in \{1, 2\}$, let M_i be the edge set of $G[V_i]$. We can route the edges of G such that each of the k edges in the cut crosses all three edges of T and each of the $m - k$ edges in $M_1 \cup M_2$ crosses exactly two

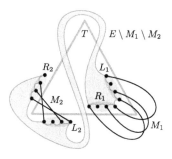

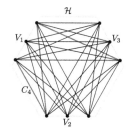

Fig. 3. Given a 3-regular graph G, a drawing of $G' = G+T$ with the maximum number of crossings yields a maximum cut of G if the edges of triangle T have much larger weight than the edges of G. The edges (in the light blue region) that cross T trice are in the cut. (Color figure online)

Fig. 4. A crossingmaximal drawing of the complete tripartite graph $K_{k,k,k}$

edges of T. Hence, max-wt-cr$(G') \geq$ max-wt-cr$(D') \geq t(3k + 2(m - k)) = c'$ as desired.

Conversely, let D' be any drawing of G' and let $c' =$ max-wt-cr(D'). We need to show that $G = G' - T$ has a cut that is crossed by at least $\lfloor c'/t \rfloor - 2m$ edges. As incident edges cannot cross, the triangle T of G' must be drawn in D' without self-crossings. Since max-cr$(G) \leq \binom{m}{2} = \binom{3n/2}{2} < 9n^2/8 = t$, we have that $x = \lfloor c'/t \rfloor$ is the number of crossings between edges of G and edges of T.

Let V_1 be the set of vertices of G in the interior of T, and let $V_2 = V \setminus V_1$. Consider the cut (V_1, V_2), and let k be the number of edges crossing this cut. Each of these k edges contributes at most 3 to x, and each of the $m - k$ edges that lie in $G[V_1]$ or $G[V_2]$ contributes at most 2 to x (as in Fig. 3). Hence, $x \leq 3k + 2(m - k) = k + 2m$, and $k \geq x - 2m = \lfloor c'/t \rfloor - 2m$ as desired.

Clearly, our reduction takes polynomial time.   □

We now set out to strengthen the result of Theorem 2; we want to show that even the maximum unweighted crossing number is hard to compute. Observe that in the above proof, the given graph G from the 3MAXCUT instance remained unweighted, but we required a heavily weighted additional triangle T. Our goal is now, essentially, to substitute T with an unweighted structure that serves the same purpose. Unfortunately, due to the large number of crossings of this new structure, we cannot make any statement about non-approximability of the unweighted case. The naïve approach of simply adding multiple unweighted triangles does not easily work since already the entanglement of the triangles among each other is non-trivial to argue.

Theorem 3. *Given a graph G, max-cr(G) is NP-complete to compute.*

Proof. The membership in $\mathcal{NP}$ follows from Theorem 2. To argue hardness, given an instance G of 3MAXCUT, we construct an unweighted graph G'—the instance for computing max-cr(G')—as the disjoint union of G and a complete tripartite graph $K := K_{k,k,k}$ with k vertices per partition set, $k > \sqrt{9/8} \cdot n$. A result of Harborth [12] yields max-cr(K) = $\binom{3k}{4} - 3\binom{k}{4} - 6k\binom{k}{3} \in \Theta(k^4)$.

We first analyze a crossing-maximal drawing of K; see Fig. 4. Consider a straight-line drawing "on a regular hexagon $\mathcal{H}$". Let V_1, V_2, V_3 be the partition sets of K and label the edges of $\mathcal{H}$ cyclically $1, 2, \ldots, 6$. Place V_i, $1 \le i \le 3$, along edge $2i$ of $\mathcal{H}$. We claim that max-cr(K) is achieved by this drawing. In fact, the arguments are analogous to the maximality of the naïve drawing for complete bipartite graphs on two layers: a 4-cycle can have at most one crossing. In the above drawing, every 4-cycle has a crossing. On the other hand, any crossing in *any* drawing of K is contained in a 4-cycle.

Intuitively, when thinking about shrinking the sides $1, 3, 5$ in $\mathcal{H}$, we obtain a drawing akin to T in the hardness proof for the maximum weighted crossing number. It remains to argue that there is an optimal drawing of full G' where K is drawn as described. Consider a drawing realizing max-cr(G') and note that any triangle in K is formed by a vertex triple, with a vertex from each partition set. Pick a triple $\tau = (v_1, v_2, v_3) \in V_1 \times V_2 \times V_3$ that induces a triangle T_τ with maximum number of crossings with G among all such triangles. Now, redraw K along T_τ according to the above drawing scheme such that, for $i = 1, 2, 3$, it holds that (a) all vertices of V_i are in a small neighborhood of v_i and (b) any edge $(w_i, w_j) \in V_i \times V_j$ for some $j \ne i$ crosses exactly the same edges of G as the edge (v_i, v_j). Our new drawing retains the same crossings within G', achieves the maximum number of crossings within K, and does not decrease the number of crossings between K and G; hence it is optimal. In this drawing, K plays the role of the heavy triangle T in the hardness proof of the weighted case, again yielding NP-hardness.   □

5   Conclusions and Open Problems

We have considered the crossing maximization problem in the topological, rectilinear, and convex settings. In particular, we disproved a conjecture of Alpert et al. [1] that the maximum crossing number in the latter two settings always coincide. We proposed the new "separated drawing" setting, and ask whether for every bipartite graph the maximum rectilinear, maximum convex, maximum separated, and maximum separated convex crossing numbers coincide. In particular, for bipartite graphs, the separation of the rectilinear and the separated convex setting is still open.

We have shown that the maximum rectilinear crossing number is APX-hard and the maximum topological crossing number is NP-hard. Is the latter also APX-hard? We have shown this to be true in the *weighted* topological case. It also remains open whether rectilinear crossing maximization is in $\mathcal{NP}$. For planar graphs, MAXCUT is tractable, and our hardness arguments no longer apply, leaving open the complexity of computing the maximum crossing number for this graph class.

Acknowledgments. This work started at the 2016 Bertinoro Workshop of Graph Drawing. We thank the organizers and other participants for discussions, in particular Michael Kaufmann. We also thank Marcus Schaefer, Gábor Tardos, and Manfred Scheucher.

References

1. Alpert, M., Feder, E., Harborth, H.: The maximum of the maximum rectilinear crossing numbers of d-regular graphs of order n. Electron. J. Combin. **16**(1), 54 (2009)
2. Bald, S., Johnson, M.P., Liu, O.: Approximating the maximum rectilinear crossing number. In: Dinh, T.N., Thai, M.T. (eds.) COCOON 2016. LNCS, vol. 9797, pp. 455–467. Springer, Cham (2016). https://doi.org/10.1007/978-3-319-42634-1_37
3. Baltzer, R.: Eine Erinnerung an Möbius und seinen Freund Weiske. Berichte über die Verhandlungen der Königlich Sächsischen Gesellschaft der Wissenschaften zu Leipzig. Mathematisch-Physische Classe **37**, 1–6 (1885)
4. Berman, P., Karpinski, M.: On some tighter inapproximability results (Extended Abstract). In: Wiedermann, J., van Emde Boas, P., Nielsen, M. (eds.) ICALP 1999. LNCS, vol. 1644, pp. 200–209. Springer, Heidelberg (1999). https://doi.org/10.1007/3-540-48523-6_17
5. Çakiroglu, O.A., Erten, C., Karatas, Ö., Sözdinler, M.: Crossing minimization in weighted bipartite graphs. J. Discret. Algorithms **7**(4), 439–452 (2009)
6. Chimani, M., Felsner, S., Kobourov, S., Ueckerdt, T., Valtr, P., Wolff, A.: On the maximum crossing number. J. Graph Algorithms Appl. **22**, 67–87 (2018)
7. Eades, P., Wormald, N.C.: Edge crossings in drawings of bipartite graphs. Algorithmica **11**, 379–403 (1994)
8. Furry, W., Kleitman, D.: Maximal rectilinear crossing of cycles. Stud. Appl. Math. **56**(2), 159–167 (1977)
9. Garey, M.R., Johnson, D.S.: Crossing number is NP-complete. SIAM J. Alg. Disc. Methods **4**, 312–316 (1983)
10. Goemans, M., Williamson, D.: Improved approximation algorithms for maximum cut and satisfiability problems using semidefinite programming. J. ACM **42**(6), 1115–1145 (1995)
11. Grünbaum, B.: Arrangements and spreads. In: CBMS Regional Conference Series in Mathematics, vol. 10. AMS, Providence, RI (1972)
12. Harborth, H.: Parity of numbers of crossings for complete n-partite graphs. Mathematica Slovaca **26**(2), 77–95 (1976)
13. Håstad, J.: Some optimal inapproximability results. J. ACM **48**(4), 798–859 (2001)
14. Kang, M., Pikhurko, O., Ravsky, A., Schacht, M., Verbitsky, O.: Obfuscated drawings of planar graphs. ArXiv (2008). http://arxiv.org/abs/0803.0858v3
15. Khot, S., Kindler, G., Mossel, E., O'Donnell, R.: Optimal inapproximability results for MAX-CUT and other 2-variable CSPs? SIAM J. Comput. **37**(1), 319–357 (2007)
16. Lovász, L., Pach, J., Szegedy, M.: On Conway's thrackle conjecture. Discret. Comput. Geom. **18**(4), 369–376 (1997)
17. Nagamochi, H.: An improved bound on the one-sided minimum crossing number in two-layered drawings. Discret. Comput. Geom. **33**(4), 569–591 (2005)
18. Piazza, B., Ringeisen, R., Stueckle, S.: Properties of nonminimum crossings for some classes of graphs. In: Proceedings of Graph Theory, Combinatorics, and Applications, pp. 975–989 (1991)

19. Ringeisen, R., Stueckle, S., Piazza, B.: Subgraphs and bounds on maximum crossings. Bull. Inst. Combin. Appl. **2**, 9–27 (1991)
20. Ringel, G.: Extremal problems in the theory of graphs. In: Proceedings of Theory of Graphs and its Applications (Smolenice 1963), pp. 85–90 (1964)
21. Schaefer, M.: The graph crossing number and its variants: a survey. Electron. J. Combin. 100 p. (2014). Dynamic Survey #DS21. http://www.combinatorics.org/ojs/index.php/eljc/article/view/DS21
22. Shahrokhi, F., Sýkora, O., Székely, L.A., Vrt'o, I.: On bipartite drawings and the linear arrangement problem. SIAM J. Comput. **30**, 1773–1789 (2001)
23. Staudacher, H.: Lehrbuch der Kombinatorik: Ausführliche Darstellung der Lehre von den kombinatorischen Operationen (Permutieren, Kombinieren, Variieren). J. Maier, Stuttgart (1893)
24. Steinitz, E.: Über die Maximalzahl der Doppelpunkte bei ebenen Polygonen von gerader Seitenzahl. Mathematische Zeitschrift **17**(1), 116–129 (1923)
25. Verbitsky, O.: On the obfuscation complexity of planar graphs. Theoret. Comput. Sci. **396**(1), 294–300 (2008)
26. Woodall, D.R.: Thrackles and deadlock. In: Welsh, D. (ed.) Proceedings of Combinatorial Mathematics and its Applications, pp. 335–347. Academic Press, Cambridge (1971)
27. Yannakakis, M.: Node- and edge-deletion NP-complete problems. In: Proceedings of 10th Annual ACM Symposium Theory Computer (STOC 1978), pp. 253–264. ACM (1978)

Approximation Results for the Incremental Knapsack Problem

Federico Della Croce[1,2]([✉]), Ulrich Pferschy[3], and Rosario Scatamacchia[1]

[1] Dipartimento di Automatica e Informatica, Politecnico di Torino,
Corso Duca degli Abruzzi 24, 10129 Torino, Italy
{federico.dellacroce,rosario.scatamacchia}@polito.it
[2] CNR, IEIIT, Torino, Italy
[3] Department of Statistics and Operations Research, University of Graz,
Universitaetsstrasse 15, 8010 Graz, Austria
pferschy@uni-graz.at

Abstract. We consider the 0–1 Incremental Knapsack Problem where the knapsack capacity grows over time periods and if an item is placed in the knapsack in a certain period, it cannot be removed afterwards. The problem calls for maximizing the sum of the profits over the whole time horizon. In this work, we manage to prove the tightness of some approximation ratios of a general purpose algorithm currently available in the literature. We also devise a Polynomial Time Approximation Scheme (PTAS) when the input value indicating the number of periods is considered as a constant. Then, we add the mild and natural assumption that each item can be packed in the first time period. For this variant, we discuss different approximation algorithms suited for any number of time periods and provide an algorithm with a constant approximation factor of $\frac{6}{7}$ for the case with two periods.

1 Introduction

We consider the 0–1 Incremental Knapsack Problem (IKP) as introduced in [2]. IKP is a generalization of the standard 0–1 Knapsack Problem (KP) ([4]) where the capacity grows over T time periods. If an item is placed in the knapsack in a certain period, it cannot be removed afterwards. The problem calls for maximizing the sum of the profits over the whole time horizon.

IKP has many real-life applications since, from a practical perspective, it is often required in resource allocation problems to deal with changes in the input conditions and/or in a multi-period optimization framework. In [2], incremental versions of maximum flow, bipartite matching, and knapsack problems are introduced. The authors in [2] discuss the complexity of these problems and show how the incremental version even of a polynomial time solvable problem, like the max flow problem, turns out to be NP–hard. General techniques to adapt the algorithms for the considered optimization problems to their respective incremental versions are discussed. Also, a general purpose approximation algorithm is introduced.

© Springer International Publishing AG, part of Springer Nature 2018
L. Brankovic et al. (Eds.): IWOCA 2017, LNCS 10765, pp. 75–87, 2018.
https://doi.org/10.1007/978-3-319-78825-8_7

In [1], it is shown that IKP is strongly NP–hard. In addition, a PTAS is derived for IKP under the assumption that T is in $O(\sqrt{\log n})$, where n is the number of items. A constant factor algorithm is also provided under mild restrictions on the growth rate of the knapsack capacity. For further details on the matter, see [1].

In this work, we prove, first, the tightness of some approximation ratios derived in [2]. Then, we devise a PTAS for IKP when the number of time periods T is a constant. While this is a stronger assumption than the one made for the PTAS in [1], our algorithm is much simpler and does not require a huge number of complicated LP models.

Finally, we consider the case where each item can be packed in the first period. Under this assumption, we give further insights into IKP and manage to derive several approximation algorithms for any T. We focus also on the variant with $T = 2$ and show an algorithm with a tight approximation factor of $\frac{6}{7}$. Due to limits in paper's length some of the results are given without proofs.

2   Notation and Problem Formulation

In IKP a set of n items is given together with a knapsack with increasing capacity values c_t over time periods $t = 1, \ldots, T$. Each item i has an integer profit $p_i \geq 0$ and an integer weight $w_i \geq 0$. If an item is placed in the knapsack, it cannot be removed at a later time. The problem calls for maximizing the total profit of the selected items without exceeding the knapsack capacity over the given time horizon.

In order to derive an ILP-formulation, we associate with each item i a $0/1$ variable x_{it} such that $x_{it} = 1$ iff item i is contained in the knapsack in period t. IKP can be formulated by the following ILP model (denoted by (IKP)):

$(\boldsymbol{IKP})$:

$$\text{maximize} \quad \sum_{t=1}^{T} \sum_{i=1}^{n} p_i x_{it} \tag{1}$$

$$\text{subject to} \quad \sum_{i=1}^{n} w_i x_{it} \leq c_t \quad t = 1, \ldots, T; \tag{2}$$

$$x_{i(t-1)} \leq x_{it} \quad i = 1, \ldots, n, \quad t = 2, \ldots, T; \tag{3}$$

$$x_{it} \in \{0, 1\} \quad i = 1, \ldots, n, \quad t = 1, \ldots, T. \tag{4}$$

The cost function (1) maximizes the sum of the profits over the time horizon. Constraints (2) guarantee that the items weights sum does not exceed capacity c_t in each period t. Constraints (3) ensure that an item chosen at time t cannot be removed afterwards. Constraints (4) indicate that all variables are binary.

We remark that the linear relaxation of model (IKP), denoted by IKP^{LP} and where constraints (4) are replaced by the inclusion in the interval $[0, 1]$, can be easily computed. In fact, it suffices to order the items by non-increasing $\frac{p_i}{w_i}$ and to fill the capacity of the knapsack in each period according to this ordering.

In the following, we will denote by z^* the optimal solution value of model (IKP) and by z^X the solution value yielded by a generic algorithm X. For each period t and the related capacity value c_t, we define the corresponding standard knapsack problem as KP_t. This means that in KP_t we consider only one of the T constraints (2). The optimal solution value of each KP_t will be denoted by z_t. Finally, we define for a generic item set S the canonical weight sum $w(S) := \sum_{j \in S} w_j$ and profit sum $p(S) := \sum_{j \in S} p_j$.

3 Approximating IKP

3.1 Approximation Ratios of a General Purpose Algorithm

In [2], a general framework for deriving approximation algorithms is provided. Following the scheme in [2] for IKP, we consider the following algorithm A. The algorithm employs an ε–approximation scheme to obtain a feasible solution for each knapsack problem KP_t. Denote the corresponding solution value by z_t^A for $t = 1, \ldots, T$. Each such solution is also a feasible solution for IKP where z_t^A is present in all successive time periods. The algorithm chooses as a solution value z^A the maximum among all these candidates, i.e.

$$z^A = \max_{t=1,\ldots,T} \{(T - t + 1) z_t^A\}. \tag{5}$$

The following Theorem which is a reformulation of Theorem 3 in [2] holds.

Theorem 1. *Algorithm A is an approximation algorithm for IKP with ratio bounded by $\frac{1-\varepsilon}{\mathcal{H}_T}$, where $\mathcal{H}_T$ is the harmonic number $1 + \frac{1}{2} + \cdots + \frac{1}{T}$.*

At the present state of the art, the tightness of the bound is an open question. We consider here the case where algorithm A solves each KP_t to optimality. Hence we have $z_t^A = z_t$ for $t = 1, \ldots, T$. And the corresponding approximation ratio is $\frac{1}{\mathcal{H}_T}$.

We prove the tightness of this bound by an alternative model. Consider an LP formulation with non-negative variables h^A and h_t associated with z^A and z_t^A respectively and a positive parameter $OPT > 0$ associated with z^*. The corresponding LP model for evaluating the worst case performance of algorithm A is as follows:

$$\text{minimize} \quad h^A \tag{6}$$

$$\text{subject to} \quad h^A \geq (T - t + 1) h_t \quad t = 1, \ldots, T; \tag{7}$$

$$\sum_{t=1}^{T} h_t \geq OPT \tag{8}$$

$$h_t \geq 0 \quad t = 1, \ldots, T. \tag{9}$$

The value of the objective function (6) provides a lower bound on the worst case performance of algorithm A. Constraints (7) guarantee that the contribution of each knapsack problem KP_t as a solution of IKP will be taken into account according to (5). Constraint (8) indicates that the sum of the optimal KPs solution values z_t^A over all T knapsack problems constitute a trivial upper bound on z^*. We will denote the optimal value of h^A and h_t ($t = 1, \ldots, T$) by h^{A^*} and h_t^* respectively. By setting parameter OPT to an arbitrary positive value, the corresponding lower bounds on the performance ratio of algorithm A for any T are given by $\frac{h^{A^*}}{OPT}$.

Model (6)–(9) allows us to prove the tightness of the approximation bound $\frac{1}{\mathcal{H}_T}$ of algorithm A.

Theorem 2. *For any value of T, if algorithm A solves to optimality each KP_t, $t = 1, \ldots, T$, the approximation ratio $\frac{1}{\mathcal{H}_T}$ of the algorithm is tight for IKP.*

Proof. We first provide a characterization of the optimal solution of model (6)–(9) and show that, for any T, the bound $\frac{h^{A^*}}{OPT}$ is actually equal to $\frac{1}{\mathcal{H}_T}$. Given the model constraints, we note that the optimal value h^{A^*} will be naturally equal to at least one of the right–hand side values $(T - t + 1)h_t^*$ of constraints (7). Also, the optimal solution will always fulfill $\sum_{t=1}^{T} h_t^* = OPT$. Suppose by contradiction that there is an optimal solution with $\sum_{t=1}^{T} h_t^* > OPT$. In such a case, we could always decrease h^{A^*} by jointly decreasing the corresponding h_t^* values (i.e. such that $h^{A^*} = (T - t + 1)h_t^*$), thus contradicting the optimality of the solution. In the optimal solution all right–hand side values of constraints (7) will reach the same value. Suppose again by contradiction that there exists an optimal solution where this structure does not hold, i.e. there are two time periods t' and t'' with $(T - t' + 1)h_{t'} = \max_t\{(T - t + 1)h_t\} > (T - t'' + 1)h_{t''}$. In this case, we could lower the objective by decreasing $h_{t'}$ and increasing $h_{t''}$ by the same value thus preserving the equality in (8) but allowing a decrease of h^A, which contradicts the claim. Based on this structural property, computing the optimal solution of the LP model amounts to solving the following system with $T + 1$ equations in $T + 1$ unknowns inducing a unique solution:

$$\begin{cases} h_t^* = \dfrac{h^{A^*}}{T-t+1} & t = 1, \ldots, T \\ \displaystyle\sum_{t=1}^{T} h_t^* = OPT \end{cases} \tag{10}$$

Indeed, by combining the first T equations with the latter one, we have that

$$\frac{h^{A^*}}{T} + \frac{h^{A^*}}{T-1} + \cdots + h^{A^*} = \mathcal{H}_T h^{A^*} = OPT \implies h^{A^*} = \frac{OPT}{\mathcal{H}_T} \tag{11}$$

that is $\frac{h^{A^*}}{OPT} = \frac{1}{\mathcal{H}_T}$. To prove the tightness of the bound $\frac{1}{\mathcal{H}_T}$, notice from (10) and (11) that we have

$$h_t^* = \frac{OPT}{\mathcal{H}_T(T - t + 1)} \quad t = 1, \ldots, T. \tag{12}$$

Then, it suffices to derive instances where the optimal solution values of KPs in each period are equal to h_t^* for $t = 1, \ldots, T$, and the optimal solution value for IKP is equal to the sum of all these solutions, namely $z^* = \sum_{t=1}^{T} h_t^*$. Such target instances can be generated by the following procedure:

1. We first represent the harmonic number $\mathcal{H}_T$ as a fraction, i.e. $\mathcal{H}_T = \frac{a}{b}$ where b is the smallest common multiple of the denominators of the fractions $\frac{1}{2} + \cdots + \frac{1}{T}$. Then, we set $OPT = a$ and solve model (6)–(9) according to (11)–(12). This setting guarantees to get integer values h_t^*.

2. Then, we generate an IKP instance with: $n = b$, $p_j = w_j = 1$ $(j = 1, \ldots, n)$, $c_t = h_t^*$ $(t = 1, \ldots, T)$. The optimal solution of each KP_t will pack items until the corresponding capacity c_t is fulfilled and thus will yield a solution value equal to h_t^*. The number of items is b because the capacity in the last period T is $c_T = h_T^* = \frac{OPT}{\mathcal{H}_T(T-T+1)} = \frac{a}{\frac{a}{b}} = b$. At the same time, the optimal solution for IKP can be obtained by progressively packing all items over time periods while filling the capacities c_t, hence $z^* = \sum_{t=1}^{T} c_t = \sum_{t=1}^{T} h_t^*$.   □

As an example of the outlined procedure, consider the case with $T = 3$ for which $\mathcal{H}_T = \frac{11}{6}$. We solve model (6)–(9) by setting $OPT = 11$. Then, the following instance with $n = 6$, $p_j = w_j = 1$ $(j = 1, \ldots, 6)$, $c_1 = 2$, $c_2 = 3$, $c_3 = 6$ is generated. The optimal solution is given by packing all items over time periods and filling the corresponding capacities $(z^* = 11)$. The optimal solutions values of the KPs are equal to 2, 3, 6 respectively. Hence we have $z^A = \max\{3 * 2, 2 * 3, 6\} = 6$ which proves the tightness of the approximation bound $\frac{1}{\mathcal{H}_3} = \frac{6}{11}$.

We remark that the bound tightness cannot be straightforwardly generalized when an ε–approximation scheme is adopted for solving each KP. We could get the ratio in Theorem 1 by solving model (6)–(9) where the term $\sum_{t=1}^{T} h_t$ in constraint (8) is divided by $(1 - \varepsilon)$. However, the generation of tight instances is strictly related to the choice of the approximation algorithm for KPs.

3.2 A PTAS When T is a Constant

Similarly to the line of reasoning for deriving PTAS's for KP (see, e.g., [5]), we propose an approximation scheme for IKP based on guessing the k items with largest profits in an optimal solution. We first define the following variant of algorithm A described in Sect. 3.1, denoted as algorithm A'. We run an FPTAS for each time period t yielding ε-approximations $z_t^{A'}$. Then we also consider an alternative solution for IKP derived by computing the optimal solution of IKP^{LP} and rounding down all fractional variables to 0. Thus, we get a feasible solution for IKP with solution value z'. Finally, we take the maximum between these $T + 1$ candidates reaching a solution value $z^{A'}$, namely

$$z^{A'} = \max\{z', \max_t\{(T - t + 1)z_t^{A'}\}\}. \tag{13}$$

Since computing z' requires $O(n \log n)$ and the running time of the FPTAS for KP can be bounded by $O(n \log(\frac{1}{\varepsilon}) + (\frac{1}{\varepsilon})^3 \log^2(\frac{1}{\varepsilon}))$ (see [3]) the overall running time of algorithm A' is $O(n \log n + T(n \log(\frac{1}{\varepsilon}) + (\frac{1}{\varepsilon})^3 \log^2(\frac{1}{\varepsilon})))$.

A useful property of algorithm A' is the following. Because of the special structure of IKP^{LP}, at most T fractional variables will be rounded down to get z'. Thus, we have that

$$z^* \leq z' + Tp_{max} \leq z^{A'} + Tp_{max} \tag{14}$$

where p_{max} is the maximum profit of any item.

The overall approximation ratio of algorithm A', denoted by ρ, can be stated by considering the solution values $z_t^{A'}$ in each time period. Hence, according to Theorem 1, we have $\rho = \frac{1-\varepsilon}{\mathcal{H}_T}$.

We can now state our approximation scheme, denoted by algorithm *Approx*, as follows:

1. We sort the items by decreasing efficiency $\frac{p_j}{w_j}$ and guess the k items with largest profits in an optimal solution. Also, we guess how these k items are distributed over the T periods. This corresponds to consider $O(n^k)$ choices for the items and $O(k^T)$ possible choices for their distributions over time. We set $k := \min\left\{n, \left\lceil\frac{T}{\varepsilon}\right\rceil\right\}$.
2. For each distribution of the items, we consider the remaining IKP instance. Let $Pr(k)$ be the overall profit contribution of the k items. We denote by c_t^R the residual capacities after the items insertion in each period t.
3. To keep the incremental structure of the problem, we set $c_t^R = \min\{c_t^R, c_{t+1}^R\}$ in decreasing order of t, i.e. for $t = T-1, \ldots, 1$. The corresponding residual IKP instance is denoted by R.
4. Given the items ordering, we apply algorithm A' to instance R getting a solution value $z_R^{A'}$. The sum $z_R^{A'} + Pr(k)$ yields the related solution value.
5. The overall best solution over all choices of k with value z^{Approx} is returned.

The following proposition holds.

Proposition 1. *The running time complexity of Algorithm Approx is polynomial in the size n of the input.*

Proof. The running time is given by the initial sorting of the items and by the running time of algorithm A' (without the sorting contribution) multiplied by $O(n^k k^T)$. Thus, the overall complexity is

$$O(n \log n + n^{\lceil\frac{T}{\varepsilon}\rceil}(\left\lceil\frac{T}{\varepsilon}\right\rceil)^T T(n \log(\frac{1}{\varepsilon}) + (\frac{1}{\varepsilon})^3 \log^2(\frac{1}{\varepsilon}))). \qquad \Box \tag{15}$$

The following theorem holds.

Theorem 3. *Algorithm Approx is a PTAS for IKP when T is a constant.*

Proof. Omitted.

We remark that [1] introduces a PTAS for T in $O(\sqrt{\log n})$ thus providing a stronger result. Still, the PTAS in [1] relies on solving a large number, namely $O(n(\frac{1}{\varepsilon} + T)^{O(\log(\frac{T}{\varepsilon})/\varepsilon^2)})$, of non-trivial LP models. Our approach instead does not require the solution of LP models and requires lower complexity. Hence, it constitutes an appealing approximation algorithm for reasonable values of T.

4   Approximation Algorithms for a Relevant IKP Variant

In this section we consider IKP under the mild assumption that each item can be packed in the first period, i.e. $w_i \le c_1, i = 1, \ldots, n$. We refer to this variant of the problem as IKP'. Let us denote by s_t the split items in the linear relaxation of each KP_t for $t = 1, \ldots, T$, which also correspond to the fractional items in the optimal solution of IKP^{LP}. We state the following algorithm H_1:

1. We first sort items by decreasing $\frac{p_i}{w_i}$ and solve IKP^{LP}.
2. We pack either items $j = 1, \ldots, s_1 - 1$ or item s_1 by taking the maximum between their profit contributions $\sum_{j=1}^{s_1-1} p_j$ and p_{s_1} in all time periods before these items are entirely packed in the optimal solution of IKP^{LP}.
3. From then on, we take the rest of the optimal solution of IKP^{LP} without any fractional item.

The following theorem holds.

Theorem 4. *Algorithm H_1 has a tight $\frac{1}{2}$-approximation ratio for IKP'.*

Proof. Consider the optimal solution values z_t of KP_t for $t = 1, \ldots, T$. From the properties of KP and since items are ordered by decreasing $\frac{p_i}{w_i}$, we have

$$\max \left\{ \sum_{j=1}^{s_t-1} p_j, p_{s_t} \right\} \ge \frac{1}{2} z_t \qquad t = 1, \ldots, T. \tag{16}$$

$$\frac{\sum_{j=1}^{s_t-1} p_j}{\sum_{j=1}^{s_t-1} w_j} \ge \frac{p_{s_t}}{w_{s_t}} \qquad t = 1, \ldots, T. \tag{17}$$

Let $\hat{t}$ be the first period when items $1, \ldots, s_1$ are fully packed in the optimal solution of IKP^{LP} (i.e. $s_{\hat{t}} \ne s_1$). Algorithm H_1 yields a solution with value

$$z^{H_1} = (\hat{t} - 1) \cdot \max \left\{ \sum_{j=1}^{s_1-1} p_j, p_{s_1} \right\} + \sum_{t=\hat{t}}^{T} \sum_{j=1}^{s_t-1} p_j. \tag{18}$$

Since inequalities $\sum_{j=1}^{s_t-1} w_j \geq \sum_{j=1}^{s_1} w_j > c_1 \geq w_i$ hold for any item i and $t \geq \hat{t}$, from (17) we get $\sum_{j=1}^{s_t-1} p_j > p_{s_t}$ for any $t \geq \hat{t}$ and thus

$$\max\left\{\sum_{j=1}^{s_t-1} p_j, p_{s_t}\right\} = \sum_{j=1}^{s_t-1} p_j \qquad t = \hat{t}, \ldots, T. \tag{19}$$

Considering (16), (18), (19) and that $\sum_{t=1}^{T} z_t$ is an upper bound on z^*, we get

$$z^{H_1} \geq \frac{1}{2}\sum_{t=1}^{\hat{t}-1} z_t + \frac{1}{2}\sum_{t=\hat{t}}^{T} z_t \geq \frac{1}{2}z^* \tag{20}$$

which shows that algorithm H_1 has a relative performance guarantee of $\frac{1}{2}$.

To prove the bound tightness, consider the following instance with $c_t = M + t$ for $t = 1, \ldots, T$ (with integer $M \gg T$) and 3 items with entries

$$p_1 = \frac{M}{2} + \delta, w_1 = \frac{M}{2}; p_2 = \frac{M}{2} + T + 1, w_2 = \frac{M}{2} + T + 1; p_3 = \frac{M}{2} - \delta, w_3 = \frac{M}{2};$$

with $\delta > 0$ being an arbitrary small number. Algorithm H_1 will select only the second item over all time periods getting a solution with value $T(\frac{M}{2} + T + 1)$. The optimal IKP' solution will pack items 1 and 3 for all periods reaching a value of $T \cdot M$. Hence, the approximation ratio of the algorithm is

$$\frac{T(\frac{M}{2} + T + 1)}{TM} = \frac{1}{2} + \frac{T+1}{M}$$

which reaches a value arbitrarily close to $\frac{1}{2}$ for large values of M.   □

We consider now an algorithm, denoted as H_2, which considers the periods from $t = T$ to $t = 1$ and computes in each step an *optimal* knapsack solution. The algorithm works as follows:

1. We solve KP_T to optimality.
2. Iteratively going back over periods $t = T - 1, \ldots, 1$, we solve to optimality the problem induced by restricting the available item set for KP_t to the set of items packed in the preceding period for $t + 1$.

The following theorem holds.

Theorem 5. *Algorithm H_2 has a tight $(\frac{1}{2} + \frac{1}{2T})$-approximation ratio for IKP'.*

Proof. Let O_t be the set of items selected in period t by the optimal IKP' solution and I_t the items packed by H_2. For any arbitrary period $t \leq T - 1$ we construct an auxiliary item set Q as follows. At first insert all items in $I_{t+1} \cap O_t$ into Q. Clearly, their total weight cannot exceed c_t. Then we continue adding items from I_{t+1} in arbitrary order as long as $w(Q) \leq c_t$. Let us denote by r the first item considered in this process violating this weight bound. Hence, in any period $t \leq T - 1$ we have that

$$w(Q) \leq c_t, \tag{21}$$
$$w(Q) + w_r > c_t. \tag{22}$$

Since by definition $w(O_t) \leq c_t$, we also have

$$w(Q) + w_r > w(O_t). \tag{23}$$

All items in I_{t+1} and thus also all items in set Q and item r are part of the optimal solution of KP_T. Therefore, with (23) the following inequality holds:

$$p(Q) + p_r \geq p(O_t) \tag{24}$$

Otherwise we could improve the optimal solution value of KP_T by replacing the items in set Q and r with the items in set O_t (also if some items of O_t are already contained in Q). We consider now the solution value contribution $z_t^{H_2}$ yielded by algorithm H_2 in period t, namely $z^{H_2} = \sum_{t=1}^{T} z_t^{H_2}$ with $z_T^{H_2}$ being the optimal solution value of KP_T. We have $z_t^{H_2} \geq p_r$ and $z_t^{H_2} \geq p(Q)$ since, by definition, item set Q constitutes a feasible solution for KP_t. Thus we get with (24) that:

$$z_t^{H_2} \geq \frac{1}{2} p(O_t) \qquad t = 1, \ldots, T - 1 \tag{25}$$

Note that if there isn't any item r, i.e. all items packed by algorithm H_2 in period $t+1$ fit in period t as well, we have $z_t^{H_2} = z_{t+1}^{H_2}$. Increasing t towards T we either find an iteration $t' > t$ where such an item r exists and thus from above $z_t^{H_2} = z_{t'}^{H_2} \geq \frac{1}{2} p(O_{t'}) \geq \frac{1}{2} p(O_t)$, or we would even get $z_t^{H_2} = z_T^{H_2} \geq p(O_T) \geq p(O_t)$.

Moreover, since $z_T^{H_2} \geq p(O_t)$ for $t = 1, \ldots, T$, the following inequality holds:

$$z_T^{H_2} \geq \frac{1}{T-1} \sum_{t=1}^{T-1} p(O_t) \tag{26}$$

Correspondingly, we have that:

$$\frac{z^{H_2}}{z^*} = \frac{\sum_{t=1}^{T} z_t^{H_2}}{\sum_{t=1}^{T} p(O_t)} \geq \frac{z_T^{H_2} + \sum_{t=1}^{T-1} z_t^{H_2}}{z_T^{H_2} + \sum_{t=1}^{T-1} p(O_t)} \geq \frac{z_T^{H_2} + \frac{1}{2} \sum_{t=1}^{T-1} p(O_t)}{z_T^{H_2} + \sum_{t=1}^{T-1} p(O_t)} \tag{27}$$

$$\geq \frac{\frac{1}{T-1} \sum_{t=1}^{T-1} p(O_t) + \frac{1}{2} \sum_{t=1}^{T-1} p(O_t)}{\frac{1}{T-1} \sum_{t=1}^{T-1} p(O_t) + \sum_{t=1}^{T-1} p(O_t)} = \frac{\frac{1}{2}(T+1)}{T} = \frac{1}{2} + \frac{1}{2T} \tag{28}$$

The first two inequalities derive from $z_T^{H_2} \geq p(O_T)$ and from (25) respectively. The last inequality is due to the fact that the value of the last term of expression (27) decreases with the decrease of $z_T^{H_2}$ which is lower bounded by (26).

To prove the tightness of the bound, we consider the following instance with capacities $c_t = M + t$ for $t = 1, \ldots, T$ (with $M \gg T$) and 3 items with entries

$$w_1 = \frac{M}{2} + T, \; p_1 = \frac{1}{2} + \delta, \; w_2 = \frac{M}{2}, \; p_2 = \frac{1}{2}, \; w_3 = M, \; p_3 = 1.$$

Algorithm H_2 will select items 1 and 2 n period T and only item 1 in the other periods. The corresponding solution value is $z^{H_2} = T(\frac{1}{2} + \delta) + \frac{1}{2}$. The optimal solution will pack only item 3 in all periods, thus, $z^* = T$. Hence, $\frac{z^{H_2}}{z^*} = \frac{1}{2} + \delta + \frac{1}{2T}$, which shows the claim as δ goes to 0. $\qquad \square$

4.1   Approximation for $T = 2$

We consider now IKP' with $T = 2$ and the following algorithm denoted as H_3.

1. We solve to optimality KP_1. We then solve KP_2 where the optimal solution set of items in KP_1 is placed inside the knapsack.
2. As an alternative solution, we first solve to optimality KP_2. Then, we consider the optimal set of items in KP_2 and solve KP_1 with these items only.
3. The best solution found is returned.

The following theorem holds.

Theorem 6. *Algorithm H_3 has a tight $\frac{6}{7}$-approximation ratio for IKP' with $T = 2$.*

Proof. Let us define the following subsets of items S_i:

- S_1: subset of items included both in the optimal solutions of KP_1 and KP_2;
- S_2: remaining items subset in the optimal solution of KP_1;
- S_3: remaining items subset not exceeding capacity c_1 in the optimal solution of KP_2;
- S_4: first item exceeding c_1 in the optimal solution of KP_2;
- S_5: remaining items subset in the optimal solution of KP_2.

The union of S_1 and S_2 gives the optimal solution set of items of KP_1. The union of S_1, S_3, S_4 and S_5 gives the optimal solution set of KP_2. Figure 1 depicts the decomposition of the optimal solution sets in each time period.

The dashed lines refer to the item in S_4 which exceeds the first capacity value. According to the above definitions, we have the following inequalities

$$w(S_1) + w(S_2) \leq c_1; \tag{29}$$

$$w(S_1) + w(S_3) \leq c_1; \tag{30}$$

$$w(S_1) + w(S_3) + w(S_4) > c_1; \tag{31}$$

$$w(S_1) + w(S_3) + w(S_4) + w(S_5) \leq c_2; \tag{32}$$

$$w(S_1) + w(S_2) + w(S_5) < c_2. \tag{33}$$

Inequality (33) derives directly from inequalities (29), (31) and (32). The optimal solution values of KP_1 and KP_2 are $z_1 = p(S_1) + p(S_2)$ and $z_2 = p(S_1) + p(S_3) + p(S_4) + p(S_5)$ respectively. Now we can state three feasible solutions reached by algorithm H_3:

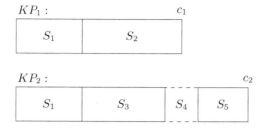

Fig. 1. Decomposition of the optimal solutions of KP_1 and KP_2.

- The optimal solution set of KP_1 in the two periods plus the additional packing of items in S_5 in the second period. The whole profit is: $2(p(S_1) + p(S_2)) + p(S_5)$;
- The optimal solution set of KP_2 in the second period with the packing of items in subsets S_1 and S_3 in the first period. The resulting profit is: $2(p(S_1) + p(S_3)) + p(S_4) + p(S_5)$;
- The optimal solution set of KP_2 in the second period with item S_4 placed in the knapsack in the first period. The profit of this solution is: $p(S_1) + p(S_3) + 2p(S_4) + p(S_5)$.

Algorithm H_3 will return a solution at least as good as the best of these solutions. In order to evaluate the worst case performance of the heuristic, we consider an LP formulation where we associate a non-negative variable h with the solution value computed by the algorithm. In addition, the profits of the subsets S_i are associated with non-negative variables $\bar{p}_i$ ($i = 1, \ldots, 5$). As in Sect. 3.1, the positive parameter OPT represents z^*. This implies the following LP model:

$$\text{minimize} \quad h \tag{34}$$
$$\text{subject to} \quad h \geq 2(\bar{p}_1 + \bar{p}_2) + \bar{p}_5 \tag{35}$$
$$h \geq 2(\bar{p}_1 + \bar{p}_3) + \bar{p}_4 + \bar{p}_5 \tag{36}$$
$$h \geq \bar{p}_1 + \bar{p}_3 + 2\bar{p}_4 + \bar{p}_5 \tag{37}$$
$$(\bar{p}_1 + \bar{p}_2) + (\bar{p}_1 + \bar{p}_3 + \bar{p}_4 + \bar{p}_5) \geq OPT \tag{38}$$
$$\bar{p}_i \geq 0 \quad i = 1, \ldots, 5. \tag{39}$$

The cost function value (34) represents a lower bound on the worst case performance of H_3. Constraints (35)–(37) guarantee that H_3 will select the best of the three feasible solutions. Constraint (38) states that the sum of the optimal values of KP_1 and KP_2 is an upper bound on z^*. Constraints (39) indicate that variables are non-negative. Setting the parameter $OPT = 1$, we get an optimal value h^* equal to $0.8571\ldots = \frac{6}{7}$. Thus, a lower bound on the performance ratio provided by algorithm H_4 is equal to $\frac{h^*}{OPT} = \frac{6}{7}$.

We can show that the ratio of $\frac{6}{7}$ is tight by considering an instance with $n = 5$, $c_1 = 3 + \delta$, $c_2 = 4$ and the following entries:

i	1		2	3	4	5
p_i	3		2	2	$1-\delta$	$1-\delta$
w_i	$3+\delta$		2	2	1	1

The optimal solution of KP_1 consists of item 1. The corresponding IKP' solution considers only item 1 in the second period with a total profit of 6. The optimal solution of KP_2 consists in packing items 2 and 3. Then either item 2 or item 3 is placed in the knapsack in the first period. The resulting profit is again 6. An optimal solution selects items 2 and 4 in the first period together with item 5 in the second period, thus $z^* = 7 - 3\delta$. The approximation ratio of algorithm H_3 is

$$\frac{\max\{6,6\}}{7-3\delta} = \frac{6}{7-3\delta}$$

which can be arbitrarily close to $\frac{6}{7}$ as the value of δ goes to 0. □

Algorithm H_3 is not polynomial since it requires the optimal solutions of KP_1 and KP_2. We could solve these knapsack problems by an ε–approximation scheme (PTAS or FPTAS) to get a polynomial running time at the cost of a decrease of the approximation bound.

The following Corollary holds.

Corollary 1. *If an ε–approximation scheme is employed for solving KP_1 and KP_2, the approximation ratio of algorithm H_3 is bounded by $\frac{6}{7}(1-\varepsilon)$.*

Proof. Omitted.

5 Conclusions

We proposed for IKP a series of results extending in different directions the contributions currently available in the literature. We proved the tightness of approximation ratios of a general purpose algorithm previously laid out and established a polynomial time approximation scheme (PTAS) when one of the problem inputs can be considered as a constant. We then focused on a restricted relevant variant of IKP which plausibly assumes the possible packing of any item since the first period. We discussed the performance of different approximation algorithms and showed an algorithm with an approximation ratio of $\frac{6}{7}$ for the variation with two time periods. In future research, we will investigate extensions of our procedures to the design of improving approximation algorithms for variants involving more than two periods. Also, since to the authors' knowledge no computational experience has been provided for IKP so far, it would also be interesting to derive new solution approaches for the problem and test their performance in solving instances after generating challenging benchmarks.

References

1. Bienstock, D., Sethuraman, J., Ye, C.: Approximation algorithms for the incremental knapsack problem via disjunctive programming (2013). arXiv:1311.4563
2. Hartline, J., Sharp, A.: An incremental model for combinatorial maximization problems. In: Àlvarez, C., Serna, M. (eds.) WEA 2006. LNCS, vol. 4007, pp. 36–48. Springer, Heidelberg (2006). https://doi.org/10.1007/11764298_4
3. Kellerer, H., Pferschy, U.: Improved dynamic programming in connection with an FPTAS for the knapsack problem. J. Comb. Optim. **8**, 5–11 (2004)
4. Kellerer, H., Pferschy, U., Pisinger, D.: Knapsack Problems. Springer, Heidelberg (2004). https://doi.org/10.1007/978-3-540-24777-7
5. Sahni, S.: Approximate algorithms for the 0–1 knapsack problem. J. ACM **22**, 115–124 (1975)

Derandomization for k-Submodular Maximization

Hiroki Oshima[(✉)]

Department of Mathematical Informatics,
Graduate School of Information Science and Technology,
The University of Tokyo, Tokyo 113-8656, Japan
hiroki_oshima@me2.mist.i.u-tokyo.ac.jp

Abstract. Submodularity is one of the most important properties of combinatorial optimization, and k-submodularity is a generalization of submodularity. Maximization of a k-submodular function is NP-hard, and approximation algorithms have been studied. Most of algorithms use randomization and achieve the approximation ratio as the expected value. For unconstrained submodular maximization, [Buchbinder and Feldman 2016] gave a derandomization scheme, and showed that randomness is not necessary for obtaining optimal results. In this paper, we extend their scheme for unconstrained k-submodular maximization and give two deterministic algorithms for k-submodular maximization. The first is for monotone k-submodular functions. We derandomize $k/(2k-1)$-approximation algorithm [Iwata, Tanigawa, and Yoshida 2016], and achieve the same approximation ratio. The second is for nonmonotone functions with $k \geq 2$. It achieves $k/(3k-2)$-approximation, and it is better than a simple greedy algorithm [Ward and Živný 2016] which achieves $1/3$-approximation for $k \geq 3$.

1 Introduction

A set function $f : 2^V \to \mathbb{R}$ is submodular if, for any $A, B \subseteq V$, $f(A) + f(B) \geq f(A \cup B) + f(A \cap B)$. Submodularity is one of the most important properties of combinatorial optimization. The rank functions of matroids and cut capacity functions of networks are submodular. Submodular functions can be seen as a discrete version of convex functions.

For submodular function minimization, Grötschel et al. [5] showed the first polynomial-time algorithm. The combinatorial strongly polynomial algorithms were shown by Iwata et al. [7] and Schrijver [11]. On the other hand, submodular function maximization is NP-hard and we can only use approximation algorithms. Let an input function for maximization be f, a maximizer of f be S^*, and an output of an algorithm be S. The approximation ratio of the algorithm is defined as $f(S)/f(S^*)$ for deterministic algorithms and $\mathbb{E}[f(S)]/f(S^*)$ for randomized algorithms. For unconstrained submodular maximization, the randomized algorithm called Double Greedy [2] achieves $1/2$-approximation. Feige et al. [3] showed $(1/2 + \epsilon)$-approximation requires exponential number of value

© Springer International Publishing AG, part of Springer Nature 2018
L. Brankovic et al. (Eds.): IWOCA 2017, LNCS 10765, pp. 88–99, 2018.
https://doi.org/10.1007/978-3-319-78825-8_8

oracle queries. This implies that, Double Greedy algorithm is one of the best algorithms in terms of the approximation ratio. Buchbinder and Feldman [1] showed a derandomized version of randomized Double Greedy algorithm, and their algorithm achieves 1/2-approximation.

k-submodularity is an extension of submodularity. It was first introduced by Huber and Kolmogolov [6]. k-submodular function is defined as below.

Definition 1 ([6]). *Let* $(k + 1)^V := \{(X_1, ..., X_k) \mid X_i \subseteq V \ (i = 1, ..., k), X_i \cap X_j = \emptyset \ (i \neq j)\}$. *A function* $f : (k+1)^V \to \mathbb{R}$ *is called k-submodular if we have*

$$f(\boldsymbol{x}) + f(\boldsymbol{y}) \geq f(\boldsymbol{x} \sqcap \boldsymbol{y}) + f(\boldsymbol{x} \sqcup \boldsymbol{y})$$

for any $\boldsymbol{x} = (X_1, ..., X_k)$, $\boldsymbol{y} = (Y_1, ..., Y_k) \in (k+1)^V$. *Note that*

$$\boldsymbol{x} \sqcap \boldsymbol{y} = (X_1 \cap Y_1, ..., X_k \cap Y_k) \text{ and}$$
$$\boldsymbol{x} \sqcup \boldsymbol{y} = (X_1 \cup Y_1 \setminus \bigcup_{i \neq 1}(X_i \cup Y_i), ..., X_k \cup Y_k \setminus \bigcup_{i \neq k}(X_i \cup Y_i)).$$

It is a submodular function if $k = 1$. It is called a bisubmodular function if $k = 2$. The rank functions of delta-matroids are bisubmodular. We can see applications of k-submodular functions in influence maximization and sensor placement [10] with maximization and computer vision [4] with minimization.

Maximization for k-submodular functions is also NP-hard and approximation algorithm have been studied. An input of the problem is a nonnegative k-submodular function. Note that, for any k-submodular function f and any $c \in \mathbb{R}$, a function $f'(\boldsymbol{x}) := f(\boldsymbol{x}) + c$ is k-submodular. An output of the problem is $\boldsymbol{x} = (X_1, ..., X_k) \in (k+1)^V$. The input function is accessed via value oracle queries. For bisubmodular functions, Iwata et al. [8] and Ward and Živný [12] showed that the algorithm for submodular functions [2] can be extended. Ward and Živný [12] analyzed an extension for k-submodular functions. They showed a randomized $1/(1 + a)$-approximation algorithm with $a = \max\{1, \sqrt{(k-1)/4}\}$ and a deterministic 1/3-approximation algorithm. Now we have a randomized 1/2-approximation algorithm shown by Iwata et al. [9]. In particular, for monotone k-submodular functions, they gave a randomized $\frac{k}{2k-1}$-approximation algorithm. (For monotone unconstrained maximization with $k = 1$, it is obvious that $X_1 = V$ is optimal solution.) They also showed any $(\frac{k+1}{2k} + \epsilon)$-approximation algorithm requires exponential number of value oracle queries.

In this paper, we extend the derandomization scheme [1], used for Double Greedy algorithm in submodular maximization, and give the scheme for k-submodular maximization. Then we show two deterministic approximation algorithms. One satisfies $\frac{k}{2k-1}$-approximation for monotone functions. This algorithm is a derandomized version of the algorithm for monotone functions [9]. The other algorithm in this paper is $\frac{k}{3k-2}$-approximation algorithm for nonmonotone functions with $k \geq 2$. First, we show a randomized $\frac{k}{3k-2}$-approximation algorithm as the base of derandomization. Then we derandomize it as in our algorithm for monotone functions.

The rest of this paper is organized as follows. In Sect. 2, we explain details of k-submodularity. In Sect. 3, we explain previous works of unconstrained k-submodular maximization. In Sect. 4, we expand the derandomization scheme [1] for k-submodular maximization. In Sect. 5, we present a deterministic algorithm for monotone functions. We present a deterministic algorithm for nonmonotone functions in Sect. 6. We conclude this paper in Sect. 7.

2 Preliminary

Define a partial order $\preceq$ on $(k+1)^V$ for $\boldsymbol{x} = (X_1, ..., X_k)$ and $\boldsymbol{y} = (Y_1, ..., Y_k)$ as follows:

$$\boldsymbol{x} \preceq \boldsymbol{y} \overset{\text{def}}{\Longleftrightarrow} X_i \subseteq Y_i \ (i = 1, ..., k).$$

Also, for $\boldsymbol{x} = (X_1, ..., X_k) \in (k+1)^V$, $e \notin \bigcup_{l=1}^{k} X_l$, and $i \in \{1, ..., k\}$, define

$$\Delta_{e,i} f(\boldsymbol{x}) = f(X_1, ..., X_{i-1}, X_i \cup \{e\}, X_{i+1}, ..., X_k) - f(X_1, ..., X_k).$$

A monotone k-submodular function is k-submodular and satisfies $f(\boldsymbol{x}) \leq f(\boldsymbol{y})$ for any $\boldsymbol{x} = (X_1, ..., X_k)$ and $\boldsymbol{y} = (Y_1, ..., Y_k)$ in $(k+1)^V$ with $\boldsymbol{x} \preceq \boldsymbol{y}$.

The property of k-submodularity can be written as another form.

Theorem 1 ([12], Theorem 7). *A function $f : (k+1)^V \to \mathbb{R}$ is k-submodular if and only if f is orthant submodular and pairwise monotone.*

Note that orthant submodularity is to satisfy

$$\Delta_{e,i} f(\boldsymbol{x}) \geq \Delta_{e,i} f(\boldsymbol{y}) \ (\boldsymbol{x}, \boldsymbol{y} \in (k+1)^V, \ \boldsymbol{x} \preceq \boldsymbol{y}, \ e \notin \bigcup_{l=1}^{k} Y_l, \ i \in \{1, ..., k\}),$$

and pairwise monotonicity is to satisfy

$$\Delta_{e,i} f(\boldsymbol{x}) + \Delta_{e,j} f(\boldsymbol{x}) \geq 0 \ (\boldsymbol{x} \in (k+1)^V, \ e \notin \bigcup_{l=1}^{k} X_l, \ i,j \in \{1, ..., k\} \ (i \neq j)).$$

To analyze k-submodular functions, it is often convenient to identify $(k+1)^V$ as $\{0, 1, ..., k\}^V$. Let $n = |V|$. An n-dimensional vector $\boldsymbol{x} \in \{0, 1, ..., k\}^V$ is associated with $(X_1, ..., X_n) \in (k+1)^V$ by $X_i = \{e \in V \mid \boldsymbol{x}(e) = i\}$.

3 Existing Algorithms

3.1 Algorithm Framework

In this section, we see the framework to maximize k-submodular functions (Algorithm 1 [9]). Iwata et al. [8] and Ward and Živný [12] used it with specific distributions.

Algorithm 1. ([9] Algorithm 1)

Input: A nonnegative k-submodular function $f : \{0, 1, ..., k\}^V \to \mathbb{R}_+$.
Output: A vector $s \in \{0, 1, ..., k\}^V$.
$s \leftarrow \mathbf{0}$.
Denote the elements of V by $e^{(1)}, ..., e^{(n)}$ ($|V| = n$).
for $j = 1, ..., n$ **do**
    Set a probability distribution $p^{(j)}$ over $\{1, ..., k\}$.
    Let $s(e^{(j)}) \in \{1, ..., k\}$ be chosen randomly with $\Pr[s(e^{(j)}) = i] = p_i^{(j)}$.
end for
return s

Now we define some variables to see Algorithm 1. Let o be an optimal solution. We consider the j-th iteration of the algorithm, and we write $s^{(j)}$ as the solution s after j-th iteration. Let other variables be as follows:

$$o^{(j)} = (o \sqcup s^{(j)}) \sqcup s^{(j)}, \quad t^{(j-1)}(e) = \begin{cases} o^{(j)}(e) & (e \neq e^{(j)}) \\ 0 & (e = e^{(j)}) \end{cases},$$

$$y_i^{(j)} = \Delta_{e^{(j)}, i} f(s^{(j-1)}), \quad a_i^{(j)} = \Delta_{e^{(j)}, i} f(t^{(j-1)}).$$

From the updating rule in the algorithm, $o^{(j)}(e) = s(e)$ for $e \in \{e^{(1)}, ..., e^{(j)}\}$, and $o^{(j)}(e) = o(e)$ for $e \in \{e^{(j+1)}, ..., e^{(n)}\}$. Algorithm 1 satisfies following lemma.

Lemma 1 ([9], Lemma 2.1). *Let $c \in \mathbb{R}_+$. Conditioning on $s^{(j-1)}$, suppose that*

$$\sum_{i=1}^{k} (a_{i^*}^{(j)} - a_i^{(j)}) p_i^{(j)} \leq c \sum_{i=1}^{k} (y_i^{(j)} p_i^{(j)}) \tag{1}$$

holds for each j with $1 \leq j \leq n$, where $i^* = o(e^{(j)})$. Then $\mathbb{E}[f(s)] \geq \frac{1}{1+c} f(o)$.

3.2 The Randomized Algorithm for Monotone Functions

In this section, we see the randomized algorithm for monotone functions [9]. We apply the distribution and their proof to our algorithms.

We show their algorithm as Algorithm 2. Algorithm 2 runs in polynomial time. The approximation ratio of Algorithm 2 satisfies Theorem 2. We rephrase the proof [9] for fitting our settings. First, we introduce Lemma 2. This lemma is used in the proof [9], and we use it for proving our theorems later.

Lemma 2 ([9], Proof of Theorem 2.2). *Let k be a positive integer, $y_i (i = 1, ..., k)$ be nonnegative, and $i_* \in \{1, ..., k\}$.*

$$\left(1 - \frac{1}{k}\right) \sum_{i=1}^{k} p_i y_i \geq \sum_{i \neq i^*} p_i y_{i^*} \tag{2}$$

is always satisfied by the distribution $\{p_i\}$ in Algorithm 2.

Algorithm 2. ([9] Algorithm 3)

Input: A monotone k-submodular function $f : \{0, 1, ..., k\}^V \to \mathbb{R}_+$.
Output: A vector $s \in \{0, 1, ..., k\}^V$.
$s \leftarrow \mathbf{0}, t \leftarrow k - 1$.
Denote the elements of V by $e^{(1)}, ..., e^{(n)}$ ($|V| = n$).
for $j = 1, ..., n$ **do**
$\quad y_i^{(j)} \leftarrow \Delta_{e^{(j)}, i} f(s) \ (1 \leq i \leq k)$.
$\quad \beta \leftarrow \sum_{i=1}^k (y_i^{(j)})^t$.
$\quad$**if** $\beta \neq 0$ **then** $p_i^{(j)} \leftarrow (y_i^{(j)})^t / \beta \ (1 \leq i \leq k)$.
$\quad$**else**
$\qquad p_1^{(j)} = 1, p_i^{(j)} = 0 \ (i = 2, ..., k)$.
$\quad$**end if**
$\quad$Let$s(e^{(j)}) \in \{1, ..., k\}$ be chosen randomly with $\Pr[s(e^{(j)}) = i] = p_i^{(j)}$.
end for
return s

Proof. We first consider the case $\beta = 0$. From the monotonicity of f, $y_i \geq a_i \geq 0$ for all $1 \leq i \leq k$. Then we have $y_i = a_i = 0$. Hence it is clear that (2) holds. Now suppose $\beta > 0$. The goal is to show

$$\left(1 - \frac{1}{k}\right) \sum_{i=1}^k y_i^{t+1} \geq \sum_{i \neq i^*} y_i^t y_{i^*}. \tag{3}$$

If $k = 1$, (3) is satisfied since $i_* = 1$ and both sides is 0. Hence we assume $k \geq 2$. Let $\gamma = (k - 1)^{\frac{1}{t}} = t^{\frac{1}{t}}$. From the weighted AM-GM inequality, $a^{\frac{1}{t+1}} b^{\frac{t}{t+1}} \leq \frac{1}{t+1} a + \frac{t}{t+1} b$ holds for all $a, b \geq 0$. By setting $a = (\gamma y_{i^*})^{t+1}$ and $b = (\sum_{i \neq i^*} y_i^t)^{(t+1)/t}$, we have

$$\sum_{i \neq i^*} y_i^t y_{i^*} = \frac{1}{\gamma} \left(\gamma y_{i^*} \cdot \sum_{i \neq i^*} y_i^t \right) \leq \frac{1}{\gamma} \left(\frac{(\gamma y_{i^*})^{t+1}}{t+1} + \frac{t}{t+1} (\sum_{i \neq i^*} y_i^t)^{\frac{t+1}{t}} \right). \tag{4}$$

From Hölder's inequality, $\sum_i a_i \leq (\sum_i a_i^{\frac{t+1}{t}})^{\frac{t}{t+1}} (\sum_i 1^{t+1})^{\frac{1}{t+1}}$ holds for any nonnegative a_i's. By setting $a_i = y_i^t$, we have

$$(\text{RHS of } (4)) \leq \frac{1}{\gamma} \left(\frac{(\gamma y_{i^*})^{t+1}}{t+1} + \frac{t}{t+1} (\sum_{i \neq i^*} y_i^{t+1}) \cdot (\sum_{i \neq i^*} 1^{t+1})^{\frac{1}{t}} \right)$$

$$= \frac{1}{\gamma} \left(\frac{(\gamma y_{i^*})^{t+1}}{t+1} + \frac{t(k-1)^{\frac{1}{t}}}{t+1} \sum_{i \neq i^*} y_i^{t+1} \right)$$

$$= \frac{\gamma^t}{t+1} \sum_i y_i^{t+1} = \left(\frac{k-1}{k} \right) \sum_i y_i^{t+1}.$$

Thus we obtain (3). $\qquad \square$

Theorem 2 ([9], Theorem 2.2). *Let o be a maximizer of a monotone k-submodular function f and let s be the output of Algorithm 2. Then $\mathbb{E}[f(s)] \geq \frac{k}{2k-1}f(o)$.*

Proof. We obtain $a_{i^*} - a_i \leq a_{i^*} \leq y_{i^*}$ from orthant submodularity and monotonicity. Then

$$\sum_{i=1}^{k} p_i(a_{i^*} - a_i) \leq \sum_{i \neq i^*} p_i y_{i^*} \tag{5}$$

is valid, regardless of the provability $\{p_i\}$.

Now we can see the validity of this theorem from Lemmas 1 and 2.   □

4   Derandomization Scheme for k-Submodular Maximization

Now we consider derandomization for k-submodular maximization. To explain our deterministic algorithms, we have to expand the derandomization scheme for submodular maximization [1] and obtain the scheme for k-submodular maximization. First, we introduce the base algorithm as Algorithm 3. The constants $c > 0$ and $\{\lambda_i(y_1, ..., y_k, l)\}$ depend on problem settings.

Let the variables be defined as follows:

$$a_i(s) = \Delta_{e^{(j)}, i} f(t_s), \ t_s(e) = \begin{cases} o[s](e) & (e \neq e^{(j)}) \\ 0 & (e = e^{(j)}) \end{cases}, \ o[s] := (o \sqcup s) \sqcup s.$$

Algorithm 3 achieves Theorem 3.

Theorem 3. *Let o be a maximizer of a monotone nonnegative k-submodular function f and let z be the output of Algorithm 3. We suppose the inequalities*

$$\sum_{i=1}^{k} p_{i,s} \lambda_i(y_1(s), ..., y_k(s), l) \geq \sum_{i=1}^{k} (a_l(s) - a_i(s)) p_{i,s} (l = 1, ..., k) \tag{10}$$

is valid and

$$c \cdot \sum_{i=1}^{k} p_{i,s} y_i(s) \geq \sum_{i=1}^{k} p_{i,s} \lambda_i(y_1(s), ..., y_k(s), l)(l = 1, ..., k) \tag{11}$$

is satisfied for all $s \in \mathcal{D}_j$ for any iterations by some randomized algorithm which follows the framework (Algorithm 1). Then $f(z) \geq \frac{1}{1+c} f(o)$.

Proof. By the assumption (11), the linear formulation (6), (7) and (8) is feasible. (For any s, we can set $p_{i,s}$ by the randomized algorithm in the assumption.) We also obtain

$$c \cdot \mathbb{E}_{s \sim \mathcal{D}_{j-1}} \left[\sum_{i=1}^{k} p_{i,s} y_i(s) \right] \geq \mathbb{E}_{s \sim \mathcal{D}_{j-1}} \left[\sum_{i=1}^{k} (a_l(s) - a_i(s)) p_{i,s} \right] \tag{12}$$

Algorithm 3. A base algorithm

Input: A (monotone/nonmonotone) k-submodular function $f : \{0, 1, ..., k\}^V \to \mathbb{R}_+$.
Output: A vector $s \in \{0, 1, ..., k\}^V$.
$\mathcal{D}_0 \leftarrow (1, \mathbf{0})$, $(\mathcal{D} = \{(p, s) \mid s \in (k+1)^V, \ 0 \leq p \leq 1\} \ (\sum_{s \in \mathcal{D}} p = 1))$.
Denote the elements of V by $e^{(1)}, ..., e^{(n)}$ $(|V| = n)$.
for $j = 1, ..., n$ **do**
    $y_i(s) \leftarrow \Delta_{e^{(j)},i} f(s)$ $(\forall s \in \text{supp}(\mathcal{D}_{j-1}) := \{s \mid (p, s) \in \mathcal{D}_{j-1}, p > 0\}$, $i \in \{1, ..., k\})$.
    Find an extreme point solution $(p_{i,s})_{i=1,...,k, \ s \in \text{supp}(\mathcal{D}_{j-1})}$ of the following linear formulation:

$$c \cdot \mathbb{E}_{s \sim \mathcal{D}_{j-1}} \left[\sum_{i=1}^{k} p_{i,s} y_i(s) \right] \geq \mathbb{E}_{s \sim \mathcal{D}_{j-1}} \left[\sum_{i=1}^{k} p_{i,s} \lambda_i(y_1(s), ..., y_k(s), l) \right] \quad (6)$$

$$(l \in \{1, ..., k\})$$

$$\sum_{i=1}^{k} p_{i,s} = 1 \quad (\forall s \in \text{supp}(\mathcal{D}_{j-1})) \quad (7)$$

$$p_{i,s} \geq 0 \quad (\forall s \in \text{supp}(\mathcal{D}_{j-1}), \ i \in \{1, ..., k\}). \quad (8)$$

(c and $\{\lambda_i(y_1, ..., y_k)\}$ depend on problem settings.)
Construct a new distribution $\mathcal{D}_j$:

$$\mathcal{D}_j \leftarrow \bigcup_{i=1}^{k} \{(p_{i,s} \cdot \text{Pr}_{\mathcal{D}_{j-1}}[s], \ s_{e^{(j)},i}) \mid s \in \text{supp}(\mathcal{D}_{j-1}), \ p_{i,s} > 0\} \quad (9)$$

$$\left(s_{e^{(j)},i}(e) = \begin{cases} s(e) & (e \neq e^{(j)}) \\ i & (e = e^{(j)}) \end{cases} \right).$$

end for
return $\text{argmax}_{s \in \text{supp}(\mathcal{D}_n)} \{f(s)\}$

---

by assumption (10). From the definition of $\{y_i\}$ and $\{a_i\}$, we can see

$$\mathbb{E}_{s \sim \mathcal{D}_{j-1}} \left[\sum_{i=1}^{k} p_{i,s} y_i(s) \right] = \mathbb{E}_{s \sim \mathcal{D}_{j-1}} \left[\sum_{i=1}^{k} \{ p_{i,s} f(s_{e^{(j)},i}) \} - f(s) \right], \quad (13)$$

$$\mathbb{E}_{s \sim \mathcal{D}_{j-1}} \left[\sum_{i=1}^{k} (a_{i^*}(s) - a_i(s)) p_{i,s} \right]$$

$$= \mathbb{E}_{s \sim \mathcal{D}_{j-1}} \left[\sum_{i=1}^{k} \{ p_{i,s} f(o[s]_{e^{(j)},i}) \} - f(o[s]) \right]. \quad (14)$$

In addition, from the construction rule of $\mathcal{D}_j$ and inequalities (12), (13) and (14), we obtain

$$c \cdot \left\{ \mathbb{E}_{s' \sim \mathcal{D}_j} \left[f(s') \right] - \mathbb{E}_{s \sim \mathcal{D}_{j-1}} \left[f(s) \right] \right\} \geq \mathbb{E}_{s \sim \mathcal{D}_{j-1}} \left[f(o[s]) \right] - \mathbb{E}_{s' \sim \mathcal{D}_j} \left[f(o[s']) \right].$$
(15)

Then, we can see

$$c \cdot \left\{ \mathbb{E}_{s' \sim \mathcal{D}_n} \left[f(s') \right] - \mathbb{E}_{s \sim \mathcal{D}_0} \left[f(s) \right] \right\} \geq \mathbb{E}_{s \sim \mathcal{D}_0} \left[f(o[s]) \right] - \mathbb{E}_{s' \sim \mathcal{D}_n} \left[f(o[s']) \right]$$

from the summation of (15). We have $\mathcal{D}_0 = \{(1, \mathbf{0})\}$ and f is nonnegative. Hence we get

$$c \cdot \mathbb{E}_{s' \sim \mathcal{D}_n} \left[f(s') \right] \geq f(o) - \mathbb{E}_{s' \sim \mathcal{D}_n} \left[f(s') \right]$$

$\square$

Algorithm 3 performs a polynomial number of value oracle queries.

Theorem 4. *Algorithm 3 returns a solution after* $O(n^2 k^2)$ *value oracle queries.*

Proof. Algorithm 3 uses the value oracle to calculate $y_i(s)$. At j-th iteration, the number of $y_i(s)$ is $k|\mathcal{D}_{j-1}|$. From (9), $|\mathcal{D}_j|$ equals the number of $p_{i,s} \neq 0$. Then we have to consider $p_{i,s} \neq 0$ at j-th iteration.

By the definition, $(p_{i,s})_{i=1,\dots,k,\ s \in \mathrm{supp}(\mathcal{D}_{j-1})}$ is an extreme point solution of (6), (7), and (8). We can see the feasible region of (6), (7), and (8) is bounded. Then some extreme point solution exists.

Let $|\mathcal{D}_{j-1}| = m$. By $(p_{i,s})_{i=1,\dots,k,\ s \in \mathrm{supp}(\mathcal{D}_{j-1})} \in \mathbb{R}^{km}$ and m equalities of (7), $km - k$ inequalities are tight at any extreme point solution. (6) have k inequalities and (8) have km inequalities. Then, at least $km - k - m$ inequalities of (8) are tight. Hence, the number of $p_{i,s} \neq 0$ is at most $m + k$.

Now we have $|\mathcal{D}_j| \leq |\mathcal{D}_{j-1}| + k$. We can also see $|\mathcal{D}_j| \leq jk + 1$. Then the number of value oracle queries is

$$\sum_{j=1}^{n} k|\mathcal{D}_{j-1}| \leq \sum_{j=1}^{n} k(jk+1).$$

$\square$

In Algorithm 3, we have to search for an extreme point solution. We can do it by solving LP for some objective function. If we use LP for our algorithm, it is polynomial-time not only for the number of queries but also for the number of operations. The simplex method is not proved to be a polynomial-time method. However, it is practical. Our algorithm needs only an extreme point solution, then if we get a basic solution, it is enough. So we can use the first phase of two-phase simplex method to find an extreme point solution.

Next we have to build a randomized algorithm and give appropriate c and $\{\lambda_i\}$ for the monotone/nonmonotone setting. To build a randomized algorithm, inequality 1 is only to consider. However, if we build it to derandomize, inequality 10 must be valid for any $\{p_i\}$, and inequality 11 must be satisfied.

An important point is using l to define $\{\lambda_i\}$. Even if the size of $\mathcal{D}_{j-1}$ is large, $i^*(= o(e^{(j)}))$ is common for all $s \in \mathcal{D}_{j-1}$. Then we reduce the number of inequalities of the linear formulation and limit increases in size of $\mathcal{D}_{j-1}$.

5   A Deterministic Algorithm for Monotone Functions

In this section, we show a polynomial-time deterministic algorithm for maximizing monotone k-submodular functions. We prove that the randomized algorithm for monotone functions [9] can be derandomized by the scheme.

Theorem 5. *If we set* $c = 1 - \frac{1}{k}$, $\lambda_l(y_1(\boldsymbol{s}), ..., y_k(\boldsymbol{s}), l) = 0$ *and* $\lambda_i(y_1(\boldsymbol{s}), ..., y_k(\boldsymbol{s}), l) = y_l$ $(i \neq l)$ *in Algorithm 3, we obtain a deterministic* $\frac{k}{2k-1}$-*approximation algorithm.*

Proof. From Lemma 2,

$$\left(1 - \frac{1}{k}\right) \sum_{i=1}^{k} p_{i,s} y_i(\boldsymbol{s}) \geq \sum_{i \neq i^*} p_{i,s} y_{i^*}(\boldsymbol{s}) \tag{2'}$$

is always satisfied by the distribution in their algorithm. And we can also obtain $a_{i^*}(\boldsymbol{s}) - a_i(\boldsymbol{s}) \leq a_{i^*}(\boldsymbol{s}) \leq y_{i^*}(\boldsymbol{s})$ from orthant submodularity and monotonicity. Then

$$\sum_{i=1}^{k} p_{i,s} (a_{i^*}(\boldsymbol{s}) - a_i(\boldsymbol{s})) \leq \sum_{i \neq i^*} p_{i,s} y_{i^*}(\boldsymbol{s}) \tag{5'}$$

is valid. The right hand side of the last inequality is the same as our setting of $\{\lambda_i\}$. Hence the setting of c and $\{\lambda_i\}$, and the randomized algorithm [9] satisfies the requirements of Theorem 3.   □

From Theorem 5, in order to maximize a monotone k-submodular function, we replace inequality (6) with (6').

$$\left(1 - \frac{1}{k}\right) \cdot \mathbb{E}_{\boldsymbol{s} \sim \mathcal{D}_{j-1}} \left[\sum_{i=1}^{k} p_{i,s} y_i(\boldsymbol{s}) \right] \geq \mathbb{E}_{\boldsymbol{s} \sim \mathcal{D}_{j-1}} \left[\sum_{i \neq l} p_{i,s} y_l(\boldsymbol{s}) \right] \tag{6'}$$

6   A Deterministic Algorithm for Nonmonotone Functions

In this section, we show a polynomial-time deterministic algorithm for maximizing nonmonotone k-submodular functions with $k \geq 2$. The derandomization scheme is the same as monotone version in the last section. First, we show a randomized algorithm. Then we derandomized it with the scheme.

6.1   The Randomized Algorithm for Derandomization

We first show a randomized $\frac{k}{3k-2}$-approximation algorithm in Algorithm 4.
   Algorithm 4 satisfies the lemma below.

Lemma 3. *Let* $\boldsymbol{o}$ *be a maximizer of a monotone nonnegative k-submodular function* f *and let* $\boldsymbol{z}$ *be the output of Algorithm 4. Then* $\mathbb{E}[f(\boldsymbol{z})] \geq \frac{k}{3k-2} f(\boldsymbol{o})$

Algorithm 4. The base randomized algorithm

Input: A k-submodular function $f : \{0, 1, ..., k\}^V \to \mathbb{R}_+$.
Output: A vector $s \in \{0, 1, ..., k\}^V$.
$s \leftarrow 0$.
Denote the elements of V by $e^{(1)}, ..., e^{(n)}$ ($|V| = n$).
for $j = 1, ..., n$ **do**
  $y_i^{(j)} \leftarrow \Delta_{e^{(j)}, i} f(s)$ $(1 \leq i \leq k)$.
  $I_+ \leftarrow \{i \mid y_i \geq 0\}, , t \leftarrow |I_+| - 1$
  $\beta \leftarrow \sum_{I_+} (y_i^{(j)})^t$.
  if $\beta > 0$ **then** $p_i^{(j)} \leftarrow (y_i^{(j)})^t / \beta$ $(i \in I_+)$, 0 $(i \notin I_+)$.
  else
    $p_1^{(j)} = 1, p_i^{(j)} = 0$ $(i = 2, ..., k)$.
  end if
  Let $s(e^{(j)}) \in \{1, ..., k\}$ be chosen randomly with $\Pr[s(e^{(j)}) = i] = p_i^{(j)}$.
end for
return s

Proof. From orthant submodularity and pairwise monotonicity, we can see $a_{i^*}(s) - a_i(s) \leq 2a_{i^*}(s) \leq 2y_{i^*}(s)$ for $i \neq i^*$. Hence

$$\sum_{i=1}^{k} p_{i,s} (a_{i^*}(s) - a_i(s)) \leq 2 \sum_{i \neq i^*} p_{i,s} y_{i^*}(s) \tag{16}$$

is satisfied. Now we prove

$$\sum_{i \neq i^*} p_{i,s} y_{i^*}(s) \leq \left(1 - \frac{1}{k}\right) \sum_{i=1}^{k} p_{i,s} y_i(s) \tag{17}$$

for Algorithm 4. If $|I_+| = k$, the validity is obvious from Lemma 2. Hence we consider the case with $|I_+| < k$.

In this case, we obtain $y_{i_-} < 0$ and $y_i > 0$ for all $i \neq i_-$ from pairwise monotonicity. And, by the algorithm, we set $p_{i_-} = 0$. So, we have to show

$$\sum_{i \neq i^*, i_-} p_{i,s} y_{i^*}(s) \leq \left(1 - \frac{1}{k}\right) \sum_{i \neq i_-} p_{i,s} y_i(s).$$

If $i^* = i_-$, the validity of the inequality is obvious. Therefore let $i^* \neq i_-$. From Lemma 2,

$$\sum_{i \neq i^*, i_-} p_{i,s} y_{i^*}(s) \leq \left(1 - \frac{1}{k-1}\right) \sum_{i \neq i_-} p_{i,s} y_i(s).$$

It is obvious that $1 - \frac{1}{k-1} \leq 1 - \frac{1}{k}$. Hence we obtain inequality (17).
From (16) and (17),

$$\sum_{i=1}^{k} p_{i,s} (a_{i^*}(s) - a_i(s)) \leq 2 \left(1 - \frac{1}{k}\right) \sum_{i=1}^{k} p_{i,s} y_i(s)$$

is satisfied by the algorithm. It is the case $c = 2(1 - 1/k)$ in Algorithm 1. Then approximation ratio is $\frac{1}{1+c} = \frac{k}{3k-2}$. □

6.2   A Deterministic Algorithm for Nonmonotone Functions

We show a polynomial-time deterministic algorithm for maximizing nonmonotone k-submodular functions. We prove that the randomized algorithm in the last section can be derandomized by the scheme.

Theorem 6. *If we set* $c = 2(1 - \frac{1}{k})$, $\lambda_l(y_1(s), ..., y_k(s), l) = 0$ *and* $\lambda_i(y_1(s), ...,$ *$y_k(s), l) = 2y_l$ $(i \neq l)$ in Algorithm 3, we obtain a deterministic $\frac{k}{3k-2}$-approximation algorithm.*

Proof. In the proof of Algorithm 4, we show

$$\sum_{i \neq i^*} p_{i,s} y_i(s) \leq \left(1 - \frac{1}{k}\right) \sum_{i=1}^{k} p_{i,s} y_i(s) \tag{18}$$

is always satisfied by the distribution in the algorithm. And we also see

$$\sum_{i=1}^{k} p_{i,s}(a_{i^*}(s) - a_i(s)) \leq 2 \sum_{i \neq i^*} p_{i,s} y_{i^*}(s) \tag{19}$$

is valid. The right hand side of the last inequality is the same as our setting of $\{\lambda_i\}$. Hence the setting of c and $\{\lambda_i\}$, and Algorithm 4 satisfy the requirements of Theorem 3. □

From Theorem 6, in order to maximize a nonmonotone k-submodular function, we replace inequality (6) with (6").

$$2\left(1 - \frac{1}{k}\right) \cdot \mathbb{E}_{s \sim \mathcal{D}_{j-1}}\left[\sum_{i=1}^{k} p_{i,s} y_i(s)\right] \geq 2 \cdot \mathbb{E}_{s \sim \mathcal{D}_{j-1}}\left[\sum_{i \neq l} p_{i,s} y_l(s)\right] \tag{6"}$$

In fact, (6") is equivalent to (6') in the monotone case.

7   Conclusion

We proposed a derandomized $\frac{k}{2k-1}$ algorithm for monotone k-submodular maximization and a derandomized $\frac{k}{3k-2}$ algorithm for nonmonotone k-submodular maximization with $k \geq 2$. For $k \geq 3$, we improved approximation ratio from $1/3$.

One of open problems is a faster method for finding an extreme point solution of the linear formulation. For submodular functions, Buchbinder and Feldman [1] showed greedy methods are effective. It is because their formulation is the form of fractional knapsack problem. Our formulation is similar to theirs, and ours can be seen as the form of an LP relaxation of multidimensional knapsack problem.

However, faster methods are not given than general LP solutions. The number of constraints in our formulation depends on k and the number of iterations. It is therefore difficult to find an extreme point faster.

For nonmonotone functions, constructing a deterministic algorithm with better approximation ratio is also an important open problem. For nonmonotone functions, we have pairwise monotonicity instead of monotonicity. In such a situation, for some i, a_i can be negative. However, if $y_j > 0$ for all j, we can't find such i. Then, if we try to use the same derandomizing method, it is difficult to construct $\{\lambda_i\}$.

References

1. Buchbinder, N., Feldman, M.: Deterministic algorithms for submodular maximization problems. In: Proceedings of the Twenty-Seventh Annual ACM-SIAM Symposium on Discrete Algorithms, pp. 392–403. SIAM (2016)
2. Buchbinder, N., Feldman, M., Seffi, J., Schwartz, R.: A tight linear time (1/2)-approximation for unconstrained submodular maximization. SIAM J. Comput. **44**(5), 1384–1402 (2015)
3. Feige, U., Mirrokni, V.S., Vondrak, J.: Maximizing non-monotone submodular functions. SIAM J. Comput. **40**(4), 1133–1153 (2011)
4. Gridchyn, I., Kolmogorov, V.: Potts model, parametric maxflow and k-submodular functions. In: Proceedings of the IEEE International Conference on Computer Vision, pp. 2320–2327 (2013)
5. Grötschel, M., Lovász, L., Schrijver, A.: The ellipsoid method and its consequences in combinatorial optimization. Combinatorica **1**(2), 169–197 (1981)
6. Huber, A., Kolmogorov, V.: Towards minimizing k-submodular functions. In: Mahjoub, A.R., Markakis, V., Milis, I., Paschos, V.T. (eds.) ISCO 2012. LNCS, vol. 7422, pp. 451–462. Springer, Heidelberg (2012). https://doi.org/10.1007/978-3-642-32147-4_40
7. Iwata, S., Fleischer, L., Fujishige, S.: A combinatorial strongly polynomial algorithm for minimizing submodular functions. J. ACM **48**(4), 761–777 (2001)
8. Iwata, S., Tanigawa, S., Yoshida, Y.: Bisubmodular function maximization and extensions. Technical Report METR 2013–16, The University of Tokyo (2013)
9. Iwata, S., Tanigawa, S., Yoshida, Y.: Improved approximation algorithms for k-submodular function maximization. In: Proceedings of the Twenty-Seventh Annual ACM-SIAM Symposium on Discrete Algorithms, pp. 404–413. SIAM (2016)
10. Ohsaka, N., Yoshida, Y.: Monotone k-submodular function maximization with size constraints. In: Advances in Neural Information Processing Systems, pp. 694–702 (2015)
11. Schrijver, A.: A combinatorial algorithm minimizing submodular functions in strongly polynomial time. J. Comb. Theory Ser. B **80**(2), 346–355 (2000)
12. Ward, J., Živný, S.: Maximizing k-submodular functions and beyond. ACM Trans. Algorithms **12**(4), 47:1–47:26 (2016)

Computational Complexity

On the Parameterized Complexity
of Happy Vertex Coloring

Akanksha Agrawal$^{(\boxtimes)}$

Department of Informatics, University of Bergen, Bergen, Norway
`akanksha.agrawal@uib.no`

Abstract. Let G be a graph, and $c : V(G) \to [k]$ be a coloring of vertices in G. A vertex $u \in V(G)$ is *happy* with respect to c if for all $v \in N_G(u)$, we have $c(u) = c(v)$, *i.e.* all the neighbors of u have color same as that of u. The problem MAXIMUM HAPPY VERTICES takes as an input a graph G, an integer k, a vertex subset $S \subseteq V(G)$, and a (partial) coloring $c : S \to [k]$ of vertices in S. The goal is to find a coloring $\tilde{c} : V(G) \to [k]$ such that $\tilde{c}|_S = c$, *i.e.* $\tilde{c}$ extends the partial coloring c to a coloring of vertices in G and the number of happy vertices in G is maximized. For the family of trees, Aravind et al. [1] gave a linear time algorithm for MAXIMUM HAPPY VERTICES for every fixed k, along with the edge variant of the problem. As an open problem, they stated whether MAXIMUM HAPPY VERTICES admits a linear time algorithm on graphs of bounded (constant) treewidth for every fixed k. In this paper, we study the problem MAXIMUM HAPPY VERTICES for graphs of bounded treewidth and give a linear time algorithm for every fixed k and (constant) treewidth of the graph. We also study the problem MAXIMUM HAPPY VERTICES with a different parameterization, which we call HAPPY VERTEX COLORING. The problem HAPPY VERTEX COLORING takes as an input a graph G, integers ℓ and k, a vertex subset $S \subseteq V(G)$, and a coloring $c : S \to [k]$. The goal is to decide if there exist a coloring $\tilde{c} : V(G) \to [k]$ such that $\tilde{c}|_S = c$ and $|H| \geq \ell$, where H is the set of happy vertices in G with respect to $\tilde{c}$. We show that HAPPY VERTEX COLORING is W[1]-hard when parameterized by ℓ. We also give a kernel for HAPPY VERTEX COLORING with $\mathcal{O}(k^2 \ell^2)$ vertices.

1 Introduction

The structure of social networks is believed to be governed by the phenomenon of *homophyly* (see Chap. 4 [5]). Similar type of people often have closely related choices. For instance, people with similar educational interests often end up joining closely related universities, people sharing similar beliefs tend to vote for the same candidate in an election, people with matching interests purchase tickets

Due to space limitations most proofs have been omitted.

The research leading to these results received funding from the European Research Council under the European Union's Seventh Framework Programme (FP/2007-2013)/ ERC Grant Agreements no. 306992 (PARAPPROX).

L. Brankovic et al. (Eds.): IWOCA 2017, LNCS 10765, pp. 103–115, 2018.
https://doi.org/10.1007/978-3-319-78825-8_9

of particular kind of movies, and so on... In fact, Li and Peng [12] showed that
several social networks admit the phenomenon of homophyly. Li and Zhang [15]
modelled the small community detection in a social network based on homophyly
law as a graph coloring problem which is described below.

Consider a graph G and a coloring $c : V(G) \rightarrow [k]$. A vertex $u \in V(G)$ is
happy in G with respect to c if for all $v \in N(u)$, we have $c(u) = c(v)$, *i.e.* all the
neighbors of u have color same as that of u. A vertex which is not happy in G
with respect to c is *unhappy*. The problem MAXIMUM HAPPY VERTICES takes
as an input a graph G, an integer k, a vertex subset $S \subseteq V(G)$, and a partial
coloring $c : S \rightarrow [k]$. The goal is to find a coloring $\tilde{c} : V(G) \rightarrow [k]$ that extends
c to a coloring of $V(G)$ such that the number of happy vertices is maximized.
We note here that Li and Zhang also studied an edge variant of the problem
(MAXIMUM HAPPY EDGES), which is not defined here since the focus of this
paper is the problem MAXIMUM HAPPY VERTICES (and its variant).

Li and Zhang [15] showed that for $k \leq 2$, both MAXIMUM HAPPY VER-
TICES and MAXIMUM HAPPY EDGES are polynomial time solvable, while
for $k \geq 3$, both the problems are NP-hard. Furthermore, they designed a
$\max\{1/k, \Omega(\Delta^{-3})\}$-approximation algorithm for MAXIMUM HAPPY VERTICES
and a $1/2$-approximation algorithm for MAXIMUM HAPPY EDGES. Here, Δ is
the maximum degree of a vertex in the input graph. Later, Zhang et al. [14]
gave improved approximation algorithms with approximation ratios $1/(1 + \Delta)$
and 0.8535 for MAXIMUM HAPPY VERTICES and MAXIMUM HAPPY EDGES,
respectively. Aravind et al. studied the problem MAXIMUM HAPPY VERTICES
and MAXIMUM HAPPY EDGES [1] on the family of trees and gave a linear time
algorithm for every fixed k.

In this paper, we study the problem MAXIMUM HAPPY VERTICES from the
viewpoint of Parameterized Complexity. A parameterized problem Q is a subset
of $\Sigma^* \times \mathbb{N}$, where Σ is a finite set of alphabets. An instance of a parameterized
problem is a tuple (I, κ), where I is a classical problem instance and κ is an
integer, which is called the *parameter*. One of the central notions in the Parameterized Complexity is *fixed-parameter tractability*, where given an instance (I, κ)
of a parameterized problem Q, the goal is to design an algorithm that runs in
time $f(\kappa)n^{\mathcal{O}(1)}$, where $n = |I|$ and $f(\cdot)$ is some computable function whose value
depends only on κ. Such an algorithm is called an FPT algorithm. A parameterized problem that admits an FPT algorithm is said to be in FPT. Another central
notion in Parameterized Complexity is *kernelization*, which mathematically captures the efficiency of a pre-processing/data reduction routine. A kernelization
algorithm for a parameterized problem Q takes as an input an instance (I, κ)
of Q and in time polynomial in $|I| + \kappa$ returns an instance (I', κ') of Q such
that $(I, \kappa) \in Q$ if and only if $(I', \kappa') \in Q$. Furthermore, $|I'| + \kappa' \leq g(\kappa)$, where
$g(\cdot)$ is some computable function whose value depends only on κ. Depending on
whether $g(\cdot)$ is a polynomial, linear or exponential function of κ, the problem
is said to admit a polynomial, linear or exponential kernel, respectively with
respect to the parameter κ. It is known that a parameterized problem is in FPT
if and only if it admits a kernel. Hereafter, whenever we talk about kernels, we

refer only to the polynomial or linear kernels. There are parameterized problems which are believed to be not in FPT under reasonable complexity-theoretic assumptions. Similar to the notion of NP-hardness and NP-hard reductions for the classical Complexity Theory, we have the notion of W[t]-hardness, where $t \in \mathbb{N}$ and parameterized reductions in the Parameterized Complexity. For more details on Parameterized Complexity we refer to the books of Downey and Fellows [7,8], Flum and Grohe [10], Niedermeier [13], and Cygan et al. [4].

Our Results. Aravind et al. [1] posed an open problem on whether MAXIMUM HAPPY VERTICES admits a linear time algorithm on graphs of bounded (constant) treewidth without any constraints on the number of pre-colored vertices. We resolve this question affirmatively, by designing an algorithm (Sect. 3) running in time $\mathcal{O}(2^{\mathcal{O}(k+w \log k)} n)$, where w is the treewidth of the input graph on n vertices, and k is the number of colors in the pre-coloring of a subset of vertices. Formally, we give an algorithm for the following problem.

MAXIMUM HAPPY VERTICES   **Parameter:** $k, \text{tw}(G)$
Input: A graph G, an integer k, a vertex subset $S \subseteq V(G)$, and a (partial) coloring $c : S \subseteq V(G) \to [k]$.
Output: An integer ℓ such that for all $\tilde{c} \in \{\hat{c} \mid \hat{c}|_S = c\}$, we have $|H_{\tilde{c}}| \leq \ell$, where $H_{\tilde{c}}$ is the set of happy vertices in G with respect to $\tilde{c}$. Furthermore, there exist $\tilde{c} \in \{\hat{c} \mid \hat{c}|_S = c\}$ such that $|H_{\tilde{c}}| = \ell$.

We note that our algorithm can be easily modified to output a coloring for which the maximum number of happy vertices, ℓ is achieved.

Next, we consider the following problem, which we call HAPPY VERTEX COLORING.

HAPPY VERTEX COLORING   **Parameter:** k, ℓ
Input: A graph G, integers k and ℓ, a vertex subset $S \subseteq V(G)$, and a (partial) coloring $c : S \to [k]$.
Question: Does there exist a coloring $\tilde{c} : V(G) \to [k]$ such that $\tilde{c}|_S = c$ and the number of happy vertices in G with respect to $\tilde{c}$ is at least ℓ?

We note that HAPPY VERTEX COLORING is para-NP-hard for $k \geq 3$, hence it is unlikely to admit an FPT algorithm when parameterized by k. This follows from the fact that MAXIMUM HAPPY VERTICES is para-NP-hard for $k \geq 3$ [15]. Another natural parameter for HAPPY VERTEX COLORING is the number of happy vertices, ℓ in the resulting coloring of vertices in the graph. In Sect. 4 we show that HAPPY VERTEX COLORING when parameterized by ℓ is W[1]-hard. We show this by giving a parameterized reduction from MULTI-COLORED INDEPENDENT SET which is known to be W[1]-hard [9].

Having known that HAPPY VERTEX COLORING when parameterized by either k or ℓ alone, is unlikely to admit an FPT algorithm, we next consider the parameter $k + \ell$. In Sect. 5 we give a kernel for HAPPY VERTEX COLORING with $\mathcal{O}(k^2 \ell^2)$ vertices.

2   Preliminaries

We denote the set of natural numbers by $\mathbb{N}$. We use $-\infty$ to denote an infinitesimals number (minus infinity) and use the convention that for any $n \in \mathbb{N}$, we have $-\infty + n = -\infty$ and $-\infty + -\infty = -\infty$. For $k \in \mathbb{N}$, by $[k]$ we denote the set $\{1, 2, \cdots, k\}$. Let $f : X \to Y$ be a function. For $y \in Y$, by $f^{-1}(y)$ we denote the set $\{x \in X \mid f(x) = y\}$. For $X' \subseteq X$, by $f|_{X'}$ we denote the function $f|_{X'} : X' \to Y$ such that $f_{X'}(x) = f(x)$, for all $x \in X'$. For an ordered set $R = X \times Y$, a function $f : R \to Z$, and an element $r = (x, y) \in R$, we slightly abuse the notation to denote $f(r) = f((x, y))$ by $f(x, y)$.

We use standard terminology from the book of Diestel [6] for graph-related terms that are not explicitly defined here. For a graph G, by $V(G)$ and $E(G)$ we denote the vertex and edge sets of G, respectively. For a graph G and a vertex $v \in V(G)$, by $N_G(v)$ we denote the set $\{u \in V(G) \mid (v, u) \in E(G)\}$ and by $N_G[v]$ we denote the set $N_G(v) \cup \{v\}$. For $S \subseteq V(G)$, by $N_G(S)$ we denote the set $(\cup_{u \in S} N_G(u)) \setminus S$. We drop the subscript G from $N_G(v)$, $N_G[v]$ and $N_G(S)$ when the context is clear. For a vertex subset $S \subseteq V(G)$, by $G[S]$ we denote the subgraph of G induced by S, i.e. the graph with vertex set S and edge set $\{(u, v) \in E(G) \mid v, u \in S\}$. By $G - S$ we denote the graph $G[V(G) \setminus S]$.

For vertices $u, v \in V(G)$, *unifying* u and v in G results in the following graph G'. We have $V(G') = (V(G) \setminus \{u, v\}) \cup \{u^\star\}$ and $E(G') = E(G[V(G) \setminus \{u, v\}]) \cup \{(u^\star, w) \mid w \in (N_G(u) \cup N_G(v)) \setminus \{u, v\}\}$ where, $u^\star$ is a vertex that is not in $V(G)$. Moreover, we refer to $u^\star$ as the resulting vertex after unification and the operation is said to unify u and v in G.

A *coloring* of a graph G with $k \in \mathbb{N}$ colors is a function $\varphi : V(G) \to [k]$. A partial coloring of G with k colors is a function $c : S \to [k]$, where $S \subseteq V(G)$. We will refer to a partial coloring as a coloring when the context is clear. A coloring $c : V(G) \to [k]$ is said to extend a partial coloring $\tilde{c} : S \to [k]$ if $c|_S = \tilde{c}$.

Definition 1. *A tree decomposition of a graph G is a pair $(\mathcal{T}, \mathcal{X} = \{X_t \mid t \in V(\mathcal{T})\})$, where an element $X \in \mathcal{X}$ is a subset of $V(G)$, called a bag, and $\mathcal{T}$ is a rooted tree satisfying the following properties:*

1. *$\cup_{X \in \mathcal{X}} X = V(G)$;*
2. *For every $(u, v) \in E(G)$, there exists $X \in \mathcal{X}$ such that $u, v \in X$;*
3. *For all $t_1, t_2, t_3 \in V(\mathcal{T})$ if t_2 lies on the unique path between t_1 and t_3 in $\mathcal{T}$ then $X_{t_1} \cap X_{t_3} \subseteq X_{t_2}$.*

Let $(\mathcal{T}, \mathcal{X})$ be a tree decomposition of a graph G. We refer to the vertices of the tree $\mathcal{T}$ as *nodes*. Note that since $\mathcal{T}$ is a rooted tree, we have a natural parent-child and ancestor-descendant relationship among nodes in $\mathcal{T}$. A *leaf* node or a *leaf* of $\mathcal{T}$ is a node with degree exactly one in $\mathcal{T}$ which is different from the root node. All the nodes of $\mathcal{T}$ which are neither the root node or a leaf will be called *non-leaf* nodes. The *width* of the tree decomposition $(\mathcal{X}, \mathcal{T})$ is defined to be $\max_{X \in \mathcal{X}}(|X| - 1)$. The *treewidth* of a graph G, denoted by $\mathsf{tw}(G)$, is the minimum of the widths of all its tree decompositions. We use the following structured tree decomposition in our algorithm.

Definition 2. *A tree decomposition* $(\mathcal{T}, \mathcal{X} = \{X_t \mid t \in V(T)\})$ *with root node as* r *of* G *is called a nice tree decomposition if the following conditions are satisfied.*

1. $X_r = \emptyset$ *and* $X_\ell = \emptyset$ *for every leaf node* ℓ *in* $\mathcal{T}$*;*
2. *Every non-leaf node* t *of* $\mathcal{T}$ *is of one of the following type:*

  - **Introduce node:** *The node* t *has exactly one child* t' *in* $\mathcal{T}$ *and* $X_t = X_{t'} \cup \{v\}$*, where* $v \notin X_{t'}$*.*
  - **Forget node:** *The node* t *has exactly one child* t' *in* $\mathcal{T}$ *and* $X_t = X_{t'} \setminus \{v\}$*, where* $v \in X_{t'}$*.*
  - **Join node:** *The node* t *has exactly two children* t_1, t_2 *in* $\mathcal{T}$ *and* $X_t = X_{t_1} = X_{t_2}$*.*

Lemma 1 ([4,11]). *If a* G *has a tree decomposition* $(\mathcal{T}, \mathcal{X})$ *of width at most* w *then there is a nice tree decomposition of* G *of width at most* w*. Moreover, given a tree decomposition* $(\mathcal{T}, \mathcal{X})$ *of* G *of width at most* w*, in time* $\mathcal{O}(w^2 \cdot \max(|V(\mathcal{T})|, |V(G)|))$ *we can compute a nice tree decomposition of* G *of width at most* w *with at most* $\mathcal{O}(w|V(G)|)$ *nodes.*

3   Algorithm for Maximum Happy Vertices on Graphs of Bounded Treewidth

In this section, we design a dynamic programming based FPT algorithm for MAXIMUM HAPPY VERTICES when parameterized by the treewidth of input graph and the number of colors in the pre-coloring of a subset of vertices.

Let $(G, k, S, c : S \to [k])$ be an instance of MAXIMUM HAPPY VERTICES, $n = |V(G)|$ and $m = |E(G)|$. Without loss of generality we assume that for each $i \in [k]$, we have $c^{-1}(i) \neq \emptyset$, otherwise we can adjust the instance appropriately by adding isolated vertices. For each $i \in [k]$, we arbitrarily choose a vertex from $c^{-1}(i)$, which we denote by $v_i^\star$. We let $S^\star = \{v_i^\star \mid i \in [k]\}$, and $\mathcal{S}^\star = (\{v_1^\star\}, \{v_2^\star\}, \cdots, \{v_k^\star\})$. We start by computing a tree decomposition $(\bar{\mathcal{T}}, \bar{\mathcal{X}})$ of width at most $w \leq 6 \cdot \mathsf{tw}(G)$ in time $\mathcal{O}(2^{\mathcal{O}(\mathsf{tw}(G))} n)$, using the algorithm by Bodlaender et al. [3]. Using Lemma 1 we compute a nice tree decomposition $(\mathcal{T}', \mathcal{X}' = \{X_t' \mid t \in V(\mathcal{T}')\})$ of G with the root node as r' and width at most w, in time $\mathcal{O}(\mathsf{tw}(G)^2 n)$. We modify the nice tree decomposition $(\mathcal{T}', \mathcal{X}')$ to obtain a more structured tree decomposition $(\mathcal{T}, \mathcal{X})$ with root node $r = r'$ as follows. We let $\mathcal{T} = \mathcal{T}'$, and $\mathcal{X} = \{X_t = X_t' \cup S^\star \mid X_t' \in \mathcal{X}'\}$, i.e. $(\mathcal{T}, \mathcal{X})$ is obtained from $(\mathcal{T}', \mathcal{X}')$ by adding all the vertices in $S^\star$ to each bag of $\mathcal{X}'$. Note that width of $(\mathcal{T}, \mathcal{X})$ is bounded by $w + k \leq 6 \cdot \mathsf{tw}(G) + k$. The purpose of adding all vertices in $S^\star$ to each bag is to ensure the subgraph induced by the subtree rooted at a node contains vertices of all k colors, which simplifies the proof. We note that the notion of introduce node, forget node, and join node naturally extends to the tree decomposition $(\mathcal{T}, \mathcal{X})$.

For a node $t \in V(\mathcal{T})$, by $\mathsf{desc}(t)$ we denote the set of nodes which are descendants of t (including t) in $\mathcal{T}$. Furthermore, for $t \in V(\mathcal{T})$, by G_t we denote the graph $G[V_t]$, where $V_t = \cup_{t' \in \mathsf{desc}(t)} X_{t'}$.

We now move to the description of the entries of the dynamic programming table. Consider a node $t \in V(\mathcal{T})$, and an ordered partition $\mathcal{P} = (P_1, P_2, \cdots P_k)$ of X_t into k sets. We call $\mathcal{P}$ a *valid* ordered partition if and only if for all $i \in [k]$, $c^{-1}(i) \cap X_t \subseteq P_i$. Note that for any valid ordered partition $\mathcal{P} = (P_1, P_2, \cdots, P_k)$, for all $i \in [k]$, we have $P_i \neq \emptyset$. This follows from the fact that $S^\star \subseteq X_t$.

For a valid ordered partition $\mathcal{P} = (P_1, P_2, \cdots P_k)$ of X_t, let $\mathcal{H} = \{(H_i, U_i) \mid H_i \uplus U_i = P_i \text{ and } i \in [k]\}$ be a set comprising of ordered pairs, which are partitions of the sets P_is into two sets. A tuple $\tau = (t, \mathcal{P}, \mathcal{H})$ is a *valid tuple* if $\mathcal{P}$ is a valid ordered partition. For a valid tuple $\tau = (t, \mathcal{P}, \mathcal{H})$, a coloring $c_\tau : V(G_t) \to [k]$ is called a τ-*good coloring* if all the following conditions are satisfied.

1. For all $i \in [k]$, we have $P_i \subseteq c_\tau^{-1}(i)$;
2. For all $i \in [k]$, all the vertices in H_i are happy in G_t with respect to c_τ;
3. $c_\tau|_{S \cap V(G_t)} = c$.

For every valid tuple $\tau = (t, \mathcal{P}, \mathcal{H})$, we have a table entry denoted by $\Pi(\tau)$ which is set to an element $z \in [|V(G_t)|] \cup \{-\infty\}$. Intuitively, $\Pi(\tau)$ is set to an element $z \in [|V(G_t)|] \cup \{-\infty\}$ which corresponds to the maximum of the number of happy vertices in G_t over all τ-*good colorings* (if it exists). Formally, the value of $\Pi(\tau)$ is determined as follows.

1. If there is no τ-good coloring of G_t then $\Pi(\tau) = -\infty$.
2. Otherwise, over all τ-*good colorings* of G_t, $\Pi(\tau)$ is set to the maximum of the number of happy vertices in $V(G_t) \setminus (\cup_{i \in [k]} U_i)$ of G_t over all such colorings.

Let $\mathbb{H}^\star$ be the set comprising of all $\mathcal{H} = \{(H_i, U_i) \mid H_i \uplus U_i = P_i^\star \text{ and } i \in [k]\}$. Observe that $\max_{\mathcal{H} \in \mathbb{H}^\star} \Pi(r, \mathcal{S}^\star, \mathcal{H})$ is exactly the number of happy vertices in G maximized over all colorings that extends c to a coloring of $V(G)$. We now move to the description on how the values of $\Pi(\cdot)$ are computed. Since we have a structured form of tree decomposition we compute the value of each of the entries at node $t \in V(\mathcal{T})$ based on the entries of its children, which will be given by the recursive formula. For leaf nodes, we compute the values directly, which corresponds to the base case for the recursive formula. Therefore, by computing the formula in a bottom-up fashion we compute the value of $\Pi(r, \mathcal{S}^\star, \mathcal{H})$, for each $\mathcal{H} \in \mathbb{H}^\star$, and hence the value of $\max_{\mathcal{H} \in \mathbb{H}^\star} \Pi(r, \mathcal{S}^\star, \mathcal{H})$. We now move to the description of computing $\Pi(\tau)$, where $\tau = (t, \mathcal{P} = (P_1, \cdots, P_k), \mathcal{H} = \{(H_i, U_i) \mid H_i \uplus U_i = P_i \text{ and } i \in [k]\})$ is a valid tuple.

Leaf node. Suppose t is a leaf node. In this case, we have $X_t = S^\star$, and $\mathcal{P} = \mathcal{S}^\star$. Note that in this case there is exactly one τ-good coloring of G_t namely, $c|_{S^\star}$. Moreover, we can find the set of happy vertices H, in G_t with respect to $c|_{S^\star}$ by looking at the adjacencies between the vertices in $S^\star$. If there exist $i \in [k]$ such that $H_i \setminus H \neq \emptyset$ then we set $\Pi(\tau) = -\infty$. Otherwise, we set $\Pi(\tau) = |H \setminus (\cup_{i \in [k]} U_i)|$. The correctness of setting the values as described is justified by the uniqueness of τ-good coloring in G_t.

Introduce node. Suppose t is an introduce node. Let t' be the unique child of t in $\mathcal{T}$, and $X_t = X_{t'} \cup \{\tilde{v}\}$, where $\tilde{v} \notin X_{t'}$. Furthermore, let P_i be the set containing $\tilde{v}$, where $i \in [k]$. Recall that by the properties of tree decomposition, there is no $u \in N_{G_t}(\tilde{v}) \setminus X_t$, i.e. all the neighbors of $\tilde{v}$ in G_t are in X_t. Let $\mathcal{P}' = (P_1, P_2, \cdots, P_i \setminus \{\tilde{v}\}, \cdots, P_k)$, and $\mathcal{H}' = (\mathcal{H} \setminus \{(H_i, U_i)\}) \cup \{(H_i \setminus \{\tilde{v}\}, U_i \setminus \{\tilde{v}\})\}$. Finally, let $\tau' = (t', \mathcal{P}', \mathcal{H}')$. Note that τ' is a valid tuple. We start by considering the following simple cases where we can immediately set the value of $\Pi(\tau)$.

Case 1 If $\tilde{v} \in H_i$ and there is $j \in [k] \setminus \{i\}$ such that $P_j \cap N_{G_t}(\tilde{v}) \neq \emptyset$ then we set $\Pi(\tau) = -\infty$ since for any τ-good coloring of G_t, $\tilde{v}$ is not a happy vertex.
Case 2 If there is $j \in [k] \setminus \{i\}$ such that $H_j \cap N_{G_t}(\tilde{v}) \neq \emptyset$ then set $\Pi(\tau) = -\infty$. The correctness of this step is justified by the fact that for any τ-good coloring c_τ, a vertex in $H_j \cap N_{G_t}(\tilde{v})$ cannot be happy in G_t with respect to c_τ.

If none of the above cases are applicable then we (recursively) set the value of $\Pi(\tau)$ as follows.

$$\Pi(\tau) = \begin{cases} 1 + \Pi(\tau') & \text{if } \tilde{v} \in H_i; \\ \Pi(\tau') & \text{if } \tilde{v} \in U_i. \end{cases} \tag{1}$$

Forget node. Suppose t is a forget node. Let t' be the unique child of t in $\mathcal{T}$ such that $X_t = X_{t'} \setminus \{\tilde{v}\}$, where $\tilde{v} \in X_{t'}$. For $i \in [k]$, let $\mathcal{P}_i = (P_1, P_2, \cdots, P_i \cup \{\tilde{v}\}, \cdots, P_k)$, i.e. the ordered partition of $X_{t'} = X_t \cup \{\tilde{v}\}$ obtained from $\mathcal{P}$ by adding $\tilde{v}$ to the set P_i. Furthermore, for the partition (H_i, U_i) of P_i in $\mathcal{H}$ let $\mathcal{H}_{i1} = (\mathcal{H} \setminus \{(H_i, U_i)\}) \cup \{(H_i \cup \{\tilde{v}\}, U_i)\}$ and $\mathcal{H}_{i2} = (\mathcal{H} \setminus \{(H_i, U_i)\}) \cup \{(H_i, U_i \cup \{\tilde{v}\})\}$, i.e. $\mathcal{H}_{i1}$ and $\mathcal{H}_{i2}$ are obtained from $\mathcal{H}$ by adding $\tilde{v}$ to the set of happy and unhappy vertices, respectively. If for some $\tilde{i} \in [k]$, we have $\tilde{v} \in c^{-1}(\tilde{i})$ then we let $\mathbb{P} = \{\mathcal{P}_{\tilde{i}}\}$, otherwise we let $\mathbb{P} = \{\mathcal{P}_i \mid i \in [k]\}$. We set the value of $\Pi(\tau)$ as follows.

$$\Pi(\tau) = \max_{\mathcal{P}_i \in \mathbb{P}, j \in [2]} \{\Pi(t', \mathcal{P}_i, \mathcal{H}_{ij})\}. \tag{2}$$

Join node. Suppose t is a join node. Let t_1, t_2 be the two children of t in $\mathcal{T}$. Recall that by the definition of nice tree decomposition we have $X_t = X_{t_1} = X_{t_2}$. We set $\Pi(t, \tilde{w}, \mathcal{P}, \mathcal{H})$ as follows.

$$\Pi(t, \tilde{w}, \mathcal{P}, \mathcal{H}) = \Pi(t_1, \tilde{w}, \mathcal{P}, \mathcal{H}) + \Pi(t_2, \tilde{w}, \mathcal{P}, \mathcal{H}) - |\cup_{i \in [k]} H_i| \tag{3}$$

Correctness. For the proof of correctness of Eqs. 1 and 2 have been omitted due to space limitations. The proof that Eq. 3 correctly computes $\Pi(\tau)$ follows from the fact that G_{t_1} and G_{t_2} are subgraphs of G_t, and in $G_t - X_t$ there is no edge in G_t between a vertex in $G_{t_1} - X_t$ and a vertex in $G_{t_2} - X_t$.

This concludes the description and the correctness proof for the recursive formulas for computing the values $\Pi(\cdot)$. We now move to the runtime analysis of the algorithm.

Runtime Analysis. Let $(G, k, S, c : S \rightarrow [k])$ be an instance of MAXIMUM HAPPY VERTICES. In time $\mathcal{O}(2^{\mathcal{O}(\text{tw}(G))}n)$, we compute a nice tree decomposition

$(\mathcal{T}', \mathcal{X}')$ of G, with r as the root node, and of width at most $w \leq 6 \cdot \mathsf{tw}(G)$. Furthermore, the number of nodes in $\mathcal{T}$ is bounded by $\mathcal{O}(wn)$. We then obtain a more structured tree decomposition $(\mathcal{T}, \mathcal{X})$, by adding $S^\star$ to each bag of $\mathcal{X}'$. For each node in $\mathcal{T}$ we have at most $k^{w+1} 2^{k+w+1}$ many table entries. Here, we get a factor of k^{w+1} in the number of table entries instead of k^{k+w+1} because for a node $t \in \mathcal{T}$, we only consider valid ordered partition of X_t, and therefore, we do not guess the set for vertices in $X_t \cap S$. Using the recursive formula we can compute each value of $\Pi(\cdot)$ in time $\mathcal{O}(2^{\mathcal{O}(k+w \log k)} n^{\mathcal{O}(1)})$. At this point of time, we cannot guarantee the runtime which linearly depends on n because we need to check the adjacency among vertices for setting the value of certain entries of the table, which using the straightforward implementation will require quadratic dependence on n. Nonetheless, we can start by computing a data structure for the graph G of treewidth at most w in time $w^{\mathcal{O}(1)} n$ that allows performing adjacency queries in time $\mathcal{O}(w)$ (for instance using [2] or Exercise 7.16 in [4]). Thus using this data structure we can compute all the entries of the table in time $\mathcal{O}(2^{\mathcal{O}(k+w \log k)} n) \in \mathcal{O}(2^{\mathcal{O}(k+\mathsf{tw}(G) \log k)} n)$, which gives us the desired running time with linear dependence on n.

Theorem 1. *Let $(G, k, S, c : S \to [k])$ be an instance of* MAXIMUM HAPPY VERTICES. *Then in time $\mathcal{O}(2^{\mathcal{O}(k+\mathsf{tw}(G) \log k)} n)$ we can find the maximum of the number of happy vertices over all colorings that extent c to a coloring of $V(G)$. Here, n is the number of vertices in G.*

We note here that using the standard backtracking technique together with the fact that we have a partition of vertices into at most k parts which extends c, we can construct a coloring which achieves the maximum number of happy vertices.

4   W[1]-hardness of Happy Vertex Coloring

In this section, we show that HAPPY VERTEX COLORING when parameterized by the number of happy vertices is W[1]-hard. We give a parameterized reduction from MULTI-COLORED INDEPENDENT SET (MIS) which is known to be W[1]-hard [9]. The problem MIS is formally defined below.

MULTI-COLORED INDEPENDENT SET (MIS)   **Parameter:** t
Input: A t-partite graph G with a partition $V_1, V_2, \cdots, V_t$ of $V(G)$.
Question: Does there exist $X \subseteq V(G)$, such that for each $i \in [t]$, $|X \cap V_i| = 1$ and $G[X]$ is an independent set?

Intuitively, given an instance $(G, V_1, V_2, \cdots, V_t)$ of MIS, for each V_i we create a vertex selection gadget, W_i which ensures that exactly one vertex from V_i can be happy in any valid coloring. Furthermore, the selected set of vertices from V_is form a set of happy vertices in the instance of HAPPY VERTEX COLORING created. We now move to the formal description of the reduction.

Let $(G, V_1, V_2, \cdots, V_t)$ be an instance of MIS. We create an instance $(G', k, \ell, S, c : S \to [k])$ of HAPPY VERTEX COLORING as follows. Let $n = |V(G)|$ and $V(G) = \{v_i \mid i \in [n]\}$. Initially, we have $V(G') = V(G)$ and $E(G') = \{(u, v) \in E(G) \mid u \in V_i, v \in V_j, i, j \in [t] \text{ and } i \neq j\}$. We now describe the vertex selection gadget W_i, for $i \in [t]$. For each $v_i \in V(G)$, we add three vertices $\tilde{v}_i, p_i, p'_i$ to $V(G')$, and add the edges $(v_i, \tilde{v}_i), (\tilde{v}_i, p_i), (\tilde{v}_i, p'_i)$, and (p_i, p'_i) to $E(G')$. Furthermore, we add $\tilde{v}_i, p_i$, and p'_i to S, and set $c(\tilde{v}_i) = i$, $c(p_i) = n + 1$, and $c(p'_i) = n + 2$. For $i \in [t]$, we add three vertices w_i, x_i, x'_i to $V(G')$ and add all the edges in $\{(u, w_i) \mid u \in V_i\} \cup \{(w_i, x_i), (w_i, x'_i), (x_i, x'_i)\}$ to $E(G')$. Furthermore, we add x_i and x'_i to S, and set $c(x_i) = n + 1$ and $c(x'_i) = n + 2$. Here, the vertices x_i and x'_i are added to ensure that w_i can never be a happy vertex in any coloring of G', and w_i is added to ensure that at most one vertex from V_i can be happy in any coloring of $V(G')$. We have $W_i = G'[V_i \cup \{\tilde{v}_j, p_j, p'_j \mid v_j \in V_i\} \cup \{w_i, x_i, x'_i\}]$. Notice that we have $S = \{\tilde{v}_j, p_j, p'_j \mid v_j \in V(G)\} \cup \{x_i, x'_i \mid i \in [t]\}$. Note that for each $u \in S$, we have described the value of $c(u)$, and we have $k = n + 2$. Finally, we set $\ell = t$, and the resulting instance of HAPPY VERTEX COLORING is $(G', k, \ell, S, c : S \to [k])$.

We state some lemmata which establish certain properties of the instance $(G', k, \ell, S, c : S \to [k])$ of HAPPY VERTEX COLORING that we created.

Lemma 2. *Let $\tilde{c}$ be a coloring that extends c to a coloring of G', and H be the set of happy vertices in G' with respect $\tilde{c}$. Then for all $u \in \{w_i \mid i \in [t]\} \cup S$ we have $u \notin H$.*

Lemma 3. *Let $\tilde{c}$ be a coloring that extends c to a coloring of G', and H be the set of happy vertices in G' with respect $\tilde{c}$. Then for all $i \in [t]$, we have $|V_i \cap H| \leq 1$.*

We now state the main lemma of this section.

Lemma 4. $(G, V_1, V_2, \cdots, V_t)$ *is a* yes *instance of* MIS *if and only if* $(G', k, \ell, S, c : S \to [k])$ *is a* yes *instance of* HAPPY VERTEX COLORING.

Theorem 2. HAPPY VERTEX COLORING *when parameterized by the number of happy vertices is* W[1]-*hard*.

5 Kernelization Algorithm for Happy Vertex Coloring

In this section, we give a polynomial kernel for the problem HAPPY VERTEX COLORING. In fact, we give a kernel for an annotated version of HAPPY VERTEX COLORING, which we call ANNOTATED HAPPY VERTEX COLORING (AHVC). The problem is formally defined below.

ANNOTATED HAPPY VERTEX COLORING (AHVC)   **Parameter:** $k + \ell$
Input: A graph G, integers k and ℓ, a vertex subsets $S, U \subseteq V(G)$, a (partial) coloring $c : S \to [k]$.
Question: Does there exist a coloring $\tilde{c} : V(G) \to [k]$ such that $\tilde{c}|_S = c$ and $|H \setminus U| \geq \ell$, where H is the set of happy vertices in G with respect to $\tilde{c}$?

Observe that HAPPY VERTEX COLORING is a special case of AHVC, where $U = \emptyset$. Moreover, given an instance $(G, k, \ell, S, U, c : S \to [k])$ of AHVC, in polynomial time we can construct an instance $(G', k', \ell, S', c' : S' \to [k'])$ of HAPPY VERTEX COLORING such that $|V(G')| \in \mathcal{O}(|V(G)|)$, $k' \in \mathcal{O}(k)$, and $|S'| \in \mathcal{O}(|S|)$ as follows. Initially, we have $G' = G$ and $c' = c$. We add two (new) vertices $u^\star, v^\star$ to $V(G')$, add the edge $(u^\star, v^\star)$ to $E(G')$, add $u^\star, v^\star$ to S, and set $c'(u^\star) = k + 1$ and $c'(v^\star) = k + 2$. Furthermore, we add the edges $\{(u, u^\star), (u, v^\star) \mid u \in U\}$ to $E(G')$ and set $k' = k + 2$. It is easy to see that $(G, k, \ell, S, U, c : S \to [k])$ is a *yes* instance of AHVC if and only if $(G', k', \ell, S', c' : S' \to [k'])$ is a *yes* instance of HAPPY VERTEX COLORING. Therefore, to design a kernel for HAPPY VERTEX COLORING with $\mathcal{O}(k^2\ell^2)$ vertices it is enough to design a kernel for AHVC with $\mathcal{O}(k^2\ell^2)$ vertices. Hereafter, the focus of this section will be to design a kernel with $\mathcal{O}(k^2\ell^2)$ vertices for AHVC.

Let $(G, k, \ell, S, U, c : S \to [k])$ be an instance of AHVC. The kernelization algorithm applies the following reduction rules in the order in which it is stated. Furthermore, at each step we assume that none of the preceding reduction rules are applicable. When none of the reduction rules are applicable we argue that we get a kernel of the desired size.

Reduction Rule 1. *If $\ell \leq 0$, then return that $(G, k, \ell, S, U, c : S \to [k])$ is a yes instance of AHVC.*

Observe that if $\ell \leq 0$, then any coloring that extends c to a coloring of $V(G)$ is a valid solution to the instance $(G, k, \ell, S, U, c : S \to [k])$ of AHVC, which implies that Reduction Rule 1 is safe.

Reduction Rule 2. *Let $v \in V(G) \setminus U$ be a vertex such that $N(v) \subseteq S$, for all $u, u' \in N(v)$ we have $c(u) = c(u')$, and one of the following conditions is satisfied. i) $v \notin S$; or ii) $c(v) = c(u)$, where $u \in N(v)$. Then delete v from G and decrease ℓ by one. The resulting instance is $(G - \{v\}, k, \ell - 1, S \setminus \{v\}, U, c|_{S \setminus \{v\}} : S \setminus \{v\} \to [k])$.*

Lemma 5. *Reduction Rule 2 is safe.*

Reduction Rule 3. *Let $v \in S \setminus U$ be a vertex such that there exists $u \in N(v) \cap S$ with $c(v) \neq c(u)$. Then add v to the set U. The resulting instance is $(G, k, \ell, S, U \cup \{v\}, c : S \to [k])$.*

The safeness of Reduction Rule 3 follows from the fact that a vertex $v \in S \setminus U$ with $u \in N(v) \cap S$ such that $c(v) \neq c(u)$ can never be a happy vertex in any coloring of G that extends c to a coloring of $V(G)$.

Reduction Rule 4. *Let $v \in V(G) \setminus U$ be a vertex such that there exists $u, u' \in N(v) \cap S$ with $c(u) \neq c(u')$. Then add v to the set U. The resulting instance is $(G, k, \ell, S, U \cup \{v\}, c : S \to [k])$.*

The safeness of Reduction Rule 4 follows from the fact that a vertex v with $u, u' \in N(v) \cap S$ such that $c(u) \neq c(u')$ can never be a happy vertex in any coloring of G that extends c to a coloring of $V(G)$.

Next we consider the following sets. For $i \in [k]$, let $U_i = \{v \in U \cap S \mid c(v) = i\}$, and $U_R = U \setminus (\cup_{i \in [k]} U_i)$. We proceed with the following reduction rules.

Reduction Rule 5. *If there exists $i \in [k]$ such that there are distinct $u, v \in U_i$ then unify u, v in G to obtain the graph G' with $u^\star$ being the vertex resulting after unification. Furthermore, let $c' : (S \setminus \{u, v\}) \cup \{u^\star\} \to [k]$ be the coloring obtained from c with $c'|_{S \setminus \{u,v\}} = c$ and $c'(u^\star) = c(u)$. The resulting instance is $(G', k, \ell, (S \setminus \{u, v\}) \cup \{u^\star\}, (U \setminus \{u, v\}) \cup \{u^\star\}, c' : (S \setminus \{u, v\}) \cup \{u^\star\} \to [k])$.*

Lemma 6. *Reduction Rule 5 is safe.*

Hereafter, we assume that Reduction Rule 5 is not applicable and hence for each $i \in [k]$, we have $|U_i| \leq 1$. Let $Z = V(G) \setminus U$. For $i \in [k]$, let $S_i^Z = \{v \in S \cap Z \mid c(v) = i\}$ and $Z_i = (Z \cap N(U_i \cup S_i^Z)) \cup S_i^Z$. Furthermore, we let $Z_R = Z \setminus (\cup_{i \in [k]} Z_i)$. Observe that for $i, j \in [k]$, $i \neq j$ we have $Z_i \cap Z_j = \emptyset$ since Reduction Rules 3 and 4 are not applicable. Also, for each $v \in Z_R$, we have $N(v) \subseteq V(G) \setminus S$. We proceed with the following reduction rules.

Reduction Rule 6. *If there exists $i \in [k]$ such that $|Z_i| \geq \ell$ then return that $(G, k, \ell, S, U, c : S \to [k])$ is a yes instance of AHVC.*

Lemma 7. *Reduction Rule 6 is safe.*

Reduction Rule 7. *If $|Z_R| \geq \ell$ then return that $(G, k, \ell, S, U, c : S \to [k])$ is a yes instance of AHVC.*

Lemma 8. *Reduction Rule 7 is safe.*

Notice that since Reduction Rule 4 is not applicable we have for each $i \in [k]$, $|U_i| = 1$. Furthermore, since Reduction Rule 6 is not applicable we have for each $i \in [k]$, $|Z_i| < \ell$, and since Reduction Rule 7 is not applicable we have $|Z_R| < \ell$. Therefore, we have $|Z \cup (\cup_{i \in [k]} U_i)| \leq k\ell + \ell - 1$. We now move to bounding the size of U_R, which will give us the desired kernel. To bound the size of U_R we employ the following marking scheme and argue that all the unmarked vertices can be deleted.

Marking Scheme for bounding $|U_R|$. We will denote the set of marked vertices by $M^\star \subseteq U_R$. For all $u, v \in V(G) \setminus U_R$ (not necessarily distinct) such that $N(u) \cap N(v) \cap U_R \neq \emptyset$, choose an arbitrary vertex in $w_{uv} \in N(u) \cap N(v) \cap U_R$ and add it to $M^\star$. That is we add a vertex in U_R to the marked set of vertices which is a common neighbor to vertices u and v.

We call a vertex in $U_R \setminus M^\star$ as an unmarked vertex. We now move to the reduction rule which deletes an unmarked vertex.

Reduction Rule 8. *If there exists $u \in U_R \setminus M^\star$ then delete u from G. The resulting instance is $(G - \{u\}, k, \ell, S, U \setminus \{u\}, c : S \to [k])$.*

Lemma 9. *Reduction Rule 8 is safe.*

Once Reduction Rule 8 is not applicable we have $|U_R| \leq \binom{|Z|}{2} + |Z|$. Therefore, when none of the Reduction Rules 1 to 8 are applicable, we get the desired kernel. Hence, we obtain the following theorem.

Theorem 3. AHVC *admits a kernel with* $\mathcal{O}(k^2\ell^2)$ *vertices, where k is the number of colors in the coloring function and ℓ is the desired number of happy vertices.*

6   Conclusion

We proved that MAXIMUM HAPPY VERTICES admits a linear time algorithm for every fixed k on graphs of bounded treewidth. It remains an interesting question whether the MAXIMUM HAPPY VERTICES problem admits an FPT algorithm when parameterized by treewidth alone. Another interesting direction of research is to study whether HAPPY VERTEX COLORING admits a linear kernel when parameterized by $k + \ell$.

Acknowledgement. The author is thankful to Saket Saurabh for helpful discussions.

References

1. Aravind, N.R., Kalyanasundaram, S., Kare, A.S.: Linear time algorithms for happy vertex coloring problems for trees. In: Mäkinen, V., Puglisi, S.J., Salmela, L. (eds.) IWOCA 2016. LNCS, vol. 9843, pp. 281–292. Springer, Cham (2016). https://doi.org/10.1007/978-3-319-44543-4_22

2. Bodlaender, H.L., Bonsma, P., Lokshtanov, D.: The fine details of fast dynamic programming over tree decompositions. In: Gutin, G., Szeider, S. (eds.) IPEC 2013. LNCS, vol. 8246, pp. 41–53. Springer, Cham (2013). https://doi.org/10.1007/978-3-319-03898-8_5

3. Bodlaender, H.L., Drange, P.G., Dregi, M.S., Fomin, F.V., Lokshtanov, D., Pilipczuk, M.: A c^k n 5-approximation algorithm for treewidth. SIAM J. Comput. **45**(2), 317–378 (2016)

4. Cygan, M., Fomin, F.V., Kowalik, L., Lokshtanov, D., Marx, D., Pilipczuk, M., Saurabh, S.: Parameterized Algorithms. Springer, Cham (2015). https://doi.org/10.1007/978-3-319-21275-3

5. David, E., Jon, K.: Networks, Crowds, and Markets: Reasoning About a Highly Connected World. Cambridge University Press, New York (2010)

6. Reinhard, D.: Graph Theory. GTM, vol. 173, 4th edn. Springer, Heidelberg (2017). https://doi.org/10.1007/978-3-662-53622-3

7. Downey, R.G., Fellows, M.R.: Parameterized Complexity. Springer, New York (1997). https://doi.org/10.1007/978-1-4612-0515-9

8. Downey, R.G., Fellows, M.R.: Fundamentals of Parameterized complexity. Springer, London (2013). https://doi.org/10.1007/978-1-4471-5559-1

9. Fellows, M.R., Hermelin, D., Rosamond, F.A., Vialette, S.: On the parameterized complexity of multiple-interval graph problems. Theor. Comput. Sci. **410**(1), 53–61 (2009)

10. Flum, J., Grohe, M.: Parameterized Complexity Theory. Texts in Theoretical Computer Science. An EATCS Series. Springer, Heidelberg (2006). https://doi.org/10.1007/3-540-29953-X
11. Kloks, T. (ed.): Treewidth: Computations and Approximations. LNCS, vol. 842. Springer, Heidelberg (1994). https://doi.org/10.1007/BFb0045375
12. Li, A., Peng, P.: The small-community phenomenon in networks. Math. Struct. Comput. Sci. **22**(3), 373–407 (2012)
13. Niedermeier, R.: Invitation to Fixed-Parameter Algorithms. Oxford Lecture Series in Mathematics and Its Applications. Oxford University Press, Oxford (2006)
14. Zhang, P., Jiang, T., Li, A.: Improved approximation algorithms for the maximum happy vertices and edges problems. In: Xu, D., Du, D., Du, D. (eds.) COCOON 2015. LNCS, vol. 9198, pp. 159–170. Springer, Cham (2015). https://doi.org/10.1007/978-3-319-21398-9_13
15. Zhang, P., Li, A.: Algorithmic aspects of homophyly of networks. Theor. Comput. Sci. **593**, 117–131 (2015)

Complexity Dichotomies for the Minimum $\mathcal{F}$-Overlay Problem

Nathann Cohen[1], Frédéric Havet[2], Dorian Mazauric[3(✉)], Ignasi Sau[4,5], and Rémi Watrigant[3]

[1] Université Paris-Sud, LRI, CNRS, Orsay, France
[2] Université Côte d'Azur, CNRS, I3S, Inria, Sophia Antipolis, Sophia Antipolis, France
[3] Université Côte d'Azur, Inria, Sophia Antipolis, Sophia Antipolis, France
`dorian.mazauric@inria.fr`
[4] Université de Montpellier, CNRS, LIRMM, Montpellier, France
[5] Departamento de Matemática, Universidade Federal do Ceará, Fortaleza, Brazil

Abstract. For a (possibly infinite) fixed family of graphs $\mathcal{F}$, we say that a graph G *overlays* $\mathcal{F}$ on a hypergraph H if $V(H)$ is equal to $V(G)$ and the subgraph of G induced by every hyperedge of H contains some member of $\mathcal{F}$ as a spanning subgraph. While it is easy to see that the complete graph on $|V(H)|$ overlays $\mathcal{F}$ on a hypergraph H whenever the problem admits a solution, the MINIMUM $\mathcal{F}$-OVERLAY problem asks for such a graph with the minimum number of edges. This problem allows to generalize some natural problems which may arise in practice. For instance, if the family $\mathcal{F}$ contains all connected graphs, then MINIMUM $\mathcal{F}$-OVERLAY corresponds to the MINIMUM CONNECTIVITY INFERENCE problem (also known as SUBSET INTERCONNECTION DESIGN problem) introduced for the low-resolution reconstruction of macro-molecular assembly in structural biology, or for the design of networks.

Our main contribution is a strong dichotomy result regarding the polynomial vs. NP-hard status with respect to the considered family $\mathcal{F}$. Roughly speaking, we show that the easy cases one can think of (*e.g.* when edgeless graphs of the right sizes are in $\mathcal{F}$, or if $\mathcal{F}$ contains only cliques) are the only families giving rise to a polynomial problem: all others are NP-complete. We then investigate the parameterized complexity of the problem and give similar sufficient conditions on $\mathcal{F}$ that give rise to W[1]-hard, W[2]-hard or FPT problems when the parameter is the size of the solution. This yields an FPT/W[1]-hard dichotomy for a relaxed problem, where every hyperedge of H must contain some member of $\mathcal{F}$ as a (non necessarily spanning) subgraph.

Keywords: Hypergraph · Minimum $\mathcal{F}$-Overlay Problem
NP-completeness · Fixed-parameter tractability

This work was partially funded by 'Projet de Recherche Exploratoire', Inria, *Improving inference algorithms for macromolecular structure determination* and ANR under contract STINT ANR-13-BS02-0007.

1 Introduction

1.1 Notation

Most notations of this paper are standard. We now recall some of them, and we refer the reader to [8] for any undefined terminology. For a graph G, we denote by $V(G)$ and $E(G)$ its respective sets of *vertices* and *edges*. The *order* of a graph G is $|V(G)|$, while its *size* is $|E(G)|$. By extension, for a hypergraph H, we denote by $V(H)$ and $E(H)$ its respective sets of *vertices* and *hyperedges*. For $p \in \mathbb{N}$, a *p-uniform* hypergraph H is a hypergraph such that $|S| = p$ for every $S \in E(H)$. Given a graph G, we say that a graph G' is a *subgraph* of G if $V(G') \subseteq V(G)$ and $E(G') \subseteq E(G)$. We say that G' is a *spanning subgraph* of G if it is a subgraph of G such that $V(G') = V(G)$. Given $S \subseteq V(G)$, we denote by $G[S]$ the graph with vertex set S and edge set $\{uv \in E(G) \mid u, v \in S\}$. We say that a graph G' is an *induced subgraph* of G if there exists $S \subseteq V(G)$ such that $G' = G[S]$. Given $S \subseteq V(G)$, we say that an edge $uv \in E(G)$ is *covered* by S if $u \in S$ or $v \in S$, and we say that $uv \in E(G)$ is *induced* by S if $\{u, v\} \subseteq S$. An *isolated* vertex of a graph is a vertex of degree 0. Finally, for a positive integer p, let $[p] = \{1, \ldots, p\}$.

1.2 Definition of the Minimum $\mathcal{F}$-Overlay problem

We define the problem investigated in this paper: MINIMUM $\mathcal{F}$-OVERLAY. Given a fixed family of graphs $\mathcal{F}$ and an input hypergraph H, we say that a graph G *overlays* $\mathcal{F}$ on H if $V(G) = V(H)$ and for every hyperedge $S \in E(H)$, the subgraph of G induced by S, $G[S]$, has a spanning subgraph in $\mathcal{F}$.

Observe that if a graph G overlays $\mathcal{F}$ on H, then the graph G with any additional edges overlays $\mathcal{F}$ on H. Thus, there exists a graph G overlaying $\mathcal{F}$ on H if and only if the complete graph on $|V(H)|$ vertices overlays $\mathcal{F}$ on H. Note that the complete graph on $|V(H)|$ vertices overlays $\mathcal{F}$ on H if and only if for every hyperedge $S \in E(H)$, there exists a graph in $\mathcal{F}$ with exactly $|S|$ vertices. It implies that deciding whether there exists a graph G overlaying $\mathcal{F}$ on H can be done in polynomial time. Hence, otherwise stated, we will always assume that there exists a graph overlaying $\mathcal{F}$ on our input hypergraph H. We thus focus on minimizing the number of edges of a graph overlaying $\mathcal{F}$ on H.

The $\mathcal{F}$-*overlay number* of a hypergraph H, denoted $\mathrm{over}_{\mathcal{F}}(H)$, is the smallest size (*i.e.*, number of edges) of a graph overlaying $\mathcal{F}$ on H.

MINIMUM $\mathcal{F}$-OVERLAY

Input: A hypergraph H, and an integer k.
Question: $\mathrm{over}_{\mathcal{F}}(H) \leq k$?

We also investigate a relaxed version of the problem, called MINIMUM $\mathcal{F}$-ENCOMPASS where we ask for a graph G such that for every hyperedge $S \in E(H)$, the graph $G[S]$ contains a (non necessarily spanning) subgraph in $\mathcal{F}$. In an analogous way, we define the $\mathcal{F}$-*encompass number*, denoted $\mathrm{encomp}_{\mathcal{F}}(H)$, of a hypergraph H.

Minimum $\mathcal{F}$-ENCOMPASS

Input: A hypergraph H, and an integer k.
Question: $\mathrm{encomp}_{\mathcal{F}}(H) \leq k$?

Observe that the MINIMUM $\mathcal{F}$-ENCOMPASS problems are particular cases of MINIMUM $\mathcal{F}$-OVERLAY problems. Indeed, for a family $\mathcal{F}$ of graphs, let $\tilde{\mathcal{F}}$ be the family of graphs containing an element of $\mathcal{F}$ as a subgraph. Then MINIMUM $\mathcal{F}$-ENCOMPASS is exactly MINIMUM $\tilde{\mathcal{F}}$-OVERLAY.

Throughout the paper, we will only consider graph families $\mathcal{F}$ whose $\mathcal{F}$-RECOGNITION problem[1] is in NP. This assumption implies that MINIMUM $\mathcal{F}$-OVERLAY and MINIMUM $\mathcal{F}$-ENCOMPASS are in NP as well (indeed, a certificate for both problems is simply a certificate of the recognition problem for every hyperedge). In particular, it is not necessary for the recognition problem to be in P as it can be observed from the family $\mathcal{F}_{Ham}$ of Hamiltonian graphs: the $\mathcal{F}$-RECOGNITION problem is NP-hard, but providing a spanning cycle for every hyperedge is a polynomial certificate and thus belongs to NP.

1.3   Related Work and Applications

MINIMUM $\mathcal{F}$-OVERLAY allows us to model lots of interesting combinatorial optimization problems of practical interest, as we proceed to discuss.

Common graph families $\mathcal{F}$ are the following: connected graphs (and more generally, ℓ-connected graphs), Hamiltonian graphs, graphs having a universal vertex (*i.e.*, having a vertex adjacent to every other vertex). When the family is the set of all connected graphs, then the problem is known as SUBSET INTERCONNECTION DESIGN, MINIMUM TOPIC-CONNECTED OVERLAY or INTERCONNECTION GRAPH PROBLEM. It has been studied by several communities in the context of designing vacuum systems [10,11], scalable overlay networks [5,14,18], reconfigurable interconnection networks [12,13], and, in variants, in the context of inferring a most likely social network [2], determining winners of combinatorial auctions [7], as well as drawing hypergraphs [3,15–17].

As an illustration, we explain in detail the importance of such inference problems for fundamental questions on structural biology [1]. A major problem is the characterization of low resolution structures of macro-molecular assemblies. To attack this very difficult question, one has to determine the plausible contacts between the subunits of an assembly, given the lists of subunits involved in all the complexes. We assume that the composition, in terms of individual subunits, of selected complexes is known. Indeed, a given assembly can be chemically split into complexes by manipulating chemical conditions. This problem can be formulated as a MINIMUM $\mathcal{F}$-OVERLAY problem, where vertices represent the subunits and hyperedges are the complexes. In this setting, an edge between two vertices represents a contact between two subunits.

Hence, the considered family $\mathcal{F}$ is the family of all trees: we want the complexes to be connected. Note that the minimal connectivity assumption avoids

---

[1] The $\mathcal{F}$-RECOGNITION problem asks, given a graph F, whether $F \in \mathcal{F}$.

speculating on the exact (unknown) number of contacts. Indeed, due to volume exclusion constraints, a given subunit cannot contact many others.

1.4   Our Contributions

In Sect. 2, we prove a strong dichotomy result regarding the polynomial vs. NP-hard status with respect to the considered family $\mathcal{F}$. Roughly speaking, we show that the easy cases one can think of (*e.g.* containing only edgeless and complete graphs) are the only families giving rise to a polynomial problem: all others are NP-complete. In particular, it implies that the MINIMUM CONNECTIVITY INFERENCE problem is NP-hard in p-uniform hypergraphs, which generalizes previous results. In Sect. 3, we then investigate the parameterized complexity of the problem and give similar sufficient conditions on $\mathcal{F}$ that gives rise to W[1]-hard, W[2]-hard or FPT problems. This yields an FPT/W[1]-hard dichotomy for MINIMUM $\mathcal{F}$-ENCOMPASS.

Due to space restrictions, proofs of results marked by ($\star$) can be found in the long version of the paper [6].

2   Complexity Dichotomy

In this section, we prove a dichotomy between families of graphs $\mathcal{F}$ such that MINIMUM $\mathcal{F}$-OVERLAY is polynomial-time solvable, and families of graphs $\mathcal{F}$ such that MINIMUM $\mathcal{F}$-OVERLAY is NP-complete.

Given a family of graphs $\mathcal{F}$ and a positive integer p, let $\mathcal{F}_p = \{F \in \mathcal{F} : |V(F)| = p\}$. We denote by K_p the complete graph on p vertices, and by $\overline{K_p}$ the edgeless graph on p vertices.

Theorem 1. *Let $\mathcal{F}$ be a family of graphs. If, for every $p > 0$, either $\mathcal{F}_p = \emptyset$ or $\mathcal{F}_p = \{K_p\}$ or $\overline{K_p} \in \mathcal{F}_p$, then MINIMUM $\mathcal{F}$-OVERLAY is polynomial-time solvable. Otherwise, it is NP-complete.*

The first part of this theorem roughly consists in analyzing the sizes of the hyperedges, and adding cliques when necessary.

Theorem 2 ($\star$). *Let $\mathcal{F}$ be a set of graphs. If, for every $p > 0$, either $\mathcal{F}_p = \emptyset$ or $\mathcal{F}_p = \{K_p\}$ or $\overline{K_p} \in \mathcal{F}_p$, then MINIMUM $\mathcal{F}$-OVERLAY is polynomial-time solvable.*

The NP-complete part requires more work. We need to prove that if there exists $p > 0$ such that $\mathcal{F}_p \neq \emptyset$, $\mathcal{F}_p \neq \{K_p\}$, and $\overline{K_p} \notin \mathcal{F}_p$, then MINIMUM $\mathcal{F}$-OVERLAY is NP-complete. Actually, it is sufficient to prove the following:

Theorem 3. *Let $p > 0$, and $\mathcal{F}_p$ be a non-empty set of graphs with p vertices such that $\mathcal{F}_p \neq \{K_p\}$ and $\overline{K_p} \notin \mathcal{F}_p$. Then MINIMUM $\mathcal{F}_p$-OVERLAY is NP-complete (when restricted to p-uniform hypergraphs).*

2.1 Prescribing Some Edges

A natural generalization of MINIMUM $\mathcal{F}$-OVERLAY is to prescribe a set E of edges to be in the graph overlaying $\mathcal{F}$ on H. We denote by $\mathrm{over}_{\mathcal{F}}(H; E)$ the minimum number of edges of a graph G overlaying $\mathcal{F}$ on H with $E \subseteq E(G)$.

PRESCRIBED MINIMUM $\mathcal{F}$-OVERLAY

Input: A hypergraph H, an integer k, and a set $E \subseteq \binom{V(H)}{2}$.
Question: $\mathrm{over}_{\mathcal{F}}(H; E) \leq k$?

In fact, in terms of computational complexity, the two problems MINIMUM $\mathcal{F}$-OVERLAY and PRESCRIBED MINIMUM $\mathcal{F}$-OVERLAY are equivalent.

Theorem 4 ($\star$). *Let $\mathcal{F}$ be a (possibly infinite) class of graphs. Then* MINIMUM $\mathcal{F}$-OVERLAY *and* PRESCRIBED MINIMUM $\mathcal{F}$-OVERLAY *are polynomially equivalent.*

2.2 Hard Sets

A set $\mathcal{F}_p$ of graphs of order p is *hard* if there is a graph J of order p and two distinct non-edges e_1, e_2 of J such that

- no subgraph of J is in $\mathcal{F}_p$ (including J itself),
- $J \cup e_1$ has a subgraph in $\mathcal{F}_p$ and $J \cup e_2$ has a subgraph in $\mathcal{F}_p$.

The graph J is called the *hyperedge graph of $\mathcal{F}_p$* and e_1 and e_2 are its two *shifting non-edges*.

For example, the set $\mathcal{F}_3 = \{P_3\}$, where P_3 is the graph with three vertices and two edges, is hard. Indeed, the graph O_3 with three vertices and one edge has no subgraph in $\mathcal{F}_3$, but adding any of the two non-edges of O_3 results in a graph isomorphic to P_3.

Lemma 1. *Let $p \geq 3$ and $\mathcal{F}_p$ be a set of graphs of order p. If $\mathcal{F}_p$ is hard, then* PRESCRIBED MINIMUM $\mathcal{F}_p$-OVERLAY *is NP-complete.*

Proof. We present a reduction from VERTEX COVER. Let J be the hyperedge graph of $\mathcal{F}_p$ and e_1, e_2 its shifting non-edges. We distinguish two cases depending on whether e_1 and e_2 are disjoint or not. The proofs of both cases are very similar, we thus omit the second case which can be found in the long version of the paper.

Case 1: e_1 and e_2 intersect. Let G be a graph. Let H_G be the hypergraph constructed as follows.

- For every vertex $v \in V(G)$ add two vertices x_v, y_v.
- For every edge $e = uv$, add a vertex z_e and three disjoint sets Z_e, Y_u^e, and Y_v^e of size $p - 3$.
- For every edge $e = uv$, create three hyperedges $Z_e \cup \{z_e, y_u, y_v\}$, $Y_u^e \cup \{x_u, y_u, z_e\}$, and $Y_v^e \cup \{x_v, y_v, z_e\}$.

We select forced edges as follows: for every edge $e = uv \in E(G)$, we force the edges of a copy of J on $Z_e \cup \{z_e, y_u, y_v\}$ with shifting non-edges $z_e y_u$ and $z_e y_v$, we force the edges of a copy of J on $Y_u^e \cup \{z_e, y_u, x_u\}$ with shifting non-edges $y_u z_e$ and $y_u x_u$, and we force the edges of a copy of J on $Y_v^e \cup \{z_e, y_v, x_v\}$ with shifting non-edges $y_v z_e$ and $y_v x_v$.

We shall prove that $\text{over}_{\mathcal{F}_p}(H_G) = |E| + \text{vc}(G) + |E(G)|$, which yields the result. Here, $\text{vc}(G)$ denotes the size of a minimum vertex cover of G.

Consider first a minimum vertex cover C of G. For every edge $e \in E(G)$, let s_e be an endvertex of e that is not in C if such vertex exists, or any endvertex of e otherwise. Set $E_G = E \cup \{x_v y_v \mid v \in C\} \cup \{z_e y_{s_e} \mid e \in E(G)\}$. One can easily check that $(V_G, E \cup E_G)$ overlays $\mathcal{F}_p$ on H_G. Indeed, for every hyperedge S of H_G, at least one of the shifting non-edges of its forced copy of J is an edge of $E \cup E_G$. Therefore $\text{over}_{\mathcal{F}_p}(H_G) \leq |E| + |E_G| = |E| + \text{vc}(G) + |E(G)|$.

Now, consider a minimum-size graph $(V_G, E \cup E_G)$ overlaying $\mathcal{F}_p$ on H_G and maximizing the edges of the form $x_u y_u$. Let $e = uv \in E(G)$. Observe that the edge $y_u y_v$ is contained in a unique hyperedge, namely $Z_e \cup \{z_e, y_u, y_v\}$. Therefore, free to replace it (if it is not in E) by $z_e y_v$, we may assume that $y_u y_v \notin E_G$. Similarly, we may assume that the edges $x_u z_e$ and $x_v z_e$ are not in E_G, and that no edge with an endvertex in $Y_u^e \cup Y_v^e \cup Z_e$ is in E_G. Furthermore, one of $x_u y_u$ and $x_v y_v$ is in E_G. Indeed, if $\{x_u y_u, x_v y_v\} \cap E_G = \emptyset$, then $\{y_u z_e, y_v z_e\} \subseteq E_G$ because E_G contains an edge included in every hyperedge. Thus replacing $y_u z_e$ by $x_u y_u$ results in another graph overlaying $\mathcal{F}_p$ on H_G with one more edge of type $x_u y_u$ than the chosen one, a contradiction.

Let $C = \{u \mid x_u y_u \in E_G\}$. By the above property, C is a vertex cover of G, so $|C| \geq \text{vc}(G)$. Moreover, E_G contains an edge in every hyperedge $Z_e \cup \{z_e, y_u, y_v\}$, and those $|E(G)|$ edges are not in $\{x_u y_u \mid u \in V(G)\}$. Therefore $|E_G| \geq |C| + |E(G)| \geq \text{vc}(G) + |E(G)|$. $\qquad\square$

Let $\mathcal{F}_p$ be a set of graphs of order p. It is *free* if there are no two distinct elements of $\mathcal{F}_p$ such that one is a subgraph of the other. The *core* of $\mathcal{F}_p$ is the free set of graphs F having no proper subgraphs in $\mathcal{F}_p$. It is easy to see that $\mathcal{F}_p$ is overlayed by a hypergraph if and only if its core does. Henceforth, we may restrict our attention to free sets of graphs.

Lemma 2. *Let $\mathcal{F}_p$ be a free set of graphs of order p. If a graph F in $\mathcal{F}_p$ has an isolated vertex and a vertex of degree 1, then $\mathcal{F}_p$ is hard.*

Proof. Let z be an isolated vertex of F, y a vertex of degree 1, and x the neighbor of y in F. The graph $J = F \setminus xy$ contains no element of $\mathcal{F}_p$ because $\mathcal{F}_p$ is free. Moreover $J \cup xy$ and $J \cup yz$ are isomorphic to F. Hence J is a hyperedge graph of $\mathcal{F}_p$. Thus, by Lemma 1, PRESCRIBED MINIMUM $\mathcal{F}_p$-OVERLAY is NP-complete. $\qquad\square$

The *star of order p*, denoted by S_p, is the graph of order p with $p-1$ edges incident to a same vertex.

Lemma 3. *Let $p \geq 3$ and let $\mathcal{F}_p$ be a free set of graphs of order p containing a subgraph of the star S_p different from $\overline{K}_p$. Then $\mathcal{F}_p$ is hard.*

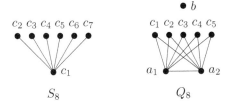

$$S_8 \qquad\qquad Q_8$$

Proof. Let S be the non-empty subgraph of S_p in $\mathcal{F}_p$. If $S \neq S_p$, then S has an isolated vertex and a vertex of degree 1, and so $\mathcal{F}_p$ is hard by Lemma 2. We may assume henceforth that $S_p \in \mathcal{F}_p$.

Let Q_p be the graph with p vertices $\{a_1, a_2, b, c_1, \ldots, c_{p-3}\}$ and edge set $\{a_1 a_2\} \cup \{a_i c_j \mid 1 \leq i \leq 2, 1 \leq j \leq p - 3\}$. Observe that Q_p does not contain S_p but $Q_p \cup a_1 b$ and $Q_p \cup a_2 b$ do. If $\mathcal{F}_p$ contains no subgraph of Q_p, then $\mathcal{F}_p$ is hard. So we may assume that $\mathcal{F}_p$ contains a subgraph of Q_p.

Let Q be the subgraph of Q_p in $\mathcal{F}_p$ that has the minimum number of triangles. If Q has a degree 1 vertex, then $\mathcal{F}_p$ is hard by Lemma 2. Henceforth we may assume that Q has no vertex of degree 1. So, without loss of generality, there exists q such that $E(Q) = \{a_1 a_2\} \cup \{a_i c_j \mid 1 \leq i \leq 2, 1 \leq j \leq q\}$.

Let $R = (Q \setminus a_1 c_1) \cup a_2 b$. Observe that $R \cup a_1 c_1$ and $R \cup a_1 b$ contain Q. If $\mathcal{F}_p$ contains no subgraph of R, then $\mathcal{F}_p$ is hard. So we may assume that $\mathcal{F}_p$ contains a subgraph R' of R. But $\mathcal{F}_p$ contains no subgraph of Q because it is free, so both $a_2 c_1$ and $a_2 b$ are in R'. In particular, c_1 and b have degree 1 in R'.

Let $T = (Q \setminus a_1 c_1)$. It is a proper subgraph of Q, so $\mathcal{F}_p$ contains no subgraph of T, because $\mathcal{F}_p$ is free. Moreover $T \cup a_1 c_1 = Q$ is in $\mathcal{F}_p$ and $T \cup a_2 b = R$ contains $R' \in \mathcal{F}_p$. Hence $\mathcal{F}_p$ is hard.   □

2.3   Proof of Theorem 3

For convenience, instead of proving Theorem 3, we prove the following statement, which is equivalent by Theorem 4.

Theorem 5. *Let p be a positive integer. Let $\mathcal{F}_p$ be a non-empty set of graphs of order p.* PRESCRIBED MINIMUM $\mathcal{F}_p$-OVERLAY *is* NP-*complete unless $\overline{K}_p \in \mathcal{F}_p$ or $\mathcal{F}_p = \{K_p\}$.*

Proof. We proceed by induction on p, the result holding trivially when $p = 1$ and $p = 2$. Assume now that $p \geq 3$. Without loss of generality, we may assume that $\mathcal{F}_p$ is a free set of graphs.

A *hypograph* of a graph G is an induced subgraph of G of order $|G| - 1$. In other words, it is a subgraph obtained by removing a vertex from G. Let $\mathcal{F}^-$ be the set of hypographs of elements of $\mathcal{F}_p$.

If $\mathcal{F}^- = \{K_{p-1}\}$, then necessarily $\mathcal{F}_p = \{K_p\}$, and PRESCRIBED MINIMUM $\mathcal{F}_p$-OVERLAY is trivially polynomial-time solvable.

If $\mathcal{F}^- \neq \{K_{p-1}\}$ and $\overline{K}_{p-1} \notin \mathcal{F}^-$, then PRESCRIBED MINIMUM $\mathcal{F}^-$-OVERLAY is NP-complete by the induction hypothesis. We shall now reduce this problem to PRESCRIBED MINIMUM $\mathcal{F}_p$-OVERLAY. Let (H^-, k^-, E^-) be an instance of PRESCRIBED MINIMUM $\mathcal{F}^-$-OVERLAY. For every hyperedge S of H^-, we create a new vertex x_S and the hyperedge $X_S = S \cup \{x_S\}$. Let H be the hypergraph defined by $V(H) = V(H^-) \cup \bigcup_{S \in E(H^-)} x_S$ and $E(H) = \{X_S \mid S \in E(H^-)\}$. We set $E = E^- \cup \bigcup_{S \in E(H^-)} \{x_S v \mid v \in S\}$.

Let us prove that $\text{over}_{\mathcal{F}_p}(H; E) = \text{over}_{\mathcal{F}^-}(H^-; E^-) + (p - 1) \cdot |S|$. Clearly, if $G^- = (V(H^-), F^-)$ overlays $\mathcal{F}^-$, then $G = (V(H), F^- \cup \bigcup_{S \in E(H^-)} \{x_S v \mid v \in S\})$ overlays $\mathcal{F}_p$. Hence $\text{over}_{\mathcal{F}_p}(H; E) \leq \text{over}_{\mathcal{F}^-}(H^-; E^-) + (p - 1) \cdot |S|$. Reciprocally, assume that G overlays $\mathcal{F}_p$. Then for each hyperedge S of H^-, the graph $G[X_S] \in \mathcal{F}_p$, and so $G[S] \in \mathcal{F}^-$. Therefore, setting the graph $G^- = G[V(H^-)]$ overlays $\mathcal{F}^-$. Moreover $E(G) \setminus E(G^-) = \bigcup_{S \in E(H^-)} \{x_S v \mid v \in S\}$. Hence $\text{over}_{\mathcal{F}_p}(H; E) \geq \text{over}_{\mathcal{F}^-}(H^-; E^-) + (p - 1) \cdot |S|$.

Assume now that $\overline{K}_{p-1} \in \mathcal{F}^-$. Then $\mathcal{F}_p$ contains a subgraph of the star S_p. If $\mathcal{F}_p$ contains $\overline{K}_p$, then PRESCRIBED MINIMUM $\mathcal{F}_p$-OVERLAY is trivially polynomial-time solvable. Henceforth, we may assume that $\mathcal{F}_p$ contains a non-empty subgraph of S_p. Thus, by Lemma 3, $\mathcal{F}_p$ is hard, and so by Lemma 1, PRESCRIBED MINIMUM $\mathcal{F}_p$-OVERLAY is NP-complete. □

3   Parameterized Analysis

We now focus on the parameterized complexity of our problems. A *parameterization* of a decision problem Q is a computable function κ that assigns an integer $\kappa(I)$ to every instance I of the problem. We say that (Q, κ) is *fixed-parameter tractable* (FPT) if every instance I can be solved in time $O(f(\kappa(I))|I|^c)$, where f is some computable function, $|I|$ is the encoding size of I, and c is some constant independent of I (we will sometimes use the $O^*(\cdot)$ notation that removes polynomial factors and additive terms). Finally, the W[i]-hierarchy of parameterized problems is typically used to rule out the existence of FPT algorithms, under the widely believed assumption that FPT $\neq$ W[1]. For more details about fixed-parameter tractability, we refer the reader to the monograph of Downey and Fellows [9].

Since MINIMUM $\mathcal{F}$-OVERLAY is NP-hard for most non-trivial cases, it is natural to ask for the existence of FPT algorithms. In this paper, we consider the so-called *standard parameterization* of an optimization problem: the size of a solution. In the setting of our problems, this parameter corresponds to the number k of edges in a solution. Hence, the considered parameter will always be k in the remainder of this section.

Similarly to our dichotomy result stated in Theorem 1, we would like to obtain necessary and sufficient conditions on the family $\mathcal{F}$ giving rise to either an FPT or a W[1]-hard problem. One step towards such a result is the following FPT-analogue of Theorem 2.

Theorem 6. *Let $\mathcal{F}$ be a family of graphs. If there is a non-decreasing function $f : \mathbb{N} \to \mathbb{N}$ such that $\lim_{n \to +\infty} f(n) = +\infty$ and $|E(F)| \geq f(|V(F)|)$ for all $F \in \mathcal{F}$, then MINIMUM $\mathcal{F}$-OVERLAY is FPT.*

Proof. Let $g : \mathbb{N} \to \mathbb{N}$ be the function that maps every $k \in \mathbb{N}$ to the smallest integer ℓ such that $f(\ell) \geq k$. Since $\lim_{n \to +\infty} f(n) = +\infty$, g is well-defined. If a hyperedge S of a hypergraph H is of size at least $g(k + 1)$, then since f is non-decreasing, over$_\mathcal{F}(H) > k$ and so the instance is negative. Therefore, we may assume that every hyperedge of H has size at most $g(k)$. Applying a simple branching algorithm (see [9]) allows us to solve the problem in time $O^*(g(k)^{O(k)})$. $\qquad\square$

Observe that if $\mathcal{F}$ is finite, setting $N = \max\{|E(F)| \mid F \in \mathcal{F}\}$, the function f defined by $f(n) = 0$ for $n \leq N$ and $f(n) = n$ otherwise satisfies the condition of Theorem 6, and so MINIMUM $\mathcal{F}$-OVERLAY is FPT. Moreover, Theorem 6 encompasses some interesting graph families. Indeed, if $\mathcal{F}$ is the family of connected graphs (resp. Hamiltonian graphs), then $f(n) = n - 1$ (resp. $f(n) = n$) satisfies the required property. Other graph families include c-vertex-connected graphs or c-edge-connected graphs for any fixed $c \geq 1$, graphs of minimum degree at least d for any fixed $d \geq 1$. In sharp contrast, we shall see in the next subsection (Theorem 7) that if, for instance, $\mathcal{F}$ is the family of graphs containing a matching of size at least c, for any fixed $c \geq 1$, then the problem becomes W[1]-hard (note that such a graph might have an arbitrary number of isolated vertices).

3.1   Negative Result

In view of Theorem 6, a natural question is to know what happens for graph families not satisfying the conditions of the theorem. Although we were not able to obtain an exact dichotomy as in the previous section, we give sufficient conditions on $\mathcal{F}$ giving rise to problems that are unlikely to be FPT (by proving W[1]-hardness or W[2]-hardness).

An interesting situation is when $\mathcal{F}$ is *closed by addition of isolated vertices*, *i.e.*, for every $F \in \mathcal{F}$, the graph obtained from F by adding an isolated vertex is also in $\mathcal{F}$. Observe that for such a family, MINIMUM $\mathcal{F}$-OVERLAY and MINIMUM $\mathcal{F}$-ENCOMPASS are equivalent, which is the reason that motivated us defining this relaxed version. We have the following result, which implies an FPT/W[1]-hard dichotomy for MINIMUM $\mathcal{F}$-ENCOMPASS.

Theorem 7. *Let $\mathcal{F}$ be a fixed family of graphs closed by addition of isolated vertices. If $\overline{K}_p \in \mathcal{F}$ for some $p \in \mathbb{N}$, then MINIMUM $\mathcal{F}$-OVERLAY is FPT. Otherwise, it is W[1]-hard parameterized by k.*

Proof. To prove the positive result, let p be the minimum integer such that $\overline{K}_p \in \mathcal{F}$. Observe that no matter the graph G, for every hyperedge $S \in E(H)$, $G[S]$ will contain $\overline{K}_{|S|}$ as a spanning subgraph, which is in $\mathcal{F}$ whenever $|S| \geq p$ (recall that $\mathcal{F}$ is closed by addition of isolated vertices). Then, a simple branching algorithm allows us to enumerate all graphs (with at least one edge) induced by hyperedges of size at most $p - 1$ in $O^*(p^{O(k)})$ time.

To prove the negative result, we use a recent result of Chen and Lin [4] stating that any constant-approximation of the parameterized DOMINATING SET is

W[1]-hard, which directly transfers to HITTING SET[2]. For an input of HITTING SET, namely a finite set U (called the *universe*), and a family $\mathcal{S}$ of subsets of U, let $\tau(U, \mathcal{S})$ be the minimum size of a set $K \subseteq U$ such that $K \cap S \neq \emptyset$ for all $S \in \mathcal{S}$ (such a set is called a *hitting set*). The result of Chen and Lin implies that the following problem is W[1]-hard parameterized by k.

GAP$_\rho$ HITTING SET

Input: A finite set U, a family $\mathcal{S}$ of subsets of U, and a positive integer k.
Question: Decide whether $\tau(U, \mathcal{S}) \leq k$ or $\tau(U, \mathcal{S}) > \rho k$.

Let F_{is} be a graph from $\mathcal{F}$ minimizing the two following criteria (in this order): number of non-isolated vertices, and minimum degree of non-isolated vertices. Let r_{is} and δ_{is} be the respective values of these criteria, $n_{is} = |V(F_{is})|$, and $m_{is} = |E(F_{is})|$. We thus have $\delta_{is} \leq r_{is}$. Let F_e be a graph in $\mathcal{F}$ with the minimum number of edges, and $n_e = |V(F_e)|$, $m_e = |E(F_e)|$.

Let $U, \mathcal{S}, k$ be an instance of GAP$_{2r_{is}}$ HITTING SET, with $U = \{u_1, \ldots, u_n\}$. We denote by H the hypergraph constructed as follows. Its vertex set is the union of:

- a set V_{is} of $r_{is} - 1$ vertices;
- a set $V_U = \bigcup_{i=1}^n V^i$, where $V^i = \{v_1^i, \ldots, v_{n_{is}-r_{is}+1}^i\}$; and
- for every $u, v \in V_{is}$, $u \neq v$, a set $V_{u,v}$ of $n_e - 2$ vertices.

Then, for every $u, v \in V_{is}$, $u \neq v$, create a hyperedge $h_{u,v} = \{u, v\} \cup V_{u,v}$ and, for every set $S \in \mathcal{S}$, create the hyperedge $h_S = V_{is} \cup \bigcup_{i:u_i \in S} V^i$. Finally, let $k' = \binom{n_{is}-1}{2} m_e + k \delta_{is}$. Since $\mathcal{F}$ is fixed, k' is a function of k only.

We shall prove that if $\tau(U, \mathcal{S}) \leq k$, then over$_\mathcal{F}(H) \leq k'$ and, conversely, if over$_\mathcal{F}(H) \leq k'$, then $\tau(U, \mathcal{S}) \leq 2r_{is}k$.

Assume first that U has a hitting set K of size at most k. For every $u, v \in V_{is}$, $u \neq v$, add to G the edges of a copy of F_e on $h_{u,v}$ with $uv \in E(G)$. This already adds $\binom{n_{is}-1}{2} m_e$ edges to G and, obviously, $G[h_{u,v}]$ contains F_e as a subgraph. Now, for every $u_i \in K$, add all edges between v_1^i and δ_{is} arbitrarily chosen vertices in V_{is}. Observe that for every $S \in \mathcal{S}$, $G[h_S]$ contains F_{is} as a subgraph, and also $|E(G)| \leq k'$.

Conversely, let G be a solution for MINIMUM $\mathcal{F}$-OVERLAY with at most k' edges. Clearly, for all $u, v \in V_{is}$, $u \neq v$, $G[V_{u,v}]$ has at least m_e edges, hence the subgraph of G induced by $V(H) \setminus V_U$ has at least $\binom{n_{is}-1}{2} m_e$ edges, and thus the number of edges of G covered by V_u is at most $k \delta_{is}$. Let K be the set of non-isolated vertices of V_U in G, and $K' = \{u_i \mid v_j^i \in K \text{ for some } j \in \{1, \ldots, n_{is}-r_{is}+1\}\}$. We claim that K' is a hitting set of $(U, \mathcal{S})$: indeed, for every $S \in \mathcal{S}$, $G[h_S]$ must contain some $F \in \mathcal{F}$ as a subgraph, but since V_{is} is composed of $r_{is} - 1$ vertices, and since F_{is} is a graph from $\mathcal{F}$ with the minimum number

---

[2] Roughly speaking, each element of the universe represents a vertex of the graph, and for each vertex, create a set with the elements corresponding to its closed neighborhood.

r_{is} of non-isolated vertices, there must exist $i \in \{1, \ldots, n\}$ such that $u_i \in S$, and $j \in \{1, \ldots, n_{is} - r_{is} + 1\}$ such that $v_j^i \in h_S \cap K$, and thus $S \cap K' \neq \emptyset$. Finally, observe that K is a set of non-isolated vertices covering $k\delta_{is}$ edges, and thus $|K| \leq 2k\delta_{is}$ (in the worst case, K induces a matching), hence we have $|K'| \leq |K| \leq 2k\delta_{is} \leq 2r_{is}k$, i.e., $\tau(U, S) \leq 2r_{is}k$, concluding the proof.   □

It is worth pointing out that the idea of the proof of Theorem 7 applies to broader families of graphs. Indeed, the required property 'closed by addition of isolated vertices' forces $\mathcal{F}$ to contain all graphs $F_{is} + \overline{K}_i$ (where $+$ denotes the disjoint union of two graphs) for every $i \in \mathbb{N}$. Actually, it would be sufficient to require the existence of a polynomial $p : \mathbb{N} \to \mathbb{N}$ such that for any $i \in \mathbb{N}$, we have $F_{is} + \overline{K}_{p(i)} \in \mathcal{F}$ (roughly speaking, for a set S of the HITTING SET instance, we would construct a hyperedge with $|V(F_{is} + \overline{K}_{p(|S|)})|$ vertices). Intuitively, most families of practical interest not satisfying such a constraint will fall into the scope of Theorem 6. Unfortunately, we were not able to obtain the dichotomy in a formal way.

Nevertheless, as explained before, this still yields an FPT/W[1]-hardness dichotomy for the MINIMUM $\mathcal{F}$-ENCOMPASS problem.

Corollary 1. *Let $\mathcal{F}$ be a fixed family of graphs. If $\overline{K}_p \in \mathcal{F}$ for some $p \in \mathbb{N}$, then* MINIMUM $\mathcal{F}$-ENCOMPASS *is* FPT. *Otherwise, it is* W[1]-*hard parameterized by k.*

We conclude this section with a stronger negative result than Theorem 7, but concerning a restricted graph family (hence both results are incomparable).

Theorem 8 (⋆). *Let $\mathcal{F}$ be a fixed graph family such that (i) $\mathcal{F}$ is closed by addition of isolated vertices; (ii) $\overline{K}_p \notin \mathcal{F}$ for every $p \geq 0$; and (iii) all graphs in $\mathcal{F}$ have the same number of non-isolated vertices. Then* MINIMUM $\mathcal{F}$-OVERLAY *is* W[2]-*hard parameterized by k.*

4   Conclusion and Future Work

Naturally, the first open question is to close the gap between Theorems 6 and 7 in order to obtain a complete FPT/W[1]-hard dichotomy for any family $\mathcal{F}$.

As further work, we are also interested in a more constrained version of the problem, in the sense that we may ask for a graph G such that for every hyperedge $S \in E(H)$, the graph $G[S]$ belongs to $\mathcal{F}$ (hence, we forbid additional edges). The main difference between MINIMUM $\mathcal{F}$-OVERLAY and this problem, called MINIMUM $\mathcal{F}$-ENFORCEMENT, is that it is no longer trivial to test for the existence of a feasible solution (actually, it is possible to prove the NP-hardness of this existence test for very simple families, *e.g.* when $\mathcal{F}$ only contains P_3, the path on three vertices). We believe that a dichotomy result similar to Theorem 1 for MINIMUM $\mathcal{F}$-ENFORCEMENT is an interesting challenging question, and will need a different approach than the one used in the proof of Theorem 5.

References

1. Agarwal, D., Caillouet, C., Coudert, D., Cazals, F.: Unveiling contacts within macro-molecular assemblies by solving minimum weight connectivity inference problems. Mol. Cell. Proteomics **14**, 2274–2284 (2015)
2. Angluin, D., Aspnes, J., Reyzin, L.: Inferring social networks from outbreaks. In: Hutter, M., Stephan, F., Vovk, V., Zeugmann, T. (eds.) ALT 2010. LNCS (LNAI), vol. 6331, pp. 104–118. Springer, Heidelberg (2010). https://doi.org/10.1007/978-3-642-16108-7_12
3. Brandes, U., Cornelsen, S., Pampel, B., Sallaberry, A.: Blocks of hypergraphs. In: Iliopoulos, C.S., Smyth, W.F. (eds.) IWOCA 2010. LNCS, vol. 6460, pp. 201–211. Springer, Heidelberg (2011). https://doi.org/10.1007/978-3-642-19222-7_21
4. Chen, Y., Lin, B.: The constant inapproximability of the parameterized dominating set problem. FOCS **2016**, 505–514 (2016)
5. Chockler, G., Melamed, R., Tock, Y., Vitenberg, R.: Constructing scalable overlays for pub-sub with many topics. In: PODC 2007, pp. 109–118. ACM (2007)
6. Cohen, N., Mazauric, D., Sau, I., Watrigant, R.: Complexity dichotomies for the minimum F-overlay problem. CoRR, abs/1703.05156 (2017)
7. Conitzer, V., Derryberry, J., Sandholm, T.: Combinatorial auctions with structured item graphs. In: AAAI 2004, pp. 212–218 (2004)
8. Diestel, R.: Graph Theory. Graduate Texts in Mathematics, vol. 173. Springer, Heidelberg (2017). https://doi.org/10.1007/978-3-662-53622-3
9. Downey, R.G., Fellows, M.R.: Fundamentals of Parameterized Complexity. Texts in Computer Science. Springer, Heidelberg (2013). https://doi.org/10.1007/978-1-4471-5559-1
10. Du, D.Z., Kelly, D.F.: On complexity of subset interconnection designs. J. Global Optim. **6**(2), 193–205 (1995)
11. Ding-Zhu, D., Miller, Z.: Matroids and subset interconnection design. SIAM J. Discrete Math. **1**(4), 416–424 (1988)
12. Fan, H., Hundt, C., Wu, Y.-L., Ernst, J.: Algorithms and implementation for interconnection graph problem. In: Yang, B., Du, D.-Z., Wang, C.A. (eds.) COCOA 2008. LNCS, vol. 5165, pp. 201–210. Springer, Heidelberg (2008). https://doi.org/10.1007/978-3-540-85097-7_19
13. Fan, H., Wu, Y.-L.: Interconnection graph problem. In: FCS 2008, pp. 51–55 (2008)
14. Hosoda, J., Hromkovic, J., Izumi, T., Ono, H., Steinov, M., Wada, K.: On the approximability and hardness of minimum topic connected overlay and its special instances. Theoret. Comput. Sci. **429**, 144–154 (2012)
15. Johnson, D.S., Pollak, H.O.: Hypergraph planarity and the complexity of drawing Venn diagrams. J. Graph Theory **11**(3), 309–325 (1987)
16. Klemz, B., Mchedlidze, T., Nöllenburg, M.: Minimum tree supports for hypergraphs and low-concurrency Euler diagrams. In: Ravi, R., Gørtz, I.L. (eds.) SWAT 2014. LNCS, vol. 8503, pp. 265–276. Springer, Cham (2014). https://doi.org/10.1007/978-3-319-08404-6_23
17. Korach, E., Stern, M.: The clustering matroid and the optimal clustering tree. Math. Program. **98**(1), 385–414 (2003)
18. Onus, M., Richa, A.W.: Minimum maximum-degree publish-subscribe overlay network design. IEEE/ACM Trans. Netw. **19**(5), 1331–1343 (2011)

Improved Complexity for Power Edge Set Problem

Benoit Darties[1]([⊠]) [iD], Annie Chateau[2], Rodolphe Giroudeau[2],
and Matthias Weller[2]

[1] Le2i FRE2005, CNRS, Arts et Métiers, University of Bourgogne Franche-Comté,
Dijon, France
benoit.darties@u-bourgogne.fr
[2] LIRMM - CNRS UMR 5506, Montpellier, France
{chateau,giroudeau,weller}@lirmm.fr

Abstract. We study the complexity of POWER EDGE SET (PES), a problem dedicated to the monitoring of an electric network. In such context we propose some new complexity results. We show that PES remains $\mathcal{NP}$-hard in planar graphs with degree at most five. This result is extended to bipartite planar graphs with degree at most six. We also show that PES is hard to approximate within a factor lower than $328/325$ in the bipartite case (resp. $17/15 - \epsilon$), unless $\mathcal{P} = \mathcal{NP}$, (resp. under $\mathcal{UGC}$). We also show that, assuming $\mathcal{ETH}$, there is no $2^{o(\sqrt{n})}$-time algorithm and no $2^{o(k)}n^{O(1)}$-time parameterized algorithm, where n is the number of vertices and k the number of PMUs placed. These results improve the current best known bounds.

1 Introduction

Monitoring the nodes of an electrical network can be carried out by means of Phasor Measurement Units (PMUs). The problem of placing an optimal number of PMUs on the nodes for complete network monitoring, is known as POWER DOMINATING SET [16]. A recent variant of the problem [15], called POWER EDGE SET (PES), is to have the PMUs on the network links rather than the nodes, considering the following two rules: (1) two endpoints of an edge bearing a PMU are monitored and (2) if one node is monitored and all but one of its neighbors are too, then the unmonitored neighbor becomes monitored. The problem of assigning a minimum number of PMUs to monitor the whole network is known to be $\mathcal{NP}$-hard in the general case but can be solved in linear time on trees [15]. In this paper, we present some new complexity results, proposing new lower bounds according to classic complexity hypotheses.

We model the electrical network by a graph $G = (V, E)$ with $|V| = n$ and $|E| = m$. We let $V(G)$ and $E(G)$ denote the respective sets of vertices and edges of G. Further, $N_G(v)$ denotes the set of neighbors of v and $d_G(v) = |N_G(v)|$ its degree in G. Finally, we let $N_G[v] := N_G(v) \cup \{v\}$ denote the *closed* neighborhood of v in G.

© Springer International Publishing AG, part of Springer Nature 2018
L. Brankovic et al. (Eds.): IWOCA 2017, LNCS 10765, pp. 128–141, 2018.
https://doi.org/10.1007/978-3-319-78825-8_11

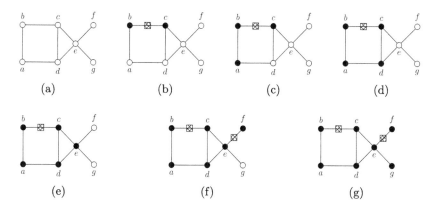

Fig. 1. Before placing any PMU (represented by crossed boxes on edges), all vertices are white (Fig. 1a). If we place a PMU on $\{b, c\}$, then $c(b) = c(c) = 1$ (black) by RULE R_1 (Fig. 1b). By applying RULE R_2 on b, we obtain $c(a) = 1$ (Fig. 1c). Then, RULE R_2 on a gives $c(d) = 1$ (Fig. 1d), and, finally, $c(e) = 1$ with RULE R_2 on c or d (Fig. 1e). The color propagation is stopped, and we need to place a second PMU. A PMU on $\{e, f\}$ implies $c(f) = 1$ by RULE R_1 (Fig. 1f) and RULE R_2 on e gives $c(g) = 1$ (Fig. 1g).

The problem POWER EDGE SET can be seen as a problem of color propagation with colors 0 (white) and 1 (black), respectively designating the states *not monitored* and *monitored* of a vertex of G. Let $G = (V, E)$ be a graph as the input of POWER EDGE SET and, for each vertex $v \in V$, let $c(v)$ be the color assigned to v (we abbreviate $\bigcup_{v \in X} c(v) =: c(X)$). Before placing the PMUs, we have $c(V) = \{0\}$. Given a set $E' \subseteq E$ of edges on which to place PMUs, colors propagate according to the following rules:

RULE R_1: if $(u, v) \in E'$, then $c(u) = c(v) = 1$ ("the endpoints of all $\{u, v\} \in E'$ are colored").

RULE R_2: for u, u' with $c(u) = 1$, $u' \in N_G(u)$ and $c(v) = 1$ for all $v \in N_G(u) \setminus \{u'\}$, then $c(u') = 1$ ("if u' is the only uncoloured neighbor of an already colored vertex u, then u' is colored" – we say that we apply RULE R_2 on u to color u', or that u' is colored by *propagation* of u).

The objective of POWER EDGE SET is to find a smallest set of edges $E' \subseteq E$ on which to place the PMUs such that c(V)=$\{1\}$ after exhaustive application of RULE R_1 and RULE R_2. We call such a set a *power edge set* of G (see Fig. 1 for a guided example of RULE R_1 and RULE R_2 on a simple graph, leading to an optimal solution with two PMUs) and we let $pmu(G)$ denote the smallest size of any power edge set.

POWER EDGE SET (PES)
Input: a graph $G = (V, E)$ and some $k \in \mathbb{N}$
Question: Is $pmu(G) \leq k$?

Previous Work. Toubaline et al. [15] propose a complexity result and an approximation threshold $1.12 - \epsilon$ for $\epsilon > 0$ based on an E-reduction from VERTEX COVER. They also propose a linear-time algorithm on trees by performing a polynomial reduction to PATH COVER. Moreover, Poirion et al. [14] develop an exact method, a linear program with binary variables, indexed on the necessary iterations using propagation RULE R_1 and RULE R_2, extended to a linear program in mixed variables, with the goal of being efficient in practice.

Our Contribution. An interesting open question stems from the assumption that power lines run in a plane or, at least in few planes or surfaces of low genus. In this work, we address this question, developing hardness results on (bipartite) planar graphs, covering both approximation and parameterized complexity. We show that PES is hard to approximate within a factor lower than $328/325$ for bipartite graphs (resp. $17/15 - \epsilon$), unless $\mathcal{P} = \mathcal{NP}$, (resp. under $\mathcal{UGC}$). We also show that, assuming $\mathcal{ETH}$, there is no $2^{o(\sqrt{n})}$-time algorithm, and no $2^{o(k)}n^{O(1)}$-time parameterized algorithm with respect to the standard parameter.

2   Preliminaries

In this section, we present some definitions and observations concerning parts of optimal solutions to PES on a graph G. We call a cycle C *ribbon* if all but exactly one vertex v of C have degree two in G and we call v the *knot* of C.

Lemma 1. *Let G be a graph, let C be a ribbon with knot v and let e be an edge of C. Then, there is an optimal power edge set S for G with $S \cap E(C) = \{e\}$.*

Proof. Suppose that no PMU is placed on the edges of C. Then, even if $c(v) = 1$, none of the neighbors of v in C can become colored and, thus, v cannot propagate on any of them. If one PMU is placed on e, we obtain $c(V(C)) = \{1\}$ by consecutive propagation of vertices of degree two.   □

Definition 1 (Passive Relay). Let G be a graph, let C be a ribbon with knot v, and let $N_G(v) \setminus V(C) = \{x, y\}$. Then, v is called *passive relay* between x and y.

If v is a passive relay between x and y, then $c(x) = 1$ implies $c(y) = 1$ by RULE R_2 applied to v. A passive relay between x and y can be built by connecting x and y to a ribbon (see Fig. 2). The interest of adding this relay lies in the fact that, by Lemma 1, any optimal power edge set intersects the ribbon, thus coloring it completely. Then, a coloration of x necessarily implies a coloration of y even if there were remaining uncolored vertices in $N_G(x)$ (and symmetrically from y to x).

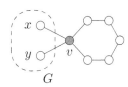

Fig. 2. A passive relay between x and y, consisting in a ribbon with knot v.

Throughout this work, we call a total order $<$ of vertices of G *valid* for any $S \subseteq E(G)$ if, for each $v \in V(G)$, there is an edge incident with v in S or there is some $u \in N_G(v)$ with $N_G[u] \leq v$ (where $\leq$ denotes the extension of $<$ by all reflexive pairs). Note that valid orders correspond to propagation processes of S in G. We also represent a total order $<$ by a sequence $(v_1, v_2, \ldots)$ such that v_i occurs before v_j in the sequence if and only if $v_i < v_j$.

3   Computational Results

In this section, we present new complexity results for PES on restricted graphs. First, we show that PES remains $\mathcal{NP}$-complete even if G is a planar graph with bounded degree at most five (Theorem 1). Then, we extend this result to planar bipartite graphs with degree at most six (Theorem 2). To prove these results, we use a reduction from 3-REGULAR PLANAR VERTEX COVER (3-RPVC) defined as follows:

3-REGULAR PLANAR VERTEX COVER (3-RPVC)
Input: a 3-regular planar graph $G = (V, E)$ and some $k \in \mathbb{N}$.
Question: Is there a size-k set $S \subseteq V$ covering E, i.e. $\forall_{e \in E}\ e \cap S \neq \emptyset$?

3-RPVC is $\mathcal{NP}$-complete [8] but admits a PTAS [1], and a $\frac{3}{2}$-approximation [3].

3.1   Hardness on Planar Graphs

First, we introduce the gadget graph H_v presented in Fig. 3:

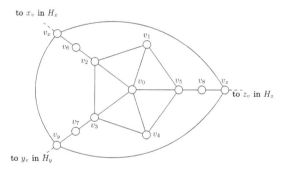

Fig. 3. Gadget H_v for a vertex v with neighbors x, y, and z.

Construction 1. *Given a vertex v of degree three with neighbors x, y, z, the gadget H_v is composed of (1) an internal 5-wheel (vertices v_0–v_5) with center v_0, (2) a set of three border-vertices, one for each neighbor of v, called v_x, v_y and v_z connected in a triangle (3) and three intermediate vertices (v_6, v_7, v_8), connected respectively to v_x and v_2, to v_y and v_3, and to v_z and v_5. The whole gadget contains 12 vertices and 19 edges.*

From any 3-regular planar graph G, we construct a planar graph G' by 1 for each $v \in V(G)$, adding H_v, and 2 for each $\{u, v\} \in E$, adding a connecting edge $\{u_v, v_u\}$, thus linking the gadgets H_u and H_v (see Fig. 6 (appendix)).

In the following, let S' be a solution to PES on G' and let $<$ be a valid order corresponding to S'.

Lemma 2. S' *contains an edge incident with v_0, v_1, or v_4 for all $v \in V(G)$.*

Proof. Towards a contradiction, assume that S' avoids all edges incident with v_0, v_1 and v_4 for some $v \in V(G)$. Then, since v_0 is neighbor of all neighbors (except v_0 itself) of v_1, we have $v_0 < v_1$ and the same holds for v_4. However, all neighbors of v_0 have either v_1 or v_4 as a neighbor (or are v_1 or v_4 themselves), implying $v_1 < v_0$ or $v_4 < v_0$, contradicting $v_0 < v_1, v_4$. ☐

Lemma 3. *For all $\{v, x\} \in E(G)$, we have $\{v_x, x_v\} \notin S'$.*

Proof. Towards a contradiction, assume that $\{v_x, x_v\} \in S'$ for some $\{x, v\} \in E(G)$. Then, we can swap $\{v_x, x_v\}$ and the edges in $S' \cap E(H_v)$ for $\{v_0, v_1\}$ and $\{v_4, v_5\}$ in S', allowing us to start $<$ with $(v_0, v_1, v_4, v_5, v_2, v_3, v_6, v_7, v_8, v_x, v_y, v_z, x_v)$ for $\{x, y, z\} = N_G(v)$. BY Lemma 2, S' did not grow larger. Further, v_x and x_v precede all $w \notin V(H_v)$ in this modified ordering, implying that it is valid for the modified power edge set. ☐

Lemma 4. *Let $v \in V(G)$ with $|S' \cap E(H_v)| = 1$, let $x \in N_G(v)$ and let $w \in \{v_0, v_1, \ldots, v_8\}$ such that w is not incident with an edge of S'. Then, $v_x < w$.*

Proof. Abbreviate $B := \{v_i \mid i \in N_G(v)\}$ and let w be chosen minimal with respect to $<$. Since w is not incident with an edge of S', there is some $u \in N_{G'}(w)$ with $N_{G'}[u] \leq w$. Assume towards a contradiction that $u \notin B$. By minimality of w, we then know that u is incident with an edge of S' and by Lemma 2, $N_{G'}[u]$ avoids B. However, since $|N_{G'}[u]| \geq 4$ for all such u, this contradicts $|S' \cap E(H_v)| = 1$. Thus, $u \in B$, implying $v_x \in N_{G'}[u]$ and $v_x < w$. ☐

Lemma 5. *Let $\{x, v\} \in E(G)$. Then $|S' \cap E(H_x)| > 1$ or $|S' \cap E(H_v)| > 1$.*

Proof. Towards a contradiction, assume that $|S' \cap E(H_x)| = |S' \cap E(H_v)| = 1$ (from Lemma 2, we know that $|S' \cap E(H_v)| \geq 1$). By symmetry, suppose that $v_x < x_v$ and note that, by Lemmas 2 and 3, v_x is not incident with an edge of S'. Thus, there is some $u \in N_{G'}(v_x)$ with $N_{G'}[u] \leq v_x$. Since $v_x < x_v$, we have $u \in V(H_v)$. By Lemma 4, we know that $u \in \{v_i \mid i \in N_G(v)\}$. However, $N_{G'}[u]$ intersects $\{v_0, v_1, \ldots, v_8\}$, contradicting Lemma 4. ☐

Theorem 1. POWER EDGE SET *is $\mathcal{NP}$-complete in planar graphs of degree at most five.*

Proof. We show that G has a size-k vertex cover if and only if the result G' of applying Construction 1 has a power edge set of size $n + k$.

"⇒": let S be a size-k vertex cover of G. We build a power edge set S' for G' as follows: for each $v \in V(G)$, add the edge $\{v_0, v_1\}$ of H_v to S' and for each $v \in S$, add the edge $\{v_4, v_5\}$ of H_v to S'. Note that $|S'| = n + k$. We construct a valid ordering $<$ of G' for S'. To this end, for each $v \in V(G)$ with (x, y, z) being an arbitrary sequence of $N_G(v)$, let

$$<_v := \begin{cases} (v_0, v_1, v_4, v_5, v_2, v_3, v_6, v_7, v_8, v_x, v_y, v_z) & \text{if } v \in S \\ (v_0, v_1, v_x, v_y, v_z, v_6, v_7, v_8, v_2, v_3, v_5, v_4) & \text{if } v \notin S. \end{cases}$$

Let $<^*$ be an arbitrary ordering of $V(G)$ such that $u <^* v$ for all $u \in S$ and $v \notin S$ and let $<$ be the result of replacing each v by the sequence $<_v$ in this ordering. Towards a contradiction, assume that $<$ is not valid for S' and let w be the first vertex of $<$ such that the subsequence of $<$ ending with w is invalid for S'. Let $v \in V(G)$ such that w is a vertex of H_v. By construction of $<_v$, this is only possible if $v \notin S$ and $w = v_x$ for some $x \in N_G(v)$. However, since S is a vertex cover, $x \in S$, implying $x <^* v$ and, thus, $V(H_x) < w$. But then, $N_{G'}[x_v] \leq v_x$ contradicting that the subsequence of S' ending with w is invalid.

"$\Leftarrow$": Let S' be a size-$(n + k)$ power edge set of G' and let $<$ be a valid total order of $V(G')$ for S'. Lemma 5 directly implies that the set $\{v \mid |S' \cap E(H_v)| > 1\}$ is a vertex cover of G and, by Claims 2, 4 and 5, its size is at most $|S'| - n = k$. □

3.2   Hardness on Bipartite Planar Graphs

In the proof of Theorem 1, the graph G' obtained by Construction 1 is not bipartite. In the following, we modify this construction to yield planar bipartite graphs while preserving large parts of the previous proof. To this end, we replace edges of odd-length cycles with a gadget (See Fig. 4) and show that this replacement does not alter the initial coloring propagation scheme in the graph.

Construction 2. *Given a graph G and an edge $e \in E(G)$, let $r(G, e)$ denote the graph $(V(G) \cup V(I(e)), E(G) \cup E(I(e)) \setminus e)$ resulting from replacing e by the gadget graph $I(e)$ in G (see Fig. 4).*

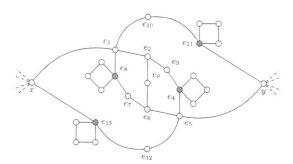

Fig. 4. Gadget graph $I(e)$ with $e = \{x, y\}$.

Note that $I(e)$ is bipartite and planar, and that the distance between x and y is even. By Lemma 1, we know that each of the four 4-cycles connected to e_4, e_8, e_{11}, and e_{13}, respectively contains a PMU. Moreover, vertex e_4 (respectively e_8, e_{11}, e_{13}) is a passive relay between e_3 and e_5 (respectively between e_1 and e_7, between e_{10} and y, between e_{12} and x). Recall that one can consider passive relays and their connected cycles as always colored.

Lemma 6. *Let G be a graph, let $e = \{x, y\} \in E(G)$, and let $G' = r(G, e)$. Then, $pmu(G) \leq k$ if and only if $pmu(G') \leq k + 4$.*

Proof. "⇒": Let F_e be a set containing one edge of each ribbon of $I(e)$, let S be a size-k power edge set for G, and let $S' := (S \setminus \{x, y\}) \cup F_e$. We suppose that $\{x, y\} \notin S$ as otherwise, $S' \cup \{x, e_1\}$ is a power edge set for G and its size is $k + 4$. Let $<$ be a valid order of G for S and let $(v_1, v_2, \ldots)$ be the sequence of $V(G)$ corresponding to $<$. From $<$, we build a valid ordering $<'$ of G' for S', thus proving that S' is a power edge set for G'. Without loss of generality, let $x < y$ and note that $(v_1, v_2, \ldots, x)$ is valid for S'. Let z be minimal with respect to $<$ such that $N_G[x] \leq z$ and let $<'$ be the result of (1) prepending the vertices of the ribbons of $I(e)$ to $<$, (2) replacing x by $(x, e_{12}, e_5, e_3, e_2)$, (3) replacing z by $(e_1, e_7, e_6, e_9, e_{10}, z)$ if $z = y$, and (4) replacing y by $(y, e_{10}, e_1, e_7, e_6, e_9)$ if $z \neq y$. Let $(v_1', v_2', \ldots)$ be the corresponding vertex sequence. Towards a contradiction, assume that there is some w such that $(v_1', v_2', \ldots, w)$ is not valid for S' and let w be minimal with respect to $<'$. As w is not incident with an edge of S', it is also not incident with an edge of S. Further, one can verify that (1)–(4) imply $w \neq e_j$ for all j and, thus, $w \in V(G)$. Since $<$ is valid for S, there is some $u \in N_G(w)$ with $N_G[u] \leq w$. First, suppose that $u = x$ and note that $x, y \leq w = z$ in this case. If $y = w = z$, then $N_G[x] \leq y$ and $N_{G'}[e_{11}] \leq' y$ by (3). Otherwise, $y < w$ and, by (4), $e_1, e_{13} <' w$, implying $N_{G'}[u] \leq' w$. Second, suppose that $u = y$. By (1) and (2), however, $e_5, e_{11} <' y <' w$, implying $N_{G'}[u] \leq' w$. Thus, $u \notin V(I(e))$, implying $N_G[u] = N_{G'}[u]$ and $N_{G'}[u] \leq' w$ as $<'$ is an extension of $<$.

"⇐": Let S' be a size-$(k+4)$ power edge set for G' and let $S_e' := S' \cap E(I(e))$. If $|S_e'| \geq 5$, then $(S \setminus E(I(e))) \cup \{\{x, y\}\}$ is clearly a power edge set for G and its size is at most k. Otherwise, $|S_e'| \leq 4$ and, by Lemma 1, S_e' consists of four edges; one in each ribbon of $I(e)$. Let $S := S' \setminus S_e'$, let $<'$ be a valid order of G' for S', and let $<$ be the restriction of $<'$ to $V(G)$. Let $(v_1', v_2', \ldots)$ and $(v_1, v_2, \ldots)$ be the sequences of $V(G')$ and $V(G)$ implied by $<'$ and $<$, respectively. Without loss of generality, let $x <' y$, implying $x < y$. By construction of $I(e)$, we observe that S_e does not propagate beyond the ribbons of $I(e)$, implying that

$$\forall_{i \in \{1,2,3,5,6,7,9,10,12\}} \ x <' e_i \quad \text{and} \quad \forall_{i \in \{1,6,7,9,10\}} \ (N_{G'}[x] \leq' e_i) \vee (y <' e_i). \quad (1)$$

We show that $<$ is valid for S. Towards a contradiction, assume that there is some $w \in V(G)$ such that $(v_1, v_2, \ldots, w)$ is not valid for S and let w be minimal with respect to $<$. Since $w \in V(G)$ and it is not incident with any edges of S, it is also not incident with any edges of S', implying that there is some $u \in V(G')$ with $N_{G'}[u] \leq' w$. First, suppose that $u \in V(G') \setminus V(G)$ and since, by (1), $w \neq x$, we have $w = y$ and $u \in \{e_5, e_{11}\}$. Thus, $N_{G'}[e_5] \leq' y$, implying $e_6 <' y$ or $N_{G'}[e_{11}] \leq' y$, implying $e_{10} <' y$. In either case, (1) implies $N_{G'}[x] \leq' y$ and, thus, $N_G[x] \leq y$. Second, suppose that $u \in V(G)$. Since $N_G[u] = N_{G'}[u]$ for all $u \in V(G) \setminus \{x, y\}$, we have $u \in \{x, y\}$ as otherwise, $N_G[u] \leq w$. If $u = y$, then $N_G[u] \leq w$ since $N_G[y] = (N_{G'}[y] \cap V(G)) \cup \{x\}$. If $u = x$ then, since $e_1 \in N_{G'}[u]$, we have $e_1 <' w$. But since $w \in N_{G'}[x]$, we have $N_{G'}[x] \not\leq' e_1$ and (1) implies $y <' e_1$. As $N_G[x] = (N_{G'}[x] \cap V(G)) \cup \{y\}$, we conclude $N_G[x] \leq w$. □

In order to show hardness on bipartite graphs, we color the vertices of the output graph G' of Construction 1 arbitrarily with two colors and replace all monochromatic edges e with $I(e)$. We can strengthen the result using the following coloring strategy. For each boundary vertex v_i of each H_v, color v_i such that $N_{G'}[v_i] \setminus \{v_6, v_7, v_8\}$ is not monochromatic and let c be the color occurring the least among $\{v_x, v_y, v_z\}$. Then, color v_0, v_6, v_7, and v_8 with c and color v_1–v_5 with the other color.

Lemma 7. *In H_v, each v_i with $i \in \{x, y, z\}$ is incident with at most two monochromatic edges.*

Proof. Let the color of v_x be blue and assume towards a contradiction that v_x is incident with at least three monochromatic edges. As $N_{G'}[v_x] \setminus \{v_6\}$ is not monochromatic, v_6 is blue. But then, blue appears least among v_x, v_y, v_z, implying that v_y and v_z are not blue. Thus, v_x is incident with at most two monochromatic edges. □

Considering Lemma 7, we observe that the graph resulting from replacing monochromatic edges of G' has maximum degree six.

Theorem 2. POWER EDGE SET *is $\mathcal{NP}$-complete in planar bipartite graphs of degree six.*

4   Some Lower Bounds

4.1   Non-approximability

In this section, we prove new approximation lower bounds for PES, improving the current best known bounds presented by Toubaline et al. [15]. First recall the definition of *L-reduction* between two difficult problems Π and Π', described by Papadimitriou and Yannakakis [13]. This reduction consists of polynomial-time computable functions f and g such that, for each instance x of Π, $f(x)$ is an instance of Π' and for each feasible solution y' for $f(x)$, $g(y')$ is a feasible solution for x. Moreover there are constants $\alpha_1, \alpha_2 > 0$ such that:

1. $OPT_{\Pi'}(f(x)) \leq \alpha_1 OPT_\Pi(x)$ and
2. $|val_\Pi(g(y')) - OPT_\Pi(x)| \leq \alpha_2 |val_{\Pi'}(y') - OPT_{\Pi'}(f(x))|$.

We use an *L*-reduction from VERTEX COVER in hypergraphs in which all edges have cardinality exactly 3.

3-UNIFORM VC (3UVC)
Input: 3-uniform hypergraph $G = (V, E)$ and some $k \geq 2$.
Question: Is there a size-k vertex set $V' \subseteq V$ covering E?

3-UNIFORM VC is hard to approximate within a factor less than $2 - \epsilon$ for all $\epsilon > 0$, unless $\mathcal{P} = \mathcal{NP}$, even if each vertex appears in at most three edges [6].

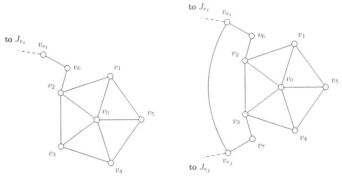

(a) H_v^1-gadget $\forall v \in V$ included into one edge e_i).

(b) H_v^2-gadget $\forall v \in V$ included into two edges e_i and e_j).

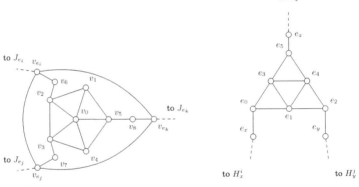

(c) H_v^3-gadget $\forall v \in V$ included into three edges e_i, e_j, e_k).

(d) J_e-gadget: $\forall e \in E$ with $e = \{x, y, z\}$

Fig. 5. Polynomial-time reduction for hypergraph G.

Theorem 3. *Under $\mathcal{UGC}$,* POWER EDGE SET *is hard to approximate within a factor of $\frac{17}{15} - \epsilon$, even on graphs of maximum degree five.*

Proof. Given an instance $I = (G, k)$ of 3-UNIFORM VC such that each vertex of G appears in at most three edges, we construct an instance $I' = (G', k + n + m)$ of PES in the following way: For each $v \in V$ included in exactly $\gamma \leq 3$ edges, we add a gadget H_v^γ given in Fig. 5a–c. Vertices $v_{e_i}, v_{e_j}, v_{e_k}$ are border-vertices for H_v^1, H_v^2, H_v^3. For each hyperedge $e = \{x, y, z\}$, we add a gadget J_e, given by Fig. 5d with border vertices e_x, e_y, e_z and we add the edges $\{x_e, e_x\}$, $\{y_e, e_y\}$, and $\{z_e, e_z\}$.

Vertex-gadgets H_v^γ are designed such that if we have $c(v_{e_j}) = 1$ for all $e_j \in E$ containing v, then placing a single PMU inside H_v^γ is sufficient to color the whole vertex-gadget. If $c(e_v) = 0$ for some e containing v, two PMUs are necessary in

H_v^γ to color the whole gadget, but this also colors e_v' for all e' containing v. Edge-gadgets J_e are designed such that if at least one border vertex e_x is colored, then only one PMU is required in J_e to color the whole edge-gadget, but this also colors v_e for all $v \in e$. Note that if there are two PMUs on any edge-gadget J_e in an optimal solution, then one can simply switch one PMU from J_e to any adjacent H_v^γ and get a solution of same cost with only one PMU per edge-gadget.

Observe that G admits a size-k vertex cover if and only if G' can be monitored with $k + n + m$ PMUs: the vertex-gadgets H_v^γ with two PMUs propagate on the border-vertices on all edge-gadgets. If we add one PMU per edge-gadget, any colored border vertex of H_v^γ propagates its color to all other border vertices. To show that the vertex-gadgets H_v^γ with two PMUs induce a vertex cover of I, suppose that there is a hyperedge $e = \{u, v, w\} \in E$ that is not covered. Then, their respective vertex gadgets contain a single PMU. Then, however, these vertex gadgets cannot be colored by RULE R_2, contradicting I' being monitored.

To show that the above constitutes an L-reduction, let f be a function transforming any instance I of 3-UNIFORM VC into an instance I' of pmuas above, let S' be any feasible solution for I', and let g be the function that transforms S' into a solution S'' that contains exactly one edge of each J_e and at least one edge of each H_v^γ, and then outputs the set of vertices v for which S'' assigns at least two PMUs to H_v^γ. First, the above argument shows that $g(S')$ is a feasible solution for 3-UNIFORM VC. Second, by construction,

$$OPT(I') = OPT(I) + n + m \tag{2}$$

and, since each vertex of I appears in at most 3 edges of I, at least one in seven vertices has to be in a vertex cover of G, implying $n/7 \le OPT(I)$. Since each vertex is incident with at most three hyperedges and each hyperedge contains exactly three vertices, Hall's theorem implies $m \le n$. We then obtain $OPT(I') \le 15 \cdot OPT(I)$. Third, by construction of g, we have

$$val(g(S')) \le val(S') - m - n \overset{(2)}{\le} val(S') - OPT(I') + OPT(I) \tag{3}$$

Thus, we constructed an L-reduction with $\alpha_1 = 15$, $\alpha_2 = 1$. Assuming $\mathcal{UGC}$, 3-UNIFORM VC is hard to approximate to a factor of $(3 - \epsilon)$ [2] and, thus

$$
\begin{aligned}
val(S') &\overset{(3)}{\ge} val(g(S')) + OPT(I') - OPT(I) \\
&\ge 3 \cdot OPT(I) + OPT(I') - OPT(I) \\
&\ge 2/15 \cdot OPT(I') + OPT(I') \\
&\ge 17/15 \cdot OPT(I') \qquad\qquad\qquad \square
\end{aligned}
$$

Theorem 4. POWER EDGE SET *on bipartite graphs of maximum degree six cannot be approximated to within a factor better than* $328/325 > 1.0092$ *unless* $\mathcal{P} = \mathcal{NP}$.

Proof. To show that the reduction from 3-RPVC presented in Construction 2 is an L-reduction, let I be an instance of 3-RPVC, let f be the described reduction and let g be the function that, given any feasible solution S' for $I' := f(I)$, transforms S' into a feasible solution S'' according to Lemmas 2, 3, 4, 5 and returns the set of vertices v such that S'' contains at least two edges more than four times the number of gadgets $I(e)$ in H_v. Let m' be the total number of edges e that are replaced by $I(e)$ by f. Using similar arguments, as in the proof of Theorem 3 we have $OPT(I') = OPT(I) + n + 4m'$ and, since the graph G of I is 3-regular, $n/2 \leq OPT(I)$ (no independent set of G can be larger than $n/2$). Additionally to the coloring scheme suggested to prove Lemma 7, we repeatedly find a H_v with at least two incident inter-gadget edges that are monochrome and swap the coloring of H_v. Then, $m' \leq 4n + m \leq 4n + m/3 = 4n + n/2$, we further have $OPT(I') \leq 39 \cdot OPT(I)$. Then, $val(S') \geq val(g(S')) + OPT(I') - OPT(I)$. Since VERTEX COVER is hard to approximate to within a factor of 1.36, even in 3-regular graphs [5,7] (unless $\mathcal{P} = \mathcal{NP}$), we conclude $val(S') \geq 328/325 OPT(I')$.

$\square$

4.2   Lower Bounds for Exact and FPT Algorithms

We propose some negative results for POWER EDGE SET about the existence of subexponential-time algorithms under $\mathcal{ETH}$ [9,10], and $\mathcal{FPT}$ Algorithms. Since the polynomial-time transformation given in the proof of Theorem 1 is linear in the number of vertices, and since 3-REGULAR PLANAR VERTEX COVER does not admit a $2^{o(\sqrt{n})}n^{O(1)}$-time algorithm [7,11], there is also no $2^{o(\sqrt{n})}n^{O(1)}$-time algorithm for POWER EDGE SET. Moreover, since the solution size k is at most n, a $2^{o(k)}n^{O(1)}$-time algorithm contradicts the non-existence (assuming $\mathcal{ETH}$) of $2^{o(n)}n^{O(1)}$-time algorithms for VERTEX COVER on planar graphs [11].

Corollary 1. *Assuming $\mathcal{ETH}$, there is no $2^{o(\sqrt{n})}n^{O(1)}$-time algorithm for* POWER EDGE SET *in planar graphs, and there is no $2^{o(k)}n^{O(1)}$-time algorithm for* POWER EDGE SET *where k is the solution size.*

5   Conclusion

In this article, we presented several new hardness results and some lowers bounds for the problem of selecting a smallest number of phasor measurement units to monitor a given (planar) network. As perspectives, it would be interesting to explore the problem on particular classes of graphs to understand to what extend the regularity of the graph, or special patterns and minors, may influence the complexity of the problem. Further, having excluded $2^{o(k)}n^{O(1)}$-time algorithms, it is also interesting to seek "the next best thing", that is, single exponential-time algorithms with respect to k as well as considering structural parameters that are independent of planarity, such as the treewidth. Finally, as the problem is hard to approximate in polynomial time, it is interesting to allow moderately exponential time, in an $\mathcal{FPT}$-approximation setting (see [4,12]).

Appendix

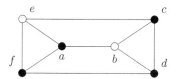

(a) A 3-regular planar graph G and an optimal solution S={a,c,d,f} to Vertex-Cover

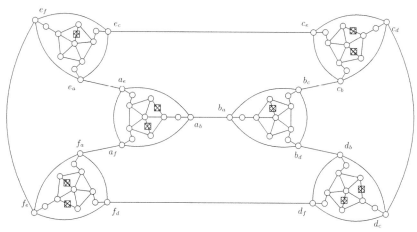

(b) The graph G' obtained from G and the solution S' obtained from S. Here PMU are placed on the edges with boxes

Fig. 6. Example of a graph constructed from an instance I of 3-RPVC (Proof of Theorem 1)

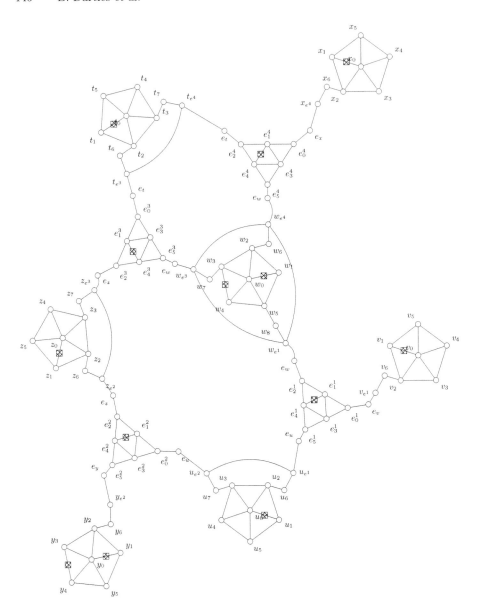

Fig. 7. Graph constructed from an instance I of ERVC with $r = 3$ (Proof of Theorem 3). The 3-uniform hypergraph from I contains 8 vertices t, u, v, w, x, y, z and the four edges $e^1 = \{u, v, w\}$, $e^2 = \{u, y, z\}$, $e^3 = \{t, w, z\}$, $e^4 = \{t, w, x\}$. An optimal solution for PES is to place PMUs on edges with a box. Vertex-Gadgets w and y are the only one with two PMU. Thus $\{w, y\}$ is a vertex cover in the hypergraph.

References

1. Baker, B.S.: Approximation algorithms for NP-complete problems on planar graphs. J. ACM **41**(1), 153–180 (1994)
2. Bansal, N., Khot, S.: Inapproximability of hypergraph vertex cover and applications to scheduling problems. In: Abramsky, S., Gavoille, C., Kirchner, C., Meyer auf der Heide, F., Spirakis, P.G. (eds.) ICALP 2010. LNCS, vol. 6198, pp. 250–261. Springer, Heidelberg (2010). https://doi.org/10.1007/978-3-642-14165-2_22
3. Bar-Yehuda, R., Even, S.: On approximating a vertex cover for planar graphs. In: Proceedings of 14th STOC, pp. 303–309 (1982)
4. Bazgan, C.: Approximation schemes and parameterized complexity. Ph.D thesis, INRIA, Orsay, France (1995)
5. Dinur, I., Safra, S.: On the hardness of approximating vertex cover. Ann. Math. **162**(1), 439–485 (2005). https://doi.org/10.4007/annals.2005.162.439. ISSN 0003-486X
6. Dinur, I., Guruswami, V., Khot, S., Regev, O.: A new multilayered PCP and the hardness of hypergraph vertex cover. SIAM J. Comput. **34**(5), 1129–1146 (2005)
7. Feige, U.: Vertex cover is hardest to approximate on regular graphs. Technical report, MCS03-15, the Weizmann Institute (2003)
8. Garey, M.R., Johnson, D.S.: Computers and Intractability: A Guide to the Theory of NP-Completeness. W. H. Freeman & Co., New York (1979)
9. Impagliazzo, R., Paturi, R.: On the complexity of k-SAT. J. Comput. Syst. Sci. **62**(2), 367–375 (2001)
10. Impagliazzo, R., Paturi, R., Zane, F.: Which problems have strongly exponential complexity? J. Comput. Syst. Sci. **63**(4), 512–530 (2001)
11. Lokshtanov, D., Marx, D., Saurabh, S.: Lower bounds based on the exponential time hypothesis. Bull. EATCS **105**, 41–72 (2011)
12. Marx, D.: Parameterized complexity and approximation algorithms. Comput. J. **51**(1), 60–78 (2008)
13. Papadimitriou, C.H., Yannakakis, M.: Optimization, approximation, and complexity classes. J. Comput. Syst. Sci. **43**(3), 425–440 (1991)
14. Poirion, P.-L., Toubaline, S., D'Ambrosio, C., Liberti, L.: The power edge set problem. Networks **68**(2), 104–120 (2016)
15. Toubaline, S., D'Ambrosio, C., Liberti, L., Poirion, P.-L., Schieber, B., Shachnai, H.: Complexité du problème power edge set. In: ROADEF 2016 (2016)
16. Yuill, W., Edwards, A., Chowdhury, S., Chowdhury, S.P.: Optimal PMU placement: a comprehensive literature review. In: 2011 IEEE Power and Energy Society General Meeting, pp. 1–8, July 2011. https://doi.org/10.1109/PES.2011.6039376

The Parameterized Complexity of Happy Colorings

Neeldhara Misra and I. Vinod Reddy[✉]

Indian Institute of Technology, Gandhinagar, Palaj 382355, India
{neeldhara.m,reddy_vinod}@iitgn.ac.in

Abstract. Consider a graph $G = (V, E)$ and a coloring c of vertices with colors from $[\ell]$. A vertex v is said to be happy with respect to c if $c(v) = c(u)$ for all neighbors u of v. Further, an edge (u, v) is happy if $c(u) = c(v)$. Given a partial coloring c of V, the Maximum Happy Vertex (Edge) problem asks for a total coloring of V extending c to all vertices of V that maximizes the number of happy vertices (edges). Both problems are known to be NP-hard in general even when $\ell = 3$, and is polynomially solvable when $\ell = 2$. In [IWOCA 2016] it was shown that both problems are polynomially solvable on trees, and for arbitrary k, it was shown that MHE is NP-hard on planar graphs and is FPT parameterized by the number of precolored vertices and branchwidth.

We continue the study of this problem from a parameterized perspective. Our focus is on both structural and standard parameterizations. To begin with, we establish that the problems are FPT when parameterized by the treewidth and the number of colors used in the precoloring, which is a potential improvement over the total number of precolored vertices. Further, we show that both the vertex and edge variants of the problem is FPT when parameterized by vertex cover and distance-to-clique parameters. We also show that the problem of maximizing the number of happy edges is FPT when parameterized by the standard parameter, the number of happy edges. We show that the maximum happy vertex (edge) problem is NP-hard on split graphs and bipartite graphs and maximum happy vertex problem is polynomially solvable on cographs.

1 Introduction

Given an undirected vertex colored graph G, we say that a vertex v in G is *happy* if v and all its neighbors have same color. Along similar lines, an edge is happy if both its endpoints have same color. Given a partially colored graph G with ℓ colors, the MAX HAPPY VERTICES (ℓ-MHV) problem is to color the remaining vertices of graph such that number of happy vertices is maximized. The MAX HAPPY EDGES (ℓ-MHE) problem is to color the remaining vertices of the graph such that number of happy edges is maximized.

The ℓ-MHE problem generalizes the MULTIWAY UNCUT problem which is defined as follows. Given a graph G and a terminal set $S = \{s_1, \cdots, s_k\} \subseteq V(G)$, the goal is to find a partition $\{V_1, \cdots, V_k\}$ of $V(G)$ such that each partition

© Springer International Publishing AG, part of Springer Nature 2018
L. Brankovic et al. (Eds.): IWOCA 2017, LNCS 10765, pp. 142–153, 2018.
https://doi.org/10.1007/978-3-319-78825-8_12

contains exactly one terminal and the number of edges with both end points present in the same V_i is maximized. The Multiway Uncut problem is a special case of ℓ-MHE problem, where each terminal has a unique precolor.

Both ℓ-MHV and ℓ-MHE are NP-hard [10] on general graphs for $\ell \geqslant 3$ and both 2-MHV and 2-MHE can be solved in polynomial time. Aravind et al. [2] showed that both the problems admit linear time algorithms on trees. Zhang and Li [10] studied both problems from the approximation point of view and given a $\max\{1/k, \Omega(d^{-3})\}$-approximation algorithm for the k-MHV problem, where d is the maximum degree of the graph, and a $1/2$-approximation algorithm for the ℓ-MHE problem.

We initiate, in this work, the study of these problems from a parameterized perspective. The problem admits several natural parameters: the number of colors (ℓ), the number of precolored vertices (say t), the number of happy vertices or edges (denoted by k, note that these parameters reflect the quality of the solution, and might hence be regarded as standard parameters), and various structural parameters. In particular, the linear time algorithms on trees prompt us to consider the question of whether the problem is FPT when parameterized by treewidth of the graph (w). The work in [2] already establishes that the problem is FPT when parameterized by treewidth and the number of precolored vertices. Since ℓ-MHV and ℓ-MHE are both NP-hard even when there are only three colors used by the precoloring, the problems are para-NP-hard by this parameter. We now proceed to describe some of the results that we obtain for various combinations of parameters.

Our Contributions. We continue the study of this problem from a parameterized perspective. Our focus is on both structural and standard parameterizations. To begin with, we establish that the problems are FPT when parameterized by the treewidth and the number of colors used in the precoloring, which is a potential improvement over the total number of precolored vertices. This follows from a MSO formulation but we also demonstrate a dynamic programming solution on nice tree decompositions.

Further, we show that both the vertex and edge variants of the problem is FPT when parameterized by vertex cover and distance-to-clique parameters. Observe that there is no exponential dependence here on the number of colors in the precoloring. We achieve this by not guessing all possible assignments of colors on the modulators, but just a wireframe of equivalence classes based on which vertices in the modulators receive the same colors, and it turns out that this coarser information is sufficient to determine an optimal coloring. We also show that the problem of maximizing the number of happy edges is FPT when parameterized by the standard parameter, the number of happy edges. This turns out to be a problem that reduces to the case of bounded vertex cover.

In the context of studying the problems on special classes of graphs, we show that the problem of maximizing the number of happy vertices is polynomially solvable on cographs. Unfortunately, our polynomial time approach for the ℓ-MHV problem does not extend in any straightforward way to the ℓ-MHE problem. On the other hand, both variants of the problem turned out to be

NP-hard on split graphs and bipartite graphs. We note that some of the results shown here, particularly relating to algorithms parameterized by the treewidth of the graph, were obtained independently in [1,3].

2   Preliminaries

In this section, we introduce the notation and the terminology that we will need to describe our algorithms. Most of our notation is standard. We use $[k]$ to denote the set $\{1, 2, \ldots, k\}$. We introduce here the most relevant definitions, and use standard notation pertaining to graph theory based on [5,6].

All our graphs will be simple and undirected unless mentioned otherwise. For a graph $G = (V, E)$ and a vertex v, we use $N(v)$ and $N[v]$ to refer to the open and closed neighborhoods of v, respectively. The *distance* between vertices u, v of G is the length of a shortest path from u to v in G; if no such path exists, the distance is defined to be ∞. A graph G is said to be *connected* if there is a path in G from every vertex of G to every other vertex of G. If $U \subseteq V$ and $G[U]$ is connected, then U itself is said to be connected in G. For a subset $S \subseteq V$, we use the notation $G \setminus S$ to refer to the graph induced by the vertex set $V \setminus S$.

Special Graph Classes. We now define some of the special graph classes considered in this paper. A graph is *bipartite* if its vertex set can be partitioned into two disjoint sets such that no two vertices in same set are adjacent. A graph is a *split graph* if its vertex set can be partitioned into a clique and an independent set. Split graphs do not contain C_4, C_5 or $2K_2$ as induced subgraphs. *Cographs* are P_4-free graphs, that is, they do not contain any induced paths on four vertices.

Parameterized Complexity. A parameterized problem denoted as $(I, k) \subseteq \Sigma^* \times \mathbb{N}$, where Σ is fixed alphabet and k is called the parameter. We say that the problem (I, k) is *fixed parameter tractable* with respect to parameter k if there exists an algorithm which solves the problem in time $f(k)|I|^{O(1)}$, where f is a computable function. For a detailed survey of the methods used in parameterized complexity, we refer the reader to the texts [5,7].

We now define the problems that we consider in this paper.

MAX HAPPY VERTICES   **Parameter:** k
Input: A graph $G = (V, E)$, a partial coloring $p : S \to [\ell]$ for some $S \subseteq V$, and a positive integer k.
Question: Is there a coloring $c : V \to [\ell]$ extending p such that G has at least k happy vertices with respect to c?

MAX HAPPY EDGES   **Parameter:** k
Input: A graph $G = (V, E)$, a partial coloring $p : S \to [\ell]$ for some $S \subseteq V$, and a positive integer k.
Question: Is there a coloring $c : V \to [\ell]$ extending p such that G has at least k happy edges with respect to c?

---

3   Structural Parameterizations

In this section, we explore the complexity of MAX HAPPY VERTICES and MAX HAPPY EDGES with respect to various structural parameterizations. A key question here is if these problems are FPT when parameterized by treewidth alone. While we do not resolve this question, we make partial progress in two ways.

First, we show that MAX HAPPY VERTICES and MAX HAPPY EDGES are both FPT when parameterized by the combined parameter $(\ell + w)$, where ℓ is the number of colors used in the precoloring and w is the treewidth of the input graph. In particular, our running time is $\mathcal{O}^*((2\ell)^{(w+1)})$. It was already known that the problem is FPT when parameterized by $(q + w)$, where q is the number of precolored vertices. It was also known that the problem admits a linear-time algorithm on trees. In general, since $\ell \leqslant q$, and the treewidth of a tree is one, our result unifies these results (although the running time we obtain on trees is quadratic, rather than linear).

Secondly, we show that MAX HAPPY VERTICES and MAX HAPPY EDGES are both FPT when parameterized by the size of the vertex cover. The running time of this algorithm has a polynomial dependence on ℓ, which is why this is not subsumed by our FPT algorithm when we parameterized by $(\ell+w)$. Along similar lines, we also show that MAX HAPPY VERTICES and MAX HAPPY EDGES are both FPT when parameterized by the size of a clique modulator.

The rest of this section is organized as follows. In the first subsection, we demonstrate that the problem is FPT parameterized by treewidth. Next, we consider the vertex cover and distance to clique parameterizations. In the last subsection, we show how MAX HAPPY EDGES admits a FPT algorithm for the standard parameter (the number of happy edges) by reducing it to the case of bounded vertex cover number.

3.1   Treewidth

Theorem 1. MAX HAPPY VERTICES *and* MAX HAPPY EDGES *are both* FPT *when parameterized by* $(\ell + w)$, *where* ℓ *is the number of colors used in the precoloring and* w *is the treewidth of input graph.*

Proof. We give two different proofs to show MAX HAPPY VERTICES and MAX HAPPY EDGES are both FPT when parameterized by $(\ell+w)$. The first proof uses a standard dynamic programming approach on nice tree decompositions (see [5] for definition). The second proof follows from an application of Courcelle's theorem [4] and the fact that the MAX HAPPY VERTICES and MAX HAPPY EDGES problems can be expressed by MSO formulas for fixed ℓ. Due to lack of space, the proof is presented in the full version of this paper [9].   □

3.2   Vertex Cover and Distance to Clique

Theorem 2. MAX HAPPY VERTICES *and* MAX HAPPY EDGES *are both* FPT *when parameterized by the size of the vertex cover of the input graph.*

Proof. We first consider the MAX HAPPY VERTICES problem. Let (G, S, ℓ, p, k) denote an instance of MAX HAPPY VERTICES. We will use Q to refer to $V \setminus S$, the set of vertices that are not precolored by p. Further, let $X \subseteq V$ be a vertex cover of G. We use d to denote $|X|$.

The algorithm begins by guessing a partition of X into at most $\min\{\ell, d\}$ parts. Note that the number of such partitions is at most d^d. Intuitively, we are guessing the behavior of an optimal coloring c when projected on the vertex cover, where our notion of behavior is given by which vertices are colored in the same way as c. We also guess the subset of vertices in the vertex cover that are happy with respect to c.

Let us formalize the notion of a behavior associated with a partition and the subset of X. To begin with, let $\chi = (X_1, \ldots, X_t)$ be a fixed partition of X, where $t \leqslant d$. For vertex $v \in X$, we abuse notation and use $\chi(v)$ to denote the index of the part that v belongs to in the partition χ. In other words, if $\chi(v) = i$, then $v \in X_i$. A pair of vertices $u, v \in X$ such that $\chi(u) = \chi(v)$ are called equivalent — we sometimes say that u is equivalent to v, or that u and v are equivalent. Finally, let $Y \subseteq X$ be a (possibly empty) subset of the vertex cover.

We say that a coloring c is valid and respects (χ, Y) if:

- c agrees with p on S,
- every vertex $v \in Y$ is happy with respect to c, and
- for all $u, v \in X$, $c(u) = c(v)$ if and only if u and v are equivalent.

Our goal now is to find a valid coloring c that respects (χ, Y). Let $\lambda(v)$ denote the set of colors employed by p in $N[v]$, in other words, $\lambda(v) := \{j \mid \exists u \in N[v], p(u) = j\}$.

It is easy to check that there exists a valid coloring c that respects (χ, Y) if and only if the following conditions, which we will refer to as $(\star)$, hold:

- For any vertex $v \in Y$, $|\lambda(v)| \leqslant 1$.
- For any pair of vertices u, v that are equivalent and $u, v \in Y$, if $\lambda(v) \neq \emptyset$ and $\lambda(v) \neq \emptyset$, $\lambda(v) = \lambda(u)$.
- For any pair of vertices u, v that are not equivalent and $u, v \in Y$, we have that $N(u) \cap N(v) = \emptyset$.
- For any vertex $v \in Y$, every vertex $u \in N(v) \cap X$ is equivalent to v.

In particular, that these conditions are necessary follow from the definition of what it means for a coloring to be valid and respect (χ, Y). The fact that they are sufficient will follow from the coloring obtained by the algorithm below.

We assume, without loss of generality, that all the conditions above are satisfied: indeed, if not, we simply reject the choice of (χ, Y). Our algorithm now proceeds to construct a coloring c of V that respects (χ, Y). In fact, among all such colorings, we will construct one that maximizes the number of happy vertices. To begin with, we initialize c to coincide with p on S. For convenience, we will use U to refer to the set of uncolored vertices with respect to c. Observe that at this stage, $U = Q$, and when the algorithm finishes, we will have $U = \emptyset$. We now proceed as follows.

Phase 1. Identifying forced colors. Let $v \in Y$ be such that $\lambda(v) \neq \emptyset$. Set $c(u) = j$ for all $u \in N[v] \cap Q$. Further, set $c(u) = j$ for all u equivalent to v in χ. At the end of this phase, let v be any vertex in Y that either has a precolored vertex in its closed neighborhood or has a precolored vertex that is equivalent to it. At the end of this phase, v is a happy vertex. Observe that c is well-defined because the conditions in $(\star)$ are true. We say that X_i is *pending* if $X_i \subseteq U$ at the end of Phase 1. If no X_i is pending, we skip directly to Phase 3.

Phase 2. Coloring the Pending X_i's. Let i be such that X_i is pending, and let $j \in [\ell]$. Let $w[i,j]$ denote the number of vertices in $(V \setminus X) \cap S$ that will be happy if all vertices of X_i are colored j. This is simply the size of the set of vertices in the independent set precolored j, whose neighborhoods lie entirely in X_i.

Consider an auxiliary weighted bipartite graph, denoted by $H = ((A,B),E)$ with edge weights given by $w : E \to [|V|]$. This graph is constructed as follows. The vertex set A contains one vertex for every X_i that is pending at the end of the first phase. The vertex set B contains a vertex corresponding to every element in $[\ell] \setminus [\cup_{v \in X} c(v)]$, that is to say, B has one vertex for every color that is not already used on vertices in X in Phase 1. The weight of the edge (a_i, b_j) is simply $w[i,j]$. We find a matching M of maximum weight in H. It is easy to see that any such matching saturates A, since $|B| \geqslant |A|$, the weights are positive, and all edges are present.

For a pending part X_i, we now color all vertices in X_i based on the matching M. In particular, if M matches a_i to b_j, then we color all vertices in X_i with color j. At the end of this phase, all vertices in X have been colored—formally, $X \cap U = \emptyset$.

Phase 3. Coloring the independent set. We say that an uncolored vertex in $V \setminus X$ is *definitely unhappy* with respect to χ if it has at least two neighbors which are not equivalent with respect to χ. If a vertex v that is definitely unhappy has a neighbor u in Y, then assign $c(u)$ to v. Observe that this coloring is well-defined, since vertices in Y that had different colors, and are hence not equivalent, have disjoint neighborhoods. Arbitrarily color all the other remaining "definitely unhappy" vertices, namely those that only have neighbors in $X \setminus Y$.

Similarly, we say that an uncolored vertex in $V \setminus X$ is *always happy* if all of its neighbors are equivalent. For a vertex v that is always happy, let j be such that all neighbors of v lie in X_j. We then color v with the same color that we used to color all vertices in X_j. This completes the description of the construction of the coloring—note that at this point, $U = \emptyset$.

To conclude, we count $k'[\chi, Y]$, the number of happy vertices with respect to the constructed coloring c. Let k^* denote $\max(k'[\chi, Y])$, where the max is taken over all χ, S for which there exists a valid coloring that respects χ, Y. The algorithm returns YES if $k^* \geqslant k$ and NO otherwise.

Proof of Correctness. [Sketch.] Let c^* be an arbitrary but fixed coloring of G that maximizes the number of happy vertices. Let χ, S be the behavior of c^* with respect to X, that is, let S be the set of happy vertices in X with respect to c^*, and let $\chi := (X_1, \ldots, X_t)$ be a partition of X based on the colors given by c^*. Let c be the coloring output by the algorithm when considering the behavior (χ, S). Note that the algorithm does output some coloring based on the characterizing nature of the conditions in $(\star)$.

It is easy to see that $c^\star$ and c agree on the colors given in Phase 1 of the algorithm. Further, note that c and $c^\star$ agree on the number (and even the subset) of happy vertices in X. Also, among all uncolored vertices of $V \setminus X$, all definitely unhappy vertices in $V \setminus X$ are not happy in $c^\star$, while all the vertices that are always happy are happy in c. Among the precolored vertices in $V \setminus X$, it can be verified that the maximum number of vertices that can be happy with respect to any coloring that respects the behavior (χ, S) is precisely the weight of the maximum matching obtained in Phase 2. Indeed, any such coloring is a matching in this auxiliary graph, and the number of happy vertices corresponds exactly to the weight of the matching. It follows that the number of happy vertices in c is at least the number of happy vertices in c^*.

Running Time Analysis. Trying all possible choices of (χ, S) requires time proportional to $O((2d)^d)$. For a fixed choice of (χ, S), all the three phases of the algorithm are straightforward to implement in polynomial time. A maximum matching can be computed in time $O(n + \sqrt{n}m)$ on bipartite graphs.

We now turn to the MAX HAPPY EDGES problem. Here the algorithm is considerably simpler. Partition E into two sets E_0 and E_1, where E_0 is the set of all edges who have both their endpoints in X, and $E_1 := E \setminus E_0$. We again guess the behavior of an optimal coloring in terms of how it partitions X into equivalence classes. For a fixed partition χ, we count all the happy edges in E_0 (note that this number does not depend on what colors are given to the parts, but merely the fact that all vertices in a part are equivalent).

Having fixed a partition, we force the colors of the parts that have precolored vertices. Construct an auxiliary bipartite graph as we did in Phase 2 of the algorithm for MAX HAPPY VERTICES, with the only difference that now the weight of the edge (a_i, b_j) is based on the number of *edges* that are made happy when all vertices X_i are colored with color j. This helps us determine a coloring of the vertices in X. Uncolored vertices in $V \setminus X$ can now be colored greedily: for an uncolored vertex $v \in V \setminus X$, let d_j denote $|N(v) \cap X_j|$, for $1 \leqslant j \leqslant t$. Note that coloring v with the same color as the one used on $\max(d_j)$ makes $\max(d_j)$ edges happy. The correctness of this approach follows from the fact that this is the best we can hope for from a coloring that is consistent with the behavior specified by χ.   □

Theorem 3. MAX HAPPY VERTICES *and* MAX HAPPY EDGES *are both* FPT *when parameterized by the size of a clique modulator of the input graph.*

Proof. Let (G, S, ℓ, p, k) denote an instance of MAX HAPPY VERTICES. We will use Q to refer to $V \setminus S$, the set of vertices that are not precolored by p. Further, let $X \subseteq V$ such that $C = G \setminus X$ is a clique. We use d to denote $|X|$.

If $\ell > d + 1$ then there exists at least two vertices u and v in C such that $p(u) \neq p(v)$, which implies no vertex of C is happy. First we guess the partition

(H, U) of X in $O(2^d)$ time, where H and U denotes the happy and unhappy vertices of X in optimal coloring c.

Let $H = (H_1, \cdots, H_t)$ be the partition of H such that all vertices in set H_i, $i \in [t]$, are colored with the same color by c. We can guess the correct partition in $O(d^d)$ time. Note that $N(H_i) \cap N(H_j) = \emptyset$: suppose $v \in N(H_i) \cap N(H_j)$ then the color of vertex v is either different from $col(H_i)$ or $col(H_j)$ which is a contradiction to the fact that all vertices in both H_i and H_j are happy.

Since H_i is happy there does not exist two vertices u and v in the set $H_i \cup N(H_i)$ such that $p(u) \neq p(v)$. For each $i \in [t]$, if at least one vertex is precolored in $H_i \cup N(H_i)$ then assign same color to all vertices of $H_i \cup N(H_i)$. For the sets $H_i \cup N(H_i)$, which do not have any precolored vertices, assign a color which is not used so far to color any H_i. At the end arbitrarily color remaining vertices. For each possible partition H and U of X and each possible partition $H_1, \cdots, H_t$ of H, count the number of happy vertices and the optimal coloring c is the coloring which maximizes the number of happy vertices.

If $\ell \leqslant d + 1$ then some of the clique vertices can be happy, but this only happens when all the clique vertices are colored by same color. Since there are at most ℓ colors we can guess the correct coloring of the clique in $O(\ell)$ time. Now it remained to color the set X, which can be done using the procedure described in the case of $l > d + 1$.

We now turn to the MAX HAPPY EDGES problem. First we give a procedure to color X when all the vertices in clique $C = G \setminus X$ are precolored and no vertex in X is precolored. Let $X = (X_1, \cdots, X_t)$ be the partition of X such that all vertices in X_i are colored in the same way by c.

Let $w[i, j]$ denote the number of edges in $G[X_i \cup C]$ that will be happy if all vertices of X_i are colored j. Consider an auxiliary weighted bipartite graph, denoted by $D = ((A, B), E)$ with edge weights given by $w : E \rightarrow [|E|]$. This graph is constructed as follows. The vertex set A contains one vertex for every set X_i and the vertex set B contains one vertex for every color used in clique. The weight of the edge (a_i, b_j) is simply $w[i, j]$. We find a matching M of maximum weight in D. It is easy to see that any such matching saturates A, since $|B| \geqslant |A|$, the weights are positive, and all edges are present. We now color all vertices in X_i based on the matching M. In particular, if M matches a_i to b_j, then we color all vertices in X_i with color j. At the end of this phase, all vertices in X have been colored. Now we are ready to describe the general case. Let C_U be the number of uncolored vertices in clique. If $|C_U| \leqslant d + 1$ then $X' = X \cup C_U$ is a vertex deletion distance to clique C' of size at most $2d + 1$, i.e., $G \setminus (X \cup C_U)$ is a clique. Since all the vertices of the clique $G \setminus (X \cup C_U)$ are precolored, The vertices of X' can be colored using the procedure described above. So without loss of generality we assume that $|C_U| > d + 1$.

Lemma 1. *If $|C_U| = n_1 > d+1$, then in any optimal coloring c all non precolored vertices in clique has to be colored with single color.*

Proof. Since ℓ colors are used in precoloring there exists a color class of size at least $\lfloor \frac{n - n_1}{\ell} \rfloor$ in C. Assigning this color to all vertices of C_U maximizes the number of happy edges, since if we color a vertex of C_U with a different color than others, then we loose at least $d + 1$ happy edges in clique C_U and can make at most d edges happy.   □

The case when $\ell \leqslant d + 1$ is easy, since we can guess optimal coloring of X in time $O(d^d)$ and then it can be easily extended to color the clique. If $\ell > d + 1$, From Lemma 1 we know that all vertices of C_U gets same color, we guess this color in $O(\ell)$ time. Now we need to color X such that the number of happy edges is maximized. This can be done by simply applying the procedure described in first case, where all vertices of clique are precolored. The running time of the algorithm is $O(\ell(d^d)(n + \sqrt{n}m)(n + m))$.   □

3.3   The Standard Parameter

We finally show that MAX HAPPY EDGES is, in fact, FPT when parameterized by the number of happy edges. Here we use the fact that if there are enough edges both of whose endpoints are uncolored, then we have a YES instance right away. If not, the number of uncolored vertices can be shown to be bounded, and since it is safe to delete edges among precolored vertices (with some bookkeeping), the problem effectively reduces to the bounded vertex cover number scenario.

Theorem 4. MAX HAPPY EDGES *is* FPT *when parameterized by* k *and* MAX HAPPY VERTICES *is* FPT *when parameterized by* k *and* ℓ.

Proof. Due to lack of space, the proof is presented in the full version of this paper [9].   □

4   Special Graph Classes

Theorem 5. MAX HAPPY VERTICES *and* MAX HAPPY EDGES *are both NP-complete on the class of bipartite and split graphs.*

Proof. The reductions for MAX HAPPY VERTICES follow by easy modifications of the reduction in [10]. Here, therefore, we only state the proofs for MAX HAPPY EDGES. We first consider the case of bipartite graphs. We reduce from MAX HAPPY EDGES on general graphs.

We let (G, ℓ, S, p, k) be an instance of MAX HAPPY EDGES. Construct a bipartite graph $(H = (A, B), E)$ as follows. For every vertex $v \in V(G)$, we introduce a vertex $a_v \in A$. For every edge $e \in E(G)$, we introduce a vertex $b_e \in B$, and if $e = (u, v)$, then b_e is adjacent to a_u and a_v. The precoloring function q mimics p on A, that is, for every $u \in S$, $q(a_u) = p(u)$. We use X to denote $\{a_u \mid u \in S\} \subseteq A$. Let $k' = m + k$. Thus our reduced instance is (H, ℓ, X, q, k').

We now argue the equivalence. First, consider the forward direction. If c is a total coloring of V that makes k edges happy, then we define a coloring c' for H as follows: color $c'(a_v) := c(v)$ for all $a_v \in A$. For every edge $e = (u, v) \in E$, color $b_e \in B$ according to $c(u)$. Note that for all edges e in G that are happy with respect c, two edges (namely (b_e, a_v) and (b_e, a_u)) are happy with respect to c'. Corresponding to all unhappy edges, H has one happy edge with respect to c'. Therefore, the total number of happy edges in c' is $2k + (m - k) = m + k$.

In the reverse direction, let c' be a coloring of H that makes at least $m + k$ edges happy. Now consider the coloring c obtained as follows: $c(u) = c'(a_u)$. We argue that at least k edges are happy in G with respect to c. Indeed, suppose not. Without loss of generality, assume that only $(k - 1)$ edges are happy with respect to c. Then in H, there are at most $(k-1)$ vertices in B that can have two happy edges incident on them, and therefore the total number of happy edges is at most $2(k - 1) + (m - k + 1) = m + k - 1$, which contradicts our assumption about the total number of happy edges in H with respect to c'.

We now turn to the case of split graphs. The construction is similar to the case of bipartite graphs. Construct a split graph $(H = (A, B), E)$ as follows. Let (G, ℓ, S, p, k) be an instance of MAX HAPPY EDGES. Let $T := \binom{m}{2} + 1$. For every vertex $v \in V(G)$, we introduce T copies of the vertex $a_v \in A$, denoted by $a_v[1], \ldots, a_v[T]$. For every edge $e \in E(G)$, we introduce a vertex $b_e \in B$, and if $e = (u, v)$, then b_e is adjacent to all copies of a_u and a_v. Finally, we add all edges among vertices in B, thereby making $H[B]$ a clique.

The precoloring function q mimics p on A across all copies, that is, for every $u \in S$, $q(a_u) = p(u)$ for all copies of a_u. We use X to denote $\{a_u \mid u \in S\} \subseteq A$. Let $k' = T(m + k)$. Thus our reduced instance is (H, ℓ, X, q, k').

We now argue the equivalence of these instances. First, consider the forward direction. If c is a total coloring of V that makes k edges happy, then we define a coloring c' for H as follows: color $c'(a_v) := c(v)$ for all copies of $a_v \in A$. For every edge $e = (u, v) \in E$, color $b_e \in B$ according to $c(u)$. Note that for all edges e in G that are happy with respect c, $2T$ edges (namely (b_e, a_v) and (b_e, a_u) across all copies) are happy with respect to c'. Corresponding to all unhappy edges, H has T happy edges with respect to c'. Therefore, the total number of happy edges in c' is at least $2Tk + T \cdot (m - k) = T \cdot (m + k)$.

In the reverse direction, , let c' be a coloring of H that makes at least $T(m+k)$ edges happy. We argue that there must be at least one copy A_i of $\{a_v \mid v \in V\}$ for which the number of happy edges with one endpoint in B and one in A_i is at least $(m + k)$. Indeed, suppose not. Then consider the following partition of the edges in H: let E_0 be all edges with both endpoints in B, and let E_i be all edges with one endpoint in B and the other endpoint in the ith copy of the vertices $\{a_v \mid v \in V\}$. For the sake of contradiction, we have assumed that the number of happy edges in E_i is less than $(m + k)$ for all $i \in [T]$. Note that the total number of edges in B is $\binom{m}{2}$. Therefore, the the number of edges happy with respect to c' is at most:

$$\binom{m}{2} + T \cdot (m + k - 1) = T \cdot (m + k) + \binom{m}{2} - T < T(m + k),$$

where the last step follows by substituting for $T = \binom{m}{2} + 1$. This leads to the desired contradiction.

Having identified one set E_i that has at least $(m + k)$ happy edges, the argument for recovering a coloring c for G that makes at least k edges happy is identical to the case of bipartite graphs. □

Theorem 6. MAX HAPPY VERTICES *is polynomial time solvable on the class of cographs.*

Proof (Sketch.) Due to lack of space, we only give a brief overview of the algorithm here and defer the details to the full version of the paper [9]. Our algorithm is a recursive routine on the modular decomposition [8] of the input graph, say G. Without loss of generality, we assume that the root r of tree M_G is a series node, otherwise, G is not connected and the number of happy vertices in G is equals to the sum of the maximum number of happy vertices in each connected component. Let the children of r be x and y. Further, let the cographs corresponding to the subtrees at x and y be G_x and G_y.

We assume that $\ell \geqslant 3$, otherwise we use the polynomial time algorithm of 2-MHV problem on general graphs to find the number of maximum happy vertices. Also, it is easy to handle the case when one of G_x or G_y has only one vertex, since such a vertex v is universal and in any optimal coloring c of G, all happy vertices are colored with the color of v. Therefore, we may assume that both G_x and G_y contains at least two vertices. This is helpful because if any part uses more than two colors in the precoloring, we may conclude that no vertex in the other part can be happy (since every vertex in G_x is adjacent to every vertex in G_y). Using this fact, algorithm now proceeds by a straightforward case analysis. □

5   Conclusion

In this paper, we study the MAX HAPPY VERTICES and MAX HAPPY EDGES problems from the parameterized perspective. We showed that Both the problems are FPT with respect to the structural parameters (a) Vertex cover (b) Distance to clique. MAX HAPPY EDGES is FPT when parameterized by number of happy edges in solution (standard parameter) and MAX HAPPY VERTICES is FPT when parameterized by number of happy vertices in the solution and the number of colors. Both MAX HAPPY VERTICES and MAX HAPPY EDGES are NP-hard on split graphs and bipartite graphs and MAX HAPPY VERTICES is polynomially solvable on cographs.

The following are some interesting open problems: Are MAX HAPPY VERTICES and MAX HAPPY EDGES are FPT when parameterized by the cluster vertex deletion number? Do the MAX HAPPY VERTICES and MAX HAPPY EDGES problems admit polynomial kernels when parameterized by either the vertex cover or the distance to clique parameters?

References

1. Agrawal, A.: On the parameterized complexity of happy vertex coloring. In: International Workshop on Combinatorial Algorithms. Springer (2017, in press)
2. Aravind, N.R., Kalyanasundaram, S., Kare, A.S.: Linear time algorithms for happy vertex coloring problems for trees. In: Mäkinen, V., Puglisi, S.J., Salmela, L. (eds.) IWOCA 2016. LNCS, vol. 9843, pp. 281–292. Springer, Cham (2016). https://doi. org/10.1007/978-3-319-44543-4_22
3. Aravind, N., Kalyanasundaram, S., Kare, A.S., Lauri, J.: Algorithms and hardness results for happy coloring problems. arXiv preprint arXiv:1705.08282 (2017)
4. Courcelle, B.: Graph rewriting: an algebraic and logic approach. In: Handbook of theoretical computer science, pp. 194–242 (1990)
5. Cygan, M., Fomin, F.V., Kowalik, Ł., Lokshtanov, D., Marx, D., Pilipczuk, M., Pilipczuk, M., Saurabh, S.: Parameterized Algorithms. Springer, Cham (2015). https://doi.org/10.1007/978-3-319-21275-3
6. Diestel, R.: Graph Theory. GTM, vol. 173. Springer, Heidelberg (2012). https:// doi.org/10.1007/978-3-662-53622-3
7. Downey, R.G., Fellows, M.R.: Fundamentals of Parameterized Complexity. TCS, vol. 4. Springer, London (2013). https://doi.org/10.1007/978-1-4471-5559-1
8. Habib, M., Paul, C.: A survey of the algorithmic aspects of modular decomposition. Comput. Sci. Rev. 4(1), 41–59 (2010)
9. Misra, N., Reddy, I.V.: The parameterized complexity of happy colorings. arXiv preprint arXiv:1708.03853 (2017)
10. Zhang, P., Li, A.: Algorithmic aspects of homophyly of networks. Theor. Comput. Sci. 593, 117–131 (2015)

Computational Complexity Relationship between Compaction, Vertex-Compaction, and Retraction

Narayan Vikas[✉]

School of Computing Science, Simon Fraser University,
Burnaby, BC V5A 1S6, Canada
vikas@cs.sfu.ca

Abstract. In this paper, we show a very close relationship between the compaction, vertex-compaction, and retraction problems for reflexive and bipartite graphs. The relationships that we present relate to a long-standing open problem concerning whether any pair of these problems are polynomially equivalent for every graph. The relationships we present also relate to the constraint satisfaction problem, providing evidence that similar to the compaction and retraction problems, it is also likely to be difficult to give a complete computational complexity classification of the vertex-compaction problem for every reflexive or bipartite graph. In this paper, we however give a complete computational complexity classification of the vertex-compaction problem for all graphs, including even partially reflexive graphs, with four or fewer vertices, by giving proofs based on mostly just knowing the computational complexity classification results of the compaction problem for such graphs determined earlier by the author. Our results show that the compaction, vertex-compaction, and retraction problems are polynomially equivalent for every graph with four or fewer vertices.

Keywords: Computational complexity · Algorithms · Graph
Partition · Colouring · Homomorphism · Retraction · Compaction
Vertex-compaction

1 Introduction

We first introduce the following definitions and problems, and then describe the motivation and results.

1.1 Definitions

A vertex v of a graph is said to have a loop if vv is an edge of the graph. A *reflexive graph* is a graph in which every vertex has a loop. An *irreflexive graph* is a graph in which no vertex has a loop. Any graph, in general, is a *partially reflexive graph*, in which its vertices may or may not have loops. Thus reflexive

© Springer International Publishing AG, part of Springer Nature 2018
L. Brankovic et al. (Eds.): IWOCA 2017, LNCS 10765, pp. 154–166, 2018.
https://doi.org/10.1007/978-3-319-78825-8_13

and irreflexive graphs are special partially reflexive graphs. A bipartite graph is irreflexive by definition.

Let G be a graph. We use $V(G)$ and $E(G)$ to denote the vertex set and the edge set of G respectively. If uv is an edge of G then vu is also an edge of G. Given an induced subgraph H of G, we denote by $G - H$, the subgraph obtained by deleting from G the vertices of H together with the edges incident with them; thus $G - H$ is a subgraph of G induced by $V(G) - V(H)$. For a vertex v of G, we define $G - v$ similarly (in the above, we have a single vertex v instead of the graph H). We say that G is *connected*, if there exists a path between every pair of vertices in G; otherwise we say that G is *disconnected*. A *component* of G is a maximal connected subgraph of G. In the following, let G and H be graphs.

A *homomorphism* $f : G \rightarrow H$, of G to H, is a mapping f of the vertices of G to the vertices of H, such that $f(g)$ and $f(g')$ are adjacent vertices of H whenever g and g' are adjacent vertices of G. If there exists a homomorphism of G to H then G is said to be *homomorphic* to H.

A *compaction* $c : G \rightarrow H$, of G to H, is a homomorphism of G to H, such that for every vertex x of H, there exists a vertex v of G with $c(v) = x$, and for every edge hh' of H, $h \neq h'$, there exists an edge gg' of G with $c(g) = h$ and $c(g') = h'$. If there exists a compaction of G to H then G is said to *compact* to H.

A *vertex-compaction* $c : G \rightarrow H$, of G to H, is a homomorphism of G to H, such that for every vertex x of H, there exists a vertex v of G with $c(v) = x$. If there exists a vertex-compaction of G to H then G is said to *vertex-compact* to H. Note that every compaction is also a vertex-compaction.

A *retraction* $r : G \rightarrow H$, of G to H, with H as an induced subgraph of G, is a homomorphism of G to H, such that $r(h) = h$, for every vertex h of H. If there exists a retraction of G to H then G is said to *retract* to H. Note that every retraction is necessarily also a compaction, and hence a vertex-compaction, but not vice versa.

For each vertex v of G, let $L(v)$ be a list of vertices of H. We denote by L the entire set of lists $L(v)$ for the vertices v of G. A *list homomorphism* $l : G \rightarrow H$, of G to H, with respect to L is a homomorphism of G to H, such that $l(v) \in L(v)$, for every vertex v of G. If $l : G \rightarrow H$ is a list homomorphism with respect to L then we say that $l : G \rightarrow H$ is a *list-L-homomorphism*. List homomorphism problems of various variety have been studied in (Feder and Hell 1998); (Feder et al. 1999) and (Vikas 2002).

An *identification* of two distinct vertices u and v of G is an execution of the following steps (1), (2), and (3), which results in a new graph: (1) For every nonloop edge uu' of G, if vu' is not an edge of G then we add the edge vu' to G (note that if uv is an edge of G then $u' = v$ and we will have the loop vv), (2) If u has a loop then v is also made to have a loop if it does not already have one, and (3) Delete the vertex u together with the edges incident with u from G.

1.2 Homomorphism, Vertex-Compaction, Compaction, and Retraction Problems

The problem of deciding the existence of a homomorphism to a fixed graph H, called the *homomorphism problem for H*, also known as the *H-colouring problem*, and denoted as *H-COL*, asks whether or not an input graph G is homomorphic to H. The problem *H-COL* is trivial and easily seen to be polynomial time solvable if H has a loop or H is bipartite. For any fixed non-bipartite irreflexive graph H, it is shown in (Hell and Nesetril 1990) that *H-COL* is NP-complete. Note that the classic k-colourability problem is a special case of the problem *H-COL* when H is the irreflexive complete graph K_k with k vertices and the input graph G is irreflexive.

The problem of deciding the existence of a vertex-compaction to a fixed graph H, called the *vertex-compaction problem for H*, and denoted as *VCOMP-H*, asks whether or not an input graph G vertex-compacts to H. Some results on the vertex-compaction problem have been described in (Vikas 2011; 2013).

The problem of deciding the existence of a compaction to a fixed graph H, called the *compaction problem for H*, and denoted as *COMP-H*, asks whether or not an input graph G compacts to H. The compaction problem is a well studied problem over last several years, and includes some popular problems. Results on the compaction problem can be found in (Vikas 1999; 2003; 2004a; 2004b; 2004c; 2005; 2011; 2013).

The problem of deciding the existence of a retraction to a fixed graph H, called the *retraction problem for H*, and denoted as *RET-H*, asks whether or not an input graph G, containing H as an induced subgraph, retracts to H. Retraction problems have been of continuing interest in graph theory for a long time and have been studied in various literature (see (Vikas 2004b) for references).

The vertex-compaction, compaction, and retraction problems are special graph colouring problems, and can also be viewed as graph partition problems, see (Vikas 2004c; 2005; 2011; 2013). Note that unlike the problem *H-COL*, the problems *VCOMP-H*, *COMP-H*, and *RET-H* are still interesting if H has a loop or H is bipartite. Some work on graph partition problems have also been studied in (Feder et al. 2003) and (Hell 2014).

1.3 Motivation and Results

It is not difficult to show that for every fixed graph H, if *RET-H* is solvable in polynomial time then *COMP-H* is also solvable in polynomial time (a polynomial transformation from *COMP-H* to *RET-H* under Turing reduction is shown in (Vikas 2004b). Is the converse true? This was also asked, in the context of reflexive graphs, by Peter Winkler in 1988 (personal communication), cf. (Vikas 2003). Thus the question is whether *RET-H* and *COMP-H* are polynomially equivalent for every fixed graph H. The answer to this is not known even when H is reflexive or bipartite. However, a very close relationship is shown between the compaction problem and the retraction problem for reflexive and bipartite graphs in (Vikas 2004b). It is shown in (Vikas 2004b) that for every fixed reflex-ive (bipartite) graph H there exists a fixed reflexive (bipartite) graph H' such that RET-H and $COMP$-H' are polynomially equivalent.

Similar to the polynomial transformation from $COMP$-H to RET-H shown in (Vikas 2004b), we give a polynomial transformation from $VCOMP$-H to RET-H under Turing reduction, for every fixed graph H. Thus if RET-H is solvable in polynomial time then $VCOMP$-H is also solvable in polynomial time for every fixed graph H, but again whether the converse is true is not known even for reflexive or bipartite graphs H. We however establish a very close relationship between the vertex-compaction problem and the retraction problem for reflexive and bipartite graphs, similar to the relationship between the compaction problem and the retraction problem established in (Vikas 2004b), showing that for every fixed reflexive (bipartite) graph H, there exists a fixed reflexive (bipartite) graph H' such that RET-H and $VCOMP$-H' are polynomially equivalent.

Similarly as in (Vikas 2004b), due to our above result and results of (Feder and Hell 1998), (Feder and Vardi 1998), and (Vikas 2004b), we establish that for every constraint satisfaction problem Π (with fixed templates), there exists a fixed reflexive (bipartite) graph H such that the constraint satisfaction problem Π, the vertex-compaction problem $VCOMP$-H, and the compaction problem $COMP$-H are polynomially equivalent.

Since it is thought to be likely difficult to determine whether every constraint satisfaction problem (with fixed templates) is polynomial time solvable or NP-complete, we thus have evidence that it is likely to be difficult to determine whether for every fixed reflexive (bipartite) graph H, the problems $VCOMP$-H and $COMP$-H are polynomial time solvable or NP-complete. Similar evidence has been shown for RET-H in (Feder and Hell 1998) in the case of fixed reflexive graphs H, and in (Feder and Vardi 1998) in the case of fixed bipartite graphs H. Issues related to the constraint satisfaction problem have also been considered in (Feder and Vardi 1998).

Although, as argued above, determining whether $VCOMP$-H, $COMP$-H, or RET-H is polynomial time solvable or NP-complete is likely to be difficult for every graph H, a complete computational complexity classification of $COMP$-H and RET-H when H has four or fewer vertices is given in (Vikas 2004c; 2005), i.e., for every graph H with at most four vertices (including when H is partially reflexive), it is determined in (Vikas 2004c; 2005) whether $COMP$-H is polyno-mial time solvable or NP-complete, and whether RET-H is polynomial time solv-able or NP-complete. As pointed out in (Vikas 2011; 2013), the computational complexity classification of $COMP$-H also holds for the problem $VCOMP$-H, for all graphs H with at most four vertices, as the proofs given in (Vikas 2004c; 2005) to determine the computational complexity classification of $COMP$-H also hold for the problem $VCOMP$-H, if we consider $VCOMP$-H instead of $COMP$-H in the proofs, for all graphs H with at most four vertices. The complexity when H is a reflexive square follows as a special case of a reflexive k-cycle, cf. (Vikas 2011; 2013). In this paper, we however give a complete computational complexity classification of $VCOMP$-H, by giving proofs based on mostly just knowing the computational complexity classification results of *COMP-H*, when *H* is any graph (including partially reflexive) having four or fewer vertices, i.e., for every graph *H* with at most four vertices, we determine whether *VCOMP-H* is polynomial time solvable or NP-complete.

The complexity classification of *VCOMP-H*, *COMP-H*, and *RET-H* do not differ for graphs *H* with at most four vertices. Thus the problems *VCOMP-H*, *COMP-H*, and *RET-H* are polynomially equivalent for all graphs *H* with four or fewer vertices. As discussed above, this relates to a long-standing open problem whether any pair of these problems are polynomially equivalent for every graph. We have various more results showing that for several graphs *H*, the problems *VCOMP-H*, *COMP-H*, and *RET-H* are polynomially equivalent, i.e., they are either all NP-complete or all polynomial time solvable. We do not know of any graph *H* for which the complexity classification of *VCOMP-H*, *COMP-H*, and *RET-H* differ.

Thus we have here two issues. One issue is concerned with the complete computational complexity classification of the vertex-compaction, compaction, and retraction problems, and the other issue is concerned with the equivalence of the vertex-compaction, compaction, and retraction problems. We have resolved fully the two issues for all graphs up to four vertices.

We have discussed above that whether *COMP-H* and *RET-H* are polynomially equivalent is an open problem, and whether *VCOMP-H* and *RET-H* are polynomially equivalent is also an open problem. The question that we consider now is whether we can say that the two very close problems *VCOMP-H* and *COMP-H* are polynomially equivalent. Similar to the proof given in (Vikas 2004b) showing a polynomial transformation from *COMP-H* to *RET-H*, we show that *VCOMP-H* polynomially transforms to *COMP-H* under Turing reduction, for every fixed graph *H*. What can we say about the converse? The answer to this again is not known even for reflexive or bipartite graphs *H*. Thus we pose the more specific open problem whether *COMP-H* polynomially transforms to *VCOMP-H* for every fixed reflexive or bipartite graph *H*. We however establish in this paper, a very close relationship between the vertex-compaction problem and the compaction problem, similar to the relationship between the compaction problem and the retraction problem established in (Vikas 2004b), showing that for every fixed reflexive (bipartite) graph *H*, there exists a fixed reflexive (bipartite) graph *H'*, such that *COMP-H'* and *VCOMP-H'* are polynomially equivalent.

In Sect. 2, we give a polynomial transformation from *VCOMP-H* to *COMP-H* and *RET-H* under Turing reduction. In Sect. 3, we give very close relationships between the vertex-compaction and compaction problems, and between the vertex-compaction and retraction problems, for reflexive and bipartite graphs. In Sect. 4, we give a complete computational complexity classification of *VCOMP-H* when *H* has four or fewer vertices.

2 A Polynomial Transformation from Vertex-Compaction to Compaction and Retraction

In this section, we give a polynomial transformation from the vertex-compaction problem to the compaction and retraction problems, similar to the polynomial transformation from the compaction problem to the retraction problem given in (Vikas 2004b). We also give an useful result relating them.

Theorem 2.1. *For every fixed graph H, the problem VCOMP-H is polynomially transformable to the problems COMP-H and RET-H under Turing reduction.*

Proof. Let H be a fixed graph, and let a graph G be an instance of *VCOMP-H*. We note from the definition of vertex-compaction that if $c : G \to H$ is a vertex-compaction then part of the requirement for c is that for all $h \in V(H)$, there exists $v \in V(G)$, with $c(v) = h$. Thus, there exists a set of $|V(H)|$ vertices in G which cover all the vertices in H under c. In other words, there exists an induced subgraph Q of G, with $|V(Q)| = |V(H)|$, such that $c : Q \to H$ is a vertex-compaction.

In general, there may exist several such above described induced subgraphs Q of G such that Q vertex-compacts to H, regardless of whether or not G vertex-compacts to H. Further, if Q vertex-compacts to H then there may exist several vertex-compactions f of Q to H. The pair (Q, f) will be used to denote such a subgraph Q of G with a vertex-compaction $f : Q \to H$. Note that for each such subgraph Q of G, there may be more than one but a fixed number of such pairs (since H is fixed). Also, we have $|V(Q)| = |V(H)|$, and since H is fixed, all such subgraphs Q of G, and all vertex-compactions f of Q to H, i.e., all the pairs (Q, f), can be found in time polynomial in the size of G.

Let β denote the number of different possible pairs (Q, f). Clearly, β is a polynomial in the size of G. Consider the ith pair (Q, f) (under an arbitrary ordering of the pairs), $1 \le i \le \beta$. We define $L_i(q) = \{f(q)\}$ (i.e., $L_i(q)$ is a singleton containing the vertex $f(q)$ of H), for all $q \in V(Q)$, and $L_i(u) = V(H)$, for all $u \in V(G) - V(Q)$. Thus we obtain the lists $L_i(v) \subseteq V(H)$, for all $v \in V(G)$, $i = 1, 2, \ldots, \beta$ in polynomial time. Clearly, G vertex-compacts to H if and only if there exists an i such that there is a list-L_i-homomorphism of G to H, $1 \le i \le \beta$.

We construct a graph G_i from G and L_i, for all $i = 1, 2, \ldots, \beta$. Let i be some fixed value, $1 \le i \le \beta$. The construction of G_i is as follows. We take a fresh copy of the graphs G and H. For each vertex u of G, if $L_i(u)$ is a singleton, say $L_i(u) = \{h\}$, for some vertex h of H, then we identify u and h. The resultant graph after identifications is our graph G_i. The graph G_i contains a copy of H as a subgraph. Further, considering the ith pair (Q, f), since Q is an induced subgraph of G and $f : Q \to H$ is a homomorphism, G_i has a copy of H as an induced subgraph. Since β is a polynomial, we have only polynomially many graphs $G_1, G_2, \ldots, G_\beta$. Thus we obtain the graphs $G_1, G_2, \ldots, G_\beta$ in polynomial time. Clearly, G vertex-compacts to H if and only if there exists an i such that G_i retracts to H, $1 \le i \le \beta$. Thus we have a polynomial transformation from *VCOMP-H* to *RET-H* under Turing reduction.

If G vertex-compacts to H then there exists an i such that G_i retracts to H, and hence G_i compacts to H, $1 \leq i \leq \beta$. Conversely, if there exists an i such that G_i compacts to H, then clearly G_i and G vertex-compacts to H, $1 \leq i \leq \beta$. Thus, G vertex-compacts to H if and only if there exists an i such that G_i compacts to H, $1 \leq i \leq \beta$. Hence, we have a polynomial transformation from VCOMP-H to COMP-H under Turing reduction.   □

Theorem 2.2. *Let G be a graph and H be fixed graph. Then there exists a graph G', such that the following statements (i), (ii), and (iii), are equivalent: (i) G vertex-compacts to H, (ii) G' compacts to H, and (iii) G' retracts to H.*

Proof. Let $G_1, G_2, \ldots, G_\beta$ be the graphs constructed in the proof of Theorem 2.1. Note that if there exists an i such that G_i compacts to H, then there exists a j (which may be same as i) such that G_j retracts to H, $1 \leq i \leq \beta$, $1 \leq j \leq \beta$. We know by definition that if there exists a j such that G_j retracts to H, then G_j also compacts to H, $1 \leq j \leq \beta$. Hence, we can find a value of i, such that G_i compacts to H if and only if G_i retracts to H. Thus, there exists an i, $1 \leq i \leq \beta$, such that the following statements (i), (ii), and (iii), are equivalent: (i) G vertex-compacts to H, (ii) G_i compacts to H, and (iii) G_i retracts to H.

We have seen in the proof of Theorem 2.1, that (i) is equivalent to (ii), and as discussed above, (ii) is equivalent to (iii). Note that we have already seen in the proof of Theorem 2.1, that (i) is equivalent to (iii) also, but the value of i may have been different for statements (ii) and (iii). The above equivalence shows that we can find a value of i that is same for both statements (ii) and (iii). We call the graph G_i in statements (ii) and (iii) as the graph G'.   □

3   Relationship between Vertex-Compaction, Compaction, and Retraction Problems for Reflexive and Bipartite Graphs

In this section, we establish a very close relationship between the vertex-compaction and compaction problems, and between the vertex-compaction and retraction problems, for reflexive and bipartite graphs.

Theorem 3.1. *For every reflexive (bipartite) graph H, there exists a reflexive (bipartite) graph H', such that RET-H, VCOMP-H', and COMP-H' are polynomially equivalent, and COMP-H is polynomially transformable to VCOMP-H'.*

Proof. The construction of the graph H' is as described in (Vikas 2004b). The equivalence of RET-H and COMP-H' for both the reflexive and bipartite cases was shown in (Vikas 2004b). We point out that the same proof given in (Vikas 2004b) analogously also holds for the equivalence of RET-H and VCOMP-H', if we consider the problem VCOMP-H' instead of COMP-H' in the proof given in (Vikas 2004b).

To prove the equivalence of RET-H and VCOMP-H', we prove the following statements (a) and (b) : (a) RET-H polynomially transforms to VCOMP-H', and (b) VCOMP-H' polynomially transforms to RET-H.

We first prove statement (a). Let a graph G, containing H as an induced subgraph, be an instance of RET-H. We construct in time polynomial in the size of G, a graph G', containing G and H' as induced subgraphs, such that the following statements (i) and (ii) are equivalent: (i) G retracts to H, and (ii) G' vertex-compacts to H'. The construction of the graph G' is as described in (Vikas 2004b), and the proof showing equivalence of statements (i) and (ii) is analogously same as given in the proof in (Vikas 2004b) to show that RET-H polynomially transforms to $COMP$-H'. Thus RET-H polynomially transforms to $VCOMP$-H'.

We now prove statement (b). Let a graph G be an instance of $VCOMP$-H'. We construct in time polynomial in the size of G, polynomial (in the size of G) number of graphs $G_1, G_2, \ldots, G_\beta$, each containing a copy of H' as an induced subgraph, and for each G_i, we construct in time polynomial in the size of G_i, a graph G'_i also containing a copy of H' as an induced subgraph, $1 \leq i \leq \beta$, such that the following statements (i), (ii), (iii), (iv), (v), and (vi) are equivalent for some value of i, $1 \leq i \leq \beta$: (i) G vertex-compacts to H', (ii) G_i retracts to H', (iii) G_i compacts to H', (iv) G'_i retracts to H', (v) G'_i compacts to H', and (vi) G'_i retracts to H.

The equivalence of (i) and (vi) would show that $VCOMP$-H' polynomially transforms to RET-H under Turing reduction, ie., statement (b) holds. The graphs $G_1, G_2, \ldots, G_\beta$ are constructed as in the proof of Theorem 2.1 in Section 2 (where we replace H by H'). As discussed there, β is a polynomial in the size of G, and the graphs $G_1, G_2, \ldots, G_\beta$ are constructed in time polynomial in the size of G. The equivalence of (i) and (ii) follows from the proof of Theorem 2.1. The equivalence of (ii) and (iii) follows from the proof of Theorem 2.2.

The construction of G'_i is analogous to the construction described in the proof given in (Vikas 2004b), to show that $COMP$-H' polynomially transforms to RET-H, where the graph G'_i is constructed from the graph G_i by identification of certain vertices of G_i. We have that (ii) is equivalent to (iv), and (iv) is equivalent to (vi), analogously as proved in (Vikas 2004b). Since (ii) and (iii) are equivalent due to Theorem 2.2 as pointed above, and the constructions are such that we reason out (iv) and (v) are equivalent, as (i) implies (iv), (iv) implies (v), and we show that (v) implies (i). Statements (iii) and (v) are additional observations. Thus, in effect, we prove that (i) is equivalent to (vi), which shows that $VCOMP$-H' polynomially transforms to RET-H under Turing reduction. Thus statement (b) holds.

Since $COMP$-H' and $VCOMP$-H' are both polynomially equivalent to RET-H, this shows that $COMP$-H' and $VCOMP$-H' are polynomially equivalent. Thus RET-H, $VCOMP$-H', and $COMP$-H' are polynomially equivalent.

We now prove that $COMP$-H polynomially transforms to $VCOMP$-H'. It is shown in (Vikas 2004b) that $COMP$-H polynomially transforms to RET-H under Turing reduction. We note from statement (a) above that RET-H polynomially transforms to $VCOMP$-H'. This implies that $COMP$-H polynomially transforms to $VCOMP$-H' under Turing reduction. □

In Theorem 3.1, we note that the graph H' is the same in all the polynomial equivalences showing the close relationships. Thus, the level of difficulty of showing polynomial equivalence between each pair of the problems *VCOMP-H*, *COMP-H*, and *RET-H*, all appears to be the same, even for reflexive or bipartite graphs H. As discussed earlier, Theorem 3.1 leads to the following theorem, providing evidence that it is likely to be difficult to determine whether for every reflexive (bipartite) graph H, the problem *VCOMP-H* is polynomial time solvable or NP-complete.

Theorem 3.2. *For every constraint satisfaction problem Π (with fixed templates), there exists a fixed reflexive (bipartite) graph H such that the constraint satisfaction problem Π is polynomially equivalent to the vertex-compaction problem VCOMP-H and the compaction problem COMP-H.*   □

4   A Complete Computational Complexity Classification of Vertex-Compaction to All Graphs with Four or Fewer Vertices

In this section, we give a complete computational complexity classification of *VCOMP-H* when H has four or fewer vertices. In (Vikas 2004c; 2005), a list of graphs H with at most four vetices is given for which *COMP-H* is shown there to be NP-complete. This list is given in Fig. 1. For rest all other graphs H with at most four vetices, not in this list, *COMP-H* is shown in (Vikas 2004c; 2005) to be polynomial time solvable, and hence *VCOMP-H* is also polynomial time solvable for these graphs H, as from Theorem 2.1, *VCOMP-H* polynomially transforms to *COMP-H*. Thus to completely classify the complexity of *VCOMP-H* for graphs H with at most four vertices, it only remains to determine the complexity of *VCOMP-H* for the list of graphs H given in Fig. 1.

The computational complexity classification of *COMP-H* is shown in (Vikas 2004c; 2005) to be same as that of *RET-H* for all graphs H with at most four vetices. As pointed out in (Vikas 2011; 2013), the computational complexity classification of *VCOMP-H* is same as that of *COMP-H*, for all graphs H with at most four vertices, as the proofs given in (Vikas 2004c; 2005) to determine the computational complexity classification of *COMP-H* also hold for *VCOMP-H*, if we consider *VCOMP-H* instead of *COMP-H* in the proofs, for all graphs H with at most four vertices. The complexity when H is a reflexive square follows as a special case of a reflexive k-cycle, cf. (Vikas 2011; 2013). We however give here proofs determining the computational complexity classification of *VCOMP-H*, mostly just by knowing results of the computational complexity classification of *COMP-H*, for all graphs H with at most four vertices.

Theorem 4.1. *VCOMP-H is NP-complete for the graphs H in Fig. 1(a), (d), (e), (f), (g), and (h).*

Proof. The problem *VCOMP-H* is clearly in NP. It is shown in (Vikas 2004c; 2005) that for each of these graphs H given in the figure, *COMP-H* is NP-complete when its input graph is connected. Let a connected graph G be an

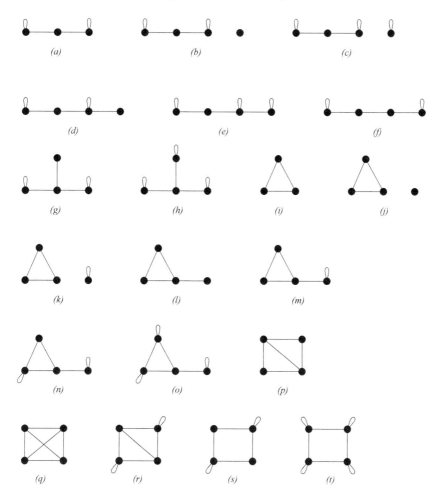

Fig. 1. List of all graphs H with at most four vertices for which $COMP$-H, RET-H, and $VCOMP$-H are NP-complete

instance of $COMP$-H. Clearly, if G compacts to H then G also vertex-compacts to H. Now suppose that G vertex-compacts to H. Since H does not have a cycle and is connected, there is a unique path between any pair of vertices in H. Hence, since G is connected, G also compacts to H. This shows that $VCOMP$-H is NP-complete.   □

Theorem 4.2. *VCOMP-H is NP-complete for the graph H in Fig. 1(s).*

Proof. Let the graph H in Fig. 1(s) be the square $h_0h_1h_2h_3h_0$ with loops on h_0 and h_2. Let S be the path $h_0h_1h_2$ with loops on h_0 and h_2. We have from Theorem 4.1 that $VCOMP$-S is NP-complete. Clearly, the problem $VCOMP$-H is in NP. To prove NP-completeness of $VCOMP$-H, we give a polynomial transformation from $VCOMP$-S to $VCOMP$-H. It is shown in (Vikas 2004c; 2005) that the input graph G for which *COMP-S* is NP-complete is a connected graph containing S as an induced subgraph. It follows from the proof of Theorem 4.1 that *VCOMP-S* is also NP-complete for the same input graph G which is connected and contains S as an induced subgraph.

Let a graph G containing S as an induced subgraph be an instance of *VCOMP-S*. We construct a graph G' by adding to G, a new vertex h_3 adjacent to h_0 and h_2 only. The resulting graph G' has G and H as induced subgraphs. We show that G vertex-compacts to S if and only if G' vertex-compacts to H. Since *VCOMP-S* is NP-complete for such an input graph G, this shows that *VCOMP-H* is also NP-complete.

Suppose first that G vertex-compacts to S. Let $c : G \to S$ be a vertex-compaction. Since h_0 and h_2 are the only vertices with a loop in S, it must be that $c(h_0) = h_0$ or h_2, and $c(h_2) = h_0$ or h_2. We define a vertex-compaction $c' : G' \to H$ as $c'(h_3) = h_3$, and $c'(v) = c(v)$, for all $v \in V(G' - h_3)$. Since h_3 is adjacent to only h_0 and h_2 in G', $c' : G' \to H$ is indeed a homomorphism and a vertex-compaction. Thus G' vertex-compacts to H.

Now suppose that G' vertex-compacts to H. Let $c' : G' \to H$ be a vertex-compaction. We first show that all the three vertices of the subgraph S of H are covered by the vertices of $G' - h_3$ under c'. Since h_0 and h_2 are the only vertices with a loop in H, it must be that $c'(h_0) = h_0$ or h_2. Thus, since h_3 is adjacent to h_0 in G', if $c'(h_3) = h_0$ or h_2 then it must be that $c'(h_0) = c'(h_3)$, and hence all the four vertices of H will be covered by the vertices of $G' - h_3$ under c'. If $c'(h_3) = h_3$ then the remaining vertices of H, i.e., all the three vertices of S, must be covered by the vertices of $G' - h_3$ under c'. If $c'(h_3) = h_1$ then due to symmetry of the vertices h_1 and h_3 in H, we can always redefine the vertex-compaction c' so that $c'(h_3) = h_3$ (if $c'(h_3) = h_1$ then for all the vertices x mapped to h_3 under c', we can redefine c' so that $c'(x) = h_1$ and $c'(h_3) = h_3$, and the redefined c' is still a vertex-compaction), and hence the other three vertices of H, i.e., the vertices of S are covered by the vertices of $G' - h_3$ under c'. Thus in all the cases, all the three vertices of the subgraph S of H are covered by the vertices of $G' - h_3$ (i.e., the graph G) under c'.

We now define a vertex-compaction $c : G \to S$ as follows: $c(v) = h_1$, if $c'(v) = h_3$, for all $v \in V(G)$, and $c(v) = c'(v)$, if $c'(v) \neq h_3$, for all $v \in V(G)$. The definition of c is same as that of c' except that we are only mapping possibly more vertices to S under c, and of course there is no need to consider h_3 under c. We showed above that all the three vertices of S are covered by the vertices of G under c'. Hence all the three vertices of S continue to be covered by the vertices of G under c also. If $c'(v) = h_3$ for some vertex $v \in V(G)$ then v is not adjacent to any vertex mapped to h_1 under c', as $c' : G' \to H$ is a homomorphism. Thus $c : G \to S$ as defined above is a homomorphism. Hence we conclude that $c : G \to S$ is a vertex-compaction. Thus G vertex-compacts to S. This completes the proof of the theorem.   □

Similarly, analogous to the above proof of Theorem 4.2, we show that *VCOMP-H* is NP-complete for the graphs H in Fig. 1(n), (o), and (r).

For the graph H in Fig. 1(m), it is shown in (Vikas 2004c; 2005) that *COMP-H* is NP-complete, and the same proof also holds if we consider the problem *VCOMP-H* instead of *COMP-H*. Thus *VCOMP-H* is NP-complete for the graph H in Fig. 1(m). The problem *VCOMP-H* is NP-complete when H is a reflexive k-cycle, for all $k \geq 4$, cf. (Vikas 2011; 2013), and hence *VCOMP-H* is NP-complete for the graph H in Fig. 1(t).

We note that a graph G is homomorphic to a graph H if and only if the disjoint union $G \cup H$ vertex-compacts to H. Thus we have a polynomial transformation from *H-COL* to *VCOMP-H*. The problem *H-COL* is shown to be NP-complete for any fixed non-bipartite irreflexive graph H in (Hell and Nesetril 1990). It follows that *VCOMP-H* is also NP-complete for any non-bipartite irreflexive graph H. Hence *VCOMP-H* is NP-complete for the graphs H in Fig. 1(i), (l), (p), and (q), as each of the graphs H given in the respective figures is a non-bipartite irreflexive graph.

The results for the compaction problem for connected and disconnected graphs given in (Vikas 2005) also hold analogously for the vertex-compaction problem, and accordingly we have the following theorem.

Theorem 4.3. *Let H be a fixed graph with components $H_1, H_2, \ldots, H_s$. If VCOMP-H_i is polynomial time solvable, for all $i = 1, 2, \ldots, s$, then VCOMP-H is also polynomial time solvable. If VCOMP-H_i is NP-complete for some i, $1 \leq i \leq s$, then VCOMP-H is also NP-complete.* □

It follows from Theorem 4.3 that *VCOMP-H* is NP-complete for the graphs H in Fig. 1(b), (c), (j), and (k), as for the first component H_1 of H in the respective figures, we have already shown that *VCOMP-H_1* is NP-complete.

We have now considered all the graphs H in Fig. 1, and shown that *VCOMP-H* is NP-complete for each of these graphs H. For all other graphs H with at most four vertices, not listed in Fig. 1, *VCOMP-H* is polynomial time solvable. Thus the problems *VCOMP-H*, *COMP-H*, and *RET-H* are polynomially equivalent for every graph H with four or fewer vertices.

References

Feder, T., Hell, P.: List homomorphisms to reflexive graphs. J. Comb. Theory Ser. B **72**, 236–250 (1998)

Feder, T., Hell, P., Huang, J.: List homomorphisms and circular arc graphs. Combinatorica **19**, 487–505 (1999)

Feder, T., Hell, P., Klein, S., Motwani, R.: List partitions. SIAM J. Discret. Math. **16**, 449–478 (2003)

Feder, T., Vardi, M.Y.: The computational structure of monotone monadic SNP and constraint satisfaction: a study through datalog and group theory. SIAM J. Comput. **28**, 57–104 (1998)

Hell, P.: Graph partitions with prescribed patterns. Eur. J. Comb. **35**, 335–353 (2014)

Hell, P., Nesetril, J.: On the complexity of H-colouring. J. Comb. Theory Ser. B **48**, 92–110 (1990)

Vikas, N.: Computational complexity of compaction to cycles. In: Proceedings of the Tenth Annual ACM-SIAM Symposium on Discrete Algorithms (SODA) (1999)

Vikas, N.: Connected and loosely connected list homomorphisms. In: Goos, G., Hartmanis, J., van Leeuwen, J., Kučera, L. (eds.) WG 2002. LNCS, vol. 2573, pp. 399–412. Springer, Heidelberg (2002). https://doi.org/10.1007/3-540-36379-3_35

Vikas, N.: Computational complexity of compaction to reflexive cycles. SIAM J. Comput. **32**, 253–280 (2003)

Vikas, N.: Computational complexity of compaction to irreflexive cycles. J. Comput. Syst. Sci. **68**, 473–496 (2004a)

Vikas, N.: Compaction, retraction, and constraint satisfaction. SIAM J. Comput. **33**, 761–782 (2004b)

Vikas, N.: Computational complexity classification of partition under compaction and retraction. In: Chwa, K.-Y., Munro, J.I.J. (eds.) COCOON 2004. LNCS, vol. 3106, pp. 380–391. Springer, Heidelberg (2004c). https://doi.org/10.1007/978-3-540-27798-9_41

Vikas, N.: A complete and equal computational complexity classification of compaction and retraction to all graphs with at most four vertices. J. Comput. Syst. Sci. **71**, 406–439 (2005)

Vikas, N.: Algorithms for partition of some class of graphs under compaction. In: Fu, B., Du, D.-Z. (eds.) COCOON 2011. LNCS, vol. 6842, pp. 319–330. Springer, Heidelberg (2011). https://doi.org/10.1007/978-3-642-22685-4_29

Vikas, N.: Algorithms for partition of some class of graphs under compaction and vertex-compaction. Algorithmica **67**, 180–206 (2013). Invited Paper

Computational Geometry

Holes in 2-Convex Point Sets

Oswin Aichholzer[1], Martin Balko[2,3]([✉]), Thomas Hackl[1],
Alexander Pilz[4], Pedro Ramos[5], Pavel Valtr[2], and Birgit Vogtenhuber[1]

[1] Institute for Software Technology, Graz University of Technology, Graz, Austria
{oaich,thackl,bvogt}@ist.tugraz.at
[2] Department of Applied Mathematics and Institute for Theoretical Computer
Science (CE-ITI), Charles University, Prague, Czech Republic
balko@kam.mff.cuni.cz
[3] Alfréd Rényi Institute of Mathematics, Hungarian Academy of Sciences,
Budapest, Hungary
[4] Department of Computer Science, ETH Zürich, Zürich, Switzerland
alexander.pilz@inf.ethz.ch
[5] Departamento de Física y Matemáticas, Universidad de Alcalá,
Alcalá de Henares, Spain
pedro.ramos@uah.es

Abstract. Let S be a set of n points in the plane in general position
(no three points from S are collinear). For a positive integer k, a k-hole
in S is a convex polygon with k vertices from S and no points of S in
its interior. For a positive integer l, a simple polygon P is l-convex if no
straight line intersects the interior of P in more than l connected components. A point set S is l-convex if there exists an l-convex polygonization
of S.

Considering a typical Erdős–Szekeres-type problem, we show that
every 2-convex point set of size n contains an $\Omega(\log n)$-hole. In comparison, it is well known that there exist arbitrarily large point sets in
general position with no 7-hole. Further, we show that our bound is tight
by constructing 2-convex point sets with holes of size at most $O(\log n)$.

Keywords: Hole · 2-convex set · Convex position · Point set
Horton set

Research supported by OEAD project CZ 18/2015 and by project no.
7AMB15A T023 of the Ministry of Education of the Czech Republic. O. Aichholzer and B. Vogtenhuber supported by ESF EUROCORES programme Euro-
GIGA - ComPoSe, Austrian Science Fund (FWF): I648-N18. M. Balko and P.
Valtr supported by grant GAUK 690214, by project CE-ITI no. P202/12/G061 of
the Czech Science Foundation GAČR, and by ERC Advanced Research Grant no
267165 (DISCONV). T. Hackl supported by Austrian Science Fund (FWF): P23629-
N18. A. Pilz supported by an Erwin Schrödinger fellowship, Austrian Science Fund
(FWF): J-3847-N35. P. Ramos supported by MINECO project MTM2014-54207,
and ESF EUROCORES programme EuroGIGA, CRP ComPoSe: MICINN Project
EUI-EURC-2011-4306, for Spain.

L. Brankovic et al. (Eds.): IWOCA 2017, LNCS 10765, pp. 169–181, 2018.
https://doi.org/10.1007/978-3-319-78825-8_14

1   Introduction

Let S be a set of n points in the plane in *general position*, i.e., the set S does not contain a collinear point triple. Throughout the whole paper we only consider point sets that are finite and in general position. A convex polygon H is a *hole* in S if its vertices are points of S and the interior of H contains no points of S. If a hole H of S has k vertices, then we say that H is a *k-hole* in S and k is the *size* of the hole. In addition, we regard single points of S as *1-holes* in S and segments determined by two points from S as *2-holes* in S. Slightly abusing the notation, we sometimes use the terms "hole" and "k-hole" also for the set of vertices of a hole and a k-hole, respectively. We remark that in some papers the definition of a hole H allows H to be non-convex.

Erdős [4] asked for the smallest integer $h(k)$ such that every set of $h(k)$ points in general position in the plane contains at least one k-hole. It is easy to check that $h(4) = 5$ and Harborth [6] showed $h(5) = 10$. After this result, the question of Erdős was settled in two phases: first, Horton showed that there are arbitrarily large point sets without 7-holes [8]. Around 25 years later, Gerken [5] and Nicolás [9] independently showed that sets with enough points always contain a 6-hole. A recent summary of known results, together with some bounds on the minimum number of k-holes in a set of n points, can be found in [3]. In the current paper, we consider this question for a restricted class of point sets.

The notion of convexity is central in discrete geometry and it has been generalized in a number of ways. Convex polygons can be characterized by looking at their intersections with straight lines: A simple polygon P is convex if and only if $P \cap \ell$ is connected for every line ℓ. Aichholzer et al. [2] extended this property to *l-convex polygons:* For a positive integer l, a simple polygon P is *l-convex* if there exists no straight line that intersects the interior of P in more than l connected components. Clearly, a convex polygon is 1-convex. An extensive study of l-convex polygons can be found in [2].

Let P be an l-convex polygon. We use ∂P to denote the boundary of P. It follows from the definition of l-convexity that every line that does not contain an edge of P intersects ∂P in at most $2l$ points. In particular, every line that does not contain an edge of a 2-convex polygon intersects its boundary in at most four points.

In [1], the notion of l-convexity was transcribed to finite point sets. A point set S is *l-convex* if there exists a polygonization $P(S)$ of S such that $P(S)$ is an l-convex polygon. Note that an l-convex polygon or point set is also $(l+1)$-convex.

The problem of deciding whether a set of n points is 2-convex can be solved in polynomial time with respect to n. Aichholzer et al. [1] provided an algorithm that solves this problem in time $O(n^{13})$. They also showed that the problem of deciding whether a point set is 3-convex is NP-complete.

In this paper, we consider the following Erdős–Szekeres-type problem for 2-convex point sets: What is the smallest number $f(k)$ such that any 2-convex point set of size $f(k)$ contains a k-hole?

We show that every 2-convex point set of size n contains a hole of size $\Omega(\log n)$, implying that $f(k)$ exists for any $k > 0$ (Sect. 3). Our proof actually yields an algorithm that, given a 2-convex set of n points, finds a hole of size $\Omega(\log n)$ in polynomial time with respect to n. Further, we show that our bound is tight by providing a construction of 2-convex point sets of size n with holes of size at most $O(\log n)$ (Sect. 4). It is natural in this context to wonder about the convexity of large sets that contain only holes of constant size. We provide an asymptotically tight lower bound $\Omega(\sqrt{n})$ on the convexity of so-called Horton sets of size n (Sect. 5).

Although some statements in the paper seem intuitively clear, despite our efforts, the presented rigorous proofs are quite technical. One reason for this is the necessity to take into account also singular cases when a line or a ray shares a whole segment with ∂P or if it touches ∂P in some point of S.

2 Properties of 2-Convex Polygons

The proof of our main result is based on the structure of 2-convex polygons shown in [2]. Let P be a simple polygon and let $\mathrm{CH}(P)$ be its convex hull. In the following, we denote a connected piecewise linear simple arc as a *chain*. We denote with $\langle p_i, \ldots, p_j \rangle$ the chain that starts at p_i and ends at p_j and that traces along ∂P in counterclockwise order. The points p_i and p_j are called the *endpoints* of $\langle p_i, \ldots, p_j \rangle$.

A *pocket* K of P is a chain $\langle p_0, \ldots, p_t \rangle$ such that its two endpoints p_0 and p_t are its only vertices of $\mathrm{CH}(P)$. The segment $p_0 p_t$ is called the *lid* of K. If a pocket consists solely of a single convex hull edge of P, we call it a *trivial pocket*.

The following three observations follow directly from the definitions. First, the lids of the pockets of P form the boundary of $\mathrm{CH}(P)$. Second, every vertex of $\mathrm{CH}(P)$ lies in exactly two pockets. And finally, if a line intersects the interior of $\mathrm{CH}(P)$ then its intersections with ∂P cannot all lie in a single pocket.

The structure of non-trivial pockets is quite simple and it will be crucial in our proof. It is outlined in the following two lemmas. In the following, let P be a set of points in general position and S be a polygonization of P.

Lemma 1 ([2, Lemma 12]). *Let $K = \langle p_0, \ldots, p_t \rangle$ be a non-trivial pocket of a 2-convex polygon between two extreme points p_0 and p_t. Then there are integers r and s with $0 \le r < s < t$ such that K consists of three chains $C_1 = \langle p_0, \ldots, p_r \rangle, C_2 = \langle p_{r+1}, \ldots, p_s \rangle, C_3 = \langle p_{s+1}, \ldots, p_t \rangle$, and two segments $p_r p_{r+1}$ and $p_s p_{s+1}$, where all vertices in C_1 and C_3 are convex vertices of P, while all vertices in C_2 are reflex; see Fig. 1.*

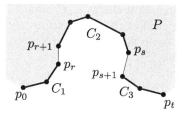

Fig. 1. A non-trivial pocket consists of three chains C_1, C_2, C_3 (drawn by thick segments) and two edges $p_r p_{r+1}$ and $p_s p_{s+1}$ (drawn by thin segments).

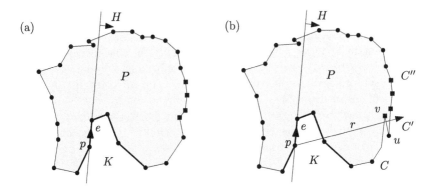

Fig. 2. (a) The order of the points in $H \cap (S \setminus K)$ along ∂P matches their radial order around p. (b) The ray r intersects K twice and each of the three chains C, C', and C'' at least once.

Lemma 2. *Let K be a non-trivial pocket and let C_1, C_2, and C_3 be the chains from Lemma 1. Then the following two statements hold for every $i \in \{1, 2, 3\}$.*

(i) *If the chain C_i contains at least three vertices, then C_i is the boundary of a convex polygon with one edge removed.*
(ii) *The convex hull of C_i is a hole in S.*

We omit the proof of Lemma 2. We say that two subsets A and B of the plane are *strictly separated* by a line ℓ, if they lie in opposite open half-planes determined by ℓ, and *weakly separated* by ℓ, if they lie in opposite closed half-planes determined by ℓ.

The following lemma is somewhat similar to Lemma 10 in [1].

Lemma 3. *Let e be an edge of a pocket K oriented according to the counterclockwise order of ∂P. Let p be the first vertex of e and let H be the right closed half-plane of e; see Fig. 2(a). Then the counterclockwise order of the points of $H \cap (S \setminus K)$ along ∂P matches their counterclockwise radial order around p.*

Proof. Suppose for contradiction that there are two points $u, v \in H \cap (S \setminus K)$ such that their counterclockwise order along ∂P does not match their counterclockwise radial order around p. Let r be any ray with apex p such that r is contained in H and the supporting line of r strictly separates u and v. Since $p \in S$ and S is in general position, such a ray r exists. We cut the chain $\partial P \setminus K$ at u and v, obtaining three chains C, C', and C''. Due to the choice of u and v, the ray r intersects each of the three chains; see Fig. 2(b). Furthermore, r intersects K in p and in one more point, which contradicts the 2-convexity of P.   □

3   The Lower Bound

In this section, we prove the following.

Theorem 1. *Every 2-convex set of n points in the plane and in general position contains a k-hole for $k \in \Omega(\log n)$.*

Throughout this section, S denotes a 2-convex set of n points in the plane in general position and P is a 2-convex polygon that is a polygonization of S.

Let us first sketch the proof of Theorem 1: If P has a large pocket, part (ii) of Lemma 2 implies the existence of a large k-hole in S. When P has no large pocket, we find a large set $Q \subset S$ of points in convex position. If Q forms a hole in S, we are done; otherwise, we can apply Lemmas 4 and 9 below to find a sufficiently large hole in S.

The *kernel* of a simple polygon P is defined as the set of points $x \in P$ such that for every point $y \in P$, the line segment xy is fully contained in P. If the kernel of P is non-empty, then P is said to be *star-shaped*.

Lemma 4. *Assume each pocket of P contains at most s points of S for some integer $s \geq 2$. Then there exists a point p in S or in the kernel of P and a set S' of at least $\frac{n}{3s} - 1$ points of S such that (1) the points of S' are strictly separated by a line from p and (2) their counterclockwise radial order around p matches their counterclockwise order along ∂P. Moreover, the points of S' appear consecutively in $S \setminus \{p\}$ both in the counterclockwise radial order around p and in the counterclockwise order of ∂P.*

Proof. Suppose first that P is star-shaped and let p be a point in the kernel of P. Consider any open half-plane H determined by a line ℓ such that p lies on ℓ and H contains at least $\frac{n-1}{2} \geq \frac{n}{3s} - 1$ points of S. The counterclockwise radial order of the points of $S \cap H$ around p matches their counterclockwise order along ∂P. The points of $S \cap H$ are strictly separated from p by a line that is parallel to ℓ and that lies in H sufficiently close to ℓ. Thus, if P is star-shaped, the lemma follows.

Suppose now that P is not star-shaped, i.e., its kernel is empty. Let $e_1, \ldots, e_n$ be the edges of P oriented according to the counterclockwise order of ∂P. For every $i \in \{1, \ldots, n\}$, let H_i be the closed left half-plane of e_i. It is well-known and not difficult to prove that the kernel of P equals the intersection $\cap_{i=1}^{n} H_i$.

Since the kernel of P is empty, Helly's theorem [7] implies that $H_a \cap H_b \cap H_c = \emptyset$ for some $a, b, c \in \{1, \ldots, n\}$. It follows that there exists an $i \in \{a, b, c\}$ such that the complement of H_i contains at least $n/3$ points of S.

Let K be the pocket that contains e_i and let p be the first vertex of e_i. Since K contains at most s points of S and two of them are the endpoints of e_i, the rays $\overrightarrow{pq}$, $q \in K \cap S$, partition the complement of H_i into at most $s - 1$ wedges. By the pigeonhole principle, one of these wedges contains a set S_0 of at least $\frac{n/3 - (s-2)}{s-1} \geq \frac{n}{3s} - 1$ points of $S \setminus K$. Due to Lemma 3, the counterclockwise order of the points of S_0 along ∂P matches their counterclockwise radial order around p.

The points of S_0 are strictly separated from p by a line that is parallel to e_i and that lies in the complement of H_i sufficiently close to e_i. This finishes the proof if P is not star-shaped. $\square$

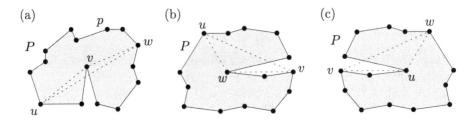

Fig. 3. Examples of reversed triples (u, v, w).

In the proof of Lemma 4, if P is star-shaped and the point p is not part of S, we can define a point set S' consisting of p and $S \cap H$, where H is the half-plane from the proof of Lemma 4. Then, it is easy to see that there is a 2-convex polygonization P' of S' in which p sees all the points in the order as they appear along $\partial P'$. Any k-hole in S' contains a $(k-1)$-hole in S. Hence the size of a largest hole in S' exceeds the size of a largest hole in S by at most 1. Thus, for simplicity, we assume that $p \in S$.

Let $\phi \subseteq S^3$ be the ternary relation representing the counterclockwise cyclic order of the vertices of P along ∂P. That is, a triple (u, v, w) of points of S is in ϕ if we can trace u, v, w in this order along ∂P in counterclockwise direction. For $u, w \in S$, an *interval* $[u, w]$ from u to w in ϕ is the set $\{v \in S : (u, v, w) \in \phi\} \cup \{u, w\}$. For each point $u \in S$, we define a linear order $<_u$ on $S \setminus \{u\}$ where $x <_u y$ if and only if $(x, y, u) \in \phi$.

The vertices of a pocket $K = \langle p_0, \dots, p_t \rangle$ of P induce a closed interval $[p_0, p_t]$ in ϕ. Consequently, ϕ induces a (counterclockwise) cyclic order of pockets of P. We choose an arbitrary pocket K_0 of P and use $K_0, K_1, \dots, K_{m-1}$ to denote this cyclic order where m is the number of pockets of P. In the rest of the section, the indices of pockets are always taken modulo m.

For $r, s \in \{0, \dots, m-1\}$, we use $[K_r, K_s]$ to denote the set consisting of pockets $K_r, K_{r+1}, \dots, K_s$. We call the set $[K_r, K_s]$ an *interval of pockets*. A *subinterval* of $[K_r, K_s]$ is any interval of pockets that can be obtained from $[K_r, K_s]$ by deleting the first i and the last j consecutive pockets of $[K_r, K_s]$ for some $i, j \in \mathbb{N}_0$.

We say that a triple $(u, v, w) \in \phi$ is *reversed* if the triangle with the vertices u, v, w traced in this order is oriented clockwise; see Fig. 3.

The following definition is crucial in our proof; see Fig. 4.

Definition 1. *For an interval of pockets $[K_r, K_s]$, a point p from $S \setminus (\cup_{i=r-1}^{s+1} K_i)$ controls $[K_r, K_s]$ if the following two conditions are satisfied:*

(C1) *If a triple (x, y, p) with $x, y \in \cup_{i=r}^s K_i$ is reversed, then x and y lie in the same pocket K and none of them is an endpoint of K,*
(C1) *$CH(\cup_{i=r}^s K_i)$ contains no point of $S \setminus (\cup_{i=r}^s K_i)$.*

Observe that if p controls $[K_r, K_s]$, then p also controls every subinterval of $[K_r, K_s]$. Further, Condition (C2) implies that p is strictly separated from the points of $[K_r, K_s]$ by a line.

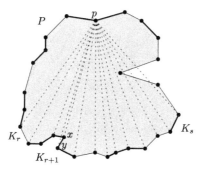

Fig. 4. The point p controls the interval $[K_r, K_s]$ of pockets. The pockets of P are drawn alternatingly along ∂P by thick and thin segments. The triple (x, y, p) is reversed.

We note that for a reversed triple (u, v, w) the point v might be a vertex of $CH(S)$; see parts (b) and (c) of Fig. 3. However, if the interval of pockets that contains the points from $[u, v]$ is controlled by some point p from S (see, for example, part (a) of Fig. 3), then v has to lie in the interior of $CH(S)$ due to Condition (C2).

Lemma 5. *Let (u, v, w) be a reversed triple of points in S and let K', K, and K'' be arbitrary pockets that contain u, v, w, respectively. Let ab be the lid of the pocket K such that $(a, v, b) \in \phi$ and such that the line $\overline{uw}$ weakly separates v from ab. Assume that $[K', K'']$ is controlled by some point $p \in S$. Then the order $<_v$ matches the radial order around v for $[u, a]$ and for $[b, w]$.*

We omit the proof of Lemma 5.

Lemma 6. *Let K_i, K_j, and K_l be pockets appearing in this order in an interval of pockets that is controlled by a point $p \in S$. Let (u, v, w) be a reversed triple of points from S such that u, v, and w lie in K_i, K_j, and K_l, respectively, and such that $\overline{uw}$ weakly separates v from the endpoints of K_j. Then v controls the intervals $[K_{i+1}, K_{j-2}]$ and $[K_{j+2}, K_{l-1}]$.*

Proof. It suffices to show that v controls $[K_{i+1}, K_{j-2}]$, as the other case is analogous. The statement is trivial for $[K_i, K_j]$ with at most three pockets, thus we assume that $[K_i, K_j]$ contains at least four pockets.

We first show that the line $\overline{uw}$ weakly separates v from $K_{i+1} \cup \cdots \cup K_{j-2}$. Let H^- and H^+ be the open half-planes determined by $\overline{uw}$ such that $v \in H^-$. It follows from the assumptions of the lemma that the endpoints of K_j are contained in $H^+ \cup \{w\}$ and thus $\overline{uw}$ intersects the pocket K_j in at least two points. The line $\overline{uw}$ (or its slight perturbation) has two additional crossings with ∂P and thus cannot intersect ∂P in any further point. Consequently, all vertices of pockets from $[K_{i+1}, K_{j-2}]$ lie in $H^+ \cup \{w\}$; see Fig. 5(a).

We now show that Conditions (C1) and (C2) are fulfilled for the point v and the interval $[K_{i+1}, K_{j-2}]$. Let ab be the lid of K_j and let C be the part of ∂P between u and a. Since v is weakly separated from C by a line, Lemma 5 implies

(a)   (b)

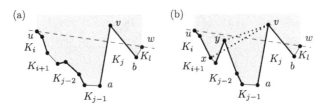

Fig. 5. Situations in the proof of Lemma 6.

that for any $x, y \in [u, a]$, (x, y, v) is not a reversed triple; see Fig. 5(b). Thus, Condition (C1) is satisfied.

Condition (C2) is also satisfied, since, by assumption, p controls the interval of pockets that contains K_i and K_j. In particular, p controls the subinterval $[K_{i+1}, K_{j-2}]$ and thus $CH(K_{i+1} \cup \cdots \cup K_{j-2})$ contains no point of $S \setminus (K_{i+1} \cup \cdots \cup K_{j-2})$. □

Lemma 7. *Let $[K_r, K_{r+3d+3}]$ be an interval of $3d+4$ pockets controlled by some point $p \in S$. Then it contains a subinterval with d pockets controlled by a point of a pocket from $[K_r, K_{r+3d+3}]$.*

The proof of Lemma 7 uses Lemma 6 and is omitted.

Lemma 8. *Let $k \geq 2$ and let $K_{i_1}, \ldots, K_{i_{k+1}}$ be distinct $k+1$ pockets traced in counterclockwise order along ∂P. Then the following two statements are true.*

(i) *If, for every $j \in \{3, \ldots, k+1\}$, we have a point $q_j \in K_{i_j}$ that controls $[K_{i_1}, K_{i_{j-1}}]$, then there is a k-hole in S.*

(ii) *If, for every $j \in \{1, \ldots, k-1\}$, we have a point $q_j \in K_{i_j}$ that controls $[K_{i_{j+1}}, K_{i_{k+1}}]$, then there is a k-hole in S.*

The proof of Lemma 8 is omitted. The following lemma is the last ingredient for our proof of Theorem 1.

Lemma 9. *For every positive integer k and every interval $[K_r, K_s]$ of pockets it holds that, if $[K_r, K_s]$ consists of at least $2 \cdot 3^{2k} - 2$ pockets and $[K_r, K_s]$ is controlled by some point of S, then $[K_r, K_s]$ contains a k-hole in S.*

Proof. Let $[K_r, K_s]$ be an interval of at least $2 \cdot 3^t - 2$ pockets for some positive integer t. Assume that there is a point of S that controls $[K_r, K_s]$. First, we show by induction on t that there are distinct pockets $K'_1, \ldots, K'_t$ in $[K_r, K_s]$ (not necessarily appearing in counterclockwise order along ∂P) such that for every $l \in \{2, \ldots, t\}$ there is a point $q_l \in K'_l$ that controls a subinterval I_l of $[K_r, K_s]$ that contains the pockets $K'_1, \ldots, K'_{l-1}$.

The statement is trivial for $t = 1$. For the induction step, we assume that $t \geq 2$. We let $d := 2 \cdot 3^{t-1} - 2$ and we note that $[K_r, K_s]$ consists of $3d + 4 = 2 \cdot 3^t - 2$ pockets. By Lemma 7, there is a point q_t contained in a pocket

K'_t from $[K_r, K_s]$ such that q_t controls a subinterval I_t of $[K_r, K_s]$ with at least d pockets. Using the induction hypothesis, it follows that I_t contains distinct pockets $K'_1, \ldots, K'_{t-1}$ such that for every $l \in \{2, \ldots, t-1\}$ there is a point $q_l \in K'_l$ that controls a subinterval I_l of I_{l+1} that contains the pockets $K_1, \ldots, K_{l-1}$. The pockets $K'_1, \ldots, K'_t$ then satisfy the statement.

Every pocket K'_l for $l \in \{2, \ldots, t\}$ either precedes I_l or succeeds I_l in the interval $[K_r, K_s]$ of pockets in the counterclockwise order of ∂P. Thus if $t \geq 2k$, we obtain distinct pockets $K_{i_1}, \ldots, K_{i_{k+1}}$ traced in this order along ∂P such that one the following two conditions holds. Either, for every $j \in \{3, \ldots, k+1\}$, there is a point $q_j \in K_{i_j}$ that controls $[K_{i_1}, K_{i_{j-1}}]$ or, for every $j \in \{1, \ldots, k-1\}$, there is a point $q_j \in K_{i_j}$ that controls $[K_{i_{j+1}}, K_{i_{k+1}}]$. In the first case, we apply part (i) of Lemma 8 and obtain a k-hole in S. In the latter case, part (ii) of Lemma 8 gives us a k-hole in S. □

Proof of Theorem 1. To show Theorem 1, we prove that in every 2-convex point set S of size n there is a k-hole for $k \geq \log n / 3$, or we have an interval with $\Omega(n / \log^3 n)$ pockets that is controlled by a point of S. In the latter case we then apply Lemma 9 and obtain a k-hole with $k \geq c \log n$ for a fixed constant $c > 0$.

First, assume that there is a pocket $K = \langle p_0, \ldots, p_t \rangle$ in P with $t \geq \log n$ in P. By Lemma 1, the pocket K can be split into three chains $C_1 = \langle p_0, p_1, \ldots, p_r \rangle$, $C_2 = \langle p_{r+1}, \ldots, p_s \rangle$, and $C_3 = \langle p_{s+1}, \ldots, p_t \rangle$ for $0 \leq r \leq s < t$, such that all vertices in C_1 and C_3 are convex in P, while all vertices in C_2 are reflex. Since K contains at least $\log n$ vertices, at least one chain C_i, for some $i \in \{1, 2, 3\}$, contains at least $\log n / 3$ vertices. By part (ii) of Lemma 2, the vertices of C_i are vertices of a k-hole for $k \geq \log n / 3$; see Fig. 6(a).

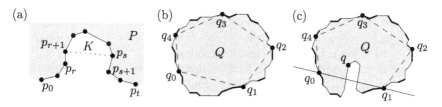

Fig. 6. (a) A large pocket gives a large hole. (b) If no point of S interferes, then Q is a hole. (c) If there is a point inside Q, then we use Lemma 6 and apply Lemma 9.

In the rest of the proof we thus assume that every pocket of P contains less than $\log n$ vertices. In particular, there are more than $n / \log n$ pockets in P and $\mathrm{CH}(S)$ has more than $n / \log n$ vertices. By Lemma 4, there are at least $z := \frac{n}{3 \log n} - 1$ points that are strictly separated by a line from a point p (that is not necessarily in S) and their counterclockwise radial order around p matches their counterclockwise order along ∂P. We call these points the *initial interval*. However, by the discussion after Lemma 4 we can assume for the following that $p \in S$. Let $q_0, \ldots, q_{\log n - 1}$ be vertices of $\mathrm{CH}(S)$ traced in counterclockwise direction along ∂P in the initial interval such that the points in each interval $[q_i, q_{i+1}]$

for $i = 0, \ldots, \log n - 1$ (indices taken modulo $\log n$) form at least $z / \log^2 n$ pockets. Clearly, if the polygon Q with the vertices $q_0, \ldots, q_{\log n-1}$ is a hole in S, then we are done; see Fig. 6(b). Otherwise there is a point q in the interior of Q and we have a reversed triple (q_i, q, q_j) for some $i, j \in \{0, \ldots, \log n - 1\}$. Let K, K', and K'' be pockets containing q_i, q, and q_j, respectively. The endpoints of K' are weakly separated from q by $\overline{q_i q_j}$, as q_i and q_j are vertices of $\mathrm{CH}(S)$; see Fig. 6(c). By Lemma 6, the point q controls the interval of pockets that are between K and K' and between K' and K''. From the choice of Q, at least one of these intervals consists of at least $z / (2 \log^2 n) = \Omega(n / \log^3 n)$ pockets. Hence, by Lemma 9, it contains a hole of size $\Omega(\log n)$. □

We note that the proof of Theorem 1 yields an algorithm that, given a 2-convex set S of n points, finds a hole of size $\Omega(\log n)$ in polynomial time with respect to n.

4   An Upper-Bound Construction

Theorem 2. *For every integer n there exists a 2-convex point set S of size n such that the largest holes in S have size $\Theta(\log n)$.*

Proof. The set is constructed recursively, following the idea shown in Fig. 7. The set S_0 is a convex pentagon with vertices $\{a, b, c, d, e\}$. For an integer $k \geq 1$ we construct $S_k := S_0 \cup L_k \cup R_k$ in the following way: L_k is a set with the same order type as S_{k-1}, located outside the convex hull of S_0 and flat enough so that no line defined by two points in L_k intersects the convex hull of S_0 (R_k is defined in an analogous way). Let $\mathcal{P}_k$ be the polygonization of S_k given by the angular order around c. Because no line intersecting more than one edge of $\mathcal{P}_k$ in L_k intersects the rest of $\mathcal{P}_k$ (and the same is true for R_k), it can be shown that the set S_k is 2-convex. We observe that $|S_k| = 5(2^{k+1} - 1)$ for every $k \geq 0$.

Let $L(S_k) := \{a\} \cup \{b\} \cup L_k$ be the left component of S_k and similarly let $R(S_k) := \{d\} \cup \{e\} \cup R_k$ be the right component of S_k.

A hole H containing points of $R(S_k)$ cannot contain points of both the left and the right component of L_k (because it would contain the point corresponding to c in L_k). The same is true for every level in the recursion and an analogous statement holds if $H \cap L(S_k) \neq \emptyset$. Hence, the number of vertices of any hole in S_k is at most linear in k. Therefore, if we set k to be the smallest integer such that $|S_k|$ is at least n, we can remove the rightmost points of S_k to obtain a set of size n and have that $|H| = O(\log n)$. Due to Theorem 1 this bound is tight. □

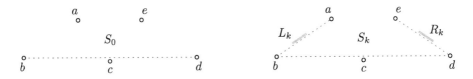

Fig. 7. Recursive operation for the construction of an upper bound example.

5   Convexity of Horton Sets

Considering holes in point sets, the Horton sets are an ubiquitous class of point sets, defined in [10,11]. A Horton set contains no 7-hole. We say that a point set U is *high above* another point set D if all points of D lie below every line through two points of U, and all points of U lie above every line through two points of D. A *Horton set* consists either of a single point or of two Horton sets U and D where U is high above D and by traversing the points by increasing x-coordinate, the points of U and D are encountered alternatingly. We show that no Horton set of size $n \geq 256$ is $\lfloor \sqrt{n}/8 - 1 \rfloor$-convex. As every set of n points is $O(\sqrt{n})$-convex [1, Theorem 2], this bound is asymptotically tight.

We use the following lemma, already used in [1] in a slightly less general form.

Lemma 10. *Let S be a set of n points for which there exists an arrangement of l lines such that no two points of S lie in the same cell of the arrangement. Then S is not $(\lceil \frac{n}{2l} \rceil - 1)$-convex.*

Proof. Every edge of a polygonization of S crosses at least one line of that arrangement, and thus there exists a line with at least $\lceil \frac{n}{l} \rceil$ crossings.   □

Using this result, we prove the following lower bound on the convexity of Horton sets.

Theorem 3. *No Horton set of size $n \geq 256$ is $\lfloor \sqrt{n}/8 - 1 \rfloor$-convex.*

Proof. We may assume that we have a Horton set of size 2^{2a} for some integer $a \geq 4$, as otherwise we may remove at most $3n/4$ rightmost points and use the fact that every subset of a k-convex point set is k-convex [1, Lemma 2].

We partition the Horton set by $2^a - 1$ vertical and $2^a - 1$ horizontal lines in the following way. We place a horizontal line strictly separating the upper from the lower Horton subset. We recursively repeat this on the two subsets defined by the line until we have placed $1 + 2 + 4 + \cdots + 2^{a-1} = 2^a - 1$ lines, thereby partitioning the original set into Horton sets of size $2^{2a}/2^a = 2^a$. Now we place a vertical line strictly separating the 2^a leftmost points from the remaining ones. Observe that any two points to the left of the vertical line are strictly separated by a horizontal line, as the points as encountered from left to right have to be alternatingly from the lower and the upper set, and this also holds for the subsets. This also holds for another vertical line strictly separating the next leftmost 2^a points and so on. Thus, after placing $2^a - 1$ vertical lines (i.e., $2^{a+1} - 2$ lines in total), we have an arrangement such that each point of the set is in a different cell.

By Lemma 10, we thus know that there is no $\lceil 2^{2a}/(2^{a+2} - 4) - 1 \rceil$-convex polygonization of a Horton set with 2^{2a} points. As every subset of a k-convex set is k-convex, we see that the original Horton set on n points is not $\lceil 2^{2a}/(2^{a+2} - 4) - 1 \rceil$-convex. Since $n \leq 2^{2a+2}$, we have

$$\left\lceil \frac{2^{2a}}{2^{a+2} - 4} - 1 \right\rceil \geq \frac{2^{2a}}{2^{a+2}} - 1 = 2^{a-2} - 1 \geq \sqrt{n}/8 - 1.$$

Since $n \geq 256$, the latter expression is at least 1, which completes the proof. □

Let $\lambda(S)$ be the minimum number of lines needed to strictly separate each pair of points of a point set S. Lemma 10 states that there is a lower bound $k \geq \Omega(|S|/\lambda(S))$ on the k-convexity of S. Since every set S is $O(\sqrt{|S|})$-convex [1, Theorem 2] and we clearly have $\lambda(S) \leq O(|S|)$, every set S is k-convex for $k \leq O(|S|^{3/2}/\lambda(S))$. To see that it is tight, consider a set of n points consisting of a set of $n/2$ points in convex position and of a Horton set on $n/2$ points.

The relation between l-convexity and the number of points needed to guarantee a k-hole seems to be an intriguing and difficult problem. We have shown that if the set is not convex, but 2-convex, n points are enough to guarantee a hole with size $\Omega(\log n)$. Moreover, if the set is not $\lfloor \sqrt{n}/8 - 1 \rfloor$-convex not even a 7-hole can be guaranteed for arbitrarily large sets (as the Horton set shows). It would be very interesting to get some bounds analogous to the bounds in [12] for the parameter $\tilde{N}(l, k)$, the minimum size of an l-convex set where the existence of a k-hole can be guaranteed. However, this seems to be hard already for the case of 3-convex point sets, as properties of 3-convex polygons analogous to the properties presented in Sect. 2 are not known.

Acknowledgments. This work was initiated during the ComPoSe Workshop on Algorithms using the Point Set Order Type held in March/April 2014 in Ratsch, Austria.

References

1. Aichholzer, O., Aurenhammer, F., Hackl, T., Hurtado, F., Pilz, A., Ramos, P., Urrutia, J., Valtr, P., Vogtenhuber, B.: On k-convex point sets. Comput. Geom. **47**(8), 809–832 (2014)
2. Aichholzer, O., Aurenhammer, F., Demaine, E.D., Hurtado, F., Ramos, P., Urrutia, J.: On k-convex polygons. Comput. Geom. **45**(3), 73–87 (2012)
3. Aichholzer, O., Fabila-Monroy, R., González-Aguilar, H., Hackl, T., Heredia, M.A., Huemer, C., Urrutia, J., Valtr, P., Vogtenhuber, B.: On k-gons and k-holes in point sets. Comput. Geom. **48**(7), 528–537 (2015)
4. Erdős, P.: Some more problems on elementary geometry. Austral. Math. Soc. Gaz. **5**(2), 52–54 (1978)
5. Gerken, T.: Empty convex hexagons in planar point sets. Discrete Comput. Geom. **39**(1–3), 239–272 (2008)
6. Harborth, H.: Konvexe Fünfecke in ebenen Punktmengen. Elem. Math. **33**(5), 116–118 (1978). In German
7. Helly, E.: Über Mengen konvexer Körper mit gemeinschaftlichen Punkten. Jahresber. Dtsch. Math. Ver. **32**, 175–176 (1923). In German
8. Horton, J.D.: Sets with no empty convex 7-gons. Canad. Math. Bull. **26**(4), 482–484 (1983)
9. Nicolás, C.M.: The empty hexagon theorem. Discrete Comput. Geom. **38**(2), 389–397 (2007)

10. Valtr, P.: Convex independent sets and 7-holes in restricted planar point sets. Discrete Comput. Geom. **7**(2), 135–152 (1992)
11. Valtr, P.: Sets in $\mathbb{R}^d$ with no large empty convex subsets. Discrete Math. **108**(1–3), 115–124 (1992)
12. Valtr, P.: A sufficient condition for the existence of large empty convex polygons. Discrete Comput. Geom. **28**(4), 671–682 (2002)

Graphs and Combinatorics

Graph Parameters and Ramsey Theory

Vadim Lozin[1,2]($\boxtimes$)

[1] Mathematics Institute, University of Warwick, Coventry CV4 7AL, UK
V.Lozin@warwick.ac.uk
[2] Lobachevsky State University of Nizhni Novgorod, Nizhny Novgorod, Russia

Abstract. Ramsey's Theorem tells us that there are exactly two minimal hereditary classes containing graphs with arbitrarily many vertices: the class of complete graphs and the class of edgeless graphs. In other words, Ramsey's Theorem characterizes the graph vertex number in terms of minimal hereditary classes where this parameter is unbounded. In the present paper, we show that a similar Ramsey-type characterization is possible for a number of other graph parameters, including neighbourhood diversity and VC-dimension.

1 Introduction

In 1930, a 26 years old British mathematician Frank Ramsey proved the following theorem, known nowadays as Ramsey's Theorem.

Theorem 1 [12]**.** *For any positive integers k, r, p, there exists a minimum positive integer $F = F(k, r, p)$ such that if the k-subsets of an F-set are colored with r colors, then there is a monochromatic p-set, i.e. a p-set all of whose k-subsets have the same color.*

It is not difficult to see that with $k = 1$ this theorem coincides with the Pigeonhole Principle. For $k = 2$, the theorem admits a nice interpretation in the terminology of graph theory, since coloring 2-subsets can be viewed as coloring the edges of a complete graph. In the case of $r = 2$ colors, the graph-theoretic interpretation of Ramsey's Theorem can be further rephrased as follows.

Theorem 2. *For any positive integer p, there is a minimum positive integer $R = R(p)$ such that every graph with at least R vertices has either a clique of size p or an independent set of size p.*

It is not difficult to see that Theorem 2 is equivalent to the following statement.

Theorem 3. *The class of complete graphs and the class of edgeless graphs are the only two minimal infinite hereditary classes of graphs.*

© Springer International Publishing AG, part of Springer Nature 2018
L. Brankovic et al. (Eds.): IWOCA 2017, LNCS 10765, pp. 185–194, 2018.
https://doi.org/10.1007/978-3-319-78825-8_15

Theorem 3 characterizes the family of hereditary classes containing graphs with a bounded number of vertices in terms of minimal "forbidden" elements, i.e. minimal hereditary classes where the vertex number is unbounded. Is it possible to find a similar characterization for other graph parameters?

The purpose of this paper is to show that Ramsey's theorem can be used to find minimal "forbidden" classes for various other parameters. For instance, directly from Ramsey's Theorem it follows that

- the class of complete graphs and the class of stars (and all their induced subgraphs) are the only two minimal hereditary classes of graphs of unbounded vertex degree.
- the class of complete graphs and the class of complete bipartite graphs are the only two minimal hereditary classes of graphs of unbounded biclique number (the size a maximum complete bipartite subgraph with equal parts).

In the subsequent sections, we show that a similar Ramsey-type characterization is possible for a number of other graph parameters. In the rest of the present section, we introduce basic definitions and notations used in the paper.

We consider only simple undirected graphs without loops and multiple edges and denote the vertex set and the edge set of a graph G by $V(G)$ and $E(G)$, respectively. If v is a vertex of G, then $N(v)$ is its neighbourhood, i.e. the set of vertices of G adjacent to v. The closed neighbourhood of v is defined and is denoted as $N[v] = N(v) \cup \{v\}$. The degree of v is $|N(v)|$.

In a graph, an *independent set* is a subset of vertices no two of which are adjacent, a *clique* is a subset of vertices every two of which are adjacent, and a *matching* is a subset of edges no two which share a vertex.

For a graph G, we denote by $\overline{G}$ the complement of G. Similarly, for a class $\mathcal{X}$ of graphs, we denote by $\overline{\mathcal{X}}$ the class of complements of graphs in $\mathcal{X}$.

Given a graph G and a subset $U \subseteq V(G)$, we denote by $G[U]$ the subgraph of G induced by U, i.e. the subgraph obtained from G by deleting all the vertices not in U. We say that a graph G contains a graph H as an induced subgraph if H is isomorphic to an induced subgraph of G. A graph G is said to be n-universal for a class of graphs $\mathcal{X}$ if G contains all n-vertex graphs from $\mathcal{X}$ as induced subgraphs.

A class $\mathcal{X}$ of graphs is *hereditary* if it is closed under taking induced subgraphs, i.e. if $G \in \mathcal{X}$ implies $H \in \mathcal{X}$ for every graph H contained in G as an induced subgraph. Two hereditary classes of particular interest in this paper are split graphs and bipartite graphs.

A graph G is a *split* graph if $V(G)$ can be partitioned into an independent set and a clique, and G is *bipartite* if $V(G)$ can be partitioned into at most two independent sets. A bipartite graph G given together with a bipartition of its vertices into independent sets A and B will be denoted $G = (A, B, E)$, in which case we will say that A and B are the color classes of G. If every vertex of A is adjacent to every vertex of B, then $G = (A, B, E)$ is *complete bipartite*. The *bipartite complement* of $G = (A, B, E)$ is the bipartite graph $G' = (A, B, E')$, where $ab \in E'$ if and only if $ab \notin E$. Clearly, by creating a clique in one of the color classes of a bipartite graph, we transform it into a split graph, and vice versa.

2 Neighbourhood Diversity

The neighbourhood diversity of a graph was introduced in [7] and was used to develop fpt-algorithms for some difficult graph problems (see e.g. [5]). This parameter can be defined as follows.

Definition 1. *Two vertices x and y are said to be similar if there is no vertex z distinguishing them i.e. if there is no vertex z adjacent to exactly one of x and y. Clearly, the similarity is an equivalence relation. The neighbourhood diversity of G is the number of similarity classes in G.*

Before we provide a Ramsey-type characterization of the neighbourhood diversity, we introduce an auxiliary notion.

Definition 2. *A skew matching in a graph G is a matching $\{x_1 y_1, \ldots, x_q y_q\}$ such that y_i is not adjacent to x_j for all $i < j$. The complement of a skew matching is a sequence of pairs of vertices that create a skew matching in the complement of G.*

Lemma 1. *For any positive integer m, there exists a positive integer $r = r(m)$ such that any bipartite graph $G = (A, B, E)$ of neighbourhood diversity r contains either a skew matching of size m or its complement.*

Proof. Define $r = 2^{2m}$ and let X be a set of pairwise non-similar vertices of size $r/2$ chosen from the same color class of G, say from A. Let y_1 be a vertex in B distinguishing the set X (i.e. y_1 has both a neighbour and a non-neighbour in X) and let us say that y_1 is *big* if the number of its neighbours in X is larger than the number of its non-neighbours, and *small* otherwise. If y_1 is small, we arbitrarily choose its neighbour in X, denote it by x_1 and remove all neighbours of y_1 from X. If y is big, we arbitrarily choose a non-neighbour of y_1 in X, denote it by x_1 and remove all non-neighbours of y_1 from X. Observe that y_1 does not distinguish the vertices in the updated set X.

We apply the above procedure to X $2m - 1$ times and obtain in this way a sequence of $2m - 1$ pairs $x_i y_i$. If m of these pairs contain small vertices y_i, then the respective pairs create a skew matching of size m. Otherwise, there is a set of m pairs containing big vertices y_i, in which case the respective pairs create the complement of a skew matching. □

Now we turn to the neighbourhood diversity and start with the bipartite case. For this, we denote by

$\mathcal{M}$ the class of graphs of vertex degree at most 1. By M_n we denote an induced matching of size n, i.e. the unique up to isomorphism graph in the class $\mathcal{M}$ with $2n$ vertices each of which has degree 1. Clearly, M_n is n-universal for graphs in $\mathcal{M}$.

$\mathcal{M}^{bc}$ the class of bipartite complements of graphs in $\mathcal{M}$. The bipartite complement of the graph M_n will be denoted M_n^{bc}. Clearly, M_n^{bc} is n-universal for graphs in $\mathcal{M}^{bc}$.

$\mathcal{Z}$ the class of *chain graphs*, i.e. bipartite graphs in which the neighbourhoods of the vertices in each part form a chain with respect to set-inclusion. By Z_n we denote a chain graph such that for each $i \in \{1, 2, \ldots, n\}$, each part of the graph contains exactly one vertex of degree i. Figure 1 represents the graph Z_n for $n = 5$. It is known [9] that Z_n is n-universal for graphs in $\mathcal{Z}$.

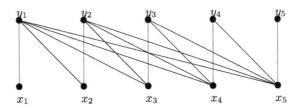

Fig. 1. The graph Z_5

Lemma 2. *For any positive integer p, there exists a positive integer $q = q(p)$ such that any bipartite graph $G = (A, B, E)$ of neighbourhood diversity q contains either an induced M_p or an induced Z_p or an induced M_p^{bc}.*

Proof. Let $m = R(p+1)$ (where R is the Ramsey number defined in Theorem 2) and $q = 2^{2m}$. According to the proof of Lemma 1, G contains a skew matching of size m or its complement. If G contains a skew matching M, we color each pair $(x_i y_i, x_j y_j)$ of edges of M $(i < j)$ in two colors as follows:

- color 1 if x_i is not adjacent to y_j,
- color 2 if x_i is adjacent to y_j.

By Ramsey's Theorem, M contains a monochromatic set M' of edges of size $p + 1$. If the color of each pair of edges in M' is

1. then M' is an induced matching of size $p + 1$,
2. then the vertices of M' induce a Z_{p+1}.

Analogously, in the case when G contains the complement of a skew matching, we find either an induced M_{p+1}^{bc} or an induced Z_p (observe that the bipartite complement of Z_{p+1} contains an induced Z_p). $\qquad\square$

Now we proceed to the general case and denote by

$\mathcal{M}^*$ the class of split graphs obtained from graphs in $\mathcal{M}$ by creating a clique in one of the color classes. The graph obtained from M_n by creating clique in one its color classes will be denoted by M_n^*. Clearly, M_n^* is n-universal for graphs in $\mathcal{M}^*$.

$\mathcal{Z}^*$ the class of split graphs obtained from graphs in $\mathcal{Z}$ by creating a clique in one of the color classes. This class is known in the literature as the class of *threshold graphs*. The graph obtained from Z_n by creating a clique in one of its color classes will be denoted Z_n^*. This graph is n-universal for threshold graphs [6].

Lemma 3. *For any positive integer p, there exists a positive integer $Q = Q(p)$ such that every graph G of neighbourhood diversity Q contains one of the following nine graphs as an induced subgraph: M_p, M_p^{bc}, Z_p, $\overline{M}_p$, $\overline{M}_p^{bc}$, $\overline{Z}_p$, M_p^*, $\overline{M}_p^*$, Z_p^*.*

Proof. Let $Q = R(q)$, where $q = 2^{2m}$ and $m = R(R(p) + 1)$ (R is the Ramsey number). We choose one vertex from each similarity class of G and find in the chosen set a subset A of vertices that form an independent set or a clique of size $q = 2^{2m}$. Let us call the vertices of A white. We denote the remaining vertices of G by B and call them black. Let G' denote the bipartite subgraph of G formed by the edges between A and B. By the choice of A, all vertices of this set have pairwise different neighbourhoods in G'. Therefore, according to the proof of Lemma 2, G' contains a subgraph G'' inducing either M_n, or M_n^{bc} or Z_n with $n = R(p)$. Among the n black vertices of G'', we can find a subset B' of vertices that form either a clique or an independent set of size p in the graph G. Then B' together with a subset of A of size p induce in G one of the nine graphs listed in the statement of the lemma.   □

Since the nine graphs of Lemma 3 are universal for their respective classes, we make the following conclusion.

Theorem 4. *There exist exactly nine minimal hereditary classes of graphs of unbounded neighbourhood diversity: $\mathcal{M}$, $\mathcal{M}^{bc}$, $\mathcal{Z}$, $\overline{\mathcal{M}}$, $\overline{\mathcal{M}}^{bc}$, $\overline{\mathcal{Z}}$, $\mathcal{M}^*$, $\overline{\mathcal{M}}^*$, $\mathcal{Z}^*$.*

3   VC-Dimension

A set system (X, S) consists of a set X and a family S of subsets of X. A subset $A \subseteq X$ is *shattered* if for every subset $B \subseteq A$ there is a set $C \in S$ such that $B = A \cap C$. The VC-dimension of (X, S) is the cardinality of a largest shattered subset of X.

The VC-dimension of a graph $G = (V, E)$ was defined in [1] as the VC-dimension of the set system (V, S), where S the family of closed neighbourhoods of vertices of G, i.e. $S = \{N[v] \; : \; v \in V(G)\}$. Let us denote the VC-dimension of G by $vc[G]$.

In this section, we characterize VC-dimension by means of three minimal hereditary classes where this parameter is unbounded. To this end, we first redefine it in terms of open neighbourhoods as follows. Let $vc(G)$ be the size of a largest set A of vertices of G such that for any subset $B \subseteq A$ there is a vertex v *outside of A* with $B = A \cap N(v)$. In other words, $vc(G)$ is the size of a largest subset of vertices shattered by *open* neighbourhoods of vertices of G.

We start by showing that the two definitions are equivalent in the sense that they both are either bounded or unbounded in a hereditary class. To prove this, we introduce the following terminology. Let A be a set of vertices which is shattered by a collection of closed neighbourhoods. For a subset $B \subseteq A$ we will denote by $v(B)$ the vertex whose neighbourhood intersect A at B. We will say that B is *closed* if $v(B)$ belongs to B, and *open* otherwise.

Lemma 4. $vc(G) \leq vc[G] \leq vc(G)(vc(G) + 1) + 1$

Proof. The first inequality is obvious. To prove the second one, let A be a subset of $V(G)$ of size $vc[G]$ which is shattered by a collection of closed neighbourhoods. If A has no closed subsets, then $vc[G] = vc(G)$. Otherwise, let B be a closed subset of A.

Assume first that $|B| = 1$. Then $B = \{v(B)\}$ and $v(B)$ is isolated in $G[A]$, i.e. it has no neighbours in A. Let C be the set of all such vertices, i.e. vertices each of which is a closed subset of A. By removing from A any vertex $x \in C$ we obtain a new set A and may assume that it has no closed subsets of size 1. Indeed, for any vertex $y \in C$ different from x, there must exist a vertex $y' \notin A$ such that $N(y') \cap A = \{x, y\}$ (since A is shattered). After the removal of x from A, we have $N(y') \cap A = \{y\}$ and hence $\{y\}$ is not a closed subset anymore. This discussion allows us to assume in what follows that A has no closed subsets of size 1, in which case we only need to show that $vc[G] \leq vc(G)(vc(G) + 1)$.

Assume now that B is a closed subset of A of size at least 2. Suppose that $B - v(B)$ contains a closed subset C, i.e. $v(C) \in C$. Observe that $v(C)$ is adjacent to $v(B)$, as every vertex of $B - v(B)$ is adjacent to $v(B)$. But then $N[v(C)] \cap A$ contains $v(B)$ contradicting the fact that $N[v(C)] \cap A = C$. This contradiction shows that every subset of $B - v(B)$ is open, i.e. $|B - v(B)| \leq vc(G)$.

The above observation allows us to apply the following procedure: as long as A contains a closed subset B with at least two vertices, delete from A all vertices of B except for $v(B)$. Denote the resulting set by A^*. Assume the procedure was applied p times and let $B_1, \ldots, B_p$ be the closed subsets it was applied to. It is not difficult to see that the set $\{v(B_1), \ldots, v(B_p)\}$ has no closed subsets and hence its size cannot be large than $vc(G)$, i.e. $p \leq vc(G)$. Combining, we conclude:

$$vc[G] = |A| \leq |A^*| + \sum_{i=1}^{p} |B_i - v(B_i)| \leq vc(G) + p \cdot vc(G) \leq vc(G)(vc(G) + 1).$$

□

This lemma allows us to assume that if A is shattered, then there is a set C *disjoint* from A such that for any subset $B \subseteq A$ there is a vertex $v \in C$ with $B = A \cap N(v)$, in which case we will say that A is shattered by C, or C shatters A.

Let $Q_n = (A, B, E)$ be the bipartite graph with $|A| = n$ and $|B| = 2^n$ such that all vertices of B have pairwise different neighbourhood in A. Also, let S_n be the split graph obtained from Q_n by creating a clique in A.

Lemma 5. *The graph Q_n is an n-universal bipartite graph, i.e. it contains every bipartite graph with n vertices as an induced subgraph.*

Proof. Let G be a bipartite graph with n vertices and with parts A and B of size n_1 and n_2, respectively. By adding at most n_2 vertices to A, we can guarantee that all vertices of B have pairwise different neighbourhoods in A. Clearly, Q_n contains the extended graph and hence it also contains G as an induced subgraph.

□

Corollary 1. *Every co-bipartite graph with at most n vertices is contained in $\overline{Q}_n$ and every split graph with at most n vertices is contained both in S_n and in $\overline{S}_n$.*

Lemma 6. *If a set A shatters a set B with $|B| = 2^n$, then B shatters a subset A^* of A with $|A^*| = n$.*

Proof. Without loss of generality we assume that B is the set of all binary sequences of length n. Then every vertex $a \in A$ defines a Boolean function of n variables (the neighbourhood of a consists of the binary sequences, where the function takes value 1). For each $i = 1, \ldots, n$, let us denote by a_i the Boolean function such that $a_i(x_1, \ldots, x_n) = 1$ if and only if $x_i = 1$. Let A' be an arbitrary subset of $A^* = \{a_1, \ldots, a_n\}$ and $\alpha = (\alpha_1, \ldots, \alpha_n)$ its characteristic vector, i.e. $\alpha_i = 1$ if and only if $a_i \in A'$. Clearly, $\alpha \in B$ and $N(\alpha) \cap A^* = A'$. Therefore, B shatters A^*.   □

Lemma 7. *For every n, there exists a $k = k(n)$ such that every graph G with $vc(G) = k$ contains one of $Q_n, \overline{Q}_n, S_n, \overline{S}_n$ as an induced subgraph.*

Proof. Define $k = R(2^{R(n)})$, where R is the Ramsey number. Since $vc(G) = k$, there are two subsets A and B of $V(G)$ such that $|A| = k$ and B shatters A. By definition of k, A must have a subset A' of size $2^{R(n)}$ which is a clique or an independent set. Clearly, B shatters A' and hence, by Lemma 6, A' shatters a subset B' of B of size $R(n)$. Then B' must have a subset B'' of size n which is either a clique or an independent set. Now $G[A' \cup B'']$ is either bipartite or co-bipartite or split graph, $|B''| = n$ and A' shatters B''. Therefore, $G[A' \cup B'']$ contains one of $Q_n, \overline{Q}_n, S_n, \overline{S}_n$ as an induced subgraph.   □

Theorem 5. *The classes of bipartite, co-bipartite and split graphs are the only three minimal hereditary classes of graphs of unbounded VC-dimension.*

Proof. Clearly these three classes have unbounded VC-dimension, since they contain $Q_n, \overline{Q}_n, S_n, \overline{S}_n$ with arbitrarily large values of n.

Now let X be a hereditary class containing none of these three classes. Therefore, there is a bipartite graph G_1, a co-bipartite graph G_2 and a split graph G_3 which are forbidden for X. Denote by n the maximum number of vertices in these graphs.

Assume that VC-dimension is not bounded for graphs in X and let $G \in X$ be a graph with $vc(G) = k$, where $k = k(n)$ is from Lemma 7. Then G contains one of $Q_n, \overline{Q}_n, S_n, \overline{S}_n$, say Q_n. Since Q_n is n-universal for bipartite graphs (Lemma 5), it contains G_1 as an induced subgraph, which is impossible because G_1 is forbidden for graphs in X. This contradiction shows that VC-dimension is bounded in the class X.   □

4   More Results and Discussion

In [4], it was shown that for every t, p, s, there exists a $z = z(t, p, s)$ such that every graph with a (not necessarily induced) matching of size at least z contains either an induced matching of size t or an induced complete bipartite graph with color classes of size p or a clique of size s. This result was used in [4] to develop fpt-algorithms for the maximum induced matching problem in special classes of graphs where the problem is W[1]-hard. Now we use this result to derive a Ramsey-type characterization of the *matching number* $\mu(G)$, i.e. the size of a maximum matching in G. It is well known that $\mu(G) \le \tau(G) \le 2\mu(G)$, where $\tau(G)$ is the *vertex cover number*, i.e. the size of a minimum vertex cover in G. Therefore, the same characterization is valid the vertex cover number.

Theorem 6. *The class of complete graphs, the class of complete bipartite graphs and the class $\mathcal{M}$ of graphs of vertex degree at most 1 are the only three minimal hereditary classes of graphs of unbounded matching number and unbounded vertex cover number.*

One more result of the same nature was proved in [2]. It states that for every t, p, s, there exists a $z = z(t, p, s)$ such that every graph with a (not necessarily induced) path of length at least z contains either an induced path of length t or an induced complete bipartite graph with color classes of size p or a clique of size s. This result was used in [2] to obtain fpt-algorithms in special classes of graphs for the k-Biclique problem, which is generally W[1]-hard [8]. Now we use this result to derive the following Ramsey-type characterization of the *path number*, i.e. the length of a longest path.

Theorem 7. *The class of complete graphs, the class of complete bipartite graphs and the class of linear forests (i.e. graphs every connected component of which is a path) are the only three minimal hereditary classes of graphs of unbounded path number.*

From the last two theorems and the remark in the introduction it follows that path number lies between matching number and biclique number in the hierarchy of graph parameters. Also, it is well known that graphs of bounded path number have bounded tree-width and that the complete graphs and complete bipartite graphs are minimal hereditary classes of unbounded tree-width. Therefore, tree-width lies between path number and biclique number. On the other hand, it is known that tree-width lies below clique-width, i.e. bounded tree-width implies bounded clique-width, which was shown in [3]. In the same paper it was also shown that bounded clique-width together with bounded biclique number imply bounded tree-width. Therefore, the family of classes of bounded tree-width is precisely the intersection of the family of classes of bounded clique-width and the family of classes of bounded biclique number.

The above discussion shows that a Ramsey-type characterization of tree-width can be derived from such a characterization for clique-width, if it exists. However, in the case of clique-width even the existence of minimal classes is not obvious. The first two such classes have been recently discovered in [11]. In spite of this progress, a complete Ramsey-type characterization of clique-width is not possible, because in the universe of hereditary classes there are areas, where minimal classes do not exist, for instance, graphs of bounded vertex degree.

There are several ways to overcome this difficulty. One of them is to reduce the universe. For instance, by reducing the universe of hereditary classes to minor-closed classes of graphs, we conclude that the class of planar graphs is the unique minimal minor-closed class of graphs of unbounded tree-width [13] and hence it is the unique minimal minor-closed class of graphs of unbounded clique-width. One more way to overcome the difficulty of non-existence of minimal classes is to employ the notion of boundary classes, which is a relaxation of the notion of minimal classes (see e.g. [10]).

Finally, instead of characterizing graph parameters in terms of "forbidden" elements (i.e. in terms of "what is not allowed") one can characterize them in terms of "what is allowed", and the results of the present paper suggest a uniform way to such characterizations. To see this, let us observe that both neighbourhood diversity and VC-dimension describe how complex the neighbourhood of a set X of vertices can be outside of X. This complexity can be described by a hypergraph whose hyperedges correspond to the neighbourhoods of vertices in X. In this terminology, neighbourhood diversity marks the jump from finitely many to infinitely many (distinct) hyperedges. Similarly, VC-dimension marks the jump from infinitely many hyperedges to *all possible* hyperedges. Between these two extremes, lies a variety of other graph parameters, such as clique-width, and exploring their neighbourhood complexity (i.e. the structure of the corresponding hypergraphs) is a very challenging task.

Acknowledgment. This work was supported by the Russian Science Foundation Grant No. 17-11-01336.

References

1. Alon, N., Brightwell, G., Kierstead, H., Kostochka, A., Winkler, P.: Dominating sets in k-majority tournaments. J. Combin. Theory Ser. B **96**, 374–387 (2006)
2. Atminas, A., Lozin, V.V., Razgon, I.: Linear time algorithm for computing a small biclique in graphs without long induced paths. In: Fomin, F.V., Kaski, P. (eds.) SWAT 2012. LNCS, vol. 7357, pp. 142–152. Springer, Heidelberg (2012). https://doi.org/10.1007/978-3-642-31155-0_13
3. Courcelle, B., Olariu, S.: Upper bounds to the clique width of graphs. Discrete Appl. Math. **101**, 77–114 (2000)
4. Dabrowski, K.K., Demange, M., Lozin, V.V.: New results on maximum induced matchings in bipartite graphs and beyond. Theor. Comput. Sci. **478**, 33–40 (2013)
5. Gargano, L., Rescigno, A.: Complexity of conflict-free colorings of graphs. Theor. Comput. Sci. **566**, 39–49 (2015)
6. Hammer, P.L., Kelmans, A.K.: On universal threshold graphs. Comb. Probab. Comput. **3**(3), 327–344 (1994)
7. Lampis, M.: Algorithmic meta-theorems for restrictions of treewidth. Algorithmica **64**(1), 19–37 (2012)
8. Lin, B.: The parameterized complexity of k-Biclique. In: Proceedings of the Twenty-Sixth Annual ACM-SIAM Symposium on Discrete Algorithms, pp. 605–615 (2015)
9. Lozin, V., Rudolf, G.: Minimal universal bipartite graphs. Ars Comb. **84**, 345–356 (2007)

10. Lozin, V.: Boundary classes of planar graphs. Comb. Probab. Comput. **17**(2), 287–295 (2008)
11. Lozin, V.: Minimal classes of graphs of unbounded clique-width. Ann. Comb. **15**(4), 707–722 (2011)
12. Ramsey, F.P.: On a problem of formal logic. Proc. Lond. Math. Soc. **30**, 264–286 (1930)
13. Robertson, N., Seymour, P.D.: Graph minors. V. Excluding a planar graph. J. Comb. Theory Ser. B **41**, 92–114 (1986)

Letter Graphs and Geometric Grid Classes of Permutations: Characterization and Recognition

Bogdan Alecu[1], Vadim Lozin[1(✉)], Viktor Zamaraev[1],
and Dominique de Werra[2]

[1] Mathematics Institute, University of Warwick, Coventry CV4 7AL, UK
V.Lozin@warwick.ac.uk
[2] Institute of Mathematics, EPFL, 1015 Lausanne, Switzerland

Abstract. In this paper, we reveal an intriguing relationship between two seemingly unrelated notions: letter graphs and geometric grid classes of permutations. We also present the first constructive polynomial-time algorithm for the recognition of 3-letter graphs.

1 Introduction

Letter graphs and geometric grid classes of permutations have been introduced independently of each other in [1,9], respectively. Nothing in the definition of these notions suggests that there is anything in common between them. The only common property is well-quasi-orderability. We believe that there is much more in common between letter graphs and geometric grid classes of permutations and that they can be closely connected through the notion of permutation graph. Speaking informally, we believe that geometric grid classes of permutations and letter graphs are two languages describing the same concept in the universe of permutations and permutation graphs, respectively. We state this formally as a conjecture as follows:

Conjecture 1. Let X be a class of permutations and $\mathcal{G}_X$ the corresponding class of permutation graphs. Then X is a geometric grid class if and only if $\mathcal{G}_X$ is a class of k-letter graphs for a finite value of k.

In this conjecture, the parameter k stands for the size of the alphabet used to describe graphs by means of letters (all definitions will be given in Sect. 2). In Sect. 3, we prove the "only if" part of the conjecture, i.e. we translate the concept of geometric grid classes of permutations to the language of letter graphs.

Well-quasi-orderability implies that geometric grid classes of permutations can be described by finitely many forbidden patterns. Similarly, for each fixed k, the class of k-letter graphs can be described by finitely many forbidden induced subgraphs. This provides a nonconstructive proof of the fact that geometric grid classes of permutations and k-letter graphs (for a fixed k) can be recognized

© Springer International Publishing AG, part of Springer Nature 2018
L. Brankovic et al. (Eds.): IWOCA 2017, LNCS 10765, pp. 195–205, 2018.
https://doi.org/10.1007/978-3-319-78825-8_16

in polynomial time. However, constructive algorithms are not available for the recognition problem, except for the 2-letter graphs and corresponding classes of permutations. As a step towards solving this problem for larger values of k, in Sect. 4 we study the class of 3-letter graphs. Our results lead to a cubic algorithm to recognize graphs in this class.

All preliminary information related to the topic of the paper can be found in Sect. 2.

2   Preliminaries

All graphs in this paper are finite, undirected, without loops and multiple edges. The vertex set and the edge set of a graph G are denoted by $V(G)$ and $E(G)$, respectively. For a vertex $x \in V(G)$ we denote by $N(x)$ the neighbourhood of x, i.e. the set of vertices of G adjacent to x. A subgraph of G induced by a subset of vertices $U \subseteq V(G)$ is denoted $G[U]$. By $\overline{G}$ we denote the complement of G.

A *clique* in a graph is a subset of pairwise adjacent vertices and an *independent set* is a subset of pairwise non-adjacent vertices. By K_n we denote the complete graph on n vertices.

Chain graphs. A graph G is a chain graph if it is bipartite and admits a bipartition $V(G) = V_1 \cup V_2$ such that for any two vertices x, y in the same part V_i either $N(x) \subseteq N(y)$ or $N(y) \subseteq N(x)$. In other words, the vertices in each part of the bipartition of G can be linearly ordered under inclusion of their neighbourhoods (form a chain). In terms of minimal forbidden induced subgraphs, the chain graphs are precisely the $2K_2$-free bipartite graphs.

Permutation graphs. Let π be a permutation of the set $\{1, 2, \ldots, n\}$. The permutation graph G_π of this permutation has $\{1, 2, \ldots, n\}$ as its vertex set with i and j being adjacent if and only if $(i - j)(\pi(i) - \pi(j)) < 0$. A graph G is a *permutation graph* if there is a permutation π such that G is isomorphic to G_π.

2.1   Letter Graphs

Let Σ be a finite alphabet and $\mathcal{P} \subseteq \Sigma^2$ a set of ordered pairs of symbols from Σ, called the *decoder*. To each word $w = w_1 w_2 \cdots w_n$ with $w_i \in \Sigma$ we associate a graph $G(\mathcal{P}, w)$, called the *letter graph* of w, by defining $V(G(\mathcal{P}, w)) = \{1, 2, \ldots, n\}$ with i being adjacent to $j > i$ if and only if the ordered pair (w_i, w_j) belongs to the decoder $\mathcal{P}$.

It is not difficult to see that every graph G is a letter graph in an alphabet of size at most $|V(G)|$ with an appropriate decoder $\mathcal{P}$. The minimum ℓ such that G is a letter graph in an alphabet of ℓ letters is the *lettericity* of G and is denoted $\ell(G)$. A graph is a k-letter graph if its lettericity is at most k.

The notion of k-letter graphs was introduced in [9] and in the same paper the author characterized k-letter graphs as follows.

Theorem 1. *A graph G is a k-letter graph if and only if*

1. *there is a partition $V_1, V_2, \ldots, V_p$ of $V(G)$ with $p \leq k$ such that each V_i is either a clique or an independent set in G, and*
2. *there is a linear ordering L of $V(G)$ such that for each pair of distinct indices $1 \leq i, j \leq p$, the intersection of $E(G)$ with $V_i \times V_j$ is one of the following four types (where L is considered as a binary relation, i.e. as a set of pairs):*
   (a) $L \cap (V_i \times V_j)$;
   (b) $L^{-1} \cap (V_i \times V_j)$;
   (c) $V_i \times V_j$;
   (d) $\emptyset$.

The notion of letter graphs is of interest for various reasons. First, some important graphs classes, such as threshold graphs [7], can be described in the terminology of letter graphs. In particular, a graph G is threshold if and only if G is a 2-letter graph representable over the alphabet $\Sigma = \{a, b\}$ with the decoder $\mathcal{P} = \{(a, a), (a, b)\}$. Second, letter graphs provide an interesting contribution to the theory of ordered graphs [8]. Third, they contribute to the rich theory of graph parameters. On the one hand, graph lettericity generalizes neighbourhood diversity (which was recently introduced in [6] to study parameterized complexity of algorithmic graph problems) in the sense that bounded neighbourhood diversity implies bounded lettericity, but not vice versa (threshold graphs have unbounded neighbourhood diversity). On the other hand, bounded lettericity implies bounded linear clique-width. Moreover, from an algorithmic point of view graphs of bounded lettericity are more attractive, since they can be recognized efficiently, while for graphs of bounded linear clique-width this question is open [4].

Finally, and perhaps most importantly, graphs of bounded lettericity (i.e. of lettericity at most k for a fixed value of k) are well-quasi-ordered by the induced subgraph relation [9]. This is a rare property of graphs, which was shown, up to date, only for some restricted graph classes [3,5]. Moreover, as was observed in [5], k-letter graphs are well-quasi-ordered under the stronger relation of labelled induced subgraphs (see [2] for more information on this topic).

2.2   Geometric Grid Classes of Permutations

The notion of geometric grid classes of permutations was introduced in [1] as follows. Suppose that M is a $0/\pm 1$ matrix. The *standard figure* of M is the set of points in $\mathbb{R}^2$ consisting of

- the increasing open line segment from $(k-1, \ell-1)$ to (k, ℓ) if $M_{k,\ell} = 1$ or
- the decreasing open line segment from $(k-1, \ell)$ to $(k, \ell-1)$ if $M_{k,\ell} = -1$.

We index matrices first by column, counting left to right, and then by row, counting bottom to top throughout. The *geometric grid class* of M, denoted by Geom(M), is then the set of all permutations that can be drawn on this figure in the following manner. Choose n points in the figure, no two on a common horizontal or vertical line. Then label the points from 1 to n from bottom to top and record these labels reading left to right. Figure 1 represents two permutations that lie, respectively, in grid classes of

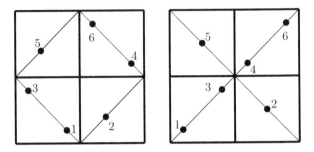

Fig. 1. The permutation 351624 on the left and the permutation 153426 on the right.

$$\begin{pmatrix} 1 & -1 \\ -1 & 1 \end{pmatrix} \text{ and } \begin{pmatrix} -1 & 1 \\ 1 & -1 \end{pmatrix}.$$

A permutation class is said to be *geometrically griddable* if it is contained in some geometric grid class. The geometrically griddable classes of permutations enjoy many nice properties. In particular, in [1] the following results have been proved.

Theorem 2. *Every geometrically griddable class of permutations is well-quasi-ordered and is in bijection with a regular language.*

3 From Geometric Grid Class of Permutations to Letter Graphs

In this section, we prove the "only if" of Conjecture 1, i.e. we prove the following result.

Theorem 3. *Let X be a class of permutations and $\mathcal{G}_X$ the corresponding class of permutation graphs. If X is a geometric grid class, then $\mathcal{G}_X$ is a class of k-letter graphs for a finite value of k.*

To prove Theorem 3, we first outline the correspondence (bijection) between a geometrically griddable class of permutations and a regular language established in Theorem 2. To this end, we need the following definition from [1].

Definition 1. We say that a $0/\pm 1$ matrix M of size $t \times u$ is a *partial multiplication matrix* if there are column and row signs

$$c_1, \ldots, c_t, r_1, \ldots, r_u \in \{1, -1\}$$

such that every entry $M_{k,\ell}$ is equal to either 0 or the product $c_k r_\ell$.

Example 1. The matrix $\begin{pmatrix} 1 & 0 & -1 \\ -1 & 1 & 0 \end{pmatrix}$ is a partial multiplication matrix. This matrix has column and row signs $c_2 = c_3 = r_1 = 1$ and $c_1 = r_2 = -1$.

The importance of this notion for the study of geometric grid classes of permutations is due to the following proposition proved in [1].

Proposition 1. *Every geometric grid class is the geometric grid class of a partial multiplication matrix.*

Let M be a $t \times u$ partial multiplication matrix with column and row signs

$$c_1, \ldots, c_t, r_1, \ldots, r_u \in \{1, -1\}$$

and let Φ_M be the standard gridded figure of M. We will interpret the signs of the columns and rows of M as the "directions" associated with the columns and rows of Φ with the following convention: $c_i = 1$ corresponds to $\rightarrow$, $c_i = -1$ corresponds to $\leftarrow$, $r_i = 1$ corresponds to $\uparrow$, and $r_i = -1$ corresponds to $\downarrow$. The standard gridded figure of the matrix $M = \begin{pmatrix} 0 & 1 & 1 \\ 1 & -1 & -1 \end{pmatrix}$ with row signs $r_1 = -1$ and $r_2 = 1$ and column signs $c_1 = -1$, $c_2 = c_3 = 1$ is represented in Fig. 2.

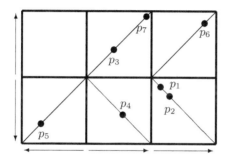

Fig. 2. A standard gridded figure of a partial multiplication matrix.

The *base point* of a cell $C_{k,\ell}$ of the figure Φ is one of the four corners of the cell, where both directions (associated with column k and row ℓ) start. For instance, in Fig. 2 the base point of the cell $C_{3,1}$ is the top-left corner.

In order to establish a bijection between $\mathrm{Geom}(M)$ and a regular language, we first fix an alphabet Σ (known as the *cell alphabet* of M) as follows:

$$\Sigma = \{a_{k\ell} : M_{k,\ell} \neq 0\}.$$

Now, let π be a permutation in $\mathrm{Geom}(M)$, i.e. a permutation represented by a set of n points in the figure Φ_M. For each point p_i of π, let d_i be the distance from the base point of the cell containing p_i to p_i. Without loss of generality, we assume that these distances are pairwise different and the points are ordered so that $0 < d_1 < d_2 < \cdots < d_n < 1$. If p_i belongs to the cell $C_{k,\ell}$ of Φ_M, we define $\phi(p_i) = a_{k\ell}$. Then $\phi(\pi) = \phi(p_1)\phi(p_2)\cdots\phi(p_n)$ is a word in the alphabet Σ, i.e. ϕ defines a mapping from $\mathrm{Geom}(M)$ to Σ^*. Figure 2 shows seven points defining the permutation 1527436. The mapping ϕ associates with this permutation a word in the alphabet Σ as follows: $\phi(1527436) = a_{31}a_{31}a_{22}a_{21}a_{11}a_{32}a_{22}$.

Conversely, let $w = w_1 \cdots w_n$ be a word in Σ^* and let $0 < d_1 < \cdots < d_n < 1$ be n distances chosen arbitrarily. If $w_i = a_{k\ell}$, we let p_i be the point on the line segment in cell $C_{k,\ell}$ at distance d_i from the base point of $C_{k,\ell}$. The n points of Φ_M constructed in this way define a permutation $\psi(w)$ in $\text{Geom}(M)$. Therefore, ψ is a mapping from Σ^* to $\text{Geom}(M)$.

This correspondence between Σ^* and $\text{Geom}(M)$ is not yet a bijection, as illustrated in Fig. 3, because the order in which the points are consecutively inserted into *independent* cells (i.e. cells which share neither a column nor a row) is irrelevant. To turn this correspondence into a bijection, we say that two words $v, w \in \Sigma^*$ are *equivalent* if one can be obtained from the other by successively interchanging adjacent letters which represent independent cells. The equivalence classes of this relation form a *trace monoid* and each element of this monoid is called a *trace*. It is known that in any trace monoid it is possible to choose a unique representative from each trace in such a way that the resulting set of representatives forms a regular language. This is the language which is in a bijection with $\text{Geom}(M)$, as was shown in [1].

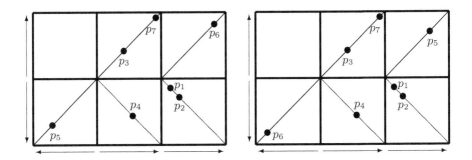

Fig. 3. Two drawings of the permutation 1527436. The drawing on the left is encoded as $a_{31}a_{31}a_{22}a_{21}a_{11}a_{32}a_{22}$ and the drawing on the right is encoded as $a_{31}a_{31}a_{22}a_{21}a_{32}a_{11}a_{22}$.

Next, we will show that the permutation graph G_π of $\pi \in \text{Geom}(M)$ is a k-letter graph with $k = |\Sigma|$. Indeed, the non-empty cells of the figure Φ_M defines a partition of the vertex set of G_π into cliques and independent sets and the word $\phi(\pi)$ defines the order of the vertex set of G_π satisfying conditions of Theorem 1. More formally, let us show that the matrix M uniquely defines a decoder $\mathcal{P} \subseteq \Sigma^2$ such that the letter graph $G(\mathcal{P}, w)$ of the word $w = \phi(\pi)$ coincides with G_π. In order to define the decoder $\mathcal{P}$, we observe that two points p_i and p_j of a permutation $\pi \in \text{Geom}(M)$ corresponds to a pair of adjacent vertices in G_π if and only if one of them lies to the left and above the second one in the figure Φ_M. Therefore, if

- $M_{k,\ell} = 1$, then the points lying in the cell $C_{k,\ell}$ form an independent set in the permutation graph of π. Therefore, we do not include the pair $(a_{k\ell}, a_{k\ell})$ in $\mathcal{P}$.

- $M_{k,\ell} = -1$, then the points lying in the cell $C_{k,\ell}$ form a clique in the permutation graph of π. Therefore, we include the pair $(a_{k\ell}, a_{k\ell})$ in $\mathcal{P}$.
- two cells $C_{k,\ell}$ and $C_{s,t}$ are independent with $k < s$ and $\ell < t$, then no point of $C_{k,\ell}$ is adjacent to any point of $C_{s,t}$ in the permutation graph of π. Therefore, we include neither $(a_{k\ell}, a_{st})$ nor $(a_{st}, a_{k\ell})$ in $\mathcal{P}$.
- two cells $C_{k,\ell}$ and $C_{s,t}$ are independent with $k < s$ and $\ell > t$, then every point of $C_{k,\ell}$ is adjacent to every point of $C_{s,t}$ in the permutation graph of π. Therefore, we include both pairs $(a_{k\ell}, a_{st})$ and $(a_{st}, a_{k\ell})$ in $\mathcal{P}$.
- two cells $C_{k,\ell}$ and $C_{s,t}$ share a column, i.e. $k = s$, then we look at the sign (direction) associated with this column and the relative position of the two cells within the column.
  - If $c_k = 1$ (i.e. the column is oriented from left to right) and $\ell > t$ (the first of the two cells is above the second one), then only the pair $(a_{k\ell}, a_{kt})$ is included in $\mathcal{P}$.
  - If $c_k = 1$ and $\ell < t$, then only the pair $(a_{kt}, a_{k\ell})$ is included in $\mathcal{P}$.
  - If $c_k = -1$ (i.e. the column is oriented from right to left) and $\ell > t$ (the first of the two cells is above the second one), then only the pair $(a_{kt}, a_{k\ell})$ is included in $\mathcal{P}$.
  - If $c_k = -1$ and $\ell < t$, then only the pair $(a_{k\ell}, a_{kt})$ is included in $\mathcal{P}$.
- two cells $C_{k,\ell}$ and $C_{s,t}$ share a row, i.e. $\ell = t$, then we look at the sign (direction) associated with this row and the relative position of the two cells within the row.
  - If $r_\ell = 1$ (i.e. the row is oriented from bottom to top) and $k < s$ (the first of the two cells is to the left of the second one), then only the pair $(a_{k\ell}, a_{s\ell})$ is included in $\mathcal{P}$.
  - If $r_\ell = 1$ and $k > s$, then only the pair $(a_{s\ell}, a_{k\ell})$ is included in $\mathcal{P}$.
  - If $r_\ell = -1$ (i.e. the row is oriented from top to bottom) and $k < s$, then only the pair $(a_{s\ell}, a_{k\ell})$ is included in $\mathcal{P}$.
  - If $r_\ell = -1$ and $k > s$, then only the pair $(a_{k\ell}, a_{s\ell})$ is included in $\mathcal{P}$.

It is now a routine task to verify that $G(\mathcal{P}, w)$ coincides with G_π.

4   Recognition of 3-Letter Graphs

To develop a recognition algorithm for 3-letter graphs, we introduce some more terminology. Let $G = (V, E)$ be a graph and A an independent set in G. We will say that a linear order $(a_1, a_2, \ldots, a_k)$ of the vertices of A is

- *increasing* if $i < j$ implies $N(a_i) \subseteq N(a_j)$,
- *decreasing* if $i < j$ implies $N(a_i) \supseteq N(a_j)$,
- *monotone* if it is either increasing or decreasing.

By definition, each part of a chain graph (i.e. a $2K_2$-free bipartite graph) admits a monotone ordering. Let $G = (A \cup B, E)$ be a chain graph given together with a bipartition $V(G) = A \cup B$ of its vertices into two independent sets. We fix an order of the parts (A is first and B is second), a decreasing order for A, an increasing order for B, and call G a *properly ordered graph*. This notion suggest an easy way of representing a $2K_2$-free bipartite graph as a 2-letter graph.

Let $G = (A \cup B, E)$ be a properly ordered $2K_2$-free bipartite graph. To represent G as a 2-letter graph, we fix the alphabet $\Sigma = \{a, b\}$ and decoder $\mathcal{P} = \{(a, b)\}$. The word ω representing G can be constructed as follows. To each vertex of A we assign letter a and to each vertex of B we assign letter b. The a letters will appear in ω in the oder in which the corresponding vertices appear in A and the b letters will appear in ω in the oder in which the corresponding vertices appear in B. The rule defining the relative positions of a vertices with respect to b vertices can be described in two different ways as follows:

R_1 an a vertex is located between the last b non-neighbour (if any) and the first b neighbour (if any),

R_2 a b vertex is located between the last a neighbour (if any) and the first a non-neighbour (if any).

It is not difficult to see that both rules R_1 and R_2 define the same word and this word represents G. Figure 4 represents the chain graph defined by the word $abababababab$.

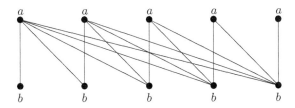

Fig. 4. A chain graph

In Sect. 4.1, we describe an efficient algorithm to recognize 3-letter graphs representable over the decoder $\{(a, b), (b, c), (c, a)\}$. For other decoders, the algorithms are similar (though not identical), so we omit them.

4.1   3-Letter Graphs with the Decoder $\{(a, b), (b, c), (c, a)\}$

Let $G = (A \cup B \cup C, E)$ be a graph whose vertex set is partitioned into three independent sets A, B, C such that

(a) $G[A \cup B]$, $G[B \cup C]$ and $G[C \cup A]$ are $2K_2$-free bipartite graphs,
(b) there are no three vertices $a \in A$, $b \in B$, $c \in C$ inducing either a triangle K_3 or an anti-triangle $\overline{K}_3$.

We call any graph satisfying (a) and (b) *nice*. Our goal is to show that a graph G is a 3-letter graph with the decoder $\{(a, b), (b, c), (c, a)\}$ if and only if it is nice. First, we prove the following lemma.

Lemma 1. *Let $G = (A \cup B \cup C, E)$ be a nice graph. Then each of the independent sets A, B and C admits a linear ordering such that all three bipartite graphs $G[A \cup B]$, $G[B \cup C]$ and $G[C \cup A]$ are properly ordered.*

Proof. We start with a proper order of $G[A \cup B]$, in which case the order of B is increasing with respect to A. Let us show that the same order of B is decreasing with respect to C.

Consider two vertices b_i and b_j of B with $i < j$, i.e. b_i precedes b_j in the linear order of B and hence $N(b_i) \cap A \subseteq N(b_j) \cap A$. To show that the linear order of B is decreasing with respect to C, assume the contrary: b_j has a neighbour $c \in C$ non-adjacent to b_i. Without loss of generality, we may suppose that the inclusion $N(b_i) \cap A \subseteq N(b_j) \cap A$ is proper, since otherwise we could change the order of the vertices b_i and b_j in B, which destroys the contradiction and keeps the graph $G[A \cup B]$ properly ordered. According to this assumption, b_j must have a neighbour $a \in A$ non-adjacent to b_i. But then either a, b_j, c induce a triangle K_3 (if a is adjacent to c) or a, b_i, c induce an anti-triangle $\overline{K}_3$ (if a is not adjacent to c). A contradiction in both cases shows that the linear order of B is decreasing with respect to C.

Similar arguments show that the order of A which is decreasing with respect to B is increasing with respect to C. Now we fix a linear order of C which is increasing with respect to B and conclude, as before, that it is decreasing with respect to A. In this way, we obtain a proper order for all three graphs $G[A \cup B]$, $G[B \cup C]$ and $G[C \cup A]$ (notice, in the last graph C is the first part and A is the second). □

Theorem 4. *A graph G is a 3-letter graph with the decoder $\{(a,b), (b,c), (c,a)\}$ if and only if it is nice.*

Proof. If G is a 3-letter graph with the decoder $\{(a,b), (b,c), (c,a)\}$, then obviously V_a (the set of vertices labelled by a), V_b and V_c are independent sets and condition (a) of the definition of nice graphs is valid for G. To show that (b) is valid, assume G contains a triangle induced by letters a, b, c. Then b must appear after a in the word representing G, and c must appear after b. But then c appears after a, in which case a is not adjacent to c, a contradiction. Similarly, an anti-triangle a, b, c is not possible and hence G is nice.

Suppose now that $G = (A \cup B \cup C, E)$ is nice. According to Lemma 1, we may assume that A, B and C are ordered in such a way that each of the three bipartite graphs $G[A \cup B]$, $G[B \cup C]$ and $G[C \cup A]$ is properly ordered.

We start by representing the graph $G[A \cup B]$ by a word ω with two letters a, b according to rules R_1 or R_2. To complete the construction, we need to place the c vertices

- among the a vertices according to rule R_1, i.e. every c vertex must be located between the last a non-neighbour a_{lnn} (if any) and the first a neighbour a_{fn} (if any),
- among the b vertices according to rule R_2, i.e. every c vertex must be located between the last b neighbour b_{ln} (if any) and the first b non-neighbour b_{fnn} (if any).

This is always possible, unless

- either a_{fn} precedes b_{ln} in ω, in which case a_{fn} is adjacent to b_{ln} and hence a_{fn}, b_{ln}, c induce a triangle K_3,
- or b_{fnn} precedes a_{lnn} in ω, in which case b_{fnn} is not adjacent to a_{lnn} and hence a_{lnn}, b_{fnn}, c induce an anti-triangle $\overline{K}_3$.

A contradiction in both case shows that ω can be extended to a word representing G. □

We now turn to the recognition of 3-letter graphs with the decoder $\{(a, b), (b, c), (c, a)\}$. If a graph G has a twin v for a vertex u (i.e. $N(v) = N(u)$), then any word representing $G - v$ can be extended to a word representing G by assigning to v the same letter as to u and placing v next to u. This observation shows that we may assume that G is twin-free.

Due to the cyclic symmetry of the alphabet, we may assume without loss of generality that

- the last letter of the word is c.

Then

- the first letter is not a, since otherwise the first and the last vertices are twins.

Assume that the first letter of the word representing G is b. Then according to the decoder

(b1) no vertex between the first b and the last c is adjacent to both of them,
(b2) every vertex non-adjacent to the first b and non-adjacent to the last c must be labelled by a,
(b3) every vertex non-adjacent to the first b and adjacent to the last c must be labelled by b,
(b4) every vertex adjacent to the first b and non-adjacent to the last c must be labelled by c.

Therefore, in order to determine whether G can be represented by a word starting with b and ending with c, we inspect every pair of adjacent vertices, assign letter b to one of them and letter c to the other, check whether the set of vertices adjacent to both of them is empty and split the remaining vertices of the graph into three subsets A, B, C according to (b2), (b3) and (b4), respectively. Finally, we verify whether the partition obtained in this way satisfies the definition of nice graphs (conditions (a) and (b)).

Finally, we determine whether G can be represented by a word starting with c. Then according to the decoder

(c1) no vertex between the first c and the last c is adjacent to both of them,
(c2) every vertex adjacent to the first c and non-adjacent to the last c must be labelled by a,
(c3) every vertex non-adjacent to the first c and adjacent to the last c must be labelled by b,

(c4) every vertex non-adjacent to the first c and non-adjacent to the last c must be labelled by c.

Therefore, in order to determine whether G can be represented by a word starting with c and ending with c, we inspect every pair of non-adjacent vertices, assign letter c to both of them, check whether the set of vertices adjacent to both of them is empty and split the remaining vertices of the graph into three subsets A, B, C according to (c2), (c3) and (c4), respectively. Finally, we verify whether the partition obtained in this way satisfies the definition of nice graphs (conditions (a) and (b)).

From the above discussion we derive the following conclusion.

Theorem 5. *The 3-letter graphs with the decoder $\{(a, b), (b, c), (c, a)\}$ can be recognized in cubic time.*

Acknowledgment. Vadim Lozin and Viktor Zamaraev acknowledge support of EPSRC, grant EP/L020408/1.

References

1. Albert, M.H., Atkinson, M.D., Bouvel, M., Ruskuc, N., Vatter, V.: Geometric grid classes of permutations. Trans. Am. Math. Soc. **365**, 5859–5881 (2013)
2. Atminas, A., Lozin, V.: Labelled induced subgraphs and well-quasi-ordering. Order **32**(3), 313–328 (2015)
3. Damaschke, P.: Induced subgraphs and well-quasi-ordering. J. Graph Theory **14**(4), 427–435 (1990)
4. Fellows, M.R., Rosamond, F.A., Rotics, U., Szeider, S.: Clique-width is NP-complete. SIAM J. Discrete Math. **23**(2), 909–939 (2009)
5. Korpelainen, N., Lozin, V.: Two forbidden induced subgraphs and well-quasi-ordering. Discrete Math. **311**(16), 1813–1822 (2011)
6. Lampis, M.: Algorithmic meta-theorems for restrictions of treewidth. Algorithmica **64**(1), 19–37 (2012)
7. Mahadev, N.V.R., Peled, U.N.: Threshold Graphs and Related Topics: Annals of Discrete Mathematics, vol. 56. North-Holland Publishing Co., Amsterdam (1995). xiv+543 pp.
8. Nešetřil, J.: On ordered graphs and graph orderings. Discrete Appl. Math. **51**(1–2), 113–116 (1994)
9. Petkovšek, M.: Letter graphs and well-quasi-order by induced subgraphs. Discrete Math. **244**, 375–388 (2002)

Fully Leafed Tree-Like Polyominoes
and Polycubes

Alexandre Blondin Massé[1]($\boxtimes$), Julien de Carufel[2], Alain Goupil[2],
and Maxime Samson[2]

[1] Laboratoire de Combinatoire et d'Informatique Mathématique,
Université du Québec à Montréal, Montreal, Canada
blondin_masse.alexandre@uqam.ca
[2] Université du Québec à Trois-Rivières, Trois-Rivières, Canada

Abstract. We present and prove recursive formulas giving the maximal
number of leaves in tree-like polyominoes and polycubes of size n. We call
these tree-like polyforms *fully leafed*. The proof relies on a combinatorial
algorithm that enumerates rooted directed trees that we call abundant.
We also show how to produce a family of fully leafed tree-like polyominoes and a family of fully leafed tree-like polycubes for each possible size,
thus gaining insight into their geometric characteristics.

1 Introduction

Polyominoes and, to a lesser extent, polycubes have been the object of important
investigations in the past 30 years either from a game theoretic or combinatorial
point of view (see [12] and references therein). Recall that a polyomino is an
edge-connected set of unit cells in the square lattice that is invariant under
translation. The 3D equivalent of a polyomino is called a polycube, which is a
face-connected set of unit cells in the cubic lattice, up to translation.

A central problem is the search for the number of polyminoes with n cells
where n is called the size of the polyomino. This problem, still open, has been
investigated from several points of view; asymptotic evaluation [15], computer
generation and counting [14,16,17], random generation [13] and combinatorial
description [2,11]. Combinatorists have also concentrated their efforts in the
description of various families of polyominoes and polycubes, such as convex
polyominoes [4], parallelogram polyominoes [1,8], tree-like polyominoes [10] and
other families [5–7].

In this paper, we are interested in two related families: *tree-like polyominoes*
and *tree-like polycubes* which are acyclic polyominoes in the graph theoretic
sense. Our main results are recursive expressions giving the maximal number of
leaves of tree-like polyominoes and polycubes of size n. A tree-like polyform of
size n is called *fully leafed* when it has the maximum possible number of leaves
among all tree-like polyforms of size n. The numbers of leaves in a fully leafed
tree-like polyomino and polycube with n cells are denoted respectively $L_2(n)$
and $L_3(n)$. The structures under investigation are similar to those solving the

© Springer International Publishing AG, part of Springer Nature 2018
L. Brankovic et al. (Eds.): IWOCA 2017, LNCS 10765, pp. 206–218, 2018.
https://doi.org/10.1007/978-3-319-78825-8_17

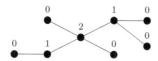

Fig. 1. The depth of the vertices in a tree.

maximum leaf spanning tree problem in grid graphs, one of the classical NP-complete problems described by Garey and Johnson in their seminal paper [9]. Both problems are concerned with the maximization of the number of leaves in subtrees, but present a fundamental difference: On one hand, spanning trees must contain all vertices of the graph while, on the other hand, induced subtrees must contain any edge of the graph between two of its vertices. To our knowledge, these new classes of polyforms present remarkable structure and properties that have not been considered yet.

2 Preliminaries

Let $G = (V, E)$ be a simple graph, $u \in V$ and $U \subseteq V$. The set of neighbors of u in G is denoted $N_G(u)$, which is naturally extended to U by defining $N_G(U) = \{u' \in N_G(u) \mid u \in U\}$. For any subset $U \subseteq V$, the *subgraph induced by U* is the graph $G[U] = (U, E \cap \mathcal{P}_2(U))$, where $\mathcal{P}_2(U)$ is the set of 2-elements subsets of V. The *extension* of $G[U]$ is defined by $\mathrm{Ext}(G[U]) = G[U \cup N_G(U)]$ and the *interior* of $G[U]$ is defined by $\mathrm{Int}(G[U]) = G[\mathrm{Int}(U)]$, where $\mathrm{Int}(U) = \{u' \in U \mid N_G(u') \subseteq U\}$. Finally, the *hull* of $G[U]$ is defined by $\mathrm{Hull}(G[U]) = \mathrm{Int}(\mathrm{Ext}(G[U]))$.

The *square lattice* is the infinite simple graph $\mathcal{G}_2 = (\mathbb{Z}^2, A_4)$, where A_4 is the *4-adjacency relation* defined by $A_4 = \{(p, p') \in \mathbb{Z}^2 \mid \mathrm{dist}(p, p') = 1\}$ where dist is the Euclidean distance of $\mathbb{R}^2$. For any $p \in \mathbb{Z}^2$, the set $c(p) = \{p' \in \mathbb{R}^2 \mid \mathrm{dist}_\infty(p, p') \leq 1/2\}$, where dist_∞ is the uniform distance of $\mathbb{R}^2$, is called the *square cell* centered in p. The function c is naturally extended to subsets of $\mathbb{Z}^2$ and subgraphs of $\mathcal{G}_2$. For any finite subset $U \subseteq \mathbb{Z}^2$, we say that $\mathcal{G}_2[U]$ is a *grounded polyomino* if it is connected. The set of all grounded polyominoes is denoted by $\mathcal{GP}$. Given two grounded polyominoes $P = \mathcal{G}_2[U]$ and $P' = \mathcal{G}_2[U']$, we write $P \equiv_t P'$ (resp. $P \equiv_i P'$) if there exists a translation $T : \mathbb{Z}^2 \to \mathbb{Z}^2$ (resp. an isometry I on $\mathbb{Z}^2$) such that $U' = T(U)$ (resp. $U' = I(U)$). A *fixed polyomino* (resp. *free polyomino*) is then an element of $\mathcal{GP}/ \equiv_t$ (resp. $\mathcal{GP}/ \equiv_i$). Clearly, any connected induced subgraph of $\mathcal{G}_2$ corresponds to exactly one set of square cells via the function c. Consequently, from now on, polyominoes will be considered as simple graphs rather than sets of edge-connected square cells. All definitions above are extended to the *cubic lattice* with the *6-adjacency relation*. Thus, we define *cubic cell*, *grounded polycube*, *fixed polycube* and *free polycube* accordingly.

Grounded polyominoes and polycubes are connected subgraphs of $\mathcal{G}_2$ and $\mathcal{G}_3$ and the terminology of graph theory becomes available. A (grounded, fixed or free) *tree-like polyomino* is therefore a (grounded, fixed or free) polyomino whose associated graph is a tree. *Tree-like polycubes* are defined similarly. Observe that

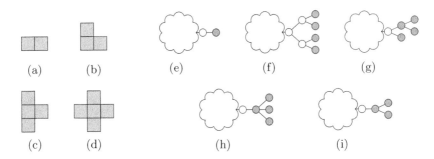

Fig. 2. Fully leafed tree-like polyominoes of size (a) 2, (b) 3, (c) 4 and (d) 5. The images (e), (f), (g), (h) and (i) depict the five cases of Lemma 3 (gray cells are removed).

if u, v are adjacent cells in a tree-like polyomino T then $\deg_T(u) + \deg_T(v) \leq 6$, which extends to $\deg_T(u) + \deg_T(v) \leq 2d + 2$ in arbitrary $\mathbb{Z}^d$ with $d \geq 2$.

Let $T = (V, E)$ be any finite simple non empty tree. We say that $u \in V$ is a *leaf* of T when $\deg_T(u) = 1$. Otherwise u is called an *inner vertex* of T. For any $d \in \mathbb{N}$, the number of vertices of degree d is denoted by $n_d(T)$ and $n(T) = |V|$ is the number of vertices of T which is also called the *size* of T. The *depth* of $u \in V$ in T, denoted by $depth_T(u)$, is defined recursively by

$$depth_T(u) = \begin{cases} 0, & \text{if } \deg_T(u) \leq 1; \\ 1 + depth_{T'}(u), & \text{otherwise,} \end{cases}$$

where T' is the tree obtained from T by removing all its leaves (see Fig. 1).

3 Fully Leafed Tree-Like Polyominoes

In this section, we describe the number of leaves of fully leafed tree-like polyominoes. For any integer $n \geq 2$, let the function $\ell_2(n)$ be defined as follows:

$$\ell_2(n) = \begin{cases} 2, & \text{if } n = 2; \\ n - 1, & \text{if } n = 3, 4, 5; \\ \ell_2(n - 4) + 2, & \text{if } n \geq 6. \end{cases} \tag{1}$$

We claim that $\ell_2(n)$ is the maximal number of leaves for a tree-like polyomino of size n so that $\ell_2(n) = L_2(n)$. The first step is straightforward.

Lemma 1. *For all $n \geq 2$, $L_2(n) \geq \ell_2(n)$.*

Proof. We build a family of tree-like polyominoes $\{T_n \mid n \geq 2\}$ whose number of leaves is given by (1). For $n = 2, 3, 4, 5$, the polyominoes T_n respectively in (a), (b), (c) and (d) of Fig. 2 satisfy (1). For $n \geq 6$, let T_n be the polyomino obtained by appending the polyomino of Fig. 2(c) to the right of T_{n-4}.

By induction on n, we have $n_1(T_n) = \ell_2(n)$ for all $n \geq 2$, since the fact that appending the T-shaped polyomino of Fig. 2(c) adds 4 cells and 3 leaves, but subtracts 1 leaf.   □

In order to prove that the family $\{T_n \mid n \geq 2\}$ described in the proof of Lemma 1 is maximal, we need the following result characterizing particular sub-trees that appear in possible counter-examples of minimum size.

Lemma 2. *Let T be a tree-like polyomino of minimum size such that $n_1(T) > \ell_2(n(T))$ and let T' be a tree-like polyomino such that $n(T') = n(T) - i$, for some $i \in \{1, 3, 4\}$. Also, let $\Delta\ell_2(1) = 0$, $\Delta\ell_2(3) = 1$ and $\Delta\ell_2(4) = 2$. Then $n_1(T) > n_1(T') + \Delta\ell_2(i)$.*

Proof. It is easy to prove by induction that for any $k \geq 2$, $\ell_2(k + i) \geq \ell_2(k) + \Delta\ell_2(i)$, where $i \in \{1, 3, 4\}$. Therefore,

$$
\begin{aligned}
n_1(T) &> \ell_2(n(T)), && \text{by assumption,} \\
&= \ell_2(n(T') + i), && \text{by definition of } T', \\
&\geq \ell_2(n(T')) + \Delta\ell_2(i), && \text{by the observation above,} \\
&\geq L_2(n(T')) + \Delta\ell_2(i), && \text{by minimality of } n(T), \\
&\geq n_1(T') + \Delta\ell_2(i), && \text{by definition of } L_2,
\end{aligned}
$$

concluding the proof.   □

We are ready to prove that the family $\{T_n \mid n \geq 2\}$ is maximal.

Lemma 3. *For all $n \geq 2$, $L_2(n) \leq \ell_2(n)$.*

Proof. Suppose, by contradiction, that T is a tree-like polyomino of minimal size such that $n_1(T) > \ell_2(n(T))$. We first show that all vertices of T of depth 1 have degree 3 or 4. Arguing by contradiction, assume that there exists a vertex u_1 of T such that $depth_T(u_1) = 1$ and $\deg_T(u_1) = 2$. Let T' be the tree-like polyomino obtained from T by removing the leaf adjacent to u_1 (see Fig. 2(e)). Then $n(T') = n(T) - 1$ and $n_1(T') = n_1(T)$, contradicting Lemma 2.

Now, we show that T cannot have a vertex of depth 2. Again by contradiction, assume that such a vertex u_2 exists. Clearly, $\deg_T(u_2) \neq 4$, otherwise u_2 would have a neighbor of depth 1 and degree 2, which was just shown to be impossible. If $\deg_T(u_2) = 3$, then we are either in case (f) or (g) of Fig. 2. In each case, let T' be the tree-like polyomino obtained by removing the four gray cells. Then $n(T') = n(T) - 4$ and $n_1(T') = n_1(T) - 2$, contradicting Lemma 2. Finally, if $\deg_T(u_2) = 2$, then either (h) or (i) of Fig. 2 holds, leading to a contradiction with Lemma 2 when removing the gray cells. Since every tree-like polyomino of size larger than 6 has at least one vertex of depth 2, the proof is completed by exhaustive inspection of all tree-like polyominoes of size at most 6.   □

Combining Lemmas 1 and 3, we have proved the following result.

Theorem 4. *For all integers $n \geq 2$, $L_2(n) = \ell_2(n)$ and the asymptotic growth of L_2 is given by $L_2(n) \sim \frac{1}{2}n$.*   □

4   Fully Leafed Polycubes

The basic concepts introduced in Sect. 3 are now extended to tree-like polycubes with additional considerations that complexify the arguments. Recall that for all integers $n \geq 2$,

$$L_3(n) = \max\{n_1(T) \mid T \text{ is a tree-like polycube of size } n\}.$$

If the construction used for polyominoes would be extended to polycubes, then the ratio $L_3(n)/n$ would lead to $4/6$ as $n \to \infty$. In this section, we show that the optimal ratio is actually $28/41$. Define the function $\ell_3(n)$ as follows:

$$\ell_3(n) = \begin{cases} f_3(n) + 1, & \text{if } n = 6, 7, 13, 19, 25; \\ f_3(n), & \text{if } 2 \leq n \leq 40 \text{ and } n \neq 6, 7, 13, 19, 25; \\ f_3(n - 41) + 28, & \text{if } 41 \leq n \leq 81; \\ \ell_3(n - 41) + 28, & \text{if } n \geq 82. \end{cases} \quad (2)$$

$$\text{where} \qquad f_3(n) = \begin{cases} \lfloor (2n + 2)/3 \rfloor, & \text{if } 0 \leq n \leq 11; \\ \lfloor (2n + 3)/3 \rfloor, & \text{if } 12 \leq n \leq 27; \\ \lfloor (2n + 4)/3 \rfloor, & \text{if } 28 \leq n \leq 40. \end{cases} \quad (3)$$

The following key observations on ℓ_3 prove to be useful.

Proposition 5. *The function ℓ_3 satisfies the following properties:*

(i) *For all positive integers k, the sequence $(\ell_3(n + k) - \ell_3(n))_{n \geq 0}$ is bounded, so that the function $\Delta \ell_3 : \mathbb{N} \to \mathbb{N}$ defined by*

$$\Delta \ell_3(i) = \liminf_{n \to \infty}(\ell_3(n + i) - \ell_3(n))$$

is well-defined.

(ii) *For any positive integers n and k, if $\ell_3(n + k) - \ell_3(n) < \Delta \ell_3(k)$, then $n \in \{6, 7, 13, 19, 25\}$.*

Proof. Omitted due to lack of space.

We now introduce rooted tree-like polycubes.

Definition 6. *A* rooted grounded tree-like polycube *is a triple $R = (T, r, \vec{u})$ such that*

(i) *$T = (V, E)$ is a grounded tree-like polycube of size at least 2;*
(ii) *$r \in V$, called the* root *of R, is a cell adjacent to at least one leaf of T;*
(iii) *$\vec{u} \in \mathbb{Z}^3$, called the* direction *of R, is a unit vector such that $r + \vec{u}$ is a leaf of T.*

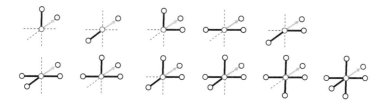

Fig. 3. Atomic tree-like polycubes up to isometry

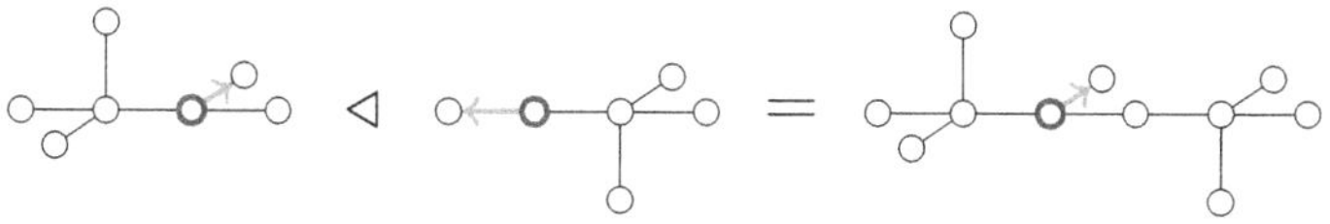

Fig. 4. A well-defined, non-final graft union of two rooted grounded tree-like polycubes

The *height* of R is the maximum length of a path from the root r to some leaf. *Rooted fixed tree-like polycubes* and *rooted free tree-like polycubes* are defined similarly. If R is a rooted, grounded or fixed, tree-like polycube, a unit vector $\vec{v} \in \mathbb{Z}^3$ is called a *free direction* of R whenever $r - \vec{v}$ is a leaf of T. A rooted grounded, fixed or free, tree-like polycube R is called *atomic* if its height is 1. The 10 atomic rooted free tree-like polycubes are illustrated in Fig. 3.

We now introduce an operation called the *graft union* of tree-like polycubes.

Definition 7 (Graft union). *Let $R = (T, r, \vec{u})$ and $R' = (T', r', \vec{u'})$ be rooted grounded tree-like polycubes such that $\vec{u'}$ is a free direction of R. The graft union of R and R', whenever it exists, is the rooted grounded tree-like polycube*

$$R \lhd R' = (\mathbb{Z}_3[V \cup \tau(V')], r, \vec{u}),$$

where V, V' are the sets of vertices of T, T' respectively and τ is the translation with respect to the vector $\vec{r'r} - \vec{u'}$.

The graft union is naturally extended to fixed and free tree-like polycubes. In the latter case however, $R \lhd R'$ is not a single rooted free tree-like polycube, but rather the set of all possible graft unions obtained from an isometry. Observe that graft union is a partial application on rooted grounded tree-like polycubes, i.e. the triple $(\mathbb{Z}_3[V \cup \tau(V')], r, \vec{u})$ is not always a rooted tree-like polycube. More precisely, the induced subgraph $\mathbb{Z}_3[V \cup \tau(V')]$ is always connected, but not always acyclic. Also, $r + \vec{u}$ needs not be a leaf. Therefore, we say that a graft union $R \lhd R'$ is

(i) *non-final* if $R \lhd R'$ is a rooted grounded tree-like polycube;
(ii) *final* if the graph $G = \mathbb{Z}_3[V \cup \tau(V')]$ is a tree-like polycube, $\vec{u'} = -\vec{u}$ and $r + \vec{u}$ is not a leaf of G;
(iii) *well-defined* if it is either non-final or final;
(iv) *invalid* otherwise.

Figure 4 illustrates a well-defined graft union of two rooted tree-like poly-cubes. The graft union interacts well with the functions $n(R)$ and $n_i(R)$ giving respectively the total number of cells and the number of cells of degree i in T.

Lemma 8. *Let R_1, R_2 be rooted grounded tree-like polycubes such that $R_1 \lhd R_2$ is well-defined. Then*

$$n_1(R_1 \lhd R_2) = n_1(R_1) + n_1(R_2) - 2,$$
$$n_i(R_1 \lhd R_2) = n_i(R_1) + n_i(R_2), \quad \text{for } i \geq 2;$$
$$n(R_1 \lhd R_2) = n(R_1) + n(R_2) - 2.$$

Proof. This is an immediate consequence of Definition 7.

We are now ready to define a family of fully leafed tree-like polycubes.

Lemma 9. *For all $n \geq 2$, $L_3(n) \geq \ell_3(n)$.*

Proof. We exhibit a family of tree-like polycubes $\{U_n \mid n \geq 2\}$ realizing ℓ_3, i.e. such that $n_1(U_n) = \ell_3(n)$ for all $n \geq 2$. First, for $n = 6, 7, 13, 19, 25$, let U_n be the tree-like polycubes depicted in Fig. 5(a), (b), (c), (d) and (e) respectively. It is easy to verify that $n_1(U_n) = \ell_3(n)$ in these cases.

Now, let $n \notin \{6, 7, 13, 19, 25\}$, let q and r be the quotient and remainder of the division of $n - 2$ by 41 and define the integers a, b, c, d, e as follow.

$$a = \chi(r \geq 10)$$
$$b = \chi(r \in \{1, 4, 7, 10, 11, 14, 17, 20, 23, 26, 27, 30, 33, 36, 39\})$$
$$c = \chi(r \in \{2, 5, 8, 12, 15, 18, 21, 24, 28, 31, 34, 37, 40\})$$
$$d = \lfloor (r - 10 \, (\chi(r \geq 10) + \chi(r \geq 26))) \, /3 \rfloor$$
$$e = \chi(r \geq 26),$$

where χ is the usual characteristic function. Let U_n be the tree-like polycube associated with a rooted grounded tree-like polycube R_n of the form

$$R_n = R_{12}^a \lhd R_{43}^q \lhd R_3^b \lhd R_4^c \lhd R_5^d \lhd R_{12}^e, \tag{4}$$

where, for $n = 3, 4, 5, 12, 14$, R_n is the rooted grounded tree-like polycube depicted in Fig. 5(f), and the exponent notation is defined by

$$R_n^k = \begin{cases} R_2, & \text{if } k = 0; \\ \rho(R_n) \lhd R_n^{k-1}, & \text{if } n = 5, 43; \\ R_n \lhd R_n^{k-1}, & \text{otherwise.} \end{cases}$$

with $\rho(R_n)$ the rotation of $90°$ of R_n about the "horizontal" axis in Fig. 5 (f) and (g) applied $k - 1$ times. In other words, when copies of R_5 and R_{43} are grafted to themselves, the new copy must be rotated by $90°$ before being grafted. We assume that the roots and directions used for the graft union are respectively as depicted in Fig. 5 (f) by red dots and blue arrows. Note also that the two rooted

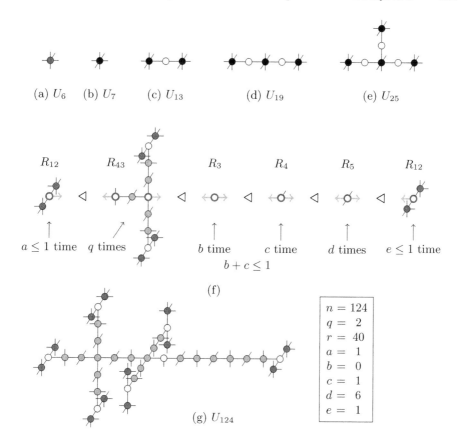

Fig. 5. Fully leafed tree-like polycubes (Color figure online)

grounded tree-like polycubes R_{12} at each end of Fig. 5 (f) are shown in the proper position up to a rotation of $90°$. Clearly, all graft unions in Eq. (4) are well-defined and it follows from Lemma 8 that $n(R_n) = 41q+10(a+e)+b+2c+3d+2 = n$ and $n_1(R_n) = 28q+7(a+e)+c+2d+2 = \ell_3(n)$ (The recursive part in the definition of $\ell_3(n)$ is straightforward since q is arbitrarily large and $n_1(R_{43}) = 28$.) Hence, for $n \geq 2$ and $n \notin \{6, 7, 13, 19, 25\}$, we obtain U_n by taking the unrooted version of R_n, concluding the proof. Figure 5(g) shows the tree-like polycube U_{124} obtained from R_{124} in Eq. (4) with $n = 124$.   □

We now introduce a notation for the operation of graft factorization of tree-like polycubes associated to the graft union.

Definition 10 (Branch). *Let $T = (V, E)$ be a tree-like polycube and r, r' two adjacent vertices of T. Let V_r and $V_{r'}$ be the set of vertices of T defined by*

(i) *$r \in V_r$, $r' \in V_{r'}$,*
(ii) *the subgraphs of T induced by V_r and $V_{r'}$ are precisely the two connected components obtained from T by removing the edge $\{r, r'\}$.*

Then the rooted tree-like polycube $B = (T[V_r \cup \{r'\}], r, \vec{rr'})$ *is called a* branch *of* T *and the rooted tree-like polycube* $B^c = (T[V_{r'} \cup \{r\}], r', \vec{r'r})$ *is called the* co-branch *of* B *in* T. *When neither* r *nor* r' *are leaves of* T, *then we say that* B *and* B^c *are* proper branches *of* T.

Proposition 11. *Let* T *be a tree-like polycube and* B *a proper branch of* T. *Then both* $B \triangleleft B^c$ *and* $B^c \triangleleft B$ *are well-defined and final, while their corresponding unrooted tree-like polycube is precisely* T.

Proof. This follows from Definitions 7 and 10.

We wish to identify branches appearing in potential counter-examples, which would need to have many leaves with respect to their number of cells.

Definition 12. *Let* R, R' *be two rooted tree-like polycubes having the same direction. We say that* R *is* substitutable *by* R' *if, for any tree-like polycube* T *containing the branch* R, $R^c \triangleleft R'$ *is well-defined.*

Definition 12 means that R can always be replaced by R' without creating a cycle whenever it appears in some tree-like polycube T. A sufficient condition for finding substitutable rooted tree-like polycubes is related to its *hull* (see the first paragraph of Sect. 2).

Proposition 13. *Let* R *be a rooted tree-like polycube and* R' *a rooted subtree of* Hull(R) *having the same root as* R. *Then* R *is substitutable by* R'.

Proof. Omitted due to lack of space.

We are now ready to classify rooted tree-like polycubes.

Definition 14. *Let* R *be a rooted tree-like polycube. We say that* R *is* abundant *if one of the following two conditions is satisfied:*

(i) R *contains exactly two cells,*
(ii) *There does not exist another abundant rooted tree-like polycube* R', *such that* R *is substitutable by* R', $n(R') < n(R)$ *and*

$$n_1(R) - n_1(R') \leq \Delta\ell_3(n(R) - n(R')) \tag{5}$$

Otherwise, we say that R *is* sparse.

The following observation is immediate.

Proposition 15.
All branches of abundant rooted tree-like polycubes are also abundant.

Proof. Omitted due to lack of space.

Algorithm 1. Generation of all abundant rooted tree-like polycubes.

```
 1: function ABUNDANTBRANCHES(h : height) : pair of maps
 2:     For i = 1, 2, . . . , h, let A[i] ← ∅ and F[i] ← ∅
 3:     A[1], F[1] ← {atomic free tree-like polycubes of size 5 and 6}
 4:     for i ← 1, 2, . . . , h do
 5:         for each atomic rooted free tree-like polycube B do
 6:             for each B′ ∈ B ◁ ∪ⱼ₌₀ⁱ⁻¹ A[j] of height i do
 7:                 if B′ is abundant then
 8:                     if B′ is final then F[i] ← F[i] ∪ B′
 9:                     else A[i] ← A[i] ∪ B′
10:                 end if
11:             end for
12:         end for
13:     end for
14:     return (A, F)
15: end function
```

Using Definition 14, one can enumerate all abundant rooted tree-like polycubes up to a given height, both final and nonfinal, using a brute-force approach as described by Algorithm 1. In Algorithm 1, for a given integer $h > 0$ and each height $i = 1, 2, \ldots, h$, the abundant final and nonfinal rooted tree-like polycubes are stored respectively in the two lists $F[i]$ and $A[i]$.

Algorithm 1 was implemented in Python and run with increasing values of h [3]. It turned out that there exists no abundant rooted tree-like polycube for $h = 11$, i.e. $|A[11]| = |F[11]| = 0$. Due to a lack of space, we cannot exhibit all abundant rooted tree-like polycubes, but we can give some examples. For instance, in Fig. 5, any rooted version of the trees U_6, U_7 and U_{12} is abundant, while rooted versions of U_3, U_4, U_5, U_{13}, U_{19}, U_{25} and U_{43} are sparse.

The following facts are directly observed by computation.

Lemma 16. *Let $T = B ◁ B^c$ be an abundant rooted tree-like polycube. Then*

(i) *The height of T is at most 10.*
(ii) *If T is final then $n_1(T) \leq \ell_3(n(T))$.*
(iii) *If $n(T) \in \{6, 7, 13, 19, 25\}$, then either B or B^c is sparse.*

Proof. Let $A = \bigcup_{i=1}^{h} A(i)$ and $F = \bigcup_{i=1}^{h} F(i)$, where $A(i)$ and $F(i)$ are respectively the sets of abundant nonfinal and final rooted tree-like polycubes computed by Algorithm 1 with $h = 11$. In particular, one notices that $|A(i)|, |F(i)| > 0$ for $1 \leq i \leq 10$, but $|A(11)| = |F(11)| = 0$ (see [3]). (i) By Proposition 15, it is immediate that $|A(i)| = 0$ implies both $|A(i+1)| = 0$ and $|F(i+1)| = 0$ for any $i \geq 1$, so the result follows. (ii) By exhaustive inspection of F. (iii) Assume by contradiction that both B and B^c are abundant. By inspecting F, we must have $T \in F$, but F does not contain any final, abundant, rooted tree-like polycube with 6, 7, 13, 19 or 25 vertices. □

The nomenclature "sparse" and "abundant" is better understood with the following lemma.

Lemma 17. *Assume that there exists a tree-like polycube T of minimum size $n(T)$ such that $n_1(T) > \ell_3(n(T))$. Then every branch of T is abundant.*

Proof. Let B be any branch of T and B^c its co-branch so that $T = B \triangleleft B^c$. Assume that B is sparse, using the substitution of B by the abundant rooted tree-like polycube B', let $T' = B' \triangleleft B^c$. Suppose first that

$$\ell_3(n(B \triangleleft B^c)) - \ell_3(n(B' \triangleleft B^c)) \geq \Delta\ell_3(n(B) - n(B')) \geq n_1(B) - n_1(B'), \quad (6)$$

where the last inequality is deduced from Inequation (5). Then

$$\begin{aligned} \ell_3(n(T)) = \ell_3(n(B \triangleleft B^c)) &\geq n_1(B) - n_1(B') + \ell_3(n(B' \triangleleft B^c)) \\ &\geq n_1(B) - n_1(B') + n_1(B' \triangleleft B^c) \\ &= n_1(B \triangleleft B^c) = n_1(T), \end{aligned}$$

contradicting the hypothesis $n_1(T) > \ell_3(n(T))$. It follows that

$$\ell_3(n(B \triangleleft B^c)) - \ell_3(n(B' \triangleleft B^c)) < \Delta\ell_3(n(B) - n(B')). \quad (7)$$

By Proposition 5(ii), this implies that $n(B' \triangleleft B^c) \in \{6, 7, 13, 19, 25\}$. Since B' is abundant, Lemma 16(iii) implies that B^c is sparse. Therefore, using the same reasoning as above, either we have Inequations (6) by swapping B and B^c, leading to a contradiction, or B^c can be substituted by some abundant branch C such that $n(B \triangleleft C) \in \{6, 7, 13, 19, 25\}$. Hence, $n(B), n(B^c) \leq 25$. To conclude, observe that B and B^c must be fully leafed: If B (or B^c) is not fully leafed, then B^c (or B) would be a counter-example of size $n(B^c) < n(T)$, contradicting our minimality assumption. Finally, exhaustive inspection of sparse and fully leafed rooted tree-like polycube of size at most 25 yields no counterexample T. □

The following fact, together with Lemma 9, leads to our main result.

Lemma 18. *For all $n \geq 2$, $L_3(n) \leq \ell_3(n)$.*

Proof. By contradiction, assume that there exists a tree-like polycube T of minimum size such that $n_1(T) > \ell_3(n(T))$. By Lemma 17, every branch of T is abundant, so that there must exist two abundant branches B and B' such that $T = B \triangleleft B'$. The result follows from Lemma 16(ii). Thus we have proved. □

Theorem 19. *For all $n \geq 2$, $L_3(n) = \ell_3(n)$ and the asymptotic growth of L_3 is given by $L_3(n) \sim \frac{28}{41}n$.*

5   Concluding Remarks

Theorems 4 and 19 give the exact values for the ratios $L_d(n)/n, d = 2, 3$ which are $1/2$ and $28/41$ respectively. For polycubes of higher dimension $d > 3$, elementary arguments allow to find lower and upper bounds for $L_d(n)/n$. Indeed, it is always possible to build such polycubes by alternating cells of degree $2d$ and 2, as was done in dimension 2, for any dimension $d > 2$ (see for example polycube U_{19} in

Fig. 5). Since this connected set of cells has ratio $L_d(n)/n = (d-1)/d$, a lower bound is deduced for $L_d(n)/n$ in all dimensions $d > 1$. Similarly the structure of the polycube U_{12} in Fig. 5 can be extended in all dimensions $d > 3$ as the graft union of three cells with respective degrees $2d - 1, 3, 2d - 1$, whose ratio $L_d(n)/n = (4d - 3)/4d$ can be exceeded neither by the ratio of any realizable triplet of cells nor by any connected set of $k > 3$ cells in dimension d. Therefore the ratio $(4d - 3)/4d$ is an upper bound for $L_d(n)/n$ in any dimension d.

Fully leafed polycubes of size n all seem to possess the same geometric shape: a connected kernel of cells of degree 4 grafted on its extremities to as many polycubes U_{43} as possible. The best ratio occurs when there are 7 cells of degree 4 between every pair of polycubes U_{43}. Future work on fully leafed polyforms will consist in the extension of the results to other regular lattices such as $\mathbb{Z}^d$, for $d > 3$, the hexagonal, the triangular lattice and also to nonperiodic lattices.

References

1. Aval, J.-C., D'Adderio, M., Dukes, M., Hicks, A., Le Borgne, Y.: Statistics on parallelogram polyominoes and a q, t-analogue of the narayana numbers. J. Comb. Theory Ser. A **123**(1), 271–286 (2014)
2. Barcucci, E., Frosini, A., Rinaldi, S.: On directed-convex polyominoes in a rectangle. Discrete Math. **298**(1–3), 62–78 (2005)
3. Blondin Massé, A.: A sagemath program to compute fully leafed tree-like polycubes (2016). https://bitbucket.org/ablondin/fully-leafed-tree-polycubes
4. Bousquet-Mélou, M., Guttmann, A.J.: Enumeration of three dimensional convex polygons. Ann. Comb. **1**(1), 27–53 (1997)
5. Bousquet-Mélou, M., Rechnitzer, A.: The site-perimeter of bargraphs. Adv. Appl. Math. **31**(1), 86–112 (1997)
6. Castiglione, G., Frosini, A., Munarini, E., Restivo, A., Rinaldi, S.: Combinatorial aspects of L-convex polyominoes. Eur. J. Comb. **28**(6), 1724–1741 (2007)
7. Champarnaud, J.-M., Dubernard, J.-P., Cohen-Solal, Q., Jeanne, H.: Enumeration of specific classes of polycubes. Electr. J. Comb. **20**(4), 26 (2013)
8. Delest, M.-P., Viennot, G.: Algebraic languages and polyominoes enumeration. Theoret. Comput. Sci. **34**, 169–206 (1984)
9. Garey, M.R., Johnson, D.S.: Computers and Intractability. Freeman, San Francisco (1979)
10. Goupil, A., Cloutier, H., Nouboud, F.: Enumeration of polyominoes inscribed in a rectangle. Discrete Appl. Math. **158**(18), 2014–2023 (2010)
11. Goupil, A., Pellerin, M.E., de Wouters d'Oplinter, J.: Partially directed snake polyominos. arXiv:1307.8432v2 (2014)
12. Guttmann, A.J.: Polygons, Polyominoes and Polycubes. Springer, Heidelberg (2009). https://doi.org/10.1007/978-1-4020-9927-4
13. Hochstättler, W., Loebl, M., Moll, C.: Generating convex polyominoes at random. Discrete Math. **153**(1–3), 165–176 (1996)
14. Jensen, I.: Enumerations of lattice animals and trees. J. Stat. Mech. **102**(18), 865–881 (2001)

15. Klarner, D.A., Rivest, R.L.: A procedure for improving the upper bound for the number of n-ominoes. Can. J. Math. **25**, 585–602 (1973)
16. Knuth, D.E.: Polynum, program available from knuth's. http://Sunburn.Stanford. EDU/~knuth/programs.html#polyominoes (1981)
17. Redelmeyer, D.H.: Counting polyominoes: yet another attack. Discrete Math. **36**(3), 191–203 (1981)

Improved Lower Bound on Broadcast Function Based on Graph Partition

Hovhannes A. Harutyunyan and Zhiyuan Li[✉]

Concordia University, Montreal, QC H3G 1M8, Canada
l_zhiyua@encs.concordia.ca

Abstract. Broadcasting is a one-to-all information spreading process in a communication network. The network is modeled as a graph. The broadcast time of a given vertex is the minimum time required to broadcast a message from that vertex to all vertices of the graph. The broadcast time of a graph is the maximum time required to broadcast from any vertex in the graph. A graph G on n vertices is a minimum broadcast graph if the broadcast time of G is the minimum possible time: $\lceil \log n \rceil$, and the number of edges in G is minimized. The broadcast function $B(n)$ denotes the number of edges in a minimum broadcast graph on n vertices. The exact value of $B(n)$ is only known for $n = 2^m$, $2^m - 2$, and some small values of $n < 64$. Finding $B(n)$ is very difficult due to the lack of tight lower bounds on $B(n)$. The existing lower bounds are based on the vertex degree of the originator vertex. However, most of the minimum broadcast graphs are not necessarily regular. In this paper, we present an improved general lower bound on $B(n)$ based on new observations about partitioning broadcast graphs.

1 Introduction

One-to-all information dissemination problem, named *broadcasting*, is one of the essential tasks in computer networks. The performance of broadcasting usually measures the overall efficiency of the network. To study the properties of broadcasting and the topology of a network, the network is defined as a simple connected graph $G = (V, E)$, where the vertex set V represents the nodes in the network, and the edge set E represents the communication links. A broadcast in a graph originates from one vertex, called the *originator*, and finishes when every vertex in the graph is informed.

In past decades, a long sequence of research papers study this topic under different models. These models differ at the number of originators, the number of destinations, the lifetime of each sender, the specific topologies of the graphs, the number of receivers at each time unit, the distance of each call, and other characteristics. In this paper, we focus on the classical model with the following assumptions.

© Springer International Publishing AG, part of Springer Nature 2018
L. Brankovic et al. (Eds.): IWOCA 2017, LNCS 10765, pp. 219–230, 2018.
https://doi.org/10.1007/978-3-319-78825-8_18

- the graph has only one originator;
- every call is a fundamental process of broadcasting and requires one time unit;
- every call is from exactly one informed vertex, the sender to one of its uninformed neighbors, the receiver.

With the assumptions above, we have the following formal definitions.

Definition 1. *A broadcast scheme is a sequence of parallel calls in a graph G originating from a vertex v. Each call, represented by a directed edge, defines a sender and a receiver. A broadcast scheme generates a broadcast tree, which is a spanning tree of the graph rooted at the originator.*

Definition 2. *Let G be a graph on n vertices and v be the broadcast originator in graph G, the broadcast time of vertex v, $b(G,v)$ defines the minimum number of time units required to broadcast from originator v in graph G. The broadcast time of graph G $b(G) = max\{b(G,v)|v \in V(G)\}$ defines the maximum of all broadcast times of any vertex in graph G.*

Since a sender can inform at most one receiver node during one time unit, the number of informed vertices at each time unit is at most doubled. Thus, $b(G) \geq \lceil \log_2 n \rceil = \lceil \log n \rceil$. For convenience, we will omit the base of all logarithms throughout this paper when the base is 2.

Definition 3. *A graph G on n vertices is a broadcast graph if $b(G) = \lceil \log n \rceil$. A broadcast graph with the minimum number of edges is a minimum broadcast graph. This minimum number of edges $B(n)$ is called the broadcast function.*

From the applications perspective minimum broadcast graphs are the cheapest graphs (with minimum number of edges) where broadcasting can be accomplished in minimum possible time.

The study of minimum broadcast graphs and broadcast function $B(n)$ has a long history. In [5], Farley et al. introduced minimum broadcast graphs, defined the broadcast function, determined the values of $B(n)$, for $n \leq 15$ and $n = 2^k$ and proved that hypercubes are minimum broadcast graphs. Khachatrian and Haroutunian [19] and independently Dinneen et al. [4] show that Knödel graphs, defined in [20], are minimum broadcast graphs on $n = 2^k - 2$ vertices. Park and Chwa prove that the recursive circulant graphs on 2^k vertices are minimum broadcast graphs [25]. The comparison of the three classes of minimum broadcast graphs mentioned above can be found in [6]. Besides these three classes, there is no other known infinite construction of minimum broadcast graphs. The values of $B(n)$ are also known for $n = 17$ [24], $n = 18, 19$ [3,30], $n = 20, 21, 22$ [23], $n = 26$ [26,31], $n = 27, 28, 29, 58, 61$ [26], $n = 30, 31$ [3], $n = 63$ [22], $n = 127$ [11] and $n = 1023, 4095$ [27].

By the difficulty of constructing minimum broadcast graphs, a long sequence of papers present different techniques to construct broadcast graphs in order to obtain upper bounds on $B(n)$. Furthermore, proving that a lower bound matches the upper bound is also extremely difficult, because most of the lower bound proofs are based on vertex degrees. However, minimum broadcast graphs except hypercubes and Knödel graphs on $2^k - 2$ vertices are not regular. Thus, in general upper bounds cannot match lower bounds.

Upper bounds on $B(n)$ are given by constructing sparse broadcast graphs. General Knödel graphs on even number of vertices give a good bound on $B(n)$ for even n. Compounding two broadcast graphs of smaller sizes is a powerful method to construct good broadcast graphs with larger size. In [1,12], compounding binomial trees, hypercubes, and Knödel graphs improves the upper bound on $B(n)$ for $2^{m-1} + 1 \leq n \leq 2^m - 2^{\frac{m}{2}}$ if the value of n is separated by intervals $[2^m - 2^{m-1} + 1, 2^m]$. The vertex addition method is also used in the constructions. In [13], authors add one vertex to Knödel graphs and improve the upper bound on $B(n)$ for even $2^m - 2^{\frac{m}{2}} + 1 \leq n \leq 2^m$. Ad hoc constructions sometimes also provides good upper bounds. This method usually constructs broadcast graphs by adding edges to a binomial tree [9,14]. The vertex deletion is studied in [3]. Several other constructions are presented in [3,8,9,14,28–30].

Lower bounds on $B(n)$ are also studied in the literature. In [8], Gargano and Vaccaro show $B(n) \geq \frac{n}{2}(\lfloor \log n \rfloor - \log(1 + 2^{\lceil \log n \rceil} - n))$, for any n. $B(n) \geq \frac{n}{2}(m - p - 1)$ is proved in [21], where m is the length of the binary representation $a_{m-1}a_{m-2}...a_1a_0$ of n and p is the index of the leftmost 0 bit. Harutyunyan and Liestman study k-broadcasting (every sender can inform at most k neighbors in each time unit) and give a lower bound on k-broadcast graph in [15]. The latter bound is the best known general lower bound for our model of broadcasting (which corresponds to the case $k = 1$ in [15]). Below we summarize this lower bound for our model.

Theorem 1. *Let* $n = 2^m - 2^k + 1 - d$, $1 \leq k \leq m - 2$ *and* $0 \leq d \leq 2^k - 1$.

$$B(n) \geq \frac{n}{2}(m - k)$$

Besides the general lower bounds, Labahn shows $B(n) \geq \frac{m^2(2^m - 1)}{2(m+1)}$ for $n = 2^m - 1$ in [22] by considering the broadcast tree rooted at a vertex with the minimum degree. Saclé follows this method and gives tight lower bounds on $B(2^m - 3)$, $B(2^m - 4)$, $B(2^m - 5)$ and $B(2^m - 6)$ in [26]. Grigoryan and Harutyunyan show a better lower bound for $n = 2^m - 2^k + 1$.

Theorem 2. [10] $B(2^m - 2^k + 1) \geq \frac{2^m - 2^k + 1}{2}(m - k + \frac{m(2k-1)-(k^2+k-1)}{m(m-1)-(k-1)})$

Better lower bounds for $n = 24, 25$ are given in [2]. Note that $23 \leq n \leq 25$ are the only values of $n \leq 32$ for which $B(n)$ is not known. For more on broadcasting and gossiping in general see the following survey papers [7,16–18]. This paper is organized as follows. Section 2 introduces some important definitions and presents three useful observations. Section 3 gives a general lower bound on $B(n)$ by using the observations. Section 4 give conclusions and future work.

2   Definitions and Observations

Definition 4. *A binomial tree* BT_m *on* 2^m *vertices of order* m *consists of*

1. *a single vertex which is also the root, if* $m = 1$;
2. *two copies of binomial trees* BT_{m-1} *having the two roots connected by an edge, if* $m > 1$.

Definition 5. *Let BT_m and BT_k are two binomial trees of order m and k respectively, and $m > k$. u is the root of BT_m. $BT_m \backslash BT_k$ is a tree obtained by removing a complete binomial tree BT_k from u in BT_m except the root u.*

Figure 1 gives an example of a binomial tree and $BT_m \backslash BT_k$ for $m = 5$ and $k = 3$.

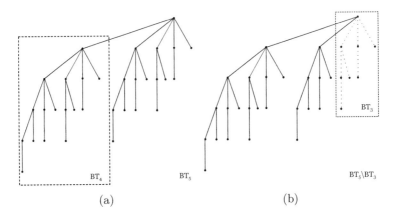

(a)   (b)

Fig. 1. (a) is an example of a binomial tree BT_5. (b) the solid edges and the associated vertices give an example of $BT_5 \backslash BT_3$.

Definition 6. *Let T be a broadcast tree of graph G originating from root u. Then $L_k(T)$, the first k broadcast level tree of T, consists of all the vertices of T which are informed in the first k time units following the broadcast scheme from originator vertex u in graph G.*

We know that any broadcast tree of graph G on n vertices is a subtree of a binomial tree $BT_{\lceil \log n \rceil}$. So, the first k broadcast level tree $L_k(T)$ is a subtree of a binomial tree BT_k. Figure 2 gives one example of a broadcast tree BT_4 and its first 3 broadcast level tree.

Let G be a minimum broadcast graph on $n = 2^m - 2^k + 1$ vertices, where $1 \le k \le m - 2$; u be a vertex of degree $m - k$ in G; T be the broadcast tree rooted at vertex u; and $L_k(T)$ be the first k broadcast level tree of T. If the neighbors of u are sorted in the decreasing order of their degrees and the i-th neighbor corresponds to the i-th branch, we have the following observations.

Observation 1. *$BT_m \backslash BT_k$ is a broadcast tree T of broadcast graph G on $n = 2^m - 2^k + 1$ vertices rooted at a vertex u of degree $m - k$.*

Proof. Graph G has $2^m - 2^k + 1$ vertices, so the broadcasting must be completed in m time units. It is clear that during this m time broadcasting from originator u in $BT_m \backslash BT_k$ there are no idle vertices (informed vertices but not transferring the message). Thus, branches of the root u are complete binomial trees BT_{m-1},

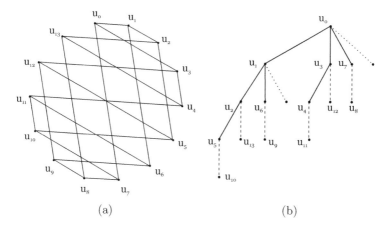

Fig. 2. (a) is a broadcast graph G on 14 vertices. (b) is a binomial tree BT_4 on 16 vertices. 14 vertices with labels among them together with the solid and the dashed edges give a broadcast tree T of G. And the solid edges form a first 3 broadcast level tree $L_3(T)$.

$BT_{m-2}, \cdots, BT_k$. There are in total $2^{m-1} + \cdots + 2^k + 1 = 2^m - 2^k + 1$ vertices, which is exactly the same as the number of vertices in G. Thus, the broadcast tree has to be $BT_m \setminus BT_k$.

Observation 2. *If $k \geq \frac{m}{2}$, the i-th branch of u has 2^{k-i} vertices in $L_k(T)$.*

Proof. If we ignore the first level (only one vertex: the root u), broadcast tree T becomes a forest of binomial trees BT_{m-1}, BT_{m-2}, $\cdots$, BT_k. So, the first k level broadcast tree $L_k(T)$ of T consists of the first $k - 1$ level broadcast tree $L_{k-1}(BT_{m-1})$ of the first branch, $L_{k-2}(BT_{m-2})$ of the second branch, and $L_{k-i}(BT_{m-i})$ of the i-th branch in general. If $k \geq \frac{m}{2}$, then the last neighbor is informed at time unit $m - k \leq k$. Thus, $L_{2k-m}(BT_{m-k})$ is a complete binomial tree BT_{2k-m}. Then, each of $L_{k-i}(BT_{m-i})$ becomes a complete binomial tree BT_{k-i}. So, there are 2^{k-i} vertices on the i-th branch.

Observation 3. *If a vertex w is in $L_k(T)$, then the corresponding vertex in broadcast tree T has degree greater than $m - k$.*

Proof. Observation 1 ensures that on i-th branch, BT_{m-i} is a complete binomial tree. So, $L_{k-i}(BT_{m-i})$ is indeed a complete binomial tree BT_{k-i} of order $k - i$, and it can be obtained by replacing every vertex in BT_{k-i} by a binomial tree BT_{m-k}. Thus, if a vertex w in $L_{k-i}(BT_{m-i})$ (which is BT_{m-k}) has degree a, then vertex w has degree $a + m - k$ in BT_{m-i} (also in broadcast tree T). Every leaf in any tree has the minimum degree 1. Therefore, any leaf in $L_k(T)$ gives the minimum degree $m - k + 1 > m - k$ in broadcast tree T. Figure 3 shows an example of broadcast tree T when $k \geq \frac{m}{2}$.

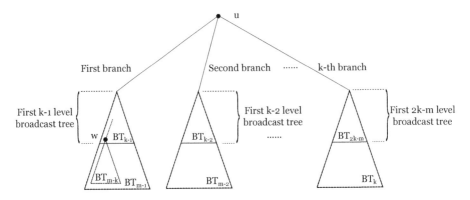

Fig. 3. An example of a broadcast tree rooted at vertex u of degree $m - k$. The triangle at the i-th branch is a binomial tree BT_{m-i}. The smaller triangle is the first $k - i$ broadcast level tree $L_{k-i}(BT_{m-i})$. And leaf w is an example of a vertex in $L_k(T)$. w has degree 1 in $L_k(T)$ and degree $m - k$ in $T - L_k(T)$. So, the degree is $m - k + 1$ in total.

3 New Lower Bound

In this section, we first give a lower bound on $B(n)$ when $n = 2^m - 2^k + 1 - d$, where $\frac{m}{2} \le k \le m - 2$ and $d = 0$. Then, we generalize the lower bound for any $0 \le d \le 2^k - 1$. That is we give a lower bound on $B(n)$ for all $2^{m-1} + 1 \le n \le 2^m - 2^{\frac{m}{2}+1} + 1$.

Theorem 3. *If $n = 2^m - 2^k + 1$ and $\frac{m}{2} \le k \le m - 2$,*

$$B(n) \ge \frac{n}{2}(m - k + \frac{1}{2} - \frac{1}{4m - 4k + 2}) + \frac{2^{k+1} - 2^{2k-m+1} - m + k}{2m - 2k + 1}$$

Proof. Observation 1 shows that the minimum degree of any vertex in G is $m - k$. So, we partition the vertices of G into V_{m-k}, the vertices of degree $m - k$; and V_{other}, other vertices. We also partition the edges into E_{m-k}, the edges connecting two vertices in V_{m-k}; E_{inter}, the edges connecting one vertex in V_{m-k} and one vertex in V_{other}; and E_{other}, the edges connecting two vertices in V_{other}. Let $v_{m-k}, v_{other}, e_{m-k}, e_{inter},$ and e_{other} be the cardinality of each of the respective sets. It is easy to see $n = v_{m-k} + v_{other}$ and $e = e_{m-k} + e_{inter} + e_{other}$.

Case 1. If there is no vertex of degree $m - k$ in graph G, then the minimum degree is $m - k + 1$, we have

$$e \ge \frac{n}{2}(m - k + 1) \tag{1}$$

Case 2. If there is a vertex of degree $m - k$ in graph G, we consider the broadcast tree T originating from such a vertex u. In order to inform all vertices in graph

G within m time units, every vertex except originator u cannot be idle during the minimum time broadcasting in G. So, the vertices informed by u (also the neighbors of u in broadcast tree T) must have degree m, $m-1$, ..., $k+1$. In other words the broadcast tree of originator u must be $BT_m \setminus BT_k$.

Since $k \geq \frac{m}{2}$, then the last neighbor of u has degree $k+1 > m-k$. Thus, there is no vertex of degree $m-k$ having a neighbor of degree $m-k$. Furthermore, if an edge is attached to a vertex of degree $m-k$, then it must be attached to a vertex of degree at least $m-k+1$.

$$e_{m-k} = 0$$

$$e_{inter} = (m-k)v_{m-k}$$

Again we consider the broadcast tree T and estimate e_{other}. By Observation 3, every vertex in the first k broadcast level tree $L_k(T)$ except the root u has degree greater than $m-k$. Thus, every edge except the ones on the first level in $L_k(T)$ has both of its endpoints of degree greater than $m-k$. And by Observation 2, $L_k(T)$ becomes a forest of $L_{k-1}(BT_{m-1})$, $\cdots$, $L_{2k-m}(BT_{m-k})$ by ignoring the root and its incident edges. Then, e_{other} can be estimated by counting the number of the edges in the forest. Therefore,

$$e_{other} \geq 2^{k+1} - 2^{2k-m+1} - (m-k)$$

Combining e_{m-k}, e_{inter}, and e_{other},

$$e = e_{m-k} + e_{inter} + e_{other}$$

$$e \geq (m-k)v_{m-k} + 2^{k+1} - 2^{2k-m+1} - (m-k)$$

$$v_{m-k} \leq \frac{e - 2^{k+1} - 2^{2k-m+1} - (m-k)}{m-k}$$

$$n - v_{m-k} \geq n - \frac{e - 2^{k+1} + 2^{2k-m+1} + (m-k)}{m-k}$$

$$v_m + \cdots + v_{m-k+1} \geq n - \frac{e - 2^{k+1} + 2^{2k-m+1} + (m-k)}{m-k} \tag{2}$$

We have the following trivial inequalities.

$$2e \geq (m-k)v_{m-k} + \cdots + mv_m$$

$$2e \geq (m-k)n + v_{m-k+1} + 2v_{m-k+2} + \cdots + kv_m$$

$$2e \geq (m-k)n + v_{m-k+1} + v_{m-k+2} + \cdots + v_m \tag{3}$$

By substituting inequality (2) we get

$$2e \geq (m-k)n + n - \frac{e - 2^{k+1} + 2^{2k-m+1} + (m-k)}{m-k}$$

$$e \geq \frac{n}{2}(m - k + \frac{1}{2} - \frac{1}{4m - 4k + 2})$$

$$+ \frac{2^{k+1} + 2^{2k-m+1} + (m-k)}{2m - 2k + 1} \tag{4}$$

Now we combine inequality (1) and inequality (4) given by the two different cases. Let RHS_1 and RHS_2 be the right hand side of the two inequalities respectively.

$$
\begin{aligned}
RHS_1 - RHS_2 &= \frac{m-k+1}{2(2m-2k+1)}n - \frac{2^{k+1}-2^{2k-m+1}-m+k}{2m-2k+1} \\
&= \frac{1}{2(2m-2k+1)}((m-k+1)(2^m-2^k+1) \\
&\quad - (2^{k+2}-2^{2k-m+2}-2m+2k)) \\
&\geq \frac{1}{2(2m-2k+1)}(3(2^{k+2}-2^k+1) \\
&\quad - (2^{k+2}-2^{2k-m+2}-2m+2k)) \\
&= \frac{1}{2(2m-2k+1)}((2^{k+3}+2^{k+1}+3) \\
&\quad - (2^{k+2}-2^{2k-m+2}-2m+2k)) \\
&> 0
\end{aligned}
$$

Thus, inequality (4) is the worst case and gives the lower bound, which completes the proof. □

By simple comparison, we can see that Theorem 3 does not give a better bound than Theorem 2. But Theorem 3 can be generalized to other n.

Theorem 4.

$$
B(n) \geq \frac{n}{2}(m-k+\frac{1}{2}+\frac{\alpha-1}{4m-4k-2\alpha+2}) + \frac{2^{k+1}-2^{2k-m+1}-m+k-d}{2m-2k+1}
$$

where

$$
\alpha = \lfloor \frac{-W_{-1}(-2^{-d-2^{2k-m+1}+2k-m+1}\ln(2))}{\ln(2)} \rfloor - d - 2^{2k-m+1}
$$

and $W_{-1}(x)$ is the lower branch of Lambert-W function.

Proof. Observation 1 is not true for general n, but in this case the minimum degree is always $m-k$. Assume a vertex r has degree $m-k-1$ in a broadcast graph G on $n = 2^m - 2^k + 1 - d$ vertices, where $0 \leq d \leq 2^k - 1$. Minimum time broadcasting from originator r informs at most 2^{m-1} vertices on the first branch, 2^{m-2} vertices on the second branch, $\cdots$, and 2^{k+1} vertices on the last branch. Together with the originator r, there are $2^m - 2^{k+1} + 1$ vertices in total, which is $2^m - 2^{k+1} + 1 > 2^m - 2^{k+1} \geq 2^m - 2^k + 1 - d$. Thus, the minimum degree has to be $m-k$. Then, we have the two cases similar to Theorem 3.

Case 1. If the minimum degree is greater than $m-k$, then

$$
e \geq \frac{n}{2}(m-k+1) \tag{5}
$$

Case 2. If the minimum degree is $m - k$, we again have e_{m-k}, e_{inter}, and e_{other} indicating the cardinality of the different edge set as in the proof of Theorem 3. However, the value of e_{m-k}, e_{inter}, and e_{other} are different after removing d vertices.

Let u be a vertex of degree $m - k$. Assume α neighbors of u become degree $m - k$ after removing d vertices.

$$e_{m-k} + e_{inter} \geq \frac{1}{2}\alpha v_{m-k} + (m - k - \alpha)v_{m-k}$$

$$= \frac{1}{2}(2m - 2k - \alpha)v_{m-k}$$

$e_{m-k} + e_{inter}$ is minimized when α is maximized, which is the worst case for the lower bound. Consider the broadcast tree T rooted at vertex u. The neighbors s_1, s_2, $\cdots$, s_{m-k} of u have degree m, $m - 1$, $\cdots$, $k + 1$ respectively. And s_i is the root of a binomial tree BT_{m-i}. To maximize α, we remove vertices and make neighbors of u of degree $m - k$ from s_{m-k} to s_1, because the last neighbor s_{m-k} has the smallest degree. So, $2k - m + 1$ neighbors of s_{m-k} are removed. $2^{2k-m+1} - 1$ vertices are removed from the last branch. And the binomial tree BT_k attached to s_{m-k} becomes $BT_k \setminus BT_{2k-m+1}$.

In general, to make s_i of degree $m - k$, $2^{k-i+1} - 1$ vertices are removed from the i-th branch. Thus, if α neighbors of u are of degree $m - k$, we need to remove $2^{2k-m+1} - 1 + 2^{2k-m+2} - 1 + \cdots + 2^{2k-m+\alpha} - 1 = 2^{2k-m+\alpha+1} - 2^{2k-m+1} - \alpha$ vertices from broadcast tree T. Since the number of removed vertices cannot exceed d, we have the following inequality:

$$d \geq 2^{2k-m+\alpha+1} - 2^{2k-m+1} - \alpha$$

$$2^{2k-m+1}2^\alpha \leq d + \alpha + 2^{2k-m+1}$$

$$2^\alpha \leq 2^{2k-m+1}\alpha + 2^{-(2k-m+1)}d + 1$$

Let $\alpha = -x - d - 2^{2k-m+1}$

$$2^{-x-d-2^{2k-m+1}} \leq -x2^{-(2k-m+1)}$$

$$-2^{-2^{2k-m+1}-d+2k-m+1} \geq x2^x$$

$$-2^{-2^{2k-m+1}-d+2k-m+1}\ln(2) \geq x\ln(2)e^{x\ln(2)} \qquad (6)$$

The right hand side of inequality (6) has the form $z \cdot e^z$. It can be solved by Lambert-W function $W(z \cdot e^z) = z$. However, $W(z)$ is a multivalued relation. $W(z)$ increases when $z \geq -\frac{1}{e}$ and $W(z) \geq -1$; while it decreases when $-\frac{1}{e} \leq z < 0$ and $W(z) \leq -1$. Let $W_0(z)$ and $W_{-1}(z)$ define the two single-valued function for the two different branches of $W(z)$ respectively. We need to estimate the value of $x\ln(2)$ to decide which single-valued function is used. We know that $\alpha \geq 0$, $0 \leq d \leq 2^k - 1$, and $\frac{m}{2} \leq k \leq m - 2$.

$$-x - d - 2^{2k-m+1} \geq 0$$
$$-x - 2^{2k-m+1} \geq 0$$
$$-x \geq 2$$
$$x \ln(2) < -1$$

Thus, $W_{-1}(z)$ is used.

$$W_{-1}(-2^{-2^{2k-m+1}-d+2k-m+1} \ln(2)) \leq x \ln(2)$$

Solve α by substitution.

$$\alpha \leq -\frac{W_{-1}(-2^{-2^{2k-m+1}-d+2k-m+1} \ln(2))}{\ln(2)} - d - 2^{2k-m+1}$$

Since α is an integer,

$$\alpha = \lfloor -\frac{W_{-1}(-2^{-2^{2k-m+1}-d+2k-m+1} \ln(2))}{\ln(2)} \rfloor - d - 2^{2k-m+1}$$

e_{other} is analyzed as in the proof of Theorem 3 by counting the number of vertices in the first k broadcast level tree $L_k(T)$. If all the removed d vertices are in $L_k(T)$, then we have a trivial bound as follows.

$$e_{other} \geq 2^{k+1} - 2^{2k-m+1} - (m - k) - d$$

Therefore, we have the following inequality

$$e \geq \frac{1}{2}(2m - 2k - \alpha)v_{m-k} + 2^{k+1} - 2^{2k-m+1} - (m - k) - d$$

After reformatting,

$$v_m + \cdots + v_{m-k+1} \geq n - \frac{2e - 2^{k+1} + 2^{2k-m+1} + (m - k) + d}{2m - 2k - \alpha}$$

Then, by substituting the inequality to $2e \geq (m - k)v_{m-k} + \cdots + mv_m$ and by the similar technique given in the proof of Theorem 3,

$$e \geq \frac{n}{2}(m - k + 1)\frac{2m - 2k - \alpha}{2m - 2k - \alpha + 1} + \frac{2^{k+1} - 2^{2k-m+1} - m + k - d}{2m - 2k + 1} \quad (7)$$

Again by the similar comparison, we can see that this bound is worse than bound 5 given in the first case. Thus, inequality (7) is the general lower bound on broadcast function, which completes the proof.   □

4   Conclusion

Simple comparison shows that Theorem 4 always gives a better bound than Theorem 1 when $d \geq 0$. In the future, we can further improve this bound, because inequality (3) unifies the coefficients of $v_{m-k+1}, \cdots v_m$ to be 1. This step over shrinks the lower bound. So, the gap between the current bound and the optimal bound may not be small.

Another work can be done in the future is a further generalization. Since Theorem 4 restricts n in the range $2^{m-1} + 1 \leq n \leq 2^m - 2^{\frac{m}{2}+1} + 1$, we can further explore the lower bound for the other side of interval $[2^{m-1} + 1, 2^m]$.

References

1. Averbuch, A., Shabtai, R.H., Roditty, Y.: Efficient construction of broadcast graphs. Discrete Appl. Math. **171**, 9–14 (2014)
2. Barsky, G., Grigoryan, H., Harutyunyan, H.A.: Tight lower bounds on broadcast function for n = 24 and 25. Discrete Appl. Math. **175**, 109–114 (2014)
3. Bermond, J.-C., Hell, P., Liestman, A.L., Peters, J.G.: Sparse broadcast graphs. Discrete Appl. Math. **36**, 97–130 (1992)
4. Dinneen, M.J., Fellows, M.R., Faber, V.: Algebraic constructions of efficient broadcast networks. In: Mattson, H.F., Mora, T., Rao, T.R.N. (eds.) AAECC 1991. LNCS, vol. 539, pp. 152–158. Springer, Heidelberg (1991). https://doi.org/10.1007/3-540-54522-0_104
5. Farley, A.M., Hedetniemi, S., Mitchell, S., Proskurowski, A.: Minimum broadcast graphs. Discrete Math. **25**, 189–193 (1979)
6. Fertin, G., Raspaud, A.: A survey on Knödel graphs. Discrete Appl. Math. **137**, 173–195 (2004)
7. Fraigniaud, P., Lazard, E.: Methods and problems of communication in usual networks. Discrete Appl. Math. **53**, 79–133 (1994)
8. Gargano, L., Vaccaro, U.: On the construction of minimal broadcast networks. Networks **19**, 673–689 (1989)
9. Grigni, M., Peleg, D.: Tight bounds on mimimum broadcast networks. SIAM J. Discrete Math. **4**, 207–222 (1991)
10. Grigoryan, H., Harutyunyan, H.A.: New lower bounds on broadcast function. In: Gu, Q., Hell, P., Yang, B. (eds.) AAIM 2014. LNCS, vol. 8546, pp. 174–184. Springer, Cham (2014). https://doi.org/10.1007/978-3-319-07956-1_16
11. Harutyunyan, H.A.: An efficient vertex addition method for broadcast networks. Internet Math. **5**(3), 197–211 (2009)
12. Harutyunyan, H.A., Li, Z.: A new construction of broadcast graphs. In: Govindarajan, S., Maheshwari, A. (eds.) CALDAM 2016. LNCS, vol. 9602, pp. 201–211. Springer, Cham (2016). https://doi.org/10.1007/978-3-319-29221-2_17
13. Harutyunyan, H.A., Li, Z.: Broadcast graphs using new dimensional broadcast schemes for Knödel graphs. In: Gaur, D., Narayanaswamy, N.S. (eds.) CALDAM 2017. LNCS, vol. 10156, pp. 193–204. Springer, Cham (2017). https://doi.org/10.1007/978-3-319-53007-9_18
14. Harutyunyan, H.A., Liestman, A.L.: More broadcast graphs. Discrete Appl. Math. **98**, 81–102 (1999)
15. Harutyunyan, H.A., Liestman, A.L.: Improved upper and lower bounds for k-broadcasting. Networks **37**, 94–101 (2001)
16. Harutyunyan, H.A., Liestman, A.L., Peters, J.G., Richards, D.: Broadcasting and gossiping. In: Handbook of Graph Theorey, pp. 1477–1494. Chapman and Hall, New York (2013)

17. Hedetniemi, S.M., Hedetniemi, S.T., Liestman, A.L.: A survey of gossiping and broadcasting in communication networks. Networks **18**, 319–349 (1988)
18. Hromkovič, J., Klasing, R., Monien, B., Peine, R.: Dissemination of information in interconnection networks (broadcasting & gossiping). In: Du, D.Z., Hsu, D.F. (eds.) Combinatorial Network Theory. Applied Optimization, vol. 1, pp. 125–212. Springer, Boston (1996). https://doi.org/10.1007/978-1-4757-2491-2_5
19. Khachatrian, L.H., Harutounian, H.S.: Construction of new classes of minimal broadcast networks. In: Conference on Coding Theory, Dilijan, Armenia, pp. 69–77 (1990)
20. Knödel, W.: New gossips and telephones. Discrete Math. **13**, 95 (1975)
21. König, J.-C., Lazard, E.: Minimum k-broadcast graphs. Discrete Appl. Math. **53**, 199–209 (1994)
22. Labahn, R.: A minimum broadcast graph on 63 vertices. Discrete Appl. Math. **53**, 247–250 (1994)
23. Maheo, M., Saclé, J.-F.: Some minimum broadcast graphs. Discrete Appl. Math. **53**, 275–285 (1994)
24. Mitchell, S., Hedetniemi, S.: A census of minimum broadcast graphs. J. Combin. Inform. Syst. Sci **5**, 141–151 (1980)
25. Park, J.-H., Chwa, K.-Y.: Recursive circulant: a new topology for multicomputer networks. In: International Symposium on Parallel Architectures, Algorithms and Networks, ISPAN 1994, pp. 73–80. IEEE (1994)
26. Saclé, J.-F.: Lower bounds for the size in four families of minimum broadcast graphs. Discrete Math. **150**, 359–369 (1996)
27. Shao, B.: On K-broadcasting in graphs. Ph.D. thesis, Concordia University (2006)
28. Ventura, J.A., Weng, X.: A new method for constructing minimal broadcast networks. Networks **23**, 481–497 (1993)
29. Weng, M.X., Ventura, J.A.: A doubling procedure for constructing minimal broadcast networks. Telecommun. Syst. **3**, 259–293 (1994)
30. Xiao, J., Wang, X.: A research on minimum broadcast graphs. Chin. J. Comput. **11**, 99–105 (1988)
31. Zhou, J.-G., Zhang, K.-M.: A minimum broadcast graph on 26 vertices. Appl. Math. Lett. **14**, 1023–1026 (2001)

Graph Colourings, Labelings and Power Domination

Orientations of 1-Factors and the List Edge Coloring Conjecture

Uwe Schauz$^{(\boxtimes)}$ (ID)

Xi'an Jiaotong-Liverpool University, Suzhou 215123, China
uwe.schauz@xjtlu.edu.cn

Abstract. As starting point, we formulate a corollary to the Quantitative Combinatorial Nullstellensatz. This corollary does not require the consideration of any coefficients of polynomials, only evaluations of polynomial functions. In certain situations, our corollary is more directly applicable and more ready-to-go than the Combinatorial Nullstellensatz itself. It is also of interest from a numerical point of view. We use it to explain a well-known connection between the sign of 1-factorizations (edge colorings) and the List Edge Coloring Conjecture. For efficient calculations and a better understanding of the sign, we then introduce and characterize the sign of single 1-factors. We show that the product over all signs of all the 1-factors in a 1-factorization is the sign of that 1-factorization. Using this result in an algorithm, we attempt to prove the List Edge Coloring Conjecture for all graphs with up to 10 vertices. This leaves us with some exceptional cases that need to be attacked with other methods.

Keywords: Combinatorial nullstellensatz · One-factorizations
Edge colorings · List edge coloring conjecture
Combinatorial algorithms

1 Introduction

Using the polynomial method, we prove the List Edge Coloring Conjecture[1] for many small graphs. This means, if such a graph G can be edge colored with k colors ($\chi'(G) \leq k$), then it can also be edge colored if the color of each edge e has to be taken from an arbitrarily chosen individual list L_e of k colors ($\chi'_\ell(G) \leq k$). There are no restriction on the lists, apart from the given cardinality k. So, in general, there are very many essentially different list assignments $e \mapsto L_e$, and brute-force attempts to find one coloring from every system of lists are computationally impossible. A way out may be found in the Combinatorial Nullstellensatz, which seems to be one of our strongest tools. It can also be used for list coloring of the vertices of a graph (see [1]), but it becomes even more powerful if applied to edge colorings of regular graphs.

---

[1] See [6, Sect. 12.20] for a discussion of the origins of this coloring conjecture.

© Springer International Publishing AG, part of Springer Nature 2018
L. Brankovic et al. (Eds.): IWOCA 2017, LNCS 10765, pp. 233–243, 2018.
https://doi.org/10.1007/978-3-319-78825-8_19

Ellingham and Goddyn [3] used it to prove the List Edge Coloring Conjecture for regular planar graphs of class 1. As, by definition, the edges of a class 1 graph G can be partitioned into $\Delta(G)$ color classes, the regular class 1 graphs are precisely the 1-factorable graphs. 1-factorable graphs, as we call regular class 1 graphs from now on, are also the first target in the current paper, but our results have implications for other graphs as well. In our previous paper [15], we could already prove the List Edge Coloring Conjecture for infinitely many 1-factorable complete graphs. There, we used a group action in connection with the Combinatorial Nullstellensatz. Häggkvist and Jansson [5] could prove the conjecture for all complete graphs of class 2. Nobody, however, has a proof for K_{16}, and 120 edges and 15 colors are completely out of reach for all known numeric methods, including the algorithms that we suggest here. That we cannot even prove the conjecture for all complete graphs shows how hard the problem is. Before this background, it is surprising that Galvin could prove the conjecture for all bipartite graphs [4]. His proof does not use the Combinatorial Nullstellensatz, but the so-called kernel method. Other methods were also used by Kahn [7], who showed that the List Edge Coloring Conjecture holds asymptotically, in some sense. Moreover, most of the mentioned results can also be generalized to edge painting [13, 14], an on-line version of list coloring that allows alterations of the lists during the coloration process.

This paper has three further sections. In Sect. 2, we formulate a corollary to the Combinatorial Nullstellensatz that does not require the consideration of any coefficients of polynomials, only evaluations of polynomial functions. There, we also explain a well-known connections between the sum of the signs over all 1-factorizations (edge colorings) of a graph and the List Edge Coloring Conjecture. In Sect. 3, we then provide another characterization of the sign. We explain how this can be used to calculate the sum of the signs over all 1-factorizations more efficiently. In Sect. 4, we explain to which conclusions this approach and our computer experiments with graphs on up to 10 vertices led.

2   A Nullstellensatz for List Colorings

We start our investigations from the following coefficient formula [12]:

Theorem 1. (Quantitative Combinatorial Nullstellensatz).
    Let $L_1, L_2, \ldots, L_n$ *be finite non-empty subsets of a field* $\mathbb{F}$, *set* $L := L_1 \times L_2 \times \ldots \times L_n$ *and define* $d := (d_1, d_2, \ldots, d_n)$ *via* $d_j := |L_j| - 1$. *For polynomials* $P = \sum_{\delta \in \mathbb{N}^n} P_\delta x^\delta \in \mathbb{F}[x_1, \ldots, x_n]$ *of total degree* $\deg(P) \le d_1 + d_2 + \cdots + d_n$, *we have*

$$P_d = \sum_{x \in L} N_L(x)^{-1} P(x),$$

where $N_L(x) = N_L(x_1, \ldots, x_n) := \prod_j N_{L_j}(x_j)$ *with* $N_{L_j}(x_j) := \prod_{\xi \in L_j \setminus x_j} (x_j - \xi) \neq 0$.

    In particular, if $\deg(P) \le d_1 + d_2 + \cdots + d_n$ *then*

$$P_d \neq 0 \implies \exists x \in L : P(x) \neq 0.$$

The implication in the second part is known as Alon's Combinatorial Nullstellensatz [2]. The coefficient P_d seems to plays a central role in the Combinatorial Nullstellensatz, but it is not really important in various applications. One may get a wrong impression form the fact that P_d is assumed as non-zero in that implication. There are applications of the theorem if the total degree $\deg(P)$ is strictly smaller than $d_1 + d_2 + \cdots + d_n$, and thus $P_d = 0$. If $P_d = 0$, then it cannot be that only one summand in the sum in that theorem is non-zero, and this mens that there cannot be only one solution to the problem that was modeled by P. So, if there exist a solution, say a trivial solution, than there must also be a second solution, a non-trivial solution. This is a very elegant line of reasoning, and it does not require us to look at the coefficient P_d at all. It is enough to know that the total degree is smaller than $d_1 + d_2 + \cdots + d_n$ and that there is a single trivial solution. Beyond that, the theorem can also be used to prove the existence of solutions to problems that do not have a trivial solution, for example the existence of a list coloring of a graph. In these cases, looking at the "leading coefficient" P_d appears to be unavoidable. However, to actually calculate P_d, usually, the best idea is to use the Quantitative Combinatorial Nullstellensatz again, just with changed lists L_j. In fact, the polynomial P can be changed, too, as long as the "leading coefficient" is not altered. So, theoretically, we can calculate P_d by applying the theorem to modified lists $\tilde{L}_j$ and a modified polynomial $\tilde{P}$. Afterwards, the theorem can then be applied a second time, to P and the original lists L_j, in order to prove the existence of a certain object. In this process, the coefficient P_d stands in the middle, playing a crucial role. The coefficient P_d, however, does not appear in the initial setting and also not in the final conclusion. Therefore, it must be possible to formulate a all-in-one ready-to-go corollary in which P_d does not occur. In providing that corollary, we free the user from the need to understand what P_d is. Of course, in its most general form, there are two polynomials P and $\tilde{P}$, and two list systems L and $\tilde{L}$, which make that corollary look more technical, but it avoids mentioning P_d and should be easier to apply in many situations:

Corollary 1. *For $j = 1, 2, \ldots, n$, let L_j and $\tilde{L}_j$ be finite non-empty subsets of a field $\mathbb{F}$ with $|L_j| = |\tilde{L}_j|$. Let N_L and $N_{\tilde{L}}$ be the corresponding coefficient functions over the cartesian products L and $\tilde{L}$ of these sets. If two polynomials $P, \tilde{P} \in \mathbb{F}[x_1, \ldots, x_n]$ of total degree at most $|L_1| + |L_2| + \cdots + |L_n| - n$ have the same homogenous component of degree $|L_1| + |L_2| + \cdots + |L_n| - n$ (or at least $\tilde{P}_d = P_d$), then*

$$\sum_{x \in \tilde{L}} N_{\tilde{L}}(x)^{-1} \tilde{P}(x) = \sum_{x \in L} N_L(x)^{-1} P(x)$$

and, in particular[2],

$$\sum_{x \in \tilde{L}} N_{\tilde{L}}(x)^{-1} \tilde{P}(x) \neq 0 \implies \exists x \in L : P(x) \neq 0.$$

---

[2] Also [14, Theorem 4.5]: $\sum N_{\tilde{L}}(x)^{-1} \tilde{P}(x) \neq 0 \implies P$ *is* $(|\tilde{L}_1|, \ldots, |\tilde{L}_n|)$*-paintable.*

We want to use this corollary to verify the existence of list colorings of graphs. Therefore, we apply the corollary to the *edge distance polynomials* P_G of graphs G. The edge distance polynomial of a multi-graph G on vertices $v_1, v_2, \ldots, v_n$ is a polynomial in the variables $x_1, x_2, \ldots, x_n$, with one variable x_i for each vertex v_i. It is defined as the product over all differences $x_i - x_j$ with $v_i v_j \in E(G)$ and $i < j$, where the factor $x_i - x_j$ occurs as many times in P as the edge $v_i v_j$ occurs in the multi-set $E(G)$. It is also called the graph polynomial and was introduced in [10]. We may view it as a polynomial over any field $\mathbb{F}$. If P_G is non-zero at a point $(x_1, x_2, \ldots, x_n)$ then the assignment $v_i \mapsto x_i$ is a proper vertex coloring of G. If the colors x_i are supposed to lie in certain lists L_i then the point $(x_1, x_2, \ldots, x_m)$ just has to be taken from the Cartesian product $L_1 \times L_2 \times \ldots \times L_m$. Here, we simple need to assume that the sets L_i lie in $\mathbb{F}$, or in an extension field of $\mathbb{F}$. This is no restriction, as one can easily embed the color lists (and their full union $\bigcup_i L_i$) into any big enough field $\mathbb{F}$. We might just take $\mathbb{F} = \mathbb{Q}$. With this ideas our corollary leads to the following more special result:

Corollary 2. *Let G be a multi-graph on the vertices $v_1, v_2, \ldots, v_n$. To each edge e, between any vertices v_i and v_j with $i < j$, choose a label a_e in a field $\mathbb{F}$ (possible $a_e = 0$) and associate the monomial $x_i - x_j - a_e$ to the edge e. Let P be the product over all these monomials. For $j = 1, 2, \ldots, n$, let L_j be a finite non-empty subset of $\mathbb{F}$, and define $\ell = (\ell_1, \ell_2, \ldots, \ell_n)$ via $\ell_j := |L_j|$. If $|E(G)| \leq \ell_1 + \ell_2 + \cdots + \ell_n - n$ then*

$$\sum_{x \in L} N_L(x)^{-1} P(x) \neq 0 \implies G \text{ is } \ell\text{-list colorable and } \ell\text{-paintable.}$$

In applications, one will often choose the a_e as zero and take the lists L_j all equal, but there are also examples where more complicated choices succeeded, as for example in the proof of the last lemma in [15]. Things can be further simplified if we examine edge colorings. In that case, one has to consider the line graph $L(G)$ of G and its edge distance polynomial $P_{L(G)}$. If G is k-regular, then $L(G)$ is the edge disjoint union of n complete graphs K_k, and $P_{L(G)}$ factors into n factors accordingly. For each vertex $v \in V(G)$ there is one complete graph K_k whose vertices are the edges $e \in E(v_j)$ incident with v. The corresponding factor of $P_{L(G)}$ is the edge distance polynomial $P_{K_k}(x_e \mid e \in E(v_j))$ of that K_k. If the k-regular graph is of class 1, i.e. if its edges can be colored with k colors, then, in the corresponding vertex colorings of $L(G)$, every color occurs one time at each vertex of that K_k. Therefore, by choosing equal lists, say all equal to $(k] := \{1, 2, \ldots, k\}$, the coefficients $N_L(x)^{-1}$ in the sum in the last corollary become all the same. More precisely, $N_L(x) = N_L(y)$ if $P_{L(G)}(x) \neq 0$ and $P_{L(G)}(y) \neq 0$. Moreover, $P_{L(G)}(x)$ assumes, up to the sign, the same value for every edge coloring $x : E(G) \to (k]$. So, in that sum, one basically only has to see which edge colorings contribute a positive sign and which ones a negative sign. This was already observed in [1]. It is easy to see that the definition of the sign given there depicts what we need, but we simplify that a bit. Basically, we only have to be able to say if two edge colorings have same or opposite sign. If $c : E \to (k]$ and $c_0 : E \to (k]$ are proper edge colorings, then $c|_{E(v)}$ and $c_0|_{E(v)}$ are bijections form the set $E(v)$ of edges at $v \in V(G)$ to $(k]$, and we set

$$\mathrm{sgn}_v(c, c_0) := \mathrm{sgn}\big((c_0|_{E(v)})^{-1} \circ c|_{E(v)}\big) \quad \text{and} \quad \mathrm{sgn}(c, c_0) := \prod_{v \in V(G)} \mathrm{sgn}_v(c, c_0),$$

$$(1)$$

where $(c_0|_{E(v)})^{-1} \circ c|_{E(v)}$ is a permutation of $E(v)$ and $\mathrm{sgn}((c_0|_{E(v)})^{-1} \circ c|_{E(v)})$ is its usual sign. We could have also defined $\mathrm{sgn}_v(c, c_0)$ as the sign of the inverse permutation $(c|_{E(v)})^{-1} \circ c_0|_{E(v)}$, or as sign of the permutations $c|_{E(v)} \circ (c_0|_{E(v)})^{-1}$ or $c_0|_{E(v)} \circ (c|_{E(v)})^{-1}$ in S_k. This is all the same. It is the right definition here, because the sign of a permutation ρ in S_k is exactly the sign of the edge distance polynomial P_{K_k} of K_k evaluated at $(\rho_1, \rho_2, \ldots, \rho_k)$,

$$\mathrm{sgn}(\rho) = \frac{P_{K_k}(\rho_1, \rho_2, \ldots, \rho_k)}{|P_{K_k}(\rho_1, \rho_2, \ldots, \rho_k)|}. \tag{2}$$

Hence, we only need to fix one edge coloring $c_0 \colon E(G) \to (k]$ and then count how many colorings $c \colon E(G) \to (k]$ are positive or negative with respect to that *reference coloring*. It is convenient to define an absolute sign $\mathrm{sgn}(c)$ through

$$\mathrm{sgn}(c) := \mathrm{sgn}(c, c_0)\, \mathrm{sgn}(c_0), \tag{3}$$

where $\mathrm{sgn}(c_0)$ is fixed given as either $+1$ or -1. In this section, however, it does not mater whether c_0 is viewed as positive or negative, and we postpone the stipulation of $\mathrm{sgn}(c_0)$ till later. With that, we arrive at [1, Corollary 3.9]:

Corollary 3. *Let $G = (V, E)$ be a k-regular graph and let $C(G)$ be the set of its proper edge colorings $c \colon E \longrightarrow (k]$. Then*

$$\sum_{c \in C(G)} \mathrm{sgn}(c) \neq 0 \implies G \text{ is } k\text{-list edge colorable and edge } k\text{-paintable.}$$

Actually, we may assume that G has even many vertices, as 1-factors and k-edge colorings only exist if there are even many vertices. If we exchange two colors in an edge coloring $c \colon E \to (k]$ of a k-regular graph G, then all the factors $\mathrm{sgn}_v(c)$ in $\mathrm{sgn}(c)$ change, but sign $\mathrm{sgn}(c)$ does not change. Therefore, it makes sense to define the sign of a 1-factorization. A 1-factorization F of G is a partition $F = \{F_1, F_2, \ldots, F_k\}$ of the edge set $E(G)$ into k 1-factors (perfect matchings). To every 1-factorization F there are $k!$ edge colorings c with F as set of fibers $c^{-1}(\{\alpha\})$. All of them have the same sign, and we define

$$\mathrm{sgn}(F) := \mathrm{sgn}(c). \tag{4}$$

With that, the last corollary can be rewritten as follows:

Corollary 4. *Let $G = (V, E)$ be a k-regular graph and let $\mathrm{OF}(G)$ be the set of 1-factorizations of G. Then*

$$\sum_{F \in \mathrm{OF}(G)} \mathrm{sgn}(F) \neq 0 \implies G \text{ is } k\text{-list edge colorable and edge } k\text{-paintable.}$$

3   Another Characterization of the Sign

In this section, G denotes a k-regular graph on the vertices $v_1, v_2, \ldots, v_{2n}$, and $F = \{F_1, F_2, \ldots, F_k\}$ denotes a 1-factorization of G. We examine the sign $\mathrm{sgn}(F)$ in more detail, starting from the following definition:

Definition 1. *Let $F_1 = \{e_1, e_2, \ldots, e_n\}$ be a 1-factor of a k-regular graph G on the vertices $v_1, v_2, \ldots, v_{2n}$. Let $1 \le i_k < j_k \le 2n$ be such that $e_k = v_{i_k} v_{j_k}$, for $k = 1, 2, \ldots, n$. We say that an edge $e_k \in F_1$ intersects another edge $e_\ell \in F_1$ if $i_k < i_\ell < j_k < j_\ell$ or $i_\ell < i_k < j_\ell < j_k$. We define*

$$\mathrm{int}(e_k, e_\ell) := \begin{cases} 1 & if\ e_k\ intersects\ e_\ell, \\ 0 & otherwise, \end{cases}$$

and set

$$\mathrm{int}(F_1) := \sum_{1 \le k < \ell \le n} \mathrm{int}(e_k, e_\ell) \quad and \quad \mathrm{sgn}(F_1) := (-1)^{\mathrm{int}(F_1)}.$$

If we position the $2n$ vertices consecutively around a cycle and draw the edges as strait lines, then an intersection is an actual intersection between lines. With this picture in mind, it is not hard to see that, if $\mathrm{int}(v_i v_j, F_1)$ denotes the number of intersections of an edge $v_i v_j \in F_1$ with other edges in F_1, then

$$\mathrm{int}(v_i v_j, F_1) \equiv j - i - 1 \pmod 2. \tag{5}$$

This, however, does not help to determine the sign $\mathrm{sgn}(F_1)$ of F_1, as

$$\sum_{e \in F_1} \mathrm{int}(e, F_1) = 2\,\mathrm{int}(F_1), \tag{6}$$

with a 2 in front of $\mathrm{int}(F_1)$. Counting the number of all intersections of each edge e is not the right approach here. We may order F_1 to $\overrightarrow{F_1} = (e_1, e_2, \ldots, e_n)$ and count only the intersections of an edge e_k with the *subsequent edges* e_ℓ, $\ell > k$. If $\mathrm{int}(e_k, \overrightarrow{F_1})$ denotes this number, then the corresponding sum yields the desired result,

$$\mathrm{int}(F_1) = \sum_{e \in F_1} \mathrm{int}(e, \overrightarrow{F_1}). \tag{7}$$

Hence,

$$\mathrm{sgn}(F_1) = \prod_{e \in F_1} \mathrm{sgn}(e, \overrightarrow{F_1}), \tag{8}$$

if we set

$$\mathrm{sgn}(e, \overrightarrow{F_1}) := (-1)^{\mathrm{int}(e, \overrightarrow{F_1})}. \tag{9}$$

This formula may be used to calculate the sign of a 1-factor in algorithms that generate a 1-factor by successively adding single edges. And, there is also an analog to Formula (5). We may just count how many of the vertices b that lie between the two ends v_{i_k} and v_{j_k} of the edge e_k are not yet matched when we add e_k to the sequence $(e_1, e_2, \ldots, e_{k-1})$. So,

$$\text{int}(v_{i_k} v_{j_k}, \overrightarrow{F_1}), \equiv \left| \{ b \mid i_k < b < j_k, \ b \notin e_1 \cup e_2 \cup \cdots \cup e_{k-1} \} \right| \quad (\text{mod } 2). \quad (10)$$

In our algorithm, we kept track of these unmatched b by using a doubly linked linear lists. From each unmatched vertex b, we have at any time a link to the unmatched vertex before b and a link to the unmatched vertex after b. Updating these links can then be done without shifting all subsequent vertices one place forward.

The next theorem shows that the signs of the 1-factors in a 1-factorization F can be used to calculate the sign of F. This can then be used in algorithms that calculate the 1-factorizations of a graph by successively adding new 1-factors. The advantage is that the sign of a 1-factor that is added at a certain point has to be calculated only once, for all the 1-factorizations that are generate afterwards, by adding more 1-factors in all possible ways. It is clear that the formula in the next theorem does not really depend on the sign of the underlying reference coloring c_0, or the equivalent *reference 1-factorization* $\{ c_0^{-1}(\{\alpha\}) \mid \alpha \in (k] \}$. But, to avoid additional minus signs in the theorem, we synchronize our different signs at this point, and define

$$\text{sgn}(c_0) := \prod_{\alpha \in (k]} (-1)^{\text{int}(c_0^{-1}(\{\alpha\}))} = (-1)^{\text{int}(c_0)} \in \{-1, +1\}, \quad (11)$$

where

$$\text{int}(c_0) := \sum_{\alpha \in (k]} \text{int}(c_0^{-1}(\{\alpha\})) \quad (12)$$

is the number of intersections between edges of equal color in c_0, if the vertices $v_1, v_2, \ldots, v_{2n}$ are arranged consecutively on a cycle and the edges are drown as strait lines. With this stipulation of the sign of the reference coloring c_0, we have the following theorem:

Theorem 2. Let $G = (V, E)$ be a k-regular graph on the vertices $v_1, v_2, \ldots, v_{2n}$, and let $F = \{F_1, F_2, \ldots, F_k\}$ be a 1-factorization of G. Then

$$\text{sgn}(F) = \prod_{i=1}^{k} \text{sgn}(F_i).$$

In other words, if $c \colon E \longrightarrow (k]$ is an edge coloring, then

$$\text{sgn}(c) = (-1)^{\text{int}(c)},$$

where $\text{int}(c)$ is the number of intersections between edges of equal color, if the vertices $v_1, v_2, \ldots, v_{2n}$ of G are arranged consecutively on a cycle and the edges are drown as strait lines.

Probably, this theorem can somehow be proven by induction. We derived it in a topological way, using Jordan's Curve Theorem. From that theorem, we know that any two closed curves on the sphere have even many intersections with each other. We also used the fact that the sign of a permutation $\rho \in S_k$ is -1 to the power of the number of *inversions* of ρ. Here, a pair $(i_1, i_2) \in (k]^2$ with $i_1 < i_2$ is an inversion of ρ if $\rho(i_1) > \rho(i_2)$. We used that this property can be characterized as intersection of strait lines in $\mathbb{R}^2$. Indeed, the pair (i_1, i_2) is an inversion if and only if the line from (i_1, h_1) to $(\rho(i_1), h_2)$ intersects with the line from (i_2, h_1) to $(\rho(i_2), h_2)$, where h_1 and h_2 are any two different real numbers.

4   The List Chromatic Index of Small Graphs

Based on Corollary 4 and the results of the previous section, we have tried to determine the list chromatic index $\chi'_\ell(G)$ of all graphs on up to 10 vertices, in an attempt to prove the List Edge Coloring Conjecture for small graphs. We implemented the approach explained in the previous sections in SageMath [11], importing regular graphs from the webpage [8] described in [9]. With that we attacked all regular graphs on 4, 6, 8 or 10 vertices. The results are shown in the first paragraph of the following subsection. We tried than to draw conclusions about the list chromatic index of all graphs with up to 10 vertices. We did this by considering embeddings into regular graphs on even many vertices. Unfortunately, there are many exceptional cases and special circumstances. We report about these difficulties, and some ideas how to overcome them, in quite a view case distinctions. It was not possible to go through all the cases and to prove the List Edge Coloring Conjecture for all graphs on up to 10 vertices. If, however, the List Edge Coloring Conjecture shall be proven for just one particular small graph, one may find a way to do so within our case distinctions.

In the following case distinctions, the word *graph* stands for connected graph, and a regular graph G is a *zero-sum graph* if the sum $\sum \text{sgn}(F)$ over all 1-factorizations $F \in \text{OF}(G)$ vanishes. We call a graph *small* if it has at most 10 vertices, and we call it *even* resp. *odd* if it has even resp. odd many vertices.

4.1   Small Even Graphs

Regular Graphs. By checking all small regular even graphs, we found only three graphs of class 2. The Petersen graph and the following two graphs:

Our main method does not apply to class 2 graphs. In these three cases, however, one can simply add a suitable 1-factor, and prove the List Edge Coloring Conjecture for the resulting graph of class 1. It is, in fact, possible to choose the 1-factor in a way that the extended graph is not a zero-sum graph. So, in the shown three cases, the List Edge Coloring Conjecture holds. Unfortunately, our method also failed in a number of other cases, where the sum $\sum \operatorname{sgn}(F)$ over all 1-factorizations $F \in OF(G)$ simple was zero. The smallest zero-sum graph is $K_{3,3}$, but this graph is bipartite. Hence, it meets the List Edge Coloring Conjecture by Galvin's Theorem [4]. On 8 vertices, there are exactly three zero-sum graphs. The complement $\overline{C_3 \cup C_5}$ of the disjoint union of a 3-cycle and a 5-cycle, and the following graph and its complement:

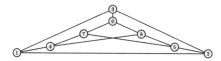

On 10 vertices there are 51 zero-sum graphs out of 164 regular class 1 graphs (1-factorable graphs). There are 5 zero-sum graphs of degree 3, 17 of degree 4, 18 of degree 5, 8 of degree 6, and 3 of degree 7. It seems that, in every small zero-sum graph, one can find a symmetry of order 2 that turns even edge coloring ($\operatorname{sgn} = +1$) into odd ones ($\operatorname{sgn} = -1$) and vice versa; which explains the vanishing sum. The most simple symmetry of this kind is given if two non-adjacent vertices of odd degree have the same neighbors, or if two adjacent vertices of even degree have the same neighbors. But, there are also more complicated cases. In the complement of the Petersen graph, for example, it is more difficult to understand how odd and even edge colorings are matched through a graph symmetry. Overall, it should be possible to proof the List Edge Coloring Conjecture for all found zero-sum graphs with other methods. Some well chosen case distinctions with respect to the color lists might suffice. This kind of reasoning, however, is usually quite tedious and depends very much on the structure of the graph.

Non-regular Graphs. If a regular graph G is of class 1 and meets the List Edge Coloring Conjecture, then every subgraph of same maximal degree still is of class 1 and still meets the List Edge Coloring Conjecture. With this argument, most non-regular small even graphs can be proven to be of class 1 and to meet the List Edge Coloring Conjecture. We just have to consider regular even extensions of same maximal degree. If an extension is still small, we may apply our findings about small regular even graphs. There are, however, three difficulties:

(i) Some small non-regular even graphs cannot be embedded into a regular graph by adding edges only, which would keep these graphs small. Several examples of this kind can be constructed from k-regular graphs ($k \geq 3$) that contain an induced path $u-v-w$ by removing the edges uv and vw, and inserting the edge uw.

(ii) The three small regular even graphs of class 2 are not suitable as regular extensions in this line of reasoning. Some of their subgraphs are actually of class 2, and we can only conclude that these class 2 subgraphs meet the List Edge Coloring Conjecture.

(iii) There are still some open cases among the small regular even class 1 graphs, for which we not yet have proven the conjecture. Circumventing these cases is not always possible, as there may not be many different ways to add edges.

4.2 Small Odd Graphs

Class 2 Graphs (including all Regular Graphs). All regular graphs of odd order
are of class 2, as no 1-factors exist. Moreover, if we start from an k-regular odd
graph and remove less than $k/2$ edges, then the graph remains in class 2, because
it is still *overfull* ($|E| > \Delta \cdot \lfloor |V|/2 \rfloor$). All graphs that we obtain in this way have
maximal degree k, which is necessarily an even number, as the initial regular
graph was odd. Odd class 2 graphs with odd maximal degree are not obtained
in this way. But, they do exist. One example is K_8 with one edge subdivided by
a new vertex, which is still overfull. To prove the List Edge Coloring Conjecture
for this graph and for all class 2 graphs G, however, we do not need to embed
G into a regular class 2 graph of same maximal degree $\Delta(G)$. To prove that
a graph G (whether of class 2 or not) has list chromatic index $\Delta(G) + 1$, we
may simply embed it into a class 1 graph whose maximal degree is $\Delta(G) + 1$.
If the List Edge Coloring Conjecture was proven for that extension graph, then
$\chi'_\ell(G) \le \Delta(G) + 1$, and then the List Edge Coloring Conjecture holds for G if G
is of class 2. We may also add vertices. In this way, most small odd graphs can
be embedded into a suitable regular graph. As in the case of even non-regular
graphs, however, there are three difficulties:

(i) Some small odd graphs cannot be embedded into a regular graph by adding
only one vertex and some edges, which would keep these graphs small. One
example of this kind is K_8 with one edge subdivided by a new vertex.

(ii) The three small regular even graphs of class 2 are not suitable as regular
extensions in this line of reasoning and must be circumvented. Since the
maximal degree can go up by one, however, there is a lot of flexibility. One
can show that the three exceptions of class 2 are not needed as extension
graphs. Still, circumventing them is an additional difficulty if one tries to
draw general conclusions.

(iii) There are still some open cases among the small regular even class 1 graphs.
If we try to embed a single small odd class 2 graph, it is often easy to
circumvent the open cases. But, in general examinations, avoiding open
cases is difficult.

Class 1 Graphs. The majority of small odd graphs are of class 1 and, in particular, non-regular. For these graphs, embedding without increasing the maximal
degree frequently works. One can try to add just one vertex and some additional
edges. In this way, the results about small even regular graphs can be applied.
As in the other case where we discussed embedding, there are three difficulties:

(i) Adding just one vertex, to stay within the small graphs, does not work if
there are not enough vertices of sub-maximal degree to which the new vertex
can be connected. In this regard, there are obviously more problematic cases
as in the discussion of small odd non-regular graphs of class 2, where we
could increase the maximal degree by one.

(ii) The three small regular even graphs of class 2 are not suitable as regular
extensions in this line of reasoning. However, if we remove just one vertex

from any of them, they remain in class 2. Hence, the three class 2 graphs do not appear as single-vertex extensions of class 1 graphs. And, if we need to add a vertex plus some edges, we may be able to circumvent these three graphs.

(iii) If we try to embed a single small odd class 1 graph, circumventing the open cases among the small regular even class 1 graphs is sometimes not possible.

References

1. Alon, N.: Restricted colorings of graphs. In: Surveys in Combinatorics. London Mathematical Society Lecture Notes Series, vol. 187, pp. 1–33. Cambridge University Press, Cambridge (1993)
2. Alon, N.: Combinatorial nullstellensatz. Comb. Probab. Comput. **8**(1–2), 7–29 (1999)
3. Ellingham, M.N., Goddyn, L.: List edge colourings of some 1-factorable multigraphs. Combinatorica **16**, 343–352 (1996)
4. Galvin, F.: The list chromatic index of a bipartite multigraph. J. Comb. Theory Ser. B **63**, 153–158 (1995)
5. Häggkvist, R., Janssen, J.: New bounds on the list-chromatic index of the complete graph and other simple graphs. Comb. Probab. Comput. **6**, 295–313 (1997)
6. Jensen, T.R., Toft, B.: Graph Coloring Problems. Wiley, New York (1995)
7. Kahn, J.: Asymptotically good list-colorings. J. Comb. Theory Ser. A **73**(1), 1–59 (1996)
8. Meringer, M.: Connected regular graphs. http://www.mathe2.uni-bayreuth.de/markus/reggraphs.html
9. Meringer, M.: Fast generation of regular graphs and construction of cages. J. Gr. Theory **30**, 137–146 (1999)
10. Petersen, J.: Die theorie der regularen graphs. Acta Math. **15**, 193–220 (1891)
11. SageMath: The sage mathematics software system (version 7.4.1). The Sage Developers (2017). http://www.sagemath.org
12. Schauz, U.: Algebraically solvable problems: describing polynomials as equivalent to explicit solutions. Electron. J. Comb. **15**, R10 (2008)
13. Schauz, U.: Mr. Paint and Mrs. Correct. Electron. J. Comb. **15**, R145 (2008)
14. Schauz, U.: A paintability version of the combinatorial Nullstellensatz, and list colorings of k-partite k-uniform hypergraphs. Electron. J. Comb. **17**(1), R176 (2010)
15. Schauz, U.: Proof of the list edge coloring conjecture for complete graphs of prime degree. Electron. J. Comb. **21**(3), 3–43 (2014)

On Solving the Queen Graph Coloring Problem

Michel Vasquez$^{(\boxtimes)}$ and Yannick Vimont

Ecole des mines d'Alès, Alès, France
{michel.vasquez,yannick.vimont}@mines-ales.fr

Abstract. The chromatic numbers of $queen - n^2$ graphs are difficult to determine when $n > 9$ and n is a multiple of 2 or 3. In previous works [6,7], we have proven that this number (denoted χ_n) is equal to n for $n \in \{12, 14, 15, 16, 18, 20, 21, 22, 24, 28, 32\}$ and that $\chi_{10} = 11$. This article describes how, by extending slightly further the previous work, the chromatic number of $queen - 26^2$ and $queen - 30^2$ can be obtained. A more general result, proving that $\chi_{2n} = 2n$ and $\chi_{3n} = 3n$ for infinitely many values of n, is then presented.

1 Introduction

Given an $n \times n$ chessboard, a queen graph is one with n^2 vertices, each of which corresponds to a square of the board. Two vertices are connected by an edge if the corresponding squares lie in the same row, column or diagonal (*both ascending and descending diagonals*); this set-up matches the rules for the queen's moves in a game of chess. The coloring problem on this graph consists of finding the minimum number of colors necessary for placing n^2 queens on the board such that no two queens of the same color can attack one another.

When the size of the chessboard is prime with 2 and 3, it is straightforward to color the graph with just n colors by using the knight's moves (*on a toroidal chessboard*) for each color position. Hence, $\chi_n = n$ for $n \in \{5, 7, 11, 13, 17, 19, ...\}$. Even recent graph coloring algorithms (see, for example, [2–5]) do not obtain optimal values for $queen - n^2$ graphs, such as $12 \leq n \leq 16$, and their best results for $queen - 12^2$, $queen - 14^2$, $queen - 15^2$ and $queen - 16^2$ are respectively 13, 15, 16 and 17 colors. These are obviously general approaches dedicated to a large set of graph coloring instances. In contrast, the present work will solely focus on specific algorithms to derive the chromatic numbers on the queen graphs.

In [6], the algorithm implemented works with the independent sets (IS) of the graph and employs an efficient backtracking procedure. Research on the queen graph chromatic number has indeed been transformed into a decision problem that requires recovering the chessboard with n IS, containing exactly n vertices each. The diagonals of the chessboard constitute cliques. After each IS enumeration, the number of non-colored vertices on each diagonal can be computed. If this number is greater than the number of remaining IS for the

L. Brankovic et al. (Eds.): IWOCA 2017, LNCS 10765, pp. 244–251, 2018.
https://doi.org/10.1007/978-3-319-78825-8_20

covering problem, then it is impossible to color the queen $- n^2$ graph with just n colors. The resulting algorithm is able to complete the search for $n = 10$ and $n = 12$ and obtain many different certificates for $\chi_{14} = 14$; however, the computational time needed to yield results on larger instances is too long.

1	2	3	4	5	6	7	8	9	10	11	12	13	14	15	16	17	18	19	20	21
10	8	1	2	15	11	9	18	12	13	17	20	6	16	5	7	3	21	14	4	19
2	14	10	12	1	7	16	5	17	11	15	9	4	3	8	13	6	19	20	21	18
3	1	2	20	18	8	11	9	14	12	13	17	15	5	4	10	21	16	7	19	6
9	7	4	1	2	3	14	13	6	17	5	18	11	20	16	15	19	12	21	10	8
20	12	11	15	16	21	19	17	1	2	7	14	10	8	6	9	18	4	3	13	5
16	6	5	8	20	4	18	7	21	19	10	1	2	9	12	11	14	13	15	17	3
14	20	7	6	10	5	21	19	13	18	9	15	16	11	3	1	2	17	8	12	4
11	4	8	16	19	15	13	20	3	6	12	10	18	2	9	21	5	14	1	7	17
18	10	21	9	3	14	17	16	15	4	8	6	5	12	11	19	1	7	13	2	20
19	9	16	11	6	12	15	21	18	5	14	4	7	17	20	3	8	2	10	1	13
15	11	19	18	9	13	2	3	4	8	6	5	20	10	21	14	12	1	17	16	7
21	18	9	14	4	17	1	11	7	16	3	8	12	15	19	20	13	10	6	5	2
5	3	6	21	11	9	12	2	10	20	1	7	19	13	17	4	16	8	18	15	14
12	21	20	19	14	2	3	1	11	9	16	13	17	18	7	5	15	6	4	8	10
4	19	12	5	7	1	8	6	16	14	18	11	9	21	13	17	10	20	2	3	15
6	16	17	3	13	20	10	15	2	7	4	21	8	19	14	12	11	9	5	18	1
8	13	18	10	17	16	5	4	20	21	19	3	14	1	2	6	7	15	11	9	12
7	17	15	13	8	19	6	12	5	1	20	2	21	4	10	18	9	3	16	14	11
13	5	14	17	12	18	4	10	8	15	21	19	3	7	1	2	20	11	9	6	16
17	15	13	7	21	10	20	14	19	3	2	16	1	6	18	8	4	5	12	11	9

Fig. 1. A certificate for $\chi_{21} = 21$ obtained by combination of rotations. Squares labeled with the same number represent independent set (or stable) of the graph *(i.e. a set of vertices without connecting edges)*. Note also that the set of vertices colored with numbers in $\{1, 9, 17, 21\}$ is invariant by rotation

To further prune this search tree, geometric characteristics are imposed on the colors output by this algorithm. The goal here is to clearly fix more than one independent set at the same time. For instance, at each search tree node, the queens of four different colors are simultaneously assigned by $\frac{\pi}{2}$ rotations. For example, in Fig. 1 solution, the colors 1, 9, 17 and 21 are simultaneously assigned. This step effectively halves the depth of the search tree explored by the algorithm. More details on this approach are given in [7]. This assumption, regarding the geometric characteristics of a possible solution, has yielded n-colorings of queen graphs with a dimension equal to $15, 16, 18, 20, 21, 22, 24, 28$ and 32. This heuristic however has its limitations, and the first certificate proving that $\chi_{26} = 26$ was found in 2005 after $1400000\,$s of computational time on a PENTIUM4 2.4 Ghz CPU.

The interested reader is invited to visit Vašek Chvátal's home page[1], which offers elegant demonstrations for queen $- 8^2$ and queen $- 9^2$, along with most of the certificates for the chromatic number values cited above.

---

[1] http://ww.cs.concordia.ca/~chvatal/queengraphs.html.

2   Additional Assumptions

The main underlying notion leading to the solution of queen -26^2 and queen -30^2 instances within reasonable computational time calls for combining geometric operators, like horizontal (H) and vertical (V) symmetries and $\frac{\pi}{2}$ rotations (R), all of which were used separately in [7].

In denoting I for the identity, we have used the following combinations: $\mathcal{C}^0 = [I]$, $\mathcal{C}^1 = [I, V]$, $\mathcal{C}^2 = [I, R^2]$, $\mathcal{C}^3 = [I, V, H, H \circ V]$ and $\mathcal{C}^4 = [I, R, R^2, R^3]$. As an example, in the certificate of queen -26^2 (Fig. 2), $\mathcal{C}^3(1) = [I(1), V(1), H(1), H \circ V(1)] = [1, 26, 18, 9]$ (1 represents either one vertex of color 1 or the n vertices of the independent set number 1). This approach is appropriate since if the square of the chessboard is free, then the other one obtained by combinations $\mathcal{C}^1$ or $\mathcal{C}^2$, or the three others obtained by combinations $\mathcal{C}^3$ or $\mathcal{C}^4$ are also free. This is obvious since: $V \circ V = I$, $R^4 = I$, $H \circ V \circ H \circ V = H \circ V \circ V \circ H = H \circ I \circ H = H \circ H = I$. Moreover, the image by symmetry (or by rotation) of an independent set remains an independent set.

The choice of the combination(s) can then be expressed mathematically by solving simple equations such as: $n = 4 \times \mathbf{r} + 2 \times \mathbf{s} + 1 \times \mathbf{i}$, where $\mathbf{r}, \mathbf{s}$ and $\mathbf{i}$ are integers: $\mathbf{r}$ represents either number of combinations $\mathcal{C}^3$ or $\mathcal{C}^4$, which implies 4 vertices; $\mathbf{s}$ represents either $\mathcal{C}^1$ or $\mathcal{C}^2$, which implies 2 vertices; and $\mathbf{i}$ corresponds to the identity $\mathcal{C}^0$, which naturally implies only 1 vertex. The objective of this game is thus to minimize the sum $\mathbf{r} + \mathbf{s} + \mathbf{i}$ under the constraint $n = 4.\mathbf{r} + 2.\mathbf{s} + \mathbf{i}$.

We can write for example $26 = 4 \times 5 + 2 \times 3$. Hence, the certificate in Fig. 2 has been obtained by enumerating 5 sets of 4 IS and then enumerating 3 sets of 2 IS, as shown in Table 1. Consequently, 8 search levels are being handled rather than 13, and a solution is found in 2200 s vs. 1400000.

Table 1. The 5 first and 3 last levels of search for the queen -26^2 coloring problem

IS1	IS2	IS3	IS4
1	$26 = V(1)$	$18 = H(1)$	$9 = H \circ V(1)$
2	$25 = V(2)$	$21 = H(2)$	$6 = H \circ V(2)$
3	$24 = V(3)$	$17 = H(3)$	$10 = H \circ V(3)$
4	$23 = V(4)$	$19 = H(4)$	$8 = H \circ V(4)$
12	$15 = V(12)$	$13 = H(12)$	$14 = H \circ V(12)$

IS1	IS2
5	$22 = V(5)$
7	$20 = V(7)$
11	$16 = V(11)$

We observe that the set of vertices colored with numbers in $\{5, 7, 11, 16, 20, 22\}$ is invariant relative to H and V (*When ordering the combinations, care must be taken to ensure that only invariant subsets of squares are being handled by isometries*).

This opportunistic approach has also provided the certificate for $\chi_{30} = 30$, with $\mathcal{C}^1$ and $\mathcal{C}^3$ as geometric operators and with settings $\mathbf{r} = 6$ and $\mathbf{s} = 3$. Again, Fig. 3 reveals that $18 = V(13)$ and $19 = V(12)$ and $24 = V(7)$: these are the six

```
 1  2  3  4  5  6  7  8  9 10 11 12 13 14 15 16 17 18 19 20 21 22 23 24 25 26
15  4  1 20  2 13 16  5 19  6  3  9 17 10 18 24 21  8 22 11 14 25  7 26 23 12
12  3  2 22  1 18  4 14  8 16 10 20 21  6  7 17 11 19 13 23  9 26  5 25 24 15
 2  1  4 15 19 24 17  9 21 11 13  5  7 20 22 14 16  6 18 10  3  8 12 23 26 25
 4 12 14  1  3 20  2 10 16 22  9 21  8 19  6 18  5 11 17 25  7 24 26 13 15 23
11 10 13  2 22  7 24  4  1 19 15 18  6 21  9 12  8 26 23  3 20  5 25 14 17 16
16  5  6 25 13  8 20 18 17 12  1  3  4 23 24 26 15 10  9  7 19 14  2 21 22 11
17 14  9 12  7 19 25 11  5 21 23 24 26  1  3  4  6 22 16  2  8 20 15 18 13 10
22 11 10 23 24 21 18 13 25 26  7  8 15 12 19 20  1  2 14  9  6  3  4 17 16  5
 8 16 20 17 26 14 23  3 12 18  6 25  5 22  2 21  9 15 24  4 13  1 10  7 11 19
24 25 23  5  6 17 26 16 15 14 19  7 18  9 20  8 13 12 11  1 10 21 22  4  2  3
 9 26  7 24  8 22 15 25 10 23 21 14 11 16 13  6  4 17  2 12  5 19  3 20  1 18
25 23 16 26 15  9 13 21  7 17  5 19 24  3  8 22 10 20  6 14 18 12  1 11  4  2
 6  8 11  9 14 26 12  2 20  3 22  4 10 17 23  5 24  7 25 15  1 13 18 16 19 21
26  9  5 10 23 11 14  6 24  8  2 15 20  7 12 25 19  3 21 13 16  4 17 22 18  1
10  6  8 16 25  3  9  7 14 15  4 22  1 26  5 23 12 13 20 18 24  2 11 19 21 17
23  7 22  3  9 15  8 17 13  1 25  6 16 11 21  2 26 14 10 19 12 18 24  5 20  4
 5 20 24  8 10  2  1 12  6  9 16 23 14 13  4 11 18 21 15 26 25 17 19  3  7 22
 3 15 26 13 20  4  6 22 11  2  8 10  9 18 17 19 25 16  5 21 23  7 14  1 12 24
 7 22 25  6 12 23 11  1  3 13 18 17 19  8 10  9 14 24 26 16  4 15 21  2  5 20
20 24 12 21 11  5 10 19 18  4 14  1 25  2 26 13 23  9  8 17 22 16  6 15  3  7
19 13 15 18 17 16 21 24 22 20 26  2 23  4 25  1  7  5  3  6 11 10  9 12 14  8
21 18 19 14  4 10  3 26  2  7 12 16 22  5 11 15 20 25  1 24 17 23 13  8  9  6
13 17 21  7 18  1 19 15 23  5 24 11  2 25 16  3 22  4 12  8 26  9 20  6 10 14
14 19 18 11 21 12  5 20  4 25 17 26  3 24  1 10  2 23  7 22 15  6 16  9  8 13
18 21 17 19 16 25 22 23 26 24 20 13 12 15 14  7  3  1  4  5  2 11  8 10  6  9
```

Fig. 2. A certificate for $\chi_{26} = 26$ obtained by setting $26 = 4 \times 5 + 2 \times 3$

```
 1  2  3  4  5  6  7  8  9 10 11 12 13 14 15 16 17 18 19 20 21 22 23 24 25 26 27 28 29 30
 3  4  6 22  2 18 16  1 14  7  5 23 11 19 21 10 12 20  8 26 24 17 30 15 13 29  9 25 27 28
 6  1  2 13  4  3 15 11 21 23 14 19  5  9  7 24 22 26 12 17  8 10 20 16 28 27 18 29 30 25
 2  7  4  3  1 13 12 10 25 22 15  8 20 26 14 17  5 11 23 16  9  6 21 19 18 30 28 27 24 29
 4  3  1  2  6  9 17 13 26 21 24 11 23 15 12 19 16  8 20  7 10  5 18 14 22 25 29 30 28 27
24 13 12 16 23  5  4  2  1 14 22  3  6 21 11 20 25 28  9 17 30 29 27 26  8 15 19 18  7 10
 5 23  7  1 18 21 19 16 20 25  4 17  2  3 22  9 28 29 14 27  6 11 15 12 10 13 30 24  8 26
25  8 21 15 24  4 26 18 12  1  9 28 29 17 20 11 14  2  3 22 30 19 13  5 27  7 16 10 23  6
21 19 23 26 25 17 28 24 29 20 18 16  1 27  9 22  4 30 15 13 11  2  7  3 14  6  5  8 12 10
26  6 22 28 27 29  7 20 10 16 17 30 12 18 13 19 24  1 14 15 21 11 23  2  4  3  9 25  5  8
29 11 27  8 10  7  9 30  5 13 28 15 19  6 17 14 25 12 16  3 18 26  1 22 24 21 23  4 20  2
14 15 24 30 28 11 29 26 27  9 12 13 25  8 10 21 23  6 18 19 22  4  5  2 20  3  1  7 16 17
18 22 25 19 14 23 27 28 24  5 10 20 30 29 16 15  2  1 11 21 26  7  3  4  8 17 12  6  9 13
12 21 30 24 22 15 18 27 23 29 28 20 26  5 11  3  2  8  4 13  6 17 16  9  7  1 10 19 25 14
11 14  5 27 13 30 10 29  6 19 16  9  3  7  8 23 24 28 22 15 12 25  2 21  1 18  4 26 17 20
27 30  8 11 12 14  3 22 16 18  6  2 10 24  5 26  7 21 29 25 13 15  9 28 17 19 20 23  1  4
13 28 14  7 29 25 30 15 19 11 26 22 21  4 23  8 27 10  9  5 20 12 16  1  6  2 24 17  3 18
19 29 15 18 30 26 11 21  7  8  4 14 22  6 25  9 17 27 28 23 24 10 20  5  1 13 16  2  3 12
30 25 19 14 21 27 22 23 11  2 13  7 15  5  3 28 26 16 24 18 29 20  8  9  4 10 17 12  6  1
22 27 11  5  3 24  2 14  8 12 21 25 18 16  1 30 15 13  6 10 19 23 17 29  7 28 26 20  4  9
23 16 29 21 11 22  5  4  3  6  1 14 24 13 19 12 18  7 17 30 25 28 27 26  9 20 10  2 15  8
28 18 26 23 15  1 21  7 22  4 19  6 17 11  2 29 20 14 25 12 27  9 24 10 30 16  8  5 13  3
15  5 28 25  7 20 23 19 13 17  2 21 22  1  4 27 30  9 10 29 14 18 12  8 11 24  6  3 26 16
 8 26 18 17 19 28 24  6  4 15 20  1  9 10 29  2 21 22 30 11 16 27 25  7  3 12 14 13  5 23
 7 12 13  6 26  8 20  9 17 30 29 10 16 28 27  4  3 15 21  2  1 14 22 11 23  5 25 18 19 24
20 10 17  9 16  2  1 12 23 28  7 27 26 25 13 18  6  5  4 24  3  8 19 30 29 15 22 14 21 11
 9 24 20 10 17 12 13  3 15 29 25  5  4 23 30  1  8 27 26  6  2 16 28 18 19 14 11  7 22 21
16 17  9 12 20 10 25 27 28 26 30 18  8  2 24  7 29 23 13  1  5  3  4  6 21 11 19 22 14 15
10 20 16 29  9 19  6 17 30 24  8 26 27 18 28  3 13  4  5 23  7  1 14 25 12 22  2 15 11 21
17  9 10 20  8 16 18  5  2  3 27 24 12 30 25  6  1 19  7  4 28 29 26 13 15 23 11 21 22 14
```

Fig. 3. A certificate for $\chi_{30} = 30$ obtained by setting $30 = 4 \times 6 + 2 \times 3$

independent sets generated using $\mathcal{C}^1$; nevertheless, $\{7, 12, 13, 18, 19, 24\}$ is stable relative to H and V.

A number of blanks have thus been filled in the chromatic number list given in Sect. 1, i.e. $\chi_n = n$ for $n \in \{12, 14, 15, 16, 18, 20, 21, 22, 24, 26, 28, 30, 32\}$. At this point however, we have reached the limit of our approach since this latest result required 28 computer units[2] (*one for each valid square of the color "1" in the second row*), while only the 6th machine produced the certificate of Fig. 3 after 1 h of computational time: the 28 *runs* took 168 h for full execution.

3   Coloring Extension

In the previous sections, assumptions were made regarding the possible characteristic solution to the queen graph coloring problem and several algorithms were implemented to solve a number of particular instances. This section will prove a construction procedure intended to solve an infinite number of instances. One common aspect of both approaches is that the algorithms have been designed from characteristics imposed by the solution these algorithms are expected to produce.

Let's start with the following remark: An independent set remains independent even after an homothetic transformation (Fig. 4).

```
 1   2   3   4   5   6   7   8   9  10  11  12
 8  11   1  10   7   4   9   6   3  12   2   5
 4   7   8  11  12  10   3   1   2   5   6   9
 6  12   5   9   3  11   2  10   4   8   1   7
 3   9   6  12   2   8   5  11   1   7   4  10
10   5   2   1   9   7   6   4  12  11   8   3
 2   4  10   7   8   1  12   5   6   3   9  11
11   8  12   6  10   9   4   3   7   1   5   2
12   6   4   8  11   3  10   2   5   9   7   1
 5   1   9   3   6   2  11   7  10   4  12   8
 9   3   7   2   1   5   8  12  11   6  10   4
 7  10  11   5   4  12   1   9   8   2   3   6

color's label = 1 + c(i, j) for 0 ≤ i, j < 12
```

```
 1 ..  2 ..  3 ..  4 ..  5 ..  6 ..  7 ..  8 ..  9 .. 10 .. 11 .. 12 ..
 ...   ...   ...   ...   ...   ...   ...   ...   ...   ...   ...   ...
 8 .. 11 ..  1  .. 10 ..  7 ..  4 ..  9 ..  6 ..  3 .. 12 ..  2 ..  5 ..
 ...   ...   ...   ...   ...   ...   ...   ...   ...   ...   ...   ...
 4 ..  7 ..  8 .. 11 .. 12 .. 10 ..  3 ..  1  ..  2 ..  5 ..  6 ..  9 ..
 ...   ...   ...   ...   ...   ...   ...   ...   ...   ...   ...   ...
 6 .. 12 ..  5 ..  9 ..  3 .. 11 ..  2 .. 10 ..  4 ..  8 ..  1  ..  7 ..
 ...   ...   ...   ...   ...   ...   ...   ...   ...   ...   ...   ...
 3 ..  9 ..  6 .. 12 ..  2 ..  8 ..  5 .. 11 ..  1  ..  7 ..  4 .. 10 ..
 ...   ...   ...   ...   ...   ...   ...   ...   ...   ...   ...   ...
10 ..  5 ..  2 ..  1  ..  9 ..  7 ..  6 ..  4 .. 12 .. 11 ..  8 ..  3 ..
 ...   ...   ...   ...   ...   ...   ...   ...   ...   ...   ...   ...
 2 ..  4 .. 10 ..  7 ..  8 ..  1  .. 12 ..  5 ..  6 ..  3 ..  9 .. 11 ..
 ...   ...   ...   ...   ...   ...   ...   ...   ...   ...   ...   ...
11 ..  8 .. 12 ..  6 .. 10 ..  9 ..  4 ..  3 ..  7 ..  1  ..  5 ..  2 ..
 ...   ...   ...   ...   ...   ...   ...   ...   ...   ...   ...   ...
12 ..  6 ..  4 ..  8 .. 11 ..  3 .. 10 ..  2 ..  5 ..  9 ..  7 ..  1  ..
 ...   ...   ...   ...   ...   ...   ...   ...   ...   ...   ...   ...
 5 ..  1  ..  9 ..  3 ..  6 ..  2 .. 11 ..  7 .. 10 ..  4 .. 12 ..  8 ..
 ...   ...   ...   ...   ...   ...   ...   ...   ...   ...   ...   ...
 9 ..  3 ..  7 ..  2 ..  1  ..  5 ..  8 .. 12 .. 11 ..  6 .. 10 ..  4 ..
 ...   ...   ...   ...   ...   ...   ...   ...   ...   ...   ...   ...
 7 .. 10 .. 11 ..  5 ..  4 .. 12 ..  1  ..  9 ..  8 ..  2 ..  3 ..  6 ..
 ...   ...   ...   ...   ...   ...   ...   ...   ...   ...   ...   ...
```

Fig. 4. Homogeneous dilation of the queen $- 12^2$ certificate to the 24×24 chessboard: the color's label in row i and column j is projected to row $2i$ and column $2j$

Next, by replacing one vertex by a square of vertices, as shown in Fig. 5, We obtain Fig. 6.

More formally, it will be proven that if $\chi_n = n$, and if p is not a multiple of 2 or 3, then $\chi_{np} = np$. The coloring formula proposed herein generalizes well-known results on the placement of the n queens [1]. Note that the particular case $\chi_{60} =$

---

[2] PENTIUM4 2.4 Ghz CPU as mentioned above.

1	2	..
8	11	..
..	..	..

1	2	3	4	5	..
4	5	1	2	3	..
2	3	4	5	1	..
5	1	2	3	4	..
3	4	5	1	2	..
..	..	..	..	..	....

color's label $= 1 + (3i + j) \equiv 5$ for $0 \leq i, j < 5$

Fig. 5. Replacing one vertex by a certificate for $\chi_5 = 5$ (*obtained by the knight's move rule on a toroidal* 5×5 *chessboard*)

1	2	3	4	5	6	7	8	9	10	11	12	13	14	15	16	17	18	19	20	21	22	23	24	25	26	27	28	29	30	31	32	33	34	35	36	37	38	39	40	41	42	43	44	45	46	47	48	49	50	51	52	53	54	55	56	57	58	59	60
3	4	5	1	2	8	9	10	6	7	13	14	15	11	12	18	19	20	16	17	23	24	25	21	22	28	29	30	26	27	33	34	35	31	32	38	39	40	36	37	43	44	45	41	42	48	49	50	46	47	53	54	55	51	52	58	59	60	56	57
5	1	2	3	4	10	6	7	8	9	15	11	12	13	14	20	16	17	18	19	25	21	22	23	24	30	26	27	28	29	35	31	32	33	34	40	36	37	38	39	45	41	42	43	44	50	46	47	48	49	55	51	52	53	54	60	56	57	58	59
2	3	4	5	1	7	8	9	10	6	12	13	14	15	11	17	18	19	20	16	22	23	24	25	21	27	28	29	30	26	32	33	34	35	31	37	38	39	40	36	42	43	44	45	41	47	48	49	50	46	52	53	54	55	51	57	58	59	60	56
4	5	1	2	3	9	10	6	7	8	14	15	11	12	13	19	20	16	17	18	24	25	21	22	23	29	30	26	27	28	34	35	31	32	33	39	40	36	37	38	44	45	41	42	43	49	50	46	47	48	54	55	51	52	53	59	60	56	57	58
36	37	38	39	40	51	52	53	54	55	1	2	3	4	5	46	47	48	49	50	31	32	33	34	35	16	17	18	19	20	41	42	43	44	45	26	27	28	29	30	11	12	13	14	15	56	57	58	59	60	6	7	8	9	10	21	22	23	24	25
38	39	40	36	37	53	54	55	51	52	3	4	5	1	2	48	49	50	46	47	33	34	35	31	32	18	19	20	16	17	43	44	45	41	42	28	29	30	26	27	13	14	15	11	12	58	59	60	56	57	8	9	10	6	7	23	24	25	21	22
40	36	37	38	39	55	51	52	53	54	5	1	2	3	4	50	46	47	48	49	35	31	32	33	34	20	16	17	18	19	45	41	42	43	44	30	26	27	28	29	15	11	12	13	14	60	56	57	58	59	10	6	7	8	9	25	21	22	23	24
37	38	39	40	36	52	53	54	55	51	2	3	4	5	1	47	48	49	50	46	32	33	34	35	31	17	18	19	20	16	42	43	44	45	41	27	28	29	30	26	12	13	14	15	11	57	58	59	60	56	7	8	9	10	6	22	23	24	25	21
39	40	36	37	38	54	55	51	52	53	4	5	1	2	3	49	50	46	47	48	34	35	31	32	33	19	20	16	17	18	44	45	41	42	43	29	30	26	27	28	14	15	11	12	13	59	60	56	57	58	9	10	6	7	8	24	25	21	22	23
16	17	18	19	20	31	32	33	34	35	36	37	38	39	40	51	52	53	54	55	56	57	58	59	60	46	47	48	49	50	11	12	13	14	15	1	2	3	4	5	6	7	8	9	10	21	22	23	24	25	26	27	28	29	30	41	42	43	44	45
18	19	20	16	17	33	34	35	31	32	38	39	40	36	37	53	54	55	51	52	58	59	60	56	57	48	49	50	46	47	13	14	15	11	12	3	4	5	1	2	8	9	10	6	7	23	24	25	21	22	28	29	30	26	27	43	44	45	41	42
20	16	17	18	19	35	31	32	33	34	40	36	37	38	39	55	51	52	53	54	60	56	57	58	59	50	46	47	48	49	15	11	12	13	14	5	1	2	3	4	10	6	7	8	9	25	21	22	23	24	30	26	27	28	29	45	41	42	43	44
17	18	19	20	16	32	33	34	35	31	37	38	39	40	36	52	53	54	55	51	57	58	59	60	56	47	48	49	50	46	12	13	14	15	11	2	3	4	5	1	7	8	9	10	6	22	23	24	25	21	27	28	29	30	26	42	43	44	45	41
19	20	16	17	18	34	35	31	32	33	39	40	36	37	38	54	55	51	52	53	59	60	56	57	58	49	50	46	47	48	14	15	11	12	13	4	5	1	2	3	9	10	6	7	8	24	25	21	22	23	29	30	26	27	28	44	45	41	42	43
26	27	28	29	30	56	57	58	59	60	21	22	23	24	25	41	42	43	44	45	11	12	13	14	15	51	52	53	54	55	6	7	8	9	10	46	47	48	49	50	16	17	18	19	20	36	37	38	39	40	1	2	3	4	5	31	32	33	34	35
28	29	30	26	27	58	59	60	56	57	23	24	25	21	22	43	44	45	41	42	13	14	15	11	12	53	54	55	51	52	8	9	10	6	7	48	49	50	46	47	18	19	20	16	17	38	39	40	36	37	3	4	5	1	2	33	34	35	31	32
30	26	27	28	29	60	56	57	58	59	25	21	22	23	24	45	41	42	43	44	15	11	12	13	14	55	51	52	53	54	10	6	7	8	9	50	46	47	48	49	20	16	17	18	19	40	36	37	38	39	5	1	2	3	4	35	31	32	33	34
27	28	29	30	26	57	58	59	60	56	22	23	24	25	21	42	43	44	45	41	12	13	14	15	11	52	53	54	55	51	7	8	9	10	6	47	48	49	50	46	17	18	19	20	16	37	38	39	40	36	2	3	4	5	1	32	33	34	35	31
29	30	26	27	28	59	60	56	57	58	24	25	21	22	23	44	45	41	42	43	14	15	11	12	13	54	55	51	52	53	9	10	6	7	8	49	50	46	47	48	19	20	16	17	18	39	40	36	37	38	4	5	1	2	3	34	35	31	32	33
11	12	13	14	15	41	42	43	44	45	26	27	28	29	30	56	57	58	59	60	6	7	8	9	10	36	37	38	39	40	21	22	23	24	25	51	52	53	54	55	1	2	3	4	5	31	32	33	34	35	16	17	18	19	20	46	47	48	49	50
13	14	15	11	12	43	44	45	41	42	28	29	30	26	27	58	59	60	56	57	8	9	10	6	7	38	39	40	36	37	23	24	25	21	22	53	54	55	51	52	3	4	5	1	2	33	34	35	31	32	18	19	20	16	17	48	49	50	46	47
15	11	12	13	14	45	41	42	43	44	30	26	27	28	29	60	56	57	58	59	10	6	7	8	9	40	36	37	38	39	25	21	22	23	24	55	51	52	53	54	5	1	2	3	4	35	31	32	33	34	20	16	17	18	19	50	46	47	48	49
12	13	14	15	11	42	43	44	45	41	27	28	29	30	26	57	58	59	60	56	7	8	9	10	6	37	38	39	40	36	22	23	24	25	21	52	53	54	55	51	2	3	4	5	1	32	33	34	35	31	17	18	19	20	16	47	48	49	50	46
14	15	11	12	13	44	45	41	42	43	29	30	26	27	28	59	60	56	57	58	9	10	6	7	8	39	40	36	37	38	24	25	21	22	23	54	55	51	52	53	4	5	1	2	3	34	35	31	32	33	19	20	16	17	18	49	50	46	47	48
46	47	48	49	50	21	22	23	24	25	6	7	8	9	10	1	2	3	4	5	41	42	43	44	45	31	32	33	34	35	26	27	28	29	30	16	17	18	19	20	56	57	58	59	60	51	52	53	54	55	36	37	38	39	40	11	12	13	14	15
48	49	50	46	47	23	24	25	21	22	8	9	10	6	7	3	4	5	1	2	43	44	45	41	42	33	34	35	31	32	28	29	30	26	27	18	19	20	16	17	58	59	60	56	57	53	54	55	51	52	38	39	40	36	37	13	14	15	11	12
50	46	47	48	49	25	21	22	23	24	10	6	7	8	9	5	1	2	3	4	45	41	42	43	44	35	31	32	33	34	30	26	27	28	29	20	16	17	18	19	60	56	57	58	59	55	51	52	53	54	40	36	37	38	39	15	11	12	13	14
47	48	49	50	46	22	23	24	25	21	7	8	9	10	6	2	3	4	5	1	42	43	44	45	41	32	33	34	35	31	27	28	29	30	26	17	18	19	20	16	57	58	59	60	56	52	53	54	55	51	37	38	39	40	36	12	13	14	15	11
49	50	46	47	48	24	25	21	22	23	9	10	6	7	8	4	5	1	2	3	44	45	41	42	43	34	35	31	32	33	29	30	26	27	28	19	20	16	17	18	59	60	56	57	58	54	55	51	52	53	39	40	36	37	38	14	15	11	12	13
6	7	8	9	10	16	17	18	19	20	46	47	48	49	50	31	32	33	34	35	36	37	38	39	40	1	2	3	4	5	56	57	58	59	60	21	22	23	24	25	26	27	28	29	30	11	12	13	14	15	41	42	43	44	45	51	52	53	54	55
8	9	10	6	7	18	19	20	16	17	48	49	50	46	47	33	34	35	31	32	38	39	40	36	37	3	4	5	1	2	58	59	60	56	57	23	24	25	21	22	28	29	30	26	27	13	14	15	11	12	43	44	45	41	42	53	54	55	51	52
10	6	7	8	9	20	16	17	18	19	50	46	47	48	49	35	31	32	33	34	40	36	37	38	39	5	1	2	3	4	60	56	57	58	59	25	21	22	23	24	30	26	27	28	29	15	11	12	13	14	45	41	42	43	44	55	51	52	53	54
7	8	9	10	6	17	18	19	20	16	47	48	49	50	46	32	33	34	35	31	37	38	39	40	36	2	3	4	5	1	57	58	59	60	56	22	23	24	25	21	27	28	29	30	26	12	13	14	15	11	42	43	44	45	41	52	53	54	55	51
9	10	6	7	8	19	20	16	17	18	49	50	46	47	48	34	35	31	32	33	39	40	36	37	38	4	5	1	2	3	59	60	56	57	58	24	25	21	22	23	29	30	26	27	28	14	15	11	12	13	44	45	41	42	43	54	55	51	52	53
51	52	53	54	55	36	37	38	39	40	56	57	58	59	60	26	27	28	29	30	46	47	48	49	50	41	42	43	44	45	16	17	18	19	20	11	12	13	14	15	31	32	33	34	35	1	2	3	4	5	21	22	23	24	25	6	7	8	9	10
53	54	55	51	52	38	39	40	36	37	58	59	60	56	57	28	29	30	26	27	48	49	50	46	47	43	44	45	41	42	18	19	20	16	17	13	14	15	11	12	33	34	35	31	32	3	4	5	1	2	23	24	25	21	22	8	9	10	6	7
55	51	52	53	54	40	36	37	38	39	60	56	57	58	59	30	26	27	28	29	50	46	47	48	49	45	41	42	43	44	20	16	17	18	19	15	11	12	13	14	35	31	32	33	34	5	1	2	3	4	25	21	22	23	24	10	6	7	8	9
52	53	54	55	51	37	38	39	40	36	57	58	59	60	56	27	28	29	30	26	47	48	49	50	46	42	43	44	45	41	17	18	19	20	16	12	13	14	15	11	32	33	34	35	31	2	3	4	5	1	22	23	24	25	21	7	8	9	10	6
54	55	51	52	53	39	40	36	37	38	59	60	56	57	58	29	30	26	27	28	49	50	46	47	48	44	45	41	42	43	19	20	16	17	18	14	15	11	12	13	34	35	31	32	33	4	5	1	2	3	24	25	21	22	23	9	10	6	7	8
56	57	58	59	60	26	27	28	29	30	16	17	18	19	20	36	37	38	39	40	51	52	53	54	55	11	12	13	14	15	46	47	48	49	50	6	7	8	9	10	21	22	23	24	25	41	42	43	44	45	31	32	33	34	35	1	2	3	4	5
58	59	60	56	57	28	29	30	26	27	18	19	20	16	17	38	39	40	36	37	53	54	55	51	52	13	14	15	11	12	48	49	50	46	47	8	9	10	6	7	23	24	25	21	22	43	44	45	41	42	33	34	35	31	32	3	4	5	1	2
60	56	57	58	59	30	26	27	28	29	20	16	17	18	19	40	36	37	38	39	55	51	52	53	54	15	11	12	13	14	50	46	47	48	49	10	6	7	8	9	25	21	22	23	24	45	41	42	43	44	35	31	32	33	34	5	1	2	3	4
57	58	59	60	56	27	28	29	30	26	17	18	19	20	16	37	38	39	40	36	52	53	54	55	51	12	13	14	15	11	47	48	49	50	46	7	8	9	10	6	22	23	24	25	21	42	43	44	45	41	32	33	34	35	31	2	3	4	5	1
59	60	56	57	58	29	30	26	27	28	19	20	16	17	18	39	40	36	37	38	54	55	51	52	53	14	15	11	12	13	49	50	46	47	48	9	10	6	7	8	24	25	21	22	23	44	45	41	42	43	34	35	31	32	33	4	5	1	2	3
21	22	23	24	25	1	2	3	4	5	41	42	43	44	45	11	12	13	14	15	26	27	28	29	30	6	7	8	9	10	51	52	53	54	55	31	32	33	34	35	46	47	48	49	50	16	17	18	19	20	56	57	58	59	60	36	37	38	39	40
23	24	25	21	22	3	4	5	1	2	43	44	45	41	42	13	14	15	11	12	28	29	30	26	27	8	9	10	6	7	53	54	55	51	52	33	34	35	31	32	48	49	50	46	47	18	19	20	16	17	58	59	60	56	57	38	39	40	36	37
25	21	22	23	24	5	1	2	3	4	45	41	42	43	44	15	11	12	13	14	30	26	27	28	29	10	6	7	8	9	55	51	52	53	54	35	31	32	33	34	50	46	47	48	49	20	16	17	18	19	60	56	57	58	59	40	36	37	38	39
22	23	24	25	21	2	3	4	5	1	42	43	44	45	41	12	13	14	15	11	27	28	29	30	26	7	8	9	10	6	52	53	54	55	51	32	33	34	35	31	47	48	49	50	46	17	18	19	20	16	57	58	59	60	56	37	38	39	40	36
24	25	21	22	23	4	5	1	2	3	44	45	41	42	43	14	15	11	12	13	29	30	26	27	28	9	10	6	7	8	54	55	51	52	53	34	35	31	32	33	49	50	46	47	48	19	20	16	17	18	59	60	56	57	58	39	40	36	37	38
41	42	43	44	45	11	12	13	14	15	31	32	33	34	35	6	7	8	9	10	1	2	3	4	5	21	22	23	24	25	36	37	38	39	40	56	57	58	59	60	51	52	53	54	55	26	27	28	29	30	46	47	48	49	50	16	17	18	19	20
43	44	45	41	42	13	14	15	11	12	33	34	35	31	32	8	9	10	6	7	3	4	5	1	2	23	24	25	21	22	38	39	40	36	37	58	59	60	56	57	53	54	55	51	52	28	29	30	26	27	48	49	50	46	47	18	19	20	16	17
45	41	42	43	44	15	11	12	13	14	35	31	32	33	34	10	6	7	8	9	5	1	2	3	4	25	21	22	23	24	40	36	37	38	39	60	56	57	58	59	55	51	52	53	54	30	26	27	28	29	50	46	47	48	49	20	16	17	18	19
42	43	44	45	41	12	13	14	15	11	32	33	34	35	31	7	8	9	10	6	2	3	4	5	1	22	23	24	25	21	37	38	39	40	36	57	58	59	60	56	52	53	54	55	51	27	28	29	30	26	47	48	49	50	46	17	18	19	20	16
44	45	41	42	43	14	15	11	12	13	34	35	31	32	33	9	10	6	7	8	4	5	1	2	3	24	25	21	22	23	39	40	36	37	38	59	60	56	57	58	54	55	51	52	53	29	30	26	27	28	49	50	46	47	48	19	20	16	17	18
31	32	33	34	35	46	47	48	49	50	51	52	53	54	55	21	22	23	24	25	16	17	18	19	20	56	57	58	59	60	1	2	3	4	5	41	42	43	44	45	36	37	38	39	40	6	7	8	9	10	11	12	13	14	15	26	27	28	29	30
33	34	35	31	32	48	49	50	46	47	53	54	55	51	52	23	24	25	21	22	18	19	20	16	17	58	59	60	56	57	3	4	5	1	2	43	44	45	41	42	38	39	40	36	37	8	9	10	6	7	13	14	15	11	12	28	29	30	26	27
35	31	32	33	34	50	46	47	48	49	55	51	52	53	54	25	21	22	23	24	20	16	17	18	19	60	56	57	58	59	5	1	2	3	4	45	41	42	43	44	40	36	37	38	39	10	6	7	8	9	15	11	12	13	14	30	26	27	28	29
32	33	34	35	31	47	48	49	50	46	52	53	54	55	51	22	23	24	25	21	17	18	19	20	16	57	58	59	60	56	2	3	4	5	1	42	43	44	45	41	37	38	39	40	36	7	8	9	10	6	12	13	14	15	11	27	28	29	30	26
34	35	31	32	33	49	50	46	47	48	54	55	51	52	53	24	25	21	22	23	19	20	16	17	18	59	60	56	57	58	4	5	1	2	3	44	45	41	42	43	39	40	36	37	38	9	10	6	7	8	14	15	11	12	13	29	30	26	27	28

Fig. 6. A certificate for $\chi_{60} = 60$, obtained by *highlighting* that of queen $-\ 12^2$ (Fig. 4) and *substituting* that of queen $-\ 5^2$ (Fig. 5)

60 has been solved (by combining the queen $-\ 12^2$ and queen $-\ 5^2$ certificates) by Gunter Stertenbrink (www.cs.concordia.ca/~chvatal/queengraphs.html).

Let's denote by (i, j) the square situated on row i and column j; moreover, $c(i, j)$ is a coloring of the queen $- n^2$ graph in n colors (e.g. see Fig. 4). If p is an integer that is not a multiple of 2 or 3, then we set $r(i, j) = (3i + j) \, modulo \, p$ (e.g. see the certificate for $\chi_5 = 5$ in Fig. 5).

The formula $R(i, j) = r(i, j) + p \times c(i/p, j/p)$ is thus a coloring of queen $- np^2$ in np colors. For one thing: $0 \le r(i, j) \le p - 1$ and $0 \le c(i/p, j/p) \le n - 1$ $\Rightarrow 0 \le R(i, j) \le p - 1 + p(n - 1) = np - 1$, which results in using np colors. Yet on the other hand, $r(i, j) < p$, $R(i, j) = R(i', j') \Leftrightarrow r(i, j) = r(i', j')$ and $c(i/p, j/p) = c(i'/p, j'/p)$. The following four equivalences are therefore derived:

1. $R(i, j) = R(i, j') \Leftrightarrow j = j' \, modulo \, p$ and $j/p = j'/p$ ($c(i, j)$ is a coloring, with positioning on the same row) $\Leftrightarrow j = j'$;
2. $R(i, j) = R(i', j) \Leftrightarrow i = i' \, modulo \, p$ (p is not a multiple of 3) and $i/p = i'/p$ ($c(i, j)$ is a coloring, with positioning on the same column) $\Leftrightarrow i = i'$;
3. $R(i, j) = R(i', j')$ and $j - i = j' - i' \Leftrightarrow i = i'$ and $j = j' \, modulo \, p$ (p is not a multiple of 2) and $i/p = i'/p$ and $j/p = j'/p$ ($c(i, j)$ is a coloring, with positioning on the same diagonal) $\Leftrightarrow i = i'$ and $j = j'$;
4. $R(i, j) = R(i', j')$ and $j + i = j' + i' \Leftrightarrow i = i'$ and $j = j' \, modulo \, p$ and $i/p = i'/p$ and $j/p = j'/p$ ($c(i, j)$ is a coloring, with positioning on the same diagonal) $\Leftrightarrow i = i'$ and $j = j'$.

Hence, a uniqueness of colors is found on the same row, column or diagonal. We have thus proven that $R(i, j)$ is a coloring of queen $- np^2$ in np colors. It can also be noted that $r(i, j)$ is a coloring of queen $- p^2$ in p colors when p is not a multiple of 2 or 3 (by simply setting $n = 1$ and $c(0, 0) = 0$).

Consequently, $\chi_{np} = np$; moreover, we can set for n any of the thirteen values given at the end of the previous section and for p any integer that is not a multiple of either 2 or 3. We then simply have to reproduce the design formed by $r(i, j)$ with n^2 translations on the groundwork $c(i, j)$. From a practical point of view, the following results have therefore been obtained: $\chi_{60} = 12 \times 5 = 60$, $\chi_{70} = 14 \times 5 = 70$, $\chi_{75} = 15 \times 5 = 75$, $\chi_{80} = 16 \times 5 = 80$, $\chi_{84} = 12 \times 7 = 84$, etc.

4   Conclusion

Below are the key features proposed by this contribution to the queen graph coloring problem:

1. transforming the search for the chromatic number into a decision problem that requires recovering the $n \times n$ chessboard with n independent sets, each containing exactly n vertices;
2. implementing an incomplete exploration of the solution space over which the distribution of colors on the chessboard verifies some of the geometric characteristics;
3. proposing a coloring composition algorithm capable of building larger certificate from two smaller ones, and proving its correctness by modular algebra.

In the first feature above, all potential solutions to $\chi_n = n$ were considered. For the second feature, strong assumptions were made regarding the geometric property of the certificates obtained by the exhaustive enumeration and the search space was reduced even further. The last feature served to impose the solution shape; in this case, the exploration ceases in favor of a linear time procedure that *prints out* the certificate for $\chi_n = n$. Under all circumstances, attention is first directed to the solution characteristics, and afterwards the algorithm design takes the corresponding constraints into account. All of these steps have yielded 14 results for the queen graph coloring problem: $\chi_{10} = 11$ and $\chi_n = n$ for $n \in \{12, 14, 15, 16, 18, 20, 21, 22, 24, 26, 28, 30, 32\}$, thus proving that **for the queen graphs, an infinite number of values of** n **multiples of** 2 **or** 3 **exist, whereby** $\chi_n = n$. Efforts are still required however to find the chromatic number of the 27×27 chessboard.

References

1. Abramson, B., Yung, M.M.: Construction through decomposition: a divide-and-conquer algorithm for the N-queens problem. In: Proceedings of 1986 ACM Fall Joint Computer Conference. ACM 1986, pp. 620–628, no. 9. IEEE Computer Society Press, Los Alamitos (1986). http://dl.acm.org/citation.cfm?id=324493.324620.
2. Caramia, M., Dell'Olmo, P.: Embedding a novel objective function in a two-phased local search for robust vertex coloring. Eur. J. Oper. Res. **189**(3), 1358–1380 (2008). https://doi.org/10.1016/j.ejor.2007.01.063. 9
3. Galinier, P., Hertz, A., Zufferey, N.: An adaptive memory algorithm for the k-coloring problem. Discrete Appl. Math. **156**(2), 267–279 (2008). https://doi.org/10.1016/j.dam.2006.07.017. Computational Methods for Graph Coloring and it's Generalizations. http://www.sciencedirect.com/science/article/pii/S0166218X07001114
4. Gualandi, S., Malucelli, F.: Exact solution of graph coloring problems via constraint programming and column generation. INFORMS J. Comput. **24**(1), 81–100 (2012). https://doi.org/10.1287/ijoc.1100.0436
5. Hansen, P., Labbé, M., Schindl, D.: Set covering and packing formulations of graph coloring: algorithms and first polyhedral results. Discrete Optim. **6**(2), 135–147 (2009). https://doi.org/10.1016/j.disopt.2008.10.004. http://www.sciencedirect.com/science/article/pii/S1572528608000716
6. Vasquez, M.: New results on the queens_n2 graph coloring problem. J. Heuristics **10**(4), 407–413 (2004). https://doi.org/10.1023/B:HEUR.0000034713.28244.e1
7. Vasquez, M., Habet, D.: Complete and incomplete algorithms for the queen graph coloring problem. In: Proceedings of the 16th European Conference on Artificial Intelligence, ECAI 2004, including Prestigious Applicants of Intelligent Systems, PAIS 2004, Valencia, Spain, 22–27 August, 2004, pp. 226–230 (2004)

Minimal Sum Labeling of Graphs

Matěj Konečný, Stanislav Kučera, Jana Novotná, Jakub Pekárek,
Štěpán Šimsa, and Martin Töpfer$^{(\boxtimes)}$

Faculty of Mathematics and Physics, Charles University, Prague, Czech Republic
matejkon@gmail.com, stanislav.kucera@outlook.com, janka.novot@seznam.cz,
edalegos@gmail.com , simsa.st@gmail.com, mtopfer@gmail.com

Abstract. A graph G is called a *sum graph* if there is a so-called *sum labeling* of G, i.e. an injective function $\ell : V(G) \to \mathbb{N}$ such that for every $u, v \in V(G)$ it holds that $uv \in E(G)$ if and only if there exists a vertex $w \in V(G)$ such that $\ell(u) + \ell(v) = \ell(w)$. We say that sum labeling ℓ is *minimal* if there is a vertex $u \in V(G)$ such that $\ell(u) = 1$. In this paper, we show that if we relax the conditions (either allow non-injective labelings or consider graphs with loops) then there are sum graphs without a minimal labeling, which partially answers the question posed by Miller in [6] and [5].

1 Introduction

An undirected graph $G = (V, E)$ is a *sum graph* if there exists an injective function $\ell : V(G) \to \mathbb{N}$ such that every pair of vertices $u \neq v \in V(G)$ is connected via an edge of G if and only if there exists a vertex $w \in V(G)$ such that $\ell(w) = \ell(u) + \ell(v)$. We call the function ℓ a *sum labeling* or *labeling function*. The value $\ell(u) + \ell(v)$ is an *edge-number* of the edge uv and it is *guaranteed* by vertex w. The *sum number* $\sigma(G)$ of a graph is defined as the least integer, such that $G + \bar{K}_{\sigma(G)}$ (G with $\sigma(G)$ additional isolated vertices) is a sum graph.

The concept of sum graphs was introduced by Harary [4] in 1990. It was further developed by Gould and Rödl [3] and Miller [6,7] who showed general upper and lower bounds on $\sigma(G)$ of order $\Omega(|E|)$ for a given general graph G and better bounds for specific classes of graphs.

For some graphs the exact sum numbers are known: $\sigma(T_n) = 1$ for trees (of order $n \geq 2$) [2], $\sigma(C_n) = 2$ for cycles ($n \geq 3$, $n \neq 4$) and $\sigma(C_4) = 3$ [4], $\sigma(K_n) = 2n - 3$ for complete graphs ($n \geq 4$) [1], $\sigma(H_{2,n}) = 4n - 5$ for cocktail party graphs ($n \geq 2$) [6] and for complete bipartite graphs [8].

In the work of Miller et al. [5,6], an open question was raised whether every sum graph has a labeling that uses number 1. Such labelings are called *minimal*. In [6], a minimal labeling of complete bipartite graphs is presented. In [5], an

All authors were supported by grant SVV-2017-260452, M. Töpfer and M. Konečný were supported by project CE-ITI P202/12/G061 of GA CR.

The full preprinted version of this paper is available at https://arxiv.org/abs/1708.00552v1.

L. Brankovic et al. (Eds.): IWOCA 2017, LNCS 10765, pp. 252–263, 2018.
https://doi.org/10.1007/978-3-319-78825-8_21

upper bound on $\sigma(G)$ for G being a disjoint union of graphs $G_1, G_2 \ldots, G_n$ is shown. If at least one of the disjoint graphs has minimal labeling then

$$\sigma(G) \leq \sum_{i=1}^{n} \sigma(G_i) - (n-1).$$

In our work, we approach sum graphs from a different perspective. Instead of grounding our research on the properties of graphs, our basic objects are sets of integers. Given an integer set M, the rules of sum labeling uniquely define a graph such that it is a sum graph and its labeling consists exactly of all the integers from the given set. From our research we provide two negative answers to questions parallel to the one raised by Miller et al.

While it is natural to require graphs to have no loops, when we construct a sum graph from an integer set, it seems more natural to allow loops (i.e. for every not necessarily distinct vertices $u, v \in V(G)$ we have $uv \in E(G)$ if and only if there exists $w \in V(G)$ such that $\ell(u) + \ell(v) = \ell(w)$). We call these graphs *sum graphs with loops*. When we say *sum graphs without loops* we refer to the previous definition, where the vertices are required to be distinct. We show the following:

Theorem 1. *There exists an infinite family of sum graphs with loops which admit no minimal labeling.*

Another relaxation of the original problem (without loops) is to allow the integer set M to be a multiset. This of course causes the labeling function to cease being injective, thus we call such graphs *non-injective sum graphs*. Nevertheless, in this approach we may consider graphs without loops and obtain the following similar result:

Theorem 2. *There exists an infinite family of non-injective sum graphs (without loops) which admit no minimal labeling.*

1.1 Preliminaries

Let G be a sum graph with some labeling ℓ. We call set $M = \{\ell(v) : v \in V(G)\}$ *label-set* of G.

For a finite multiset of natural numbers $S \subset \mathbb{N}$, let G_S be the graph with elements of S being its vertices and for every $u, v \in S$ let there be an edge $uv \in E(G_S)$ if and only if $u + v \in S$. Depending on context, we sometimes allow G_S to have loops. We say that set S *induces* a graph G_S. Let f denote the natural bijection of vertices of G_S and integers in S. For any integer $i \in S$ we denote $\psi(i)$ the subset of $V(G_S)$ such that a vertex $v \in \psi(i)$ if and only if $f(v) = i$. We say that an integer i *induces* a vertex v if $v \in \psi(i)$.

We say that two vertices u, v of a graph G without loops are *equivalent* if it holds that $N(u) \backslash \{v\} = N(v) \backslash \{u\}$.

Lemma 1. *Let G be a graph with labeling ℓ and let u, v be two of its vertices. If $\ell(u) = \ell(v)$, then u and v are equivalent.*

Proof. Suppose $\ell(u) = \ell(v)$. Consider any $w \in V(G)$ other than u and v. The edge-numbers of wv and wu are the same, so either both edges are present or none of them is. Since we do not consider loops, the only remaining edge to consider is uv. If $uv \notin E(G)$, then clearly $N(u) = N(v)$. If $uv \in E(G)$, then all neighbors of v are also neighbors of u except u itself and vice versa. In both cases, u and v are equivalent.

To get an analogous definition for a graph with loops one would not exclude the vertices v and u from the neighborhoods. This rather subtle difference actually makes dealing with sum graphs with loops much easier.

We define an operation that removes loops from sum graphs. Consider a sum graph G with loops. Let us take one by one each vertex v with a loop and choose any integer $k \geq 2$ (independently for each vertex). We replace v with a clique K_k and connect all neighbors of v to all vertices from K_k. We denote set of all possible results of this operation by $\mathcal{C}(G)$, and denote $\mathcal{C}^k(G)$ the unique result where we fix all values k from the construction to a given fixed value. Let G be induced by a multiset M, we may equivalently define $\mathcal{C}(G)$ as all graphs induced by all possible multisets obtained from the set M via raising the multiplicity of membership of any $i \in M$ such that $2i \in M$.

2   Sum Graphs with Loops

This section serves as an introduction to the topic of this paper. We show that there is a sum graph with loops that admits no minimal sum labeling. Though this is weaker result than Theorem 1, we provide a direct proof without usage of complex tools. The full proof of Theorem 1 is given later as a consequence of Theorem 2.

For a proof, we use the graph induced by the set $\{2, 3, 4, 6, 7\}$ (see Fig. 1). This specific graph was chosen based on a result of a computer experiment as the smallest graph induced by an arithmetic sequence with difference 1 starting from 2 with one element missing such that no minimal labeling was found. The proof goes through several cases and is available in the full version of this paper.

Theorem 3. *There exists a sum graph with loops that admits no minimal labeling.*

Although Theorem 3 does not require the labeling function to be injective, the theorem holds also under the constraint of injectivity. The labeling used to induce the graph is certainly injective, and the theorem shows that there exists no minimal labeling; thus, in particular no injective minimal labeling.

Fig. 1. The sum graph induced by the set $\{2, 3, 4, 6, 7\}$.

It is easy to observe that sets $\{1, 2, \ldots, k\}$ induce graphs with the maximum number of edges out of all sum graphs with loops on the same number of vertices. In fact, it can be shown that any graph with the same number of vertices and edges is necessarily isomorphic to this graph. Based on this observation, it seems reasonable to assume that the set $\{2, 3, \ldots, k-1, k\}$ might induce a graph with specific structure and possibly exclude all labelings with 1 as 1 was removed from the inducing set. This idea does not hold, as the same graph is induced for example by the set $\{1, 2, \ldots, k, 3k\}$.

However, the situation seems to change dramatically once we remove one more value. Let us call *gap-graphs* of size k all graphs induced by the set obtained from $\{2, \ldots, k\}$ by removing one element we call a *gap*. Hence, a gap-graph of size k has $k - 2$ vertices. While for small values of k some gap-graphs have a minimal labeling, we conjecture that for $k \geq 10$ none of the gap-graphs of size k with gap i such that $3 < i < k$ has a minimal labeling. While we do not prove this conjecture, it serves as a basic inspiration for our main result and as a consequence of our main result, we prove Theorem 1 by providing a partial proof of the conjecture for sufficiently large n and a specific choice of gap.

3 Non-injective Sum Graphs Without Loops

In this section, we construct graphs with (non-injective) sum labelings such that they do not admit minimal labelings.

Based on the result from the previous section and the conjecture about gap-graphs, it is a natural question whether we can modify the gap-graph idea to remove all loops and keep the desired properties that may prevent existence of minimal labelings. Let us consider the gap-graph G induced by the set $\{2, 3, 4, 6, 7\}$ from Theorem 1, and its loopless modification $H = \mathcal{C}^2(G)$. The sum graph H is induced by a multiset $\{2, 2, 3, 3, 4, 6, 7\}$ by definition. Unfortunately a graph isomorphic to H is induced by a multiset $\{1, 5, 2, 2, 4, 6, 9\}$ (with the same ordering of vertices). This numbering can be naturally extended to a (non-injective) labeling of any graph from $\mathcal{C}(G)$. While this gives us a negative result, we show that for large enough gap-graphs and at least some choice of gap, the construction $\mathcal{C}$ does in fact guarantee all produced graphs to admit no minimal labeling.

We first develop some tools applicable (up to minor adjustments) to all flavors of sum graphs (injective, non-injective, with or without loops). Namely, that labels of each sum labeling of a sum graph can be described as a set of arithmetic sequences. We show how several simultaneous description of this form limit each other and in doing so reflect some structural aspects of underlying sum graphs directly into their label-sets.

The graphs we work with are based on arithmetic sequence of integers. We define graph A_n as the unique sum graph with loops induced by the set $\{2, 3, \ldots, n-1, n, n+2\}$, in other words by a set of all integers from 2 up to $n + 2$ without the second-last value $n + 1$. Note that graph A_n has exactly n vertices.

3.1   Sequence Description

Let us denote an arithmetic sequence $(a, a+d, a+2d, \ldots, a+jd)$ with difference d as $(id + a)_i$ where we always consider i going from 0 up to some unspecified integer. We also generally refer to sequences with difference d as d-sequences.

Let us fix a vertex v and call it *generator*. The set of *terminals* associated with this generator is defined as $V(G) \backslash N(v)$. Note that v has exactly $n - deg(v)$ terminals. It is an important observation that unless v has a loop, it is its own terminal. We say that a terminal w is a *proper terminal* of v if $w \neq v$, and is *improper terminal* otherwise.

Lemma 2. *Let G be a sum graph without loops, let ℓ be a fixed labeling of G, let v be a fixed generator and let w be a proper terminal of v. Then there is no vertex in G labeled $\ell(v) + \ell(w)$.*

Proof. For contradiction, let a vertex $u \in V(G)$ be labeled $\ell(v) + \ell(w)$. Then the edge-number of the edge vw is guaranteed by u, thus $vw \in E(G)$. This however makes w a neighbor of v and we reach a contradiction.

The previous lemma does not hold for improper terminal v, as the edge used in the proof would be a loop and thus would not be an edge of G even though it is technically guaranteed. In case of sum graphs with loops this issue does not arise.

Lemma 3. *Let G be a sum graph with fixed labeling ℓ, let M denote the label-set of G, let v be a fixed generator and let u be a non-terminal of v. Then there exists a sequence $S = (i\ell(v) + \ell(u))_i \subseteq M$ such that its last element is a label of a proper terminal of v.*

Proof. Since u is a non-terminal of v, there exists an edge uv with edge-number $\ell(v) + \ell(u)$. Clearly, this edge is guaranteed by some vertex u_1 labeled $\ell(v) + \ell(u)$. The vertex u_1 cannot be v as $\ell(u_1) > \ell(v)$. If u_1 is a proper terminal, then $S = (\ell(u), \ell(u) + \ell(v))$ and we are done. Otherwise, u_1 is a non-terminal and we iterate the previous argument, building a sequence S' from u_1 and setting $S = (\ell(u)).S'$ where the dot operation denotes sequence concatenation.

Lemma 4. *Let G be a sum graph, and let v be a fixed generator with k terminals and label $\ell(v) = g$. Then for every labeling ℓ and associated label-set M, the following holds:*

1. *The label-set M can be described as a union of at most k distinct arithmetic sequences with difference g.*
2. *The last element of each of the sequences is a label of a terminal.*
3. *Each label of a proper terminal is the last element of one of the sequences.*
4. *If v has j non-equivalent terminals and the description of M has j sequences, then one of the sequences is a singleton sequence (g).*

Proof. Let u be a non-terminal of v such that it has the lowest label of all non-neighbors. From Lemma 3 we have a sequence S_1 such that it covers some elements of M and ends in a terminal. If $M\backslash S$ contains non-terminals, we iterate by using the Lemma 3 with the lowest remaining non-terminal label. If $M\backslash S$ contains no further non-terminals and is non-empty, then we create a singleton sequence $(\ell(w))$ for each proper terminal w. Clearly the only element of M that can remain not-covered is $\ell(v)$, as v is improper terminal. In such case we create one more singleton sequence $(\ell(v))$. Suppose $S_1 = (ig + a)_i$ and $S_2 = (ig + b)_i$, for some integers a, b, are two constructed non-singleton sequences (S_1 was constructed first). Since both S_1 and S_2 have the same difference, $a \notin S_2$ and $a < b$ by the choice of a and b, both sequences are distinct. Clearly, no singleton sequence shares an element with any other sequence. Since there are exactly k terminals, each sequence ends with a terminal label and all sequences are distinct, we get that there are at most k sequences in total. Naturally, M is in the union of all of the constructed sequences and the sequences contain no extra elements. This proves points 1 and 2.

From Lemma 2, we have that if a sequence contains the label of a proper terminal, the hypothetical next element of the sequence is not a label of any vertex. Thus labels of proper terminals can only be the last elements in sequences, which proves the point 3.

Let v have j non-equivalent terminals. From Lemma 1 we have that there are j distinct labels of terminals in M. If there are j sequences in the description of M via generator v, we deduce from the previous points that each of the distinct labels is the last element of a distinct sequence. In particular, the label g must be the last in a sequence as v is its own (improper) terminal. Since all labels are positive and the difference of each sequence is g, the label g must form a singleton sequence (g).

Note that some labels generated by a sequence may be labels of several vertices, if these are equivalent.

Let v be a vertex of a fixed graph G. Let v-*cover* denote the set of sequences covering the label-set M of G as described in Lemma 4. Each such set is associated with one difference value. If α is this difference value (i.e. $\alpha = \ell(v)$), then we also reference to such a cover as α-*cover*.

3.2   Cover Merging

While the results of the previous section do not give particularly strong results when all the degrees are low (in respect to the number of vertices), it does give strong limits on potential labelings once we have one or more vertices with almost full degree. In this section, we expand our tools to impose additional constraints when applying the previous results to multiple vertices simultaneously.

Let G be a graph with fixed proper labeling and let M be its label-set. Then each vertex g_i induces a cover C_i of M, as described in Lemma 4. Each such cover C_i is associated with a difference d_i ($d_i = \ell(g_i)$) and a number of terminals of g_i (including g_i) denoted t_i.

We say that two covers C_i and C_j are *mergeable* if the number of non-equivalent vertices in G (and thus also the size of any label-set of G) is at least $2 \cdot (t_i - 1) \cdot (t_j - 1) + 3$. A proof of the following lemma is available in the full version of this paper.

Lemma 5. *Let C_i and C_j be mergeable covers, then there exists a pair of sequences S_i, S_j from covers C_i resp. C_j such that they share at least three elements.*

Lemma 6. *For any fixed mergeable covers C_1, C_2 there exist positive integers j, k, such that $GCD(j, k) = 1$, $j \leq t_1 - 1$, $k \leq t_2 - 1$ and the equation $k \cdot d_1 = j \cdot d_2$ holds.*

Proof. Let us fix two mergeable covers C_1 and C_2. Let S_1, S_2 denote sequences such that $S_1 \in C_1$, $S_2 \in C_2$ and $|S_1 \cap S_2| \geq 3$. Let x_0, x_1, x_2 be the three smallest elements of $S_1 \cap S_2$ so that $x_0 < x_1 < x_2$. Since all of them come from both sequences with no holes, we may denote the distance between elements $m = x_1 - x_0 = x_2 - x_1$.

As both x_1, x_2 belong to both sequences, it must hold that $m = k \cdot d_1 = j \cdot d_2$ for some positive integers j, k such that x_2 is k-th element following x_1 in S_1 and also j-th element following x_1 in S_2.

Suppose $d_1 = d_2$, then $k = j = 1 \leq t_1 - 1, t_2 - 1$ and the lemma holds trivially. We may assume $d_1 \neq d_2$. From definition of a single sequence, S_1 contains all k possible elements from interval $(x_1, x_2]$. Similarly, S_2 contains all j elements from $(x_1, x_2]$. If $GCD(k, j) > 1$ then there is some $m_0 < m$ such that $x_1 + m_0$ is an element of both sequences which would contradict the minimality of x_1, x_2. More generally, if any two elements x_3, x_4 of S_1 such that $x_1 < x_3 < x_4 \leq x_2$ belonged to the same sequence $S_0 \in C_2$, then $x_1 + (x_4 - x_3) \in S_1 \cap S_2$ which would contradict the minimality of x_2. Analogously, we would reach a contradiction if any such x_3, x_4 belonged to any $S_0' \in C_1$. Thus the k elements of S_1 from the interval $(x_1, x_2]$ fall into distinct sequences from C_2 and we have $k \leq t_2$ as t_2 limits the number of sequences in C_2. Analogously we get $j \leq t_1$.

From Lemma 4, we know that if the k elements of S_1 fall into t_1 distinct sequences from C_2, then one of them has to be the singleton sequence (d_2). Recall that we chose x_1, x_2 from $S_1 \cap S_2$ so that they are preceded by some element x_0 from $S_1 \cap S_2$. This means that $x_1 > d_2$ and thus the singleton sequences from C_2 cannot, in fact, play any role. Thus, the limit on the number of sequences involved can be further reduced to $k \leq t_2 - 1$. Symmetrically, we obtain that $j \leq t_1 - 1$.

Lemma 7. *For any fixed mergeable covers C_1, C_2 of a graph G without loops, let $d_1 = 1$. Then $d_2 \leq t_2 - 1$.*

Proof. Consider the equation from Lemma 6. If $d_1 = 1$ then $k = j \cdot d_2$. From the same lemma we also know that $GCD(j, k) = 1$, so necessarily $j = 1$ and $k = d_2$. Finally, we also have inequality $k \leq t_2 - 1$ which together give $d_2 \leq t_2 - 1$.

3.3   Vertices $\psi(2)$ and $\psi(3)$ in $\mathcal{C}(A_n)$

In this section, we explore the exact structure of terminals of graphs from $\mathcal{C}(A_n)$. Based on a number of simple properties we show that given large enough n, no graph from $\mathcal{C}(A_n)$ admits label 1 on any vertex from $\psi(2)$ and $\psi(3)$.

For a vertex v of a fixed labeling, let $\tau(v)$ denote the set of proper terminals of v. Lemma 8 summarizes some basic observations. The proof is omitted and is available in the full version of this paper.

Lemma 8. *For any graph $G \in \mathcal{C}(A_n)$ such that $n \geq 39$, let us fix arbitrary vertices v_i such that $v_i \in \psi(i)$ for values of i from 2 to 6. Then for any integers j, k such that $2 \leq j, 2 \leq k \leq 6$ all of the following holds:*

1. *Vertex v_k has exactly k proper terminals (and $k + 1$ terminals in total), all of which have distinct labels.*
2. *If $j \leq 3$, then the vertices v_j and v_k are mergeable.*
3. *If labeling is minimal and either $\ell(v_2)$ or $\ell(v_3)$ equals to 1, then $\ell(v_k) \leq k$.*
4. *There is exactly one proper terminal in the intersection of all $\tau(v_k)$, and this terminal has the highest label in the graph.*
5. *If j is not the highest label, then either $j \notin \tau(v_2)$ or $j \notin \tau(v_3)$.*
6. *$\tau(v_k) \subset \tau(v_j)$ whenever $j \geq k + 2$.*
7. *None of the chosen vertices is a proper terminal of any of the other chosen vertices and thus $\ell(v_j) + \ell(v_k) \in M$ for any $j \neq k$.*

For convenience, we extend the meaning of *terminal*. We characterize possible labelings of graphs in terms of presence or absence of values in respect to the label-set of the graph. For a vertex v we say an integer value k is a (proper) *terminal* for v as a shortcut for the fact that there exists a vertex w, which is a (proper) terminal for v and $\ell(w) = k$. We only deal with terminals of vertices from $\psi(2), \ldots, \psi(6)$ whose all proper terminals are non-equivalent and thus have distinct labels. Hence the integer terminals and the vertex terminals are in one-to-one correspondence for these vertices.

Let G be a sum graph with a minimal labeling ℓ. Let M be the label-set of G in respect to ℓ and let $v \in V(G)$ be such that $\ell(v) = 1$. According to Lemma 4, the vertex v describes M as a union of several distinct integer intervals separated by some values that are not elements of M.

We say that an interval is *long* if its first six labels and its last six labels do not intersect. In particular, its last six labels are strictly bigger than 6. The property of the long intervals we want to use is that for any v such that $v \in \psi(i)$ for $i \leq 6$, any terminal among the last six elements of a long interval is always a proper terminal of v. Proofs of Lemmas 9 and 10 are very similar, thus the proof of Lemma 9 is available in the full version of this paper.

Lemma 9. *For any graph $G \in \mathcal{C}(A_n)$ such that $n \geq 39$, there is no labeling such that $\ell(v) = 1$ for any $v \in \psi(2)$.*

Lemma 10. *For any graph $G \in \mathcal{C}(A_n)$ such that $n \geq 39$, there is no labeling such that $\ell(v) = 1$ for any $v \in \psi(3)$.*

Proof. Let us fix arbitrary vertices v_i such that $v_i \in \psi(i)$ for values of i from 2 to 6 and let us denote their labels as $\alpha := \ell(v_2), \beta := \ell(v_3), \gamma := \ell(v_4), \delta := \ell(v_5), \epsilon := \ell(v_6)$. For contradiction let $\beta = 1$. We use the observations from the previous Lemmas 8 and 7 to reach contradiction.

As v_3 has four terminals, M composes of at most three non-trivial intervals with possible trivial interval $\{1\}$, let X, Y, Z denote the intervals other than $\{1\}$ so that $X < Y < Z$. Let k be the last label of X or Y.

As v_2 has only two proper terminals, $\alpha \leq 2$ and thus $\alpha = 2$. Since k is a terminal for v_3, it is not a terminal for v_2. Together, we get that $k + 2$ is a label and so both gaps between the intervals X, Y, Z have size exactly 1.

Let us focus on the interval X and let k from now on denote its last label. From k being a terminal for v_3 we have that k is also a terminal for v_5 and v_6. Thus there are at least three distinct non-labels strictly between k and $k + 7$, and so the third closest non-label following k is at most $k + 6$. As there are at most two of them in the gaps separating the three intervals, we have that the sum of lengths of Y and Z is at most 3 (together with 3 non-labels summing up to 6). From this we get that X is long.

Since $\alpha = 2$, there is one terminal for v_2 in X. If Y or Z has length two, then the first element is also a terminal for v_2. Together with the highest label in M we would reach a contradiction with v_2 having only two proper terminals.

We have that $Y = \{k + 2\}$ and $Z = \{k + 4\}$. The label k is a terminal for v_3 and label $k - 1$ is a terminal for v_2. Both are also terminals for v_5. Therefore, we need to set δ so that both $k + \delta$ and $k - 1 + \delta$ fall into $\{k + 1, k + 3, k + 5\}$. But there is no such value and we reach a contradiction.

3.4   Smallest Labels of $\mathcal{C}(A_n)$

In this section we give limitations on ordering of labels in general case. We apply this together with our previously acquired knowledge to further limit the position of label 1 in graphs from $\mathcal{C}(A_n)$.

Lemma 11. *Let G be a sum graph, let v_1, v_2 be equivalent vertices of G with k terminals. Let ℓ be any labeling of G and M the label-set associated with ℓ. Then $\ell(v_1)$ is one of the k smallest labels in M. Furthermore, if $\ell(v_1) = \ell(v_2)$ then $\ell(v_1)$ is one of the $k - 1$ smallest labels in M.*

Proof. Consider v_1 and let us count the number of labels from M not appearing on any neighbor of v_1. The vertex v_1 has exactly k terminals including itself. Therefore there are at least $|M| - k$ labels appearing on the neighbors of v_1 and so the edges incident with v_1 carry at least $|M| - k$ distinct edge-numbers expressed as a sum of $\ell(v_1)$ and a positive label of one of the neighbors. Together we get that M contains at least $|M| - k$ values strictly greater than $\ell(v_1)$.

Assume that $\ell(v_1) = \ell(v_2)$. We can improve the previous argument by the fact that the label of v_1 is present on a neighbor of v_1 (namely v_2). This improves the bound on labels in M strictly greater than $\ell(v_1)$ to at least $|M| - (k - 1)$.

Consider graph A_n for $n \geq 39$ with some labeling ℓ, let α denote a label of some vertex from $\psi(2)$ and let β denote a label of some vertex from $\psi(3)$. Recall Lemma 8 giving explicit amounts of terminals of vertices from $\psi(2)$ and $\psi(3)$. As a corollary of the previous Lemma 11 we have that α is among the 3 smallest labels from M and β is among the 4 smallest labels from M. And as we already know from Lemmas 9 and 10, $\alpha, \beta \neq 1$.

Recall Lemma 6, which gives us limitations on mutual relations in between labels of mergeable vertices. As we know that $\psi(2)$ are mergeable with $\psi(3)$, we know that one of the following must hold: $\alpha = 2\beta$, $2\alpha = \beta$, $3\alpha = \beta$ or $3\alpha = 2\beta$.

We use these facts to show that Lemma 11 can be applied to vertices from $\psi(2)$ and $\psi(3)$ in its stronger form, thus fully determining the smallest three labels in minimal labelings of graphs from $\mathcal{C}(A_n)$.

Lemma 12. *In any minimal labeling ℓ of a graph $G \in \mathcal{C}(A_n)$ such that $n \geq 39$, $\ell(v_1) = \ell(v_2)$ for any $v_1, v_2 \in \psi(2)$.*

Proof. For contradiction, let $v_1, v_2 \in \psi(2)$ have distinct labels α_1, α_2. Without loss of generality, $\alpha_1 < \alpha_2$. We use the observations from Lemma 8. As v_1 is mergeable with v_2 and each has three terminals, Lemma 6 implies that necessarily $2\alpha_1 = \alpha_2$.

Let β denote a label of any vertex from $\psi(3)$. From mergeability, both values α_1, α_2 relate to β, from Lemma 6. The only values for any of the two alphas are $2\beta, \frac{1}{2}\beta, \frac{1}{3}\beta, \frac{2}{3}\beta$. Since $2\alpha_1 = \alpha_2$, the only two suitable values are $\alpha_1 = \frac{1}{3}\beta$ and thus $\alpha_2 = \frac{2}{3}\beta$.

Consider the α_2-cover of the label-set M. Since both α_1 and α_2 are among the three smallest elements of M, according to Lemma 11, we know that the two proper α_2-sequences start with elements 1 and α_1, as these are smaller than α_2. The value α_2 is not an element of either of the two sequences, thus $\{\alpha_2\}$ forms an improper sequence and consequently $2\alpha_2$ is not a label. From the last point of Lemma 8, the value $\alpha_1 + \beta = 4\alpha_1 = 2\alpha_2$ is a label and we reach a contradiction.

Corollary 1. *For any minimal labeling ℓ of A_n, $\ell(v) < \ell(w)$ for any $v \in \psi(2)$ and $w \in \psi(3)$.*

Proof. The previous Lemma 12 guarantees the additional condition of Lemma 11 on labels of vertices from $\psi(2)$, thus $\ell(v)$ is at most second smallest label in A_n. As 1 is also a label, $\ell(v)$ is exactly the second smallest and thus $1 \neq \ell(w) > \ell(v)$.

Lemma 13. *In any minimal labeling ℓ of a graph $G \in \mathcal{C}(A_n)$ such that $n \geq 39$, $\ell(v_1) = \ell(v_2)$ for any $v_1, v_2 \in \psi(3)$.*

The proof of Lemma 13 is analogous to the proof of Lemma 12 and is available in the full version of this paper. We are now ready to prove the last limitation on the placement of label 1 in minimal labelings of graphs from $\mathcal{C}(A_n)$ in order to exclude all possible minimal labelings.

Lemma 14. *In any labeling ℓ of a graph $G \in \mathcal{C}(A_n)$ such that $n \geq 39$, if $v \in \psi(i)$ and $\ell(v) = 1$, then $i \leq 3$.*

Proof. Let α be a label of a vertex $v \in \psi(2)$ and let β be a label of a vertex $w \in \psi(3)$. From Lemmas 12 and 13, we know that these values are uniquely determined by ℓ. For contradiction, let $\alpha, \beta > 1$. Let M denote the label-set of G. From the previous Lemmas 12 and 13, we have the additional conditions to apply Lemma 11 to v and w in its stronger form. Together we have that $\alpha < \beta$ (corollary of Lemma 12) and labels $1, \alpha, \beta$ are the three smallest labels in M.

Let x be a vertex such that $\ell(x) = 1$. Suppose x is not a terminal for v. Then $\alpha + 1$ is a label. Since β is the first label greater than α, we have $\beta = \alpha + 1$. From the fact that $\alpha > 1$ and β is not a multiple of α and vice versa, the mergeability of v and w leaves only one possible relation, $3\alpha = 2\beta$. We conclude that $\alpha = 2$ and $\beta = 3$. Consider any β-sequence, any of its consecutive elements fall into distinct α-sequences. As there exists at least one β-sequence with at least 8 elements, the two proper α-sequences must overlap to satisfy the last 4 elements, none of which can fall into the improper α-sequence. Let k be a label of a proper terminal of v. As x is induced by a number of size at least 4, any proper terminal label of v is also a terminal label of x. Thus, $k + 1$ is not a label. This means that the other α-sequence not terminating in k cannot extend over $k + 1$ and thus, either ends before k or begins after $k+1$. Applying the same argument to the other sequence, we get that the two α-sequences must not overlap as none can extend over the last element of the other. This is a contradiction and x must be a proper terminal for v.

Since 1 is a proper terminal of v, the α-cover of M has only one non-trivial sequence with the first element k (for some yet unknown integer k). Values β and $\beta + \alpha$ are elements of M with difference α, and thus are consecutive elements of the only proper α-sequence. Since the only terminal shared between v and w has the biggest label in M, x is not a terminal of w and thus $\beta + 1$ is a label and must belong to the proper α-sequence (as $\beta > \alpha$). Together, we have that the proper α-sequence contains elements β and $\beta + 1$; thus, the difference α must equal to 1, which is a contradiction.

3.5   Results

Proof (of Theorem 2). Let G be any graph from $\mathcal{C}(A_n)$ where $n \geq 39$ and let ℓ be any labeling of G. For contradiction, let $\ell(v) = 1$ for some vertex v. Clearly $v \in \psi(i)$ for some integer i, $1 < i \leq n + 2$. As shown by Lemmas 9 and 10, $i > 3$. The Lemma 14, based on the previously mentioned lemmas, shows the complementary fact that $i \leq 3$. This is of course a contradiction.

The constant 39 is an artifact of used methods and may in fact be much smaller. While minimal labeling exists for graphs from $\mathcal{C}(A_6)$, based on a computer search we conjecture that there is in fact no minimal labeling for any graph from $\mathcal{C}(A_n)$ for any $n \geq 7$.

Proof (of Theorem 1). Let G be a graph A_n where $n \geq 39$. For contradiction let G admit a minimal labeling. We replace each vertex of G with loop by a clique, obtaining a graph $H \in \mathcal{C}(G)$, and assign all the new vertices from each new clique the label of the original vertex this clique replaces. We have a graph H with sum labeling using the same labels as the labeling of G. Hence if G admits a minimal sum labeling then H must also admit a minimal sum labeling, which is a contradiction with Theorem 2.

4 Conclusion

We have shown that the set $\{2, 3, .., n - 1, n, n + 2\}$ for $n \geq 39$ induces a family of sum graphs with loops which admit no minimal labeling. Furthermore the loops can be replaced by a cliques of sizes at least two and we obtain an infinite family of non-injective sum graphs (without loops) which also admit no minimal labeling.

The constant 39 is an artefact of used methods and it might be much smaller. While minimal labeling exists for graphs from $\mathcal{C}(A_6)$, based on a computer search, we put forward the following conjecture:

Conjecture 1. Let $G \in \mathcal{C}(A_n)$ such that $n \geq 7$, then G allows no minimal labeling.

Further computer experiments indicate that it is not necessary to restrict to the omission of the second-last element from the sequence $\{2, 3, .., n - 1, n, n + 1, n + 2\}$. Thus, we put forward the following conjecture regarding sum graphs with loops:

Conjecture 2. Let G be a gap-graph of size k, where $k \geq 10$, with gap i such that $3 < i < k$, then G admits no minimal labeling.

Acknowledgements. This paper is the output of the 2016 Problem Seminar. We would like to thank Jan Kratochvíl and Jiří Fiala for their guidance, help and tea.

References

1. Bergstrand, D., Harary, F., Hodges, K., Jennings, G., Kuklinski, L., Wiener, J.: The sum number of a complete graph. Bull. Malaysian Math. Soc. **12**, 25–28 (1989)
2. Ellingham, M.N.: Sum graphs from trees. Ars Combin. **35**, 335–349 (1993)
3. Gould, R.J., Rödl, V.: Bounds on the number of isolated vertices in sum graphs, graph theory. Graph Theory Combin. Appl. **1**, 553–562 (1991)
4. Harary, F.: Sum graphs and difference graphs. Congr. Numer. **72**, 101–108 (1990)
5. Miller, M., Ryan, J.F., Smyth, W.F.: The sum number of a disjoint union of graphs (2003)
6. Miller, M., Ryan, J.F., Smyth, W.F.: The sum number of the cocktail party graph. Bull. Inst. Combin. Appl. **22**, 79–90 (1998)
7. Nagamochi, H., Miller, M.: Bounds on the number of isolates in sum graph labeling. Discrete Math. **240**(1–3), 175–185 (2001)
8. Pyatkin, A.V.: New formula for the sum number for the complete bipartite graphs. Discrete Math. **239**, 155–160 (2001)

On the Power Domination Number
of de Bruijn and Kautz Digraphs

Cyriac Grigorious[1], Thomas Kalinowski[2,3] ⓘ, and Sudeep Stephen[3,4(✉)]

[1] Graduate School, King's College London, London, UK
cyriac.grigorious@kcl.ac.uk
[2] School of Science and Technology, University of New England, Armidale, Australia
tkalinow@une.edu.au
[3] School of Mathematical and Physical Sciences, The University of Newcastle,
Callaghan, Australia
[4] School of Mathematical Sciences,
National Institute of Science Education and Research, Bhubaneswar, India
sudeep.stephen@niser.ac.in

Abstract. Let $G = (V, A)$ be a directed graph, and let $S \subseteq V$ be a set of
vertices. Let the sequence $S = S_0 \subseteq S_1 \subseteq S_2 \subseteq \cdots$ be defined as follows:
S_1 is obtained from S_0 by adding all out-neighbors of vertices in S_0. For
$k \geqslant 2$, S_k is obtained from S_{k-1} by adding all vertices w such that for
some vertex $v \in S_{k-1}$, w is the unique out-neighbor of v in $V \setminus S_{k-1}$.
We set $M(S) = S_0 \cup S_1 \cup \cdots$, and call S a *power dominating set* for G
if $M(S) = V(G)$. The minimum cardinality of such a set is called the
power domination number of G. In this paper, we determine the power
domination numbers of de Bruijn and Kautz digraphs.

Keywords: Power domination · de Bruijn digraph · Kautz digraph

1 Introduction

Let $G = (V, A)$ be a directed graph. For a vertex $i \in V$ let $N^{\text{in}}(i)$ and $N^{\text{out}}(i)$
denote its in- and out-neighborhood, respectively, i.e.,

$$N^{\text{in}}(i) = \{j \in V \ : \ (j, i) \in A\}, \qquad N^{\text{out}}(i) = \{j \in V \ : \ (i, j) \in A\}.$$

For a node set S, we use the corresponding notation

$$N^{\text{in}}(S) = \bigcup_{i \in S} N^{\text{in}}(i), \qquad N^{\text{out}}(S) = \bigcup_{i \in S} N^{\text{out}}(i).$$

Let G be a directed graph and S a subset of its vertices. Then we denote the
set monitored by S with $M(S)$ and define it as $M(S) = S_0 \cup S_1 \cup \cdots$ where the
sequence $S_0, S_1, \ldots$ of vertex sets is defined by $S_0 = S$, $S_1 = N^{\text{out}}(S)$, and

$$S_k = S_{k-1} \cup \left\{ w \ : \ \{w\} = N^{\text{out}}(v) \cap (V \setminus S_{k-1}) \text{ for some } v \in S_{k-1} \right\}.$$

© Springer International Publishing AG, part of Springer Nature 2018
L. Brankovic et al. (Eds.): IWOCA 2017, LNCS 10765, pp. 264–272, 2018.
https://doi.org/10.1007/978-3-319-78825-8_22

A set S is called a *power dominating set* of G if $M(S) = V(G)$ and the minimum cardinality of such a set is called the *power domination number* denoted as $\gamma_p(G)$.

The undirected version of the power domination problem was introduced in [11]. The problem was inspired by a problem in electric power systems concerning the placements of phasor measurement units. The directed version of the power domination problem was introduced as a natural extension in [1] where a linear time algorithm was presented for digraphs whose underlying undirected graph has bounded treewidth. Good literature reviews on the power domination problem can be found in [7,8,18].

A closely related concept is zero forcing which was introduced for undirected graphs by the *AIM Minimum Rank – Special Graphs Work Group* in [2] as a tool to bound the minimum rank of matrices associated with the graph G. This notion was extended to digraphs with loops in [4] with the same motivation. For a red/blue coloring of the vertex set of a digraph G with loops, consider the following color-change rule: a red vertex w is converted to blue if it is the only red out-neighbor of some vertex u. We say u forces w and denote this by $u \to w$. A vertex set $S \subseteq V$ is called *zero-forcing* if, starting with the vertices in S blue and the vertices in the complement $V \setminus S$ red, all the vertices can be converted to blue by repeatedly applying the color-change rule. The minimum cardinality of a zero-forcing set for the digraph G is called the *zero-forcing number* of G, denoted by $Z(G)$. Since its introduction the zero-forcing number has been studied for its own sake as an interesting graph invariant [3,5,6,10,16]. In [12], the *propagation time* of a graph is introduced as the number of steps it takes for a zero forcing set to turn the entire graph blue. Physicists have independently studied the zero forcing parameter, referring to it as the *graph infection number*, in conjunction with the control of quantum systems [17].

Recently, Dong *et al.* [9] investigated the domination number of generalized de Bruijn and Kautz digraphs. Kuo and Wu [15] gave an upper bound for power domination in undirected de Bruijn and Kautz graphs. In this paper we study the directed versions, i.e., the zero forcing number and power domination number of de Bruijn and Kautz digraphs. Due to their attractive connectivity features these digraphs have been widely studied as a topology for interconnection networks [13], and some generalizations of these digraphs were proposed [14].

Section 2 contains some notation and precise statements of our main result. In Sect. 3 we determine the power domination number and zero forcing number for de Bruijn digraphs. In Sect. 4 we determine the power domination number and zero forcing number for Kautz digraphs.

2   Notations and Main Result

We give an interpretation of the power domination problem and zero forcing problem as a set cover problem. We call a vertex set W *strongly critical* if there is no vertex in G which has exactly one out neighbor in W. We call a vertex set W *weakly critical* if there is no vertex outside W which has exactly one out-neighbor in W. If W is strongly (weakly) critical, but no proper subset of W is strongly (weakly) critical, then we call W *minimal strongly (weakly) critical*.

Note that a vertex set S is a zero forcing set if and only if $S \cap W \neq \emptyset$ for every strongly critical set $W \subseteq V$. Similarly, S is a power dominating set if and only if $(S \cup N^{\text{out}}(S)) \cap W \neq \emptyset$ for every weakly critical set $W \subseteq V$, and therefore

$$Z(G) = \min\{|S| \ : \ S \cap W \neq \emptyset \text{ for every strongly critical set } W \subseteq V\},$$

$$\gamma_p(G) = \min\{|S| \ : \ (S \cup N_G^{\text{out}}(S)) \cap W \neq \emptyset \text{ for every weakly critical set } W \subseteq V\}.$$

For an integer $d \geqslant 2$, let $\mathbb{Z}_d = \{0, 1, \ldots, d-1\}$ denote the cyclic group of order d. The de Bruijn digraph, denoted $B(d, n)$, with parameters $d \geqslant 2$ and $n \geqslant 2$ is defined to be the graph $G = (V, A)$ with vertex set V and arc set A where

$$V = \mathbb{Z}_d^n = \{(a_1, \ldots, a_n) \ : \ a_i \in \mathbb{Z}_d \text{ for } i = 1, \ldots, n\},$$
$$A = \{((a_1, a_2, \ldots, a_n), (a_2, \ldots, a_n, b)) \ : \ (a_1, a_2, \ldots, a_n) \in V, \ b \in \mathbb{Z}_d\}.$$

The Kautz digraph, denoted $K(d, n)$, with parameters $d \geqslant 2$ and $n \geqslant 2$ is defined to be the graph $G = (V, A)$ with vertex set V and arc set A where

$$V = \{(a_1, \ldots, a_n) \ : \ a_i \in \mathbb{Z}_{d+1}, \ a_i \neq a_{i+1}\}$$
$$A = \{((a_1, a_2, \ldots, a_n), (a_2, \ldots, a_n, b)) \ : \ (a_1, a_2, \ldots, a_n) \in V, \ b \in \mathbb{Z}_{d+1} \setminus \{a_n\}\}.$$

Our main results are the following theorems.

Theorem 1. *Let G be a de Bruijn digraph with parameters $d, n \geqslant 2$. Then the zero forcing number and power domination number of G are $(d-1)d^{n-1}$ and $(d-1)d^{n-2}$, respectively.*

Theorem 2. *Let G be a Kautz digraph with parameters $d \geqslant 2$ and $n \geqslant 3$. Then the zero forcing number and power domination number of G are $(d-1)(d+1)d^{n-2}$ and $(d-1)(d+1)d^{n-3}$, respectively.*

3   The Power Domination Number of de Bruijn Digraphs

In this section we prove Theorem 1. Let us define the sets

$$X(a_1, \ldots, a_{n-1}) = \{(a_1, \ldots, a_{n-1}, \alpha) \ : \ \alpha \in \mathbb{Z}_d\}$$

which partition the vertex set V into d^{n-1} sets of size d. Furthermore, $N^{\text{out}}(v) = X(a_1, \ldots, a_{n-1})$ for every vertex v of the form $(\alpha, a_1, a_2, \ldots, a_{n-1})$.

Lemma 1. *Let G be a de Bruijn digraph with parameters $d, n \geqslant 2$. Then $Z(G) \geqslant (d-1)d^{n-1}$.*

Proof. Every 2-element subset of each of the sets $X(a_1, \ldots, a_{n-1})$ is strongly critical, and therefore, any zero forcing set S needs to intersect $X(a_1, \ldots, a_{n-1})$ in at least $d-1$ elements, and the result follows.

Lemma 2. *Let G be a de Bruijn digraph with parameters $d, n \geqslant 2$. Then $Z(G) \leqslant (d-1)d^{n-1}$.*

Proof. Consider the vertex set $S = \{(a_1,\ldots,a_{n-1},a_n) \in V \ : \ a_1 \neq a_n\}$. To show that S is a zero forcing set, it is sufficient to verify that each vertex $v = (a_1,\ldots,a_{n-1},a_n)$ is either in S or is the unique out-neighbor in $V \setminus S$ for some vertex w. If $a_1 \neq a_n$, then $v \in S$. If $a_1 = a_n$, then for any vertex of the form $w = (\beta,a_1,\ldots,a_{n-1})$, v is the only neighbor of w in $V \setminus S$.

Lemmas 1 and 2 imply the first statement of Theorem 1. In order to prove the second part of this theorem we recall that $S \subseteq V$ is a power dominating set if and only if $S \cup N^{\text{out}}(S)$ intersects every weakly critical set. In particular, it is necessary that $|(S \cup N^{\text{out}}(S)) \cap X(a_1,\ldots,a_{n-1})| \geqslant d - 1$ for every $(a_1,\ldots,a_{d-1}) \in \mathbb{Z}_d^{n-1}$.

Lemma 3. *Let G be a de Bruijn digraph with parameters $d, n \geqslant 2$. Then every power dominating set has size at least $(d-1)d^{n-2}$.*

Proof. Let S be a power dominating set, suppose $|S| < (d-1)d^{n-2}$ and set $Z = S \cup N^{\text{out}}(S)$. We have

$$(Z \setminus S) \cap X(a_1,\ldots,a_{n-1}) \neq \emptyset \implies X(a_1,\ldots,a_{n-1}) \subseteq Z.$$

For $k = 0, 1, \ldots, d$, we set $\alpha_k = \#\{(a_1,\ldots,a_{n-1}) \ : \ |S \cap X(a_1,\ldots,a_{n-1})| = k\}$, and get

$$|S| = \alpha_1 + 2\alpha_2 + \cdots + (d-1)\alpha_{d-1} + d\alpha_d.$$

Now let $I_0 = \{(a_1,\ldots,a_{n-1}) \ : \ X(a_1,\ldots,a_{n-1}) \subseteq Z\}$. Then

$$|I_0| \leqslant |S| + \alpha_d = \alpha_1 + 2\alpha_2 + \cdots + (d-1)\alpha_{d-1} + (d+1)\alpha_d.$$

For $(a_1,\ldots,a_{n-1}) \notin I_0$ we must have $|Z \cap X(a_1,\ldots,a_{n-1})| = d - 1$, and this implies that $|S \cap X(a_1,\ldots,a_{n-1})| = d - 1$. We conclude $|I_0| + \alpha_{d-1} \geqslant d^{n-1}$. Therefore

$$\alpha_1 + 2\alpha_2 + \cdots + (d-2)\alpha_{d-2} + d\alpha_{d-1} + (d+1)\alpha_d \geqslant d^{n-1},$$

and together with $|S| < (d-1)d^{n-2}$ this yields

$$\alpha_{d-1} + \alpha_d > d^{n-1} - (d-1)d^{n-2} = d^{n-2}.$$

But then $|S| \geqslant (d-1)(\alpha_{d-1} + \alpha_d) > (d-1)d^{n-2}$, which is the required contradiction.

We define a set $S \subseteq V$ by

$$S = \begin{cases} \{(0,1),(0,2),\ldots,(0,d-1)\} & \text{if } n = 2, \\ \{(a_1,a_2,a_3) \in V \ : \ a_2 = a_1, a_3 \neq a_1\} & \text{if } n = 3, \\ \{(a_1,\ldots,a_n) \in V \ : \ a_{n-1} = a_1 + a_{n-2}, a_n \neq a_1 + a_2 + a_{n-2}\} & \text{if } n \geqslant 4. \end{cases}$$

$$(1)$$

Note that $|S| = (d-1)d^{n-2}$. The construction of the set S defined in (1) can be visualized by arranging the vertices of G in a $d^2 \times d^{n-2}$-array where the

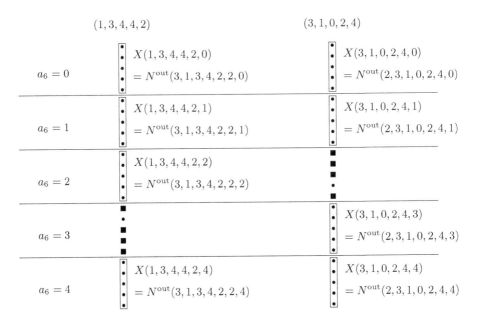

Fig. 1. Illustration of the construction of the power dominating set S for $d = 5$ and $n = 7$. For the two columns $(a_1, \ldots, a_5) = (1, 3, 4, 4, 2)$ and $(a_1, \ldots, a_5) = (3, 1, 0, 2, 4)$ we show the elements of S (black squares), and we indicate for the sets $X(a_1, \ldots, a_6)$ (enclosed by rectangles) the elements of S having them as their out-neighbourhood.

rows are indexed by pairs (a_{n-1}, a_n) and the columns are indexed by $(n - 2)$-tuples $(a_1, \ldots, a_{n-2})$. Then column $(a_1, \ldots, a_{n-2})$ is the union of the d sets $X(a_1, \ldots, a_{n-2}, a_{n-1})$ over $a_{n-1} \in \mathbb{Z}_d$, and the set S contains $d-1$ elements from each column. More precisely, the intersection of S with column $(a_1, \ldots, a_{n-2})$ is

$$X(a_1, \ldots, a_{n-2}, a_1 + a_{n-2}) \setminus \{(a_1, \ldots, a_{n-2}, a_1 + a_{n-2}, a_1 + a_2 + a_{n-2})\}.$$

In Fig. 1 this is illustrated for two columns with $d = 5$ and $n = 7$.

Lemma 4. *The set S defined in (1) is a power dominating set for G.*

Proof. For $Z = S \cup N^{\text{out}}(S)$ it is sufficient to show that $|Z \cap X(a_1, \ldots, a_{n-1})| \geqslant d - 1$ for every $(a_1, \ldots, a_{n-1})$. We provide the full argument for $n \geqslant 4$ (the cases $n = 2$ and $n = 3$ are easy to check).

Case 1. If $a_{n-1} = a_1 + a_{n-2}$, then by (1),

$$S \cap X(a_1, \ldots, a_{n-1}) = \{(a_1, \ldots, a_n) \; : \; a_n \in \mathbb{Z}_d \setminus \{a_1 + a_2 + a_{n-2}\}\},$$

hence $|Z \cap X(a_1, \ldots, a_{n-1})| \geqslant |S \cap X(a_1, \ldots, a_{n-1})| = d - 1$.

Case 2. If $a_{n-1} \neq a_1 + a_{n-2}$, then $X(a_1, \ldots, a_{n-1}) \subseteq Z$ because

$$X(a_1, \ldots, a_{n-1}) = N^{\text{out}}((a_{n-2} - a_{n-3}, a_1, a_2, \ldots, a_{n-1}))$$

and $(a_{n-2} - a_{n-3}, a_1, a_2, \ldots, a_{n-1}) \in S$.

The second part of Theorem 1 follows from Lemmas 3 and 4.

4   The Power Domination Number of Kautz Digraphs

In this section we prove Theorem 2. Let us define the sets

$$X(a_1,\ldots,a_{n-1}) = \{(a_1,\ldots,a_{n-1},a_n) \ : \ a_n \in \mathbb{Z}_{d+1} \setminus \{a_{n-1}\}\}$$

for $(a_1,\ldots,a_{n-1}) \in \mathbb{Z}_{d+1}^{n-1}$ with $a_i \neq a_{i+1}$ for all i. These sets partition the vertex set V into $(d+1)d^{n-2}$ sets of size d. Furthermore, $N^{\text{out}}(v) = X(a_1,\ldots,a_{n-1})$ for every vertex v of the form $(a_0,a_1,a_2,\ldots,a_{n-1})$.

Lemma 5. *Let G be a Kautz digraph with parameters $d, n \geqslant 2$. Then $Z(G) \geqslant (d-1)(d+1)d^{n-2}$.*

Proof. Every 2-element subset of each of the sets $X(a_1,\ldots,a_{n-1})$ is strongly critical, and therefore, any zero forcing set S needs to intersect $X(a_1,\ldots,a_{n-1})$ in at least $d-1$ elements, and the result follows.

Lemma 6. *Let G be a Kautz digraph with parameters $d, n \geqslant 2$. Then $Z(G) \leqslant (d-1)(d+1)d^{n-2}$.*

Proof. Consider the vertex set

$$S = \begin{cases} \{(a_1,a_2) \in V \ : \ a_2 \neq a_1 + 1\} & \text{if } n = 2, \\ \{(a_1,\ldots,a_n) \in V \ : \ a_n \neq a_{n-2}\} & \text{if } n \geqslant 3. \end{cases}$$

We have $|S| = (d-1)(d+1)d^{n-2}$, and to show that S is a zero forcing set, it is sufficient to verify that each vertex $v = (a_1,\ldots,a_{n-1},a_n)$ is either in S or is the unique out-neighbor in $V \setminus S$ for some vertex w.

Case $n = 2$. If $a_2 \neq a_1 + 1$, then $v \in s$. If $a_2 = a_1 + 1$, then for any vertex of the form $w = (\beta, a_1)$, v is the only neighbor of w in $V \setminus S$.

Case $n \geqslant 3$. If $a_n \neq a_{n-2}$, then $v \in S$. If $a_n = a_{n-2}$, then for any vertex of the form $w = (\beta, a_1, \ldots, a_{n-1})$, v is the only neighbor of w in $V \setminus S$.

Lemmas 5 and 6 imply the first statement of Theorem 2.

Lemma 7. *Let G be a Kautz digraph with parameters $d \geqslant 2$ and $n \geqslant 3$. Then, every power dominating set has size at least $(d-1)(d+1)d^{n-3}$.*

Proof. Let S be a power dominating set, suppose $|S| < (d-1)(d+1)d^{n-3}$ and set $Z = S \cup N^{\text{out}}(S)$. We have

$$(Z \setminus S) \cap X(a_1,\ldots,a_{n-1}) \neq \emptyset \implies X(a_1,\ldots,a_{n-1}) \subseteq Z.$$

For $k = 0, 1, \ldots, d$, we set $\alpha_k = \#\{(a_1,\ldots,a_{n-1}) \ : \ |S \cap X(a_1,\ldots,a_{n-1})| = k\}$, and get

$$|S| = \alpha_1 + 2\alpha_2 + \cdots + (d-1)\alpha_{d-1} + d\alpha_d.$$

Now let $I_0 = \{(a_1,\ldots,a_{n-1}) \ : \ X(a_1,\ldots,a_{n-1}) \subseteq Z\}$. Clearly,

$$|I_0| \leqslant |S| + \alpha_d = \alpha_1 + 2\alpha_2 + \cdots + (d-1)\alpha_{d-1} + (d+1)\alpha_d.$$

For $(a_1, \ldots, a_{n-1}) \notin I_0$ we must have $|Z \cap X(a_1, \ldots, a_{n-1})| = d - 1$ because Z intersects every weakly critical set. This implies that $|S \cap X(a_1, \ldots, a_{n-1})| = d - 1$, and we conclude $|I_0| + \alpha_{d-1} \geqslant (d+1)d^{n-2}$. Therefore

$$\alpha_1 + 2\alpha_2 + \cdots + (d-2)\alpha_{d-2} + d\alpha_{d-1} + (d+1)\alpha_d \geqslant (d+1)d^{n-2},$$

and together with $|S| < (d-1)(d+1)d^{n-3}$ this yields

$$\alpha_{d-1} + \alpha_d > (d+1)d^{n-2} - (d-1)(d+1)d^{n-3} = (d+1)d^{n-3}.$$

But then $|S| \geqslant (d-1)(\alpha_{d-1} + \alpha_d) > (d-1)(d+1)d^{n-3}$, which is the required contradiction.

We define a set $S \subseteq V$ by

$$S = \begin{cases} \{(0,1), (0,2), \ldots, (0,d)\} & \text{if } n = 2, \\ \{(a_1, a_2, a_3) \in V \; : \; a_2 = a_1 + 1, \, a_3 \neq a_1 + 2\} & \text{if } n = 3, \\ \{(a_1, a_2, a_3, a_4) \in V \; : \; a_3 = a_1, \, a_4 \neq a_2\} & \text{if } n = 4, \\ \{(a_1, \ldots, a_n) \in V \; : \; ((a_{n-2}, a_{n-1}) = (a_1, a_2) \wedge a_n \neq a_3) \vee (a_{n-1} = a_1 \wedge a_n \neq a_2)\} & \text{if } n \geqslant 5. \end{cases} \tag{2}$$

Lemma 8. $|S| = \begin{cases} d & \text{if } n = 2, \\ (d-1)(d+1)d^{n-3} & \text{if } n \geqslant 3. \end{cases}$

Proof. For $n \leqslant 4$ this is easy to check. For $n \geqslant 5$ we proceed by the following argument. We consider the partition $S = S_1 \cup S_2$ where

$$S_1 = \{(a_1, \ldots, a_n) \in S \; : \; a_{n-3} = a_1\}, \quad S_2 = \{(a_1, \ldots, a_n) \in S \; : \; a_{n-3} \neq a_1\}.$$

Let s_k be the number of words $a_1 \ldots a_k$ over the alphabet $\mathbb{Z}_{d+1}$ which satisfy $a_k = a_1$ and $a_i \neq a_{i+1}$ for all $i \in \{1, \ldots, k-1\}$. Then $s_2 = 0$ and $s_k = (d+1)d^{k-2} - s_{k-1}$ for $k \geqslant 3$. It follows by induction on k that $s_k = d^{k-1} - (-1)^k d$. Every vector $(a_1, \ldots, a_{n-3}) \in \mathbb{Z}_{d+1}^{n-3}$ with $a_i \neq a_{i+1}$ and $a_{n-3} = a_1$ can be extended to an element of S_1 by choosing $a_{n-2} \in \mathbb{Z}_{d+1} \backslash \{a_1\}$, $a_{n-1} = a_1$ and $a_n \in \mathbb{Z}_{d+1} \backslash \{a_1, a_2\}$, hence

$$|S_1| = s_{n-3}d(d-1) = \left(d^{n-4} - (-1)^{n-3}d\right)d(d-1).$$

If $a_{n-3} \neq a_1$, then we can choose $(a_{n-2}, a_{n-1}) = (a_1, a_2)$ and $a_n \in \mathbb{Z}_{d+1} \backslash \{a_2, a_3\}$, or $a_{n-2} \in \mathbb{Z}_{d+1} \backslash \{a_1, a_{n-3}\}$, $a_{n-1} = a_1$ and $a_n = \mathbb{Z}_{d+1} \backslash \{a_1, a_2\}$, hence

$$\begin{aligned} |S_2| &= \left[(d+1)d^{n-4} - s_{n-3}\right]\left[(d-1) + (d-1)^2\right] \\ &= \left[(d+1)d^{n-4} - d^{n-4} + (-1)^{n-3}d\right]d(d-1) \\ &= \left[d^{n-3} + (-1)^{n-3}d\right]d(d-1). \end{aligned}$$

Finally,

$$|S| = |S_1| + |S_2| = d(d-1)\left[d^{n-4} - (-1)^{n-3}d + d^{n-3} + (-1)^{n-3}d\right] = (d+1)(d-1)d^{n-3}.$$

Lemma 9. *The set S defined in (2) is a power dominating set for $G = K(d, n)$.*

Proof. For $Z = S \cup N^{\text{out}}(S)$ it is sufficient to show that $|Z \cap X(a_1, \ldots, a_{n-1})| \geqslant d - 1$ for every $(a_1, \ldots, a_{n-1})$. We provide the full argument for $n \geqslant 5$ (the cases $n = 2$, $n = 3$ and $n = 4$ are easy to check).

Case 1. If $a_{n-2} = a_1$ and $a_{n-1} = a_2$, then

$$|S \cap X(a_1, \ldots, a_{n-1})| = |\{(a_1, \ldots, a_n) \; : \; a_n \in \mathbb{Z}_{d+1} \setminus \{a_2, a_3\}\}| = d - 1,$$

and the claim follows from $Z \supseteq S$.

Case 2. If $a_{n-2} = a_1$ and $a_{n-1} \neq a_2$, then $X(a_1, \ldots, a_{n-1}) \subseteq Z$ because

$$X(a_1, \ldots, a_{n-1}) = N^{\text{out}}((a_{n-3}, a_1, a_2, \ldots, a_{n-1}))$$

and $(a_{n-3}, a_1, a_2, \ldots, a_{n-1}) \in S$.

Case 3. If $a_{n-2} \neq a_1$ and $a_{n-1} = a_2$, then $X(a_1, \ldots, a_{n-1}) \subseteq Z$ because

$$X(a_1, \ldots, a_{n-1}) = N^{\text{out}}((a_{n-2}, a_1, a_2, \ldots, a_{n-1}))$$

and $(a_{n-2}, a_1, a_2, \ldots, a_{n-1}) \in S$.

Case 4. If $a_{n-2} \neq a_1$ and $a_{n-1} = a_1$, then

$$|S \cap X(a_1, \ldots, a_{n-1})| = |\{(a_1, \ldots, a_n) \; : \; a_n \in \mathbb{Z}_{d+1} \setminus \{a_1, a_2\}\}| = d - 1,$$

and the claim follows from $Z \supseteq S$.

Case 5. If $a_{n-2} \neq a_1$ and $a_{n-1} \notin \{a_1, a_2\}$, then $X(a_1, \ldots, a_{n-1}) \subseteq Z$ because

$$X(a_1, \ldots, a_{n-1}) = N^{\text{out}}((a_{n-2}, a_1, a_2, \ldots, a_{n-1}))$$

and $(a_{n-2}, a_1, a_2, \ldots, a_{n-1}) \in S$.

The second part of Theorem 2 follows from Lemmas 7, 8 and 9.

5   Conclusion

In this paper, we have determined the zero forcing number and power domination number of de Bruijn and Kautz digraphs. There are many variants of de Bruijn and Kautz digraphs introduced and studied over the years, one of them being generalized de Bruijn digraphs $GB(d, n)$ and generalised Kautz digraphs $GK(d, n)$ which can be defined as follows:

$$V(GB(d, n)) = \{0, 1, \ldots, n - 1\},$$
$$A(GB(d, n)) = \{(x, y) \; : \; y \equiv dx + i \pmod{n}, \; 0 \leqslant i \leqslant d - 1\},$$
$$V(GK(d, n)) = \{0, 1, \ldots, n - 1\},$$
$$A(GK(d, n)) = \{(x, y) \; ; \; y \equiv -dx - i \pmod{n}, \; 1 \leqslant i \leqslant d\}.$$

We leave it as an open problem to determine the power domination number of generalised de Bruijn and Kautz digraphs.

Acknowledgement. The authors would like to thank Dr. Joe Ryan for his valuable comments and suggestions to improve the quality of the paper.

References

1. Aazami, A., Stilp, K.: Approximation algorithms and hardness for domination with propagation. SIAM J. Discret. Math. **23**, 1382–1399 (2009)
2. AIM Minimum Rank – Special Graphs Work Group: Zero forcing sets and the minimum rank of graphs. Linear Algebra Appl. **428**(7), 1628–1648 (2008)
3. Barioli, F., Barrett, W., Fallat, S.M., Hall, H.T., Hogben, L., Shader, B., van den Driessche, P., van der Holst, H.: Zero forcing parameters and minimum rank problems. Linear Algebra Appl. **433**(2), 401–411 (2010)
4. Barioli, F., Fallat, S.M., Hall, H.T., Hershkowitz, D., Hogben, L., van der Holst, H., Shader, B.: On the minimum rank of not necessarily symmetric matrices: a preliminary study. Electron. J. Linear Algebra **18**, 126–145 (2009)
5. Barioli, F., Barrett, W., Fallat, S.M., Hall, H.T., Hogben, L., Shader, B., van den Driessche, P., van der Holst, H.: Parameters related to tree-width, zero forcing, and maximum nullity of a graph. J. Graph Theory **72**(2), 146–177 (2012)
6. Berman, A., Friedland, S., Hogben, L., Rothblum, U.G., Shader, B.: An upper bound for the minimum rank of a graph. Linear Algebra Appl. **429**(7), 1629–1638 (2008)
7. Chang, G.J., Dorbec, P., Montassier, P., Raspaud, A.: Generalized power domination of graphs. Discret. Appl. Math. **160**, 1691–1698 (2012)
8. Dorbec, P., Mollard, M., Klavzar, S., Spacapan, S.: Power domination in product graphs. SIAM J. Discret. Math. **22**, 554–567 (2008)
9. Dong, Y., Shan, E., Kang, L.: Constructing the minimum dominating sets of generalized de Bruijn digraphs. Discret. Math. **338**, 1501–1508 (2015)
10. Edholm, C.J., Hogben, L., Huynh, M., LaGrange, J., Row, D.D.: Vertex and edge spread of zero forcing number, maximum nullity, and minimum rank of a graph. Linear Algebra Appl. **436**(12), 4352–4372 (2012)
11. Haynes, T.W., Hedetniemi, S.M., Hedetniemi, S.T., Henning, M.A.: Domination in graphs applied to electric power networks. SIAM J. Discret. Math. **15**(4), 519–529 (2002)
12. Hogben, L., Huynh, M., Kingsley, N., Meyer, S., Walker, S., Young, M.: Propagation time for zero forcing on a graph. Discret. Appl. Math. **160**(13–14), 1994–2005 (2012)
13. Huang, J., Xu, J.M.: The bondage numbers of extended de Bruijn and Kautz digraphs. Comput. Math. Appl. **51**, 1137–1147 (2006)
14. Imase, M., Itoh, M.: A design for directed graphs with minimum diameter. IEEE Trans. Comput. **32**, 782–784 (1983)
15. Kuo, J., Wu, W.L.: Power domination in generalized undirected de Bruijn graphs and Kautz graphs. Discrete Math. Algorithm. Appl. **07**, 2961–2973 (2015)
16. Lu, L., Wu, B., Tang, Z.: Proof of a conjecture on the zero forcing number of a graph. arXiv:1507.01364 (2015)
17. Severini, S.: Nondiscriminatory propagation on trees. J. Phys. A: Math. Theor. **41**(48), 482002 (2008)
18. Stephen, S., Rajan, B., Ryan, J., Grigorious, C., William, A.: Power domination in certain chemical structures. J. Discret. Algorithms **33**, 10–18 (2015)

Heuristics

A Multi-start Heuristic for Multiplicative Depth Minimization of Boolean Circuits

Sergiu Carpov[✉], Pascal Aubry, and Renaud Sirdey

CEA, LIST, Point Courrier 172, 91191 Gif-sur-Yvette Cedex, France
sergiu.carpov@cea.fr

Abstract. In this work we propose a multi-start heuristic which aims at minimizing the multiplicative depth of boolean circuits. The multiplicative depth objective is encountered in the field of homomorphic encryption where ciphertext size depends on the number of consecutive multiplications. The heuristic is based on rewrite operators for multiplicative depth-2 paths. Even if the proposed rewrite operators are simple and easy to understand the experimental results show that they are rather powerful. The multiplicative depth of the benchmarked circuits was hugely improved. In average the obtained multiplicative depths were lower by more than 3 times than the initial ones. The proposed rewrite operators are not limited to boolean circuits and can also be used for arithmetic circuits.

1 Introduction and Related Works

An *encryption scheme* describes the way of encrypting and decrypting plaintext messages such that finding which is the plaintext message from encrypted data (denoted *ciphertext* in what follows) is either very hard or even impossible without a secret. An encryption scheme is said to be *homomorphic* when some operations on plaintext messages can be done homomorphically, that is directly in the space of ciphertexts (and without decrypting them). When addition and multiplication operations are supported, the homomorphic encryption (HE) scheme is functionally complete. Since the seminal work of Gentry [9], introducing the first practical (to some extent) homomorphic encryption many other simpler and more efficient schemes have been proposed [5,6]. A HE scheme with a binary plaintext space allows to execute any boolean circuit directly over encrypted data.

A noise component is added to the ciphertext during the encryption for security reasons. The noise component is a common characteristic for HE schemes. Each new homomorphic operation applied on the ciphertexts increases the noise component in the resulting ciphertext. After a (predefined) number of homomorphic operations the noise is so large that the correctness of the decryption

This work has been supported in part by the French's FUI project CRYPTOCOMP and by the European Union's H2020 Programme under grant agreement number 727528 (project KONFIDO).

L. Brankovic et al. (Eds.): IWOCA 2017, LNCS 10765, pp. 275–286, 2018.
https://doi.org/10.1007/978-3-319-78825-8_23

cannot be not ensured anymore. Usually the noise growth induced by the addition operation is smaller than the noise growth induced by the multiplication operation. That is why many authors consider only the *multiplicative depth*[1] of evaluated circuits when HE schemes are parametrized. In order to support the evaluation of larger multiplicative depth circuits, for an equivalent security level, the ciphertext sizes must be increased and respectively the cost of homomorphic operations increases also. Another solution to this problem is to use ciphertext bootstrapping [10]. The bootstrapping procedure takes a noisy ciphertext as input and executes homomorphically the HE scheme decryption. The noise of the resulting "bootstrapped" ciphertext is lower than the noise of the input ciphertext.

Obtaining low multiplicative depth circuits is a major issue in the practical use of homomorphic encryption. With every new multiplicative level the HE scheme parameters increase in size. Therefore the execution time of the whole boolean circuit increases accordingly. Many works found in the literature treat the problem of boolean circuit optimization for hardware targets or more generally the problem of hardware synthesis. We refer to the open-source software system used for hardware synthesis ABC [3]. It is an open-source environment providing implementations of the state-of-the-art circuit optimization algorithms. The most common objectives used in hardware synthesis are circuit area and circuit depth (latency). To the best of our knowledge none of these algorithms were designed for multiplicative depth minimization.

Cryptographic literature mostly focused on the minimization of multiplicative complexity of circuits, i.e. the number of AND gates in circuits where the XOR gates are for free [4,11,14]. Minimization of boolean circuit depth was discussed in [7]. The authors introduced circuit depth minimization techniques in the context of secure multi-party computation. No distinction is made between multiplicative and additive gates. The authors of Armadillo compilation chain [8] studied the use of ABC tools for minimizing the multiplicative depth of boolean circuits in the context of a compilation chain targeting homomorphic execution.

Several works [2,12,13] study the minimization of bootstrappings in boolean circuits problem. Bootstrapping is a computational heavy procedure. It is straightforward to see that minimizing the number of bootstraps in a homomorphic evaluation of a circuit increases its execution performance. Although the problem we study in this paper shares the same goal (i.e. increase the homomorphic execution performance of boolean circuits) the employed methods to achieve it are orthogonal.

In this work we introduce and study multiplicative depth-2 path rewrite operators which decrease the multiplicative depth of a boolean circuit. We furthermore propose a heuristic method which makes use of these operators. The goal of the heuristic is to minimize the boolean circuit multiplicative depth. The paper is structured as follows, in Sect. 2 are described the proposed circuit rewrite

---

[1] Multiplicative depth is the number of sequential homomorphic multiplications which can be done on freshly encrypted ciphertexts in order to be able to decrypt and retrieve the result of multiplications.

operators and the heuristic itself, Sect. 3 presents experimental results we have performed and finally Sect. 4 concludes the paper and discuss some perspectives.

2   Multiplicative Depth Minimization Multi-start Heuristic

2.1   Preliminary Definitions

A boolean circuit is a directed acyclic graph $C = (V, E)$ with a set of nodes V and a set of edges E. Circuit nodes represent boolean functions (gates) and circuit edges are connections between nodes. The set of nodes can be split into 2 independent sub-sets:

- Nodes without a predecessor define circuit inputs. An input node can be either a boolean input variable or a boolean constant (e.g. logic "0" or logic "1" inputs).
- Nodes each representing a gate applying a basic boolean function to the values of its predecessors. The input degree of gates is 2. A sub-set of gate nodes represent circuit outputs. Without loss of generality we suppose that the set of output nodes is the same as the set of nodes with zero output degree. In this work we suppose that the boolean circuits use AND and XOR operators only. The set $\{AND, XOR\}$ together with the constant "1" is functionally complete [15]. This means that any boolean function can be expressed using these operators.

   We denote by pred: $V \rightarrow 2^V$ and succ: $V \rightarrow 2^V$ the functions giving the set of predecessors, respectively successors, of a node $v \in V$ in a boolean circuit C.

   The number of successively executed AND operators, also called *multiplicative depth*, influences the parameters of a HE scheme. The minimization of the multiplicative depth allows not only to obtain smaller ciphertext sizes but also to minimize[2] the overall execution time of the boolean circuit. Let us define a function $d: V \rightarrow \{0, 1\}$ which returns one for AND nodes and zero otherwise. Only the nodes for which $d(v) = 1$ influence circuit multiplicative depth.

   Let $l: V \rightarrow \mathbb{N}$ be a function which gives the *multiplicative depth* of circuit nodes. The multiplicative depth of node v is equal to the maximal number of AND gates on any path beginning in an input node and ending in node v. The multiplicative depths for circuit nodes are computed recursively using relation:

$$l(v) = \begin{cases} 0 & \text{if } |\text{pred}(v)| = 0, \\ \max_{u \in \text{pred}(v)} l(u) + d(v) & \text{otherwise.} \end{cases}$$

   Let $r: V \rightarrow \mathbb{N}$ be a function which gives the *reverse multiplicative depth* of circuit nodes. The reverse multiplicative depth of node v is the maximal number

---

[2] As we shall further see, more precisely it depends on the relative computational cost of circuit AND gates with respect to scheme multiplicative depth.

of AND gates on any path beginning in a successor of node v and ending in an output node. It is somewhat equivalent to the multiplicative depth function except that it does not include the depth due to the node itself. The reverse multiplicative depths for circuit nodes are computed recursively using:

$$r(v) = \begin{cases} 0 & \text{if } |\text{succ}(v)| = 0, \\ \max_{u \in \text{succ}(v)} (r(u) + d(u)) & \text{otherwise.} \end{cases}$$

The overall multiplicative depth of a circuit C is the maximal multiplicative depth of its nodes:

$$l^{\max} = \max_{v \in V} l(v) = \max_{v \in V} r(v).$$

The *critical nodes* of a circuit C are the nodes for which relation (1) is verified. We denote *critical circuit* the sub-circuit containing all the critical nodes of a circuit C. A *critical path* is a path in this circuit.

$$l(v) + r(v) = l^{\max}, \ v \in V \tag{1}$$

2.2 Multiplicative Depth-2 Path Rewrite Operators

The multiplicative depth of a boolean circuit equals to the multiplicative depth of its critical part. Decreasing the multiplicative depth of the critical part will necessarily decrease the overall circuit multiplicative depth. In this section we introduce two rewrite operators which when applied to the critical part of a boolean circuit potentially minimize the multiplicative depth of a circuit. The idea behind these operators is to rewrite critical paths of multiplicative depth 2 in such a way that the overall multiplicative depth decreases. We firstly describe an operator which rewrites simple paths composed of two AND gates only. Afterwards, a second rewrite operator is described which allows to obtain such a simple path (from two AND gates) from any path of multiplicative depth 2. Additionally, we introduce the conditions these paths should verify so that the multiplicative depth is lowered after the rewrite operators are applied.

Let P denote the set of all critical paths beginning and ending in an AND gate and containing exactly 2 AND nodes, i.e. the set of paths of multiplicative depth 2. A path $p \in P$ contains at least 2 nodes: 2 AND gates separated by zero or more XOR gates. Figure 1 illustrates such a critical path.

Let us examine a critical path p of length 2, i.e. $p = (v_1, v_{|p|})$ where v_1 and $v_{|p|}$ are AND gates. Such a path is shown on the left-hand side of Fig. 2. Path p can be rewritten using AND associativity rule: $(x \cdot y) \cdot z = x \cdot (y \cdot z)$. The right-hand side of Fig. 2 illustrates the circuit part obtained after this rewrite operator is applied to path p. Rewritten path multiplicative depth decreases only if the multiplicative depth of nodes y and z are less than the multiplicative depth of node x, i.e. $l(y) < l(x)$ and $l(z) < l(x)$. In this case the multiplicative depth of gate $v_{|p|}$ decreases by one, from $l(x) + 2$ to $l(x) + 1$. The number of AND gates in the resulting circuit either increases by one or rests the same if node v_1 does not have other successors than node $v_{|p|}$.

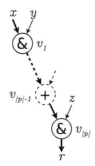

Fig. 1. Critical path of multiplicative depth 2. Thick edges represent the critical path.

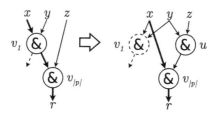

Fig. 2. Length 2 critical path $(v_1, v_{|p|})$ rewrite operator. Dotted line AND gate v_1 is kept only if needed.

In case of critical paths of length larger than 2, the inner XOR gates prevents the direct use of the rewrite operator defined above. A second rewrite operator allows to move an AND gate up the critical path by one place. We call it *AND gate move up* operator. This operator uses XOR distributivity rule: $(x \oplus y) \cdot z = (x \cdot z) \oplus (y \cdot z)$. An illustration of initial and resulting paths after the application of this operator is shown in Fig. 3. In the resulting circuit the number of AND gates increases by one and potentially the number of XOR gates increases by one also.

Suppose that we need to move up an AND gate over a path containing k XOR gates. Let $(((x \oplus y_1) \oplus \ldots) \oplus y_k) \cdot z$ be the formula of this circuit. The direct application of the AND gate move up operator adds an AND gate for each XOR gate on the path. Observing that the initial formula can be rewritten as $(x \oplus (y_1 \oplus \ldots \oplus y_k)) \cdot z$ (XOR associativity) we can transform it into $(x \cdot z) \oplus (y_1 \oplus \ldots \oplus y_k) \cdot z$. This new formulation is functionally equivalent to the one obtained using direct application of AND gate move up operator except that the number of additional AND gates is only one.

Let $p = (v_1, v_2, \ldots, v_{|p|})$ be the critical path illustrated in Fig. 1, we recall that v_1 and $v_{|p|}$ are AND gates. The AND gate move up operator is used to move node $v_{|p|}$ next to node v_1. Afterwards, a critical path of length 2 is obtained, which is rewritten using the first operator. Condition (2) insures that the multiplicative depth of the rewritten node $v_{|p|}$ decreases. It is equivalent to the condition defined earlier for length 2 paths.

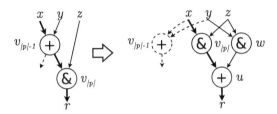

Fig. 3. AND gate move up operator. Dotted line XOR gate $v_{|p|-1}$ is kept only if needed.

$$\min_{u \in \text{pred}(v)} l(u) < l(v_1) - 1, \; v \in \left\{ v_1, v_{|p|} \right\} \tag{2}$$

We shall note that the overall boolean circuit multiplicative depth does not necessarily decrease after the above defined rewrite operators are applied, as the critical circuit can contain several parallel critical paths. All these critical paths have to be rewritten in order to decrease the overall circuit multiplicative depth by one.

2.3 Multi-start Heuristic

In this section we introduce a multi-start heuristic which uses rewrite operators defined above in order to minimize the multiplicative depth of a boolean circuit. Algorithm 1 is a priority based heuristic which rewrites critical paths of multiplicative depth 2. The path to rewrite is chosen using a priority function (introduced later). The algorithm stops either when a termination condition (e.g. time, number of iterations) is verified or when there are no more reducible critical paths, i.e. paths which respect condition (2). If the set of critical paths P is not empty, the algorithm chooses a path from it according to a priority function $prior\_func$ and rewrites this path using operators presented in previous section. If the multiplicative depth of the obtained circuit lowers then this new circuit is memorized as output circuit (variable C_{out}).

In order to decrease the overall multiplicative depth of a boolean circuit by one, all the parallel critical paths of this circuit must be rewritten. As we have observed empirically, the decrease of circuit multiplicative depth makes the new critical circuit wider and wider, that is to say the number of parallel critical paths increases. Respectively, the number of newly added gates (due to rewrite operators) increase in a non-linear way in the worst-case scenarios.

From the perspective of boolean circuit homomorphic execution, the minimization of multiplicative depth is beneficial (in terms of execution time) if the number of additional AND gates does not exceed a threshold. This threshold is defined by the ratio between the AND gate execution time at the previous multiplicative level and the AND gate execution time at the current multiplicative level. In order to obtain the best boolean circuit for homomorphic execution one can either stop when the number of newly added AND gates exceeds this threshold or store all the obtained circuits C_{out} (algorithm line 11) and choose afterwards the circuit for which the homomorphic execution time is minimal.

Algorithm 1. Multiplicative depth minimization heuristic.
___
Input: C – input boolean circuit
Input: $prior\_func$ – priority function
Output: C_{out} – multiplicative depth optimized boolean circtuit
1: $C_{out} \leftarrow C$
2: **while** termination conditions are not verified **do**
3:      $P \leftarrow$ critical paths of multiplicative depth 2 from circuit C
4:      $P \leftarrow$ filter paths $p \in P$ respecting condition (2)
5:      **if** $|P| = 0$ **then**
6:          **break**
7:      **end if**
8:      $p \leftarrow prior\_func(P)$ ▷ get highest priority path
9:      $p \leftarrow$ rewrite multiplicative depth-2 path p
10:     **if** $l_{\max}(C_{out}) > l_{\max}(C)$ **then**
11:         $C_{out} \leftarrow C$
12:     **end if**
13: **end while**
___

We introduce several functions which prioritize the path selection (ties are broken randomly):

- multiplicative depth of first path node: increasing order (d) and decreasing order (D),
- total number of critical predecessors of all path nodes: increasing order (i) and decreasing order (I),
- total number of critical successors of all path nodes: increasing order (o) and decreasing order (O),
- total number of critical predecessors and successors of all path nodes: increasing order (p) and decreasing order (P),
- critical path length: increasing order (l) and decreasing order (L).

Additionally to *non-random priority* functions[3] we have implemented a *random priority* function. Using different random seeds we obtain various search space explorations.

The *multi-start heuristic* consists in executing Algorithm 1 several times with different priority functions. In our experimentations we test two versions of the multi-start heuristic. In the first version an input circuit is optimized one time for each non-random priority function (10 executions) and in the second one the input circuit is optimized 10 times using random priority with different seeds. In both cases the best obtained solution (minimal multiplicative depth and minimal number of AND gates in case of equal multiplicative depths) is kept as multi-start heuristic result. In the next section we present the results of the experimentations we have performed for both multi-start algorithm versions.

---

[3] By abuse of language we denote so the above defined priority functions.

3   Experimental Results

Boolean circuits from the EPFL Combinational Benchmark Suite were used for experimentations. This set of benchmarks contains exclusively combinational circuits. Three types of circuits are provided: arithmetic, random/control and very large (multi-million gate designs). Please refer to [1] for more details about these benchmarks. In our experiments we have used only the first two types of benchmarks[4]: 10 arithmetic and 10 random/control circuits. Before using the benchmarks we have optimized and mapped them with ABC commands `resyn2` and `map`. The last command was used to obtain boolean circuits with AND and XOR gates only. Table 1 shows the characteristics of the obtained benchmarks after these commands were performed.

The heuristic described in the previous section was implemented in C language. The binary uses ABC as helper library. The two versions of the multi-start

Table 1. EPFL Combinational Benchmark Suite characteristics after initial optimization with ABC.

Circuit name	#input	#output	×depth	#AND
adder	256	129	255	509
bar	135	128	12	3141
div	128	128	4253	25219
hyp	256	128	24770	120203
log2	32	32	341	20299
max	512	130	204	2832
multiplier	128	128	254	14389
sin	24	25	161	3699
sqrt	128	64	4968	15571
square	64	128	247	9147
arbiter	256	129	87	11839
ctrl	7	26	8	108
cavlc	10	11	16	658
dec	8	256	3	304
i2c	147	142	15	1161
int2float	11	7	15	213
mem_ctrl	1204	1231	110	44795
priority	128	8	203	676
router	60	30	21	167
voter	1001	1	36	4229

---

[4] We assume that multi-million gate designs are out of reach for homomorphic execution, at least for the current state of HE schemes.

heuristic were executed on each benchmark circuit. Algorithm 1 execution terminates early if either the number of iterations is greater than 2 times the number of AND gates in the input circuit or the execution time exceeds 1 h. A middle-end server with AMD Opteron 6172 processors (2.1 GHz) was used as execution platform.

Table 2. Best obtained solutions for heuristic aggregated by priority function (non-random and random). Bold font is used to emphasize the best solution. The best solution considers the multiplicative depth as well as the number of AND gates.

Circuit	Initial		Non-random				Random		
	×depth	#AND	×depth	#AND	Ratio	Priority	×depth	#AND	Ratio
adder	255	509	12	911	21.2	P	**11**	1125	23.2
bar	12	3141	12	3141	1.0	-	12	3141	1.0
div	4253	25219	1852	29329	2.3	l	**1463**	31645	2.9
hyp	24770	120203	24563	120293	1.0	P	**24562**	120307	1.0
log2	341	20299	**141**	27362	2.4	p	150	22266	2.3
max	204	2832	27	4751	7.6	P	**27**	**4660**	7.6
multiplier	254	14389	60	21884	4.2	p	59	17942	4.3
sin	161	3699	**76**	5922	2.1	p	81	4473	2.0
sqrt	4968	15571	**4225**	18435	1.2	i	4391	16785	1.1
square	247	9147	**28**	10478	8.8	d, i	29	9731	8.5
arbiter	87	11839	42	8652	2.1	P	42	**8582**	2.1
ctrl	8	108	**5**	**109**	1.6	L	5	109	1.6
cavlc	16	658	**9**	669	1.8	D, I, o	10	658	1.6
dec	3	304	3	304	1.0	-	3	304	1.0
i2c	15	1161	**8**	**1185**	1.9	D, o	8	1185	1.9
int2float	15	213	**8**	216	1.9	D, o	9	214	1.7
mem_ctrl	110	44795	45	54889	2.4	p	**45**	**49175**	2.4
priority	203	676	102	1121	2.0	l	**102**	**1106**	2.0
router	21	167	11	261	1.9	o	11	**204**	1.9
voter	36	4229	**30**	**4288**	1.2	P	30	4340	1.2

Obtained results are shown in Table 2. The solutions for first version (column "non-random") and second version (column "random") of multi-start heuristic are illustrated in this table. The initial characteristics of circuits are also recalled (column "initial"). The notations we use are the multiplicative depth ("×depth"), the number of AND gates ("#AND"), the ratio between the multiplicative depth of the input circuit and the optimized one ("ratio") and the non-random priority for which the best solution was obtained ("priority").

The best solution (in terms of multiplicative depth and number of AND gates) was obtained using a non-random priority in 9 cases and using a random priority in 11 cases. For the ctrl and i2c benchmarks both heuristic versions obtained the same result. Multiplicative depth of the obtained circuits is significantly smaller when compared to the multiplicative depth of input circuits. In average the multiplicative depth decreases by more than 3 times. As expected, the price to pay for a smaller multiplicative depth is an increase in the number of AND gates (approximatively 1.2 times more in average).

The most substantial decrease is obtained for the adder benchmark, which is the usual 128-bit ripple carry adder. The proposed heuristic achieves an impressive result being able to transform a ripple carry adder with multiplicative depth 255 into "some sort of" carry-lookahead adder with a multiplicative depth 11 only.

The multiplicative depth of 2 benchmark circuits was not improved by the heuristic. In both cases the heuristic was not able to find any reducible multiplicative depth-2 paths. The dec benchmark (a 8 to 256 decoder) was already at its lowest possible multiplicative depth. As for the bar circuit (barrel shifter) we suppose that the proposed rewrite operators are too weak in terms of expressive power and more complex rewrite operators (e.g. circuit cone rewrite operators) are needed for dealing with such type of circuits.

There is not a single priority function which performs well (i.e. for which the best solution is found) on all benchmarks. The best solutions for 13 benchmarks are found using two priority functions: the total number of critical predecessors and successors of all path nodes (p,P), total number of critical successors of all path nodes in increasing order (o). We assume that each priority function performs well for a specific topology of boolean circuits.

The heuristic finished early because of time limit in the case of 5 benchmarks: div, hyp, sqrt, arbiter and mem_ctrl. The obtained multiplicative depths for these benchmarks are not the lowest possible ones. Allocating more execution time to heuristic will potentially increase the quality of presented results. We have rerun the tests for the 5 benchmarks with time limit increased to 2 h. The multiplicative

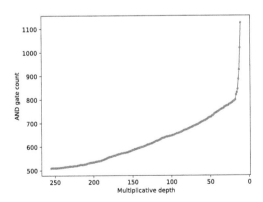

Fig. 4. Number of AND gates as a function of the multiplicative depth for the benchmark adder. Final multiplicative depth is 11.

depth further lowered for div (from 1463 to 675), hyp (from 24562 to 24417), sqrt (from 4225 to 3709), arbiter (from 42 to 11) and mem_ctrl (from 45 to 43) benchmarks. The exploration did not finish for 3 benchmarks: div, hyp and sqrt.

In order to see how the multiplicative depth influences the number of AND gates we have saved all the intermediary circuits obtained during heuristic execution. The heuristic was executed on the adder circuit. The random priority function (for which the smallest depth circuit was obtained in previous experiments) was used. The dependence between the number of AND gates and the multiplicative depth of intermediary circuits is illustrated in Fig. 4. We can see that the number of AND gates increases faster when the multiplicative depth is smaller. Moreover this increase is exponential for the last circuits (smallest multiplicative depth ones).

4   Conclusions and Perspectives

In this work we have proposed and studied a multi-start heuristic for minimizing the multiplicative depth of boolean circuits. The heuristic uses rewrite operators for boolean circuit critical paths. As a function of the used priority functions several versions of the multi-start heuristic have been studied. We have tested heuristic's performance on a set of circuits found in the literature. In average the multiplicative depth of benchmarked circuits was lowered by more than 3 times by the proposed heuristic. In perspective we envisage to study more elaborate heuristics together with new priority functions.

The optimization method described in this paper can also be applied to arithmetic circuit. An arithmetic circuit is a generalization of boolean circuits where instead of binary field operations higher degree field/ring operations are used. An arithmetic circuit is functionally complete when defined over addition and multiplication operations. It is easy to see that the optimization algorithm proposed in this paper together with rewrite operators can also be directly applied to arithmetic circuits and how to do so.

References

1. Amarú, L., Gaillardon, P.-E., De Micheli, G.: The EPFL combinational benchmark suite. In: Proceedings of the 24th International Workshop on Logic & Synthesis (IWLS) (2015)
2. Benhamouda, F., Lepoint, T., Mathieu, C., Zhou, H.: Optimization of bootstrapping in circuits. In: SODA, pp. 2423–2433. SIAM (2017)
3. Berkeley Logic Synthesis and Verification Group. ABC: A System for Sequential Synthesis and Verification, Release 30308. http://www.eecs.berkeley.edu/~alanmi/abc/
4. Boyar, J., Peralta, R.: Concrete multiplicative complexity of symmetric functions. In: Královič, R., Urzyczyn, P. (eds.) MFCS 2006. LNCS, vol. 4162, pp. 179–189. Springer, Heidelberg (2006). https://doi.org/10.1007/11821069_16

5. Brakerski, Z.: Fully homomorphic encryption without modulus switching from classical GapSVP. In: Safavi-Naini, R., Canetti, R. (eds.) CRYPTO 2012. LNCS, vol. 7417, pp. 868–886. Springer, Heidelberg (2012). https://doi.org/10.1007/978-3-642-32009-5_50

6. Brakerski, Z., Gentry, C., Vaikuntanathan, V.: (Leveled) fully homomorphic encryption without bootstrapping. In: Proceedings of the 3rd Innovations in Theoretical Computer Science Conference, ITCS 2012, pp. 309–325 (2012)

7. Buescher, N., Holzer, A., Weber, A., Katzenbeisser, S.: Compiling low depth circuits for practical secure computation. In: Askoxylakis, I., Ioannidis, S., Katsikas, S., Meadows, C. (eds.) ESORICS 2016. LNCS, vol. 9879, pp. 80–98. Springer, Cham (2016). https://doi.org/10.1007/978-3-319-45741-3_5

8. Carpov, S., Dubrulle, P., Sirdey, R.: Armadillo: a compilation chain for privacy preserving applications. In: Proceedings of the 3rd International Workshop on Security in Cloud Computing, SCC 2015, pp. 13–19 (2015)

9. Gentry, C.: Fully homomorphic encryption using ideal lattices. In: Proceedings of the Forty-First Annual ACM Symposium on Theory of Computing, STOC 2009, pp. 169–178 (2009)

10. Gentry, C., Halevi, S., Smart, N.P.: Better bootstrapping in fully homomorphic encryption. In: Fischlin, M., Buchmann, J., Manulis, M. (eds.) PKC 2012. LNCS, vol. 7293, pp. 1–16. Springer, Heidelberg (2012). https://doi.org/10.1007/978-3-642-30057-8_1

11. Kolesnikov, V., Sadeghi, A.-R., Schneider, T.: Improved garbled circuit building blocks and applications to auctions and computing minima. In: Garay, J.A., Miyaji, A., Otsuka, A. (eds.) CANS 2009. LNCS, vol. 5888, pp. 1–20. Springer, Heidelberg (2009). https://doi.org/10.1007/978-3-642-10433-6_1

12. Lepoint, T., Paillier, P.: On the minimal number of bootstrappings in homomorphic circuits. In: Adams, A.A., Brenner, M., Smith, M. (eds.) FC 2013. LNCS, vol. 7862, pp. 189–200. Springer, Heidelberg (2013). https://doi.org/10.1007/978-3-642-41320-9_13

13. Paindavoine, M., Vialla, B.: Minimizing the number of bootstrappings in fully homomorphic encryption. In: Dunkelman, O., Keliher, L. (eds.) SAC 2015. LNCS, vol. 9566, pp. 25–43. Springer, Cham (2016). https://doi.org/10.1007/978-3-319-31301-6_2

14. Schneider, T., Zohner, M.: GMW vs. Yao? Efficient secure two-party computation with low depth circuits. In: Sadeghi, A.-R. (ed.) FC 2013. LNCS, vol. 7859, pp. 275–292. Springer, Heidelberg (2013). https://doi.org/10.1007/978-3-642-39884-1_23

15. Wernick, W.: Complete sets of logical functions. Trans. Am. Math. Soc. **51**(1), 117–132 (1942)

The School Bus Routing Problem: An Analysis and Algorithm

Rhydian Lewis[1(✉)], Kate Smith-Miles[2], and Kyle Phillips[3]

[1] School of Mathematics, Cardiff University, Cardiff, Wales
LewisR9@cf.ac.uk
[2] School of Mathematical Sciences, Monash University, Melbourne, Australia
kate.smith-miles@monash.edu
[3] Visible Services and Transport, Vale of Glamorgan Council, Barry, Wales
kwphillips@valeofglamorgan.gov.uk

Abstract. In this paper we analyse a flexible real world-based model for designing school bus transit systems and note a number of parallels between this and other well-known combinatorial optimisation problems including the vehicle routing problem, the set covering problem, and one-dimensional bin packing. We then describe an iterated local search algorithm for this problem and demonstrate the sort of solutions that we can expect with different types of problem instance.

1 Introduction

Vehicle routing problems (VRPs) involve identifying routes for a fleet of vehicles that are to serve a set of customers. Often they are expressed using an edge-weighted directed graph $G = (V, E)$, where the vertex set $V = \{v_0, v_1, \ldots, v_n\}$ represents a single depot and n customers (v_0 and $v_1, \ldots, v_n$ respectively), and the weighting function $w(u, v)$ gives the travel distance (or travel time) between each pair of vertices $u, v \in V$.

Since the work of Dantzig and Ramser in the late 1950s [4], a multitude of VRP formulations have been considered in the literature [7]. These include using time-windows for visiting certain customers, placing limitations on the lengths of individual routes, the partitioning of customers into pick-up and delivery locations, and the dynamic recalibration of routes subject to the arrival of new customer requests during the transportation period [11].

Solutions to most VRP problems can be expressed by a set of routes $\mathcal{R} = \{R_1, \ldots, R_k\}$ using one vehicle per-route. In the *classical* VRP, each route should be a simple cycle in G such that:

$$R_i \cap R_j = \{v_0\} \quad \forall R_i, R_j \in \mathcal{R} \tag{1}$$

$$\bigcup_{i=1}^{k} R_i = V \tag{2}$$

These constraints specify that each customer should be assigned to exactly one route, and that all routes should start and end at the depot v_0. A variation on

L. Brankovic et al. (Eds.): IWOCA 2017, LNCS 10765, pp. 287–298, 2018.
https://doi.org/10.1007/978-3-319-78825-8_24

this is the *open* VRP in which, instead of cycles, all routes must be simple paths containing v_0 as one terminal vertex, meaning that routes either start or end at the depot, but not both [8].

In the *time constrained* VRP, extra realism is added by specifying that the total weight of edges in each route should be less than a given maximum—e.g., to ensure that driving time regulations are obeyed. In the *capacitated* VRP, meanwhile, maximum capacities are specified for each vehicle, and weights are also added to the vertices $v_1, \ldots, v_n$ in G. These vertex weights represent the size of the items being delivered to each customer, and we require the total size of items delivered by each vehicle to not exceed its maximum capacity.

The *split-delivery* VRP extends the capacitated VRP by relaxing Constraint (1) to simply: $v_0 \in R, \forall R \in \mathcal{R}$. This allows more than one vehicle to visit a customer and therefore permits a delivery to be made in many parts. Unlike the capacitated VRP, this relaxation also allows the minimum number of routes/vehicles in a solution to meet the lower bound of $\lceil (\sum_{i=1}^{n} w(v_i)) / C \rceil$, where $w(v)$ gives the weight of a vertex and C is the maximum capacity of the vehicles [15].

Objective functions for the VRP can depend on many real world factors. Most commonly we seek to minimise the number of vehicles used, the total length of the routes, or some combination of the two. In other cases we might also be concerned with the waiting times of customers, the obeying of time windows, avoiding traffic jams, or meeting individual drivers' needs. A useful survey presenting a taxonomy of the various types of VRP can be found in [5].

In this paper we look at the problem of arranging school bus transport. This problem is often cited as a type of VRP applicable in the real-world, though historically it has been less studied than other variants. One reason for this is that school transport solutions usually only involve visiting a subset of the available stopping points (bus stops); hence the issue of choosing *which* bus stops to visit adds an extra layer of complexity to the problem. Indeed, Park and Kim [10] note that bus stop selection is often omitted in the VRP literature altogether. One notable exception to this is due to Schittekat et al. [14], who use a problem based on the requirements of the Belgian school system; however, their formulation involves assumptions not considered here, most notably their limitation that bus stops can only be visited by a maximum of one vehicle in a solution.

The problem considered here is rather generic and was originally supplied by the third author of this paper, whose organisation is responsible for arranging school transport in the south of Wales (population 2.2m). Like many countries, school transport in Wales is organised by local government and then run by private bus companies. A few months before the start of the school year, a list of addresses is compiled containing all school children eligible for school transport (usually those who are in the school's catchment area but not within a reasonable walking distance). Each school is then considered individually, and a set of suitable bus routes are drawn up to serve all qualifying students. These routes are then put out for public tender, with bus companies bidding for the contracts. A yearly contract for a 70-seat bus typically ranges from GBP£25,000 to £35,000, though these costs can increase further for longer journeys and for

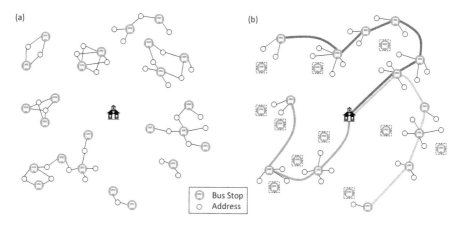

Fig. 1. (a) Example problem instance; (b) Example solution using $k = 3$ routes. Bus stops with dotted outlines are not used (i.e., are not members of V_1').

routes requiring a chaperone (i.e., routes with young children). It is therefore critical to try to reduce the number of buses used by each school. Note, however, that government guidelines also specify that journeys should not be too lengthy (less than 45 min for under-11 s, and one hour for under-18 s), though exceptions can be made for schools with very large catchment areas.

2   Problem Definition

The school bus routing problem (SBRP) considered here can be more formally stated using two sets of vertices. The first vertex set V_1 contains one school, v_0, and n bus stops $v_1, \ldots, v_n$. An edge set E_1 then contains directed edges between each $u, v \in V_1$ in each direction, making the graph (V_1, E_1) a complete digraph. A nonnegative weighting function $t(u, v)$ is also used to define the shortest driving time between each vertex pair.

The second vertex set V_2 defines the set of student *addresses*, with the weight $s(v) \in \mathbb{Z}^+$ of each $v \in V_2$ giving the number of students at this address requiring school transport. As part of the problem, a parameter m_w is defined stating the maximum distance that students are expected to walk from their home address to a bus stop. A second set of edges is thus used to signify bus stops within walking distance of each address: $E_2 = \{\{u, v\} : u \in V_2 \wedge v \in V_1 \wedge w(u, v) \le m_w\}$, where $w(u, v)$ gives the shortest walking distance between each $u \in V_2$ and $v \in V_1$. Also, students living within m_e distance units of their school are not considered eligible for school transport; consequently, $w(u, v_0) \ge m_e \ \forall u \in V_2$.

The graph $(V_1 - \{v_0\}, V_2, E_2)$ therefore constitutes an undirected bipartite graph with potentially many components, as illustrated in Fig. 1(a). Note that if there exists an address $u \in V_2$ with just one incident edge $\{u, v\} \in E_2$, then the bus stop $v \in V_1$ is *compulsory*, since it must be included in a solution in order to satisfy the needs of address u. We can also assume that $(V_1 - \{v_0\}, V_2, E_2)$

contains no isolated vertices: such vertices in $(V_1 - \{v_0\})$ would give a bus stop with no address within walking distance and can therefore be removed from the problem; isolated vertices in V_2 define an address with no suitable bus stop, making the problem unsolvable (in practice, an additional bus stop would need to be added to serve such an address).

A *feasible* solution to the SBRP is a set of routes $\mathcal{R} = \{R_1, \ldots, R_k\}$ in which each route $R \in \mathcal{R}$ is a simple path served by a single bus of capacity m_c. Each bus then travels to the school v_0 after visiting the terminal vertex on its path. The following constraints need to be satisfied.

$$\bigcup_{i=1}^{k} R_i = V_1' \tag{3}$$

$$\forall u \in V_2 \quad \exists v \in V_1' : \{u, v\} \in E_2 \tag{4}$$

$$s(R) \leq m_c \quad \forall R \in \mathcal{R} \tag{5}$$

$$t(R) \leq m_t \quad \forall R \in \mathcal{R} \tag{6}$$

Here, V_1' is a subset of $(V_1 - \{v_0\})$ that should satisfy Constraint (4): that is, for each address $u \in V_2$, the set V_1' should contain at least one bus stop within walking distance. Constraint (5) then specifies that the total number of students boarding the bus on a route R, denoted by $s(R)$, does not exceed the maximum bus capacity m_c. Similarly, Constraint (6) states that the total journey time $t(R)$ of each route should not exceed the stated time limit m_t. The aim is to then produce a feasible solution that minimises the number of routes k. An example solution to this problem is shown in Fig. 1(b). Note that these constraints allow bus stops to be included in more than one route, as is the case in the diagram. We call these bus stops *multistops*, their presence allowing different bus routes to split and merge as needed.

In our algorithm, our strategy is to relax Constraint (6) while ensuring that (3)–(5) are always satisfied. In doing so, a number of assumptions are made. First, students are always assigned to the bus stop in V_1' closest to their home. Second, all students are given bus passes that only allow them to travel on one particular route. This avoids situations where too many students might board a bus at a multistop, thereby making it too full to serve students at a later non-multistop. Third, solutions only concern buses travelling *to* school. After-school routes are assumed to follow the same paths in reverse, with any discrepancies in travel time due to one-way streets, etc. not being considered.

Our final assumption involves the use *dwell times* within a journey. These measure the time spent servicing each bus stop, including decelerating, opening doors, loading passengers, and rejoining the traffic stream. Dwell times are influenced by many factors including the number of boarding passengers, the size and position of the doors, the age of the passengers, and traffic density. Commonly, simple linear models $y = a + bx$ are used to estimate a dwell time y, where x gives the number of boarding passengers, b gives the boarding time per-passenger, and a captures all remaining delays. We follow this approach here:

Definition 1. *The journey time* $t(R)$ *of a route* $R = (u_1, u_2, \ldots, u_l) \in \mathcal{R}$ *is calculated,*

$$t(R) = \left(\sum_{i=1}^{l-1} t(u_i, u_{i+1})\right) + t(u_l, v_0) + \left(\sum_{i=1}^{l} a + b \cdot s(u_i, R)\right), \qquad (7)$$

where $s(u_i, R)$ *denotes the number of students boarding the bus on route* R *at bus stop* u_i.

In our case we use the values $a = 15$ and $b = 5$ (seconds), which are consistent with those recommended in [2,12,16].

3   Problem Analysis

In this section we now make some observations about the complexity of the SBRP and its underlying subproblems.

Theorem 1. *The task of finding a feasible solution with a minimum number of routes is NP-hard.*

Proof. Let (V_1, V_2, E_2) be a graph such that $\deg(u) = 1 \; \forall u \in V_2$. This means that, for all bus stops $v \in (V_1 - \{v_0\})$, (a) v is compulsory and must appear in at least one route, and (b) the number of boarding students is fixed at $\sum_{\forall u \in \Gamma(v)} s(u)$. This also implies the dwell times at each bus stop are fixed. This special case is equivalent to the NP-hard time-constrained capacitated split-delivery VRP, itself a generalisation of the NP-hard time-constrained VRP.

A similar proof of NP-hardness considers a generalisation of the above in which each component in $(V_1 - \{v_0\}, V_2, E_2)$ is a complete bipartite graph. In this case, all students can be assigned to a bus by including in V_1' exactly one bus stop from each component, making the problem a multi-vehicle version of the NP-hard generalised travelling salesman problem.

As stated, the primary aim in the SBRP is to minimise the number of routes (buses) being used in a solution. It is therefore desirable to fill buses where possible, bringing parallels with the NP-hard bin-packing problem [6]. Indeed, if multistops were not permitted in a solution, then the identification of a solution using k routes while obeying Constraints (3)–(5) would result in a one-dimensional bin-packing problem with bin capacity m_c and item sizes equal to the number of students boarding at each bus stop. As noted, multistops *are* permitted in this SBRP meaning that students boarding at a particular bus stop can be assigned to different routes if needed (or, equivalently, items in the corresponding packing problem can be split across different bins). This allows us to produce a solution $\mathcal{R} = \{R_1, \ldots, R_k\}$ satisfying constraints (3)–(5) that meets the lower bound of $k = \lceil (\sum_{i=1}^{n} s(v_i))/m_c \rceil$, though of course these routes could be rather long.

From a different perspective, the issue of choosing the subset V_1' of bus stops to include in a solution is closely related to the set covering problem. Recall that set covering involves taking a "universe" $U = \{1, 2, ..., n\}$ and a set S whose elements are subsets of the universe, and seeks to find the smallest subset $S' \subseteq S$ whose union equals the universe. For example, given $U = \{1, 2, 3, 4\}$ and $S = \{\{1\}, \{1, 2\}, \{1, 3\}, \{3, 4\}, \{4\}\}$ the optimal solution is $S' = \{\{1, 2\}, \{3, 4\}\}$, containing just two elements.

Definition 2. $S' \subseteq S$ *is a* complete *covering if and only if* $\bigcup_{s \in S'} = U$. *A* minimal *covering is a complete covering in which the removal of any element in* S' *results in an incomplete covering.*

According to Definition 2, an optimal solution to a set covering problem is a minimum cardinality solution among all minimal coverings. Note that while the task of identifying an optimal solution is NP-hard [6], the identification of minimal coverings is easily carried out in polynomial time. For example, starting with the complete covering $S' = S$, at each step we might simply remove any element $s \in S'$ for which $(S' - \{s\})$ is still a complete covering, repeating until S' is minimal.

With regards to the SBRP, using the bipartite graph $(V_1 - \{v_0\}, V_2, E_2)$, let S be the set whose elements correspond to the addresses within walking distance of each bus stop, $S = \{\Gamma(v) : v \in (V_1 - \{v_0\})\}$. According to Constraint (4), all addresses in a feasible solution must be served by a bus stop; hence, the task of identifying a subset $V_1' \subseteq V_1$ meeting this criterion is equivalent to the problem of finding a complete covering of the universe V_2 using the set S.

Theorem 2. *Consider the SBRP in which multistops are not permitted (i.e.,* $R_i \cap R_j = \emptyset \; \forall R_i, R_j \in \mathcal{R}$*), and let* (V_1, E_1) *be a graph whose pairwise distances satisfy the triangle inequality. Now let* $\mathcal{R} = \{R_1, \ldots, R_k\}$ *be a solution satisfying Constraints (3)–(5) that has the minimum total journey time* $\sum_{i=1}^{k} t(R_i)$. *Then the subset of bus stops* V_1' *used in* $\mathcal{R}$ *corresponds to a minimal covering* S'.

Proof. The removal of any bus stop $v \in V_1'$ corresponds to the removal of the element $\Gamma(v)$ in S' which, by definition, results in an incomplete covering and violation of Constraint (4). Conversely, the addition of an extra element $\Gamma(v)$ to S' will result in a complete but non-minimal covering; however, this corresponds to the addition of an extra bus stop v in at least one route in $\mathcal{R}$ which, due to the triangle inequality, will maintain or increase the total journey time of $\mathcal{R}$.

Note that this theorem does not hold when multistops are permitted in a solution. This is because the addition of an extra bus stop may allow a route to be shortened by redirecting it from a multistop and then through this new stop. The triangle inequality is also necessary, though it is acceptable here, being satisfied by both real-world road maps (where minimum distances/times between each pair of locations are used) and Euclidean graphs. Indeed, because of the extra delays incurred by dwell times in the SBRP, this inequality can be strengthened to $\forall u_1, u_2, u_3 \in V_1, \; t(u_1, u_2) + t(u_2, u_3) > t(u_1, u_3)$.

4 Algorithm Description

As noted, our strategy for this problem is to use a fixed number of routes k and allow the violation of Constraint (6) while ensuring that the remaining constraints (3)–(5) are always satisfied. Specialised operators are then used to try to shorten the routes in a solution such that Constraint (6) also becomes satisfied, giving a feasible solution. If this cannot be achieved at a certain computation limit, k is increased by one, and the algorithm is repeated. Initially, k is set to the lower bound $\lceil (\sum_{i=1}^{n} s(v_i)) / m_c \rceil$. An alternative approach would be to allow our search operators to alter k and then use its value as part of the objective function. However, evidence from the literature for similar partition-based problems suggests the former to usually be a more suitable approach [9,13].

Our approach is based on iterated local search using a solution space $\mathcal{V}'_1$ that contains all bus stop subsets $V'_1 \subseteq V_1$ corresponding to minimal coverings. To begin, a member of $\mathcal{V}'_1$ is generated and improved via a local search routine (Sect. 4.2). Upon termination of this routine, a new member of $\mathcal{V}'_1$ is then generated by "kicking" the incumbent solution (Sect. 4.3), and re-running the local search. This is repeated until a stopping criterion is met (see Sect. 5).

4.1 Initial Solution and Cost Function

An initial solution $\mathcal{R} = \{R_1, \ldots, R_k\}$ is constructed by first generating a subset of bus stops V'_1 corresponding to a minimal covering. In our case, this is achieved using the well-known greedy heuristic of Chvatal [3], followed by the removal of randomly selected bus stops (if necessary) until the covering is seen to be minimal.

Having generated V'_1, the number of students boarding at each bus stop $s(v)$, $v \in V'_1$ is calculated. A variant of the first-fit descending heuristic for bin packing is then used to assign bus stops to routes. Specifically, at each step, the bus stop v with the largest number of boarding students is chosen and assigned to any route (vehicle) seen to have sufficient capacity. If no such route exists, then the route with the largest spare capacity x is chosen and v is assigned to this route along with x students. This has the effect of creating a multistop, since a copy of bus stop v, along with its remaining students will also need to be assigned to a different route in a subsequent iteration.

The above process produces a solution $\mathcal{R}$ obeying Constraints (3)–(5). It is then evaluated according to an objective function $f(\mathcal{R}) = \sum_{i=1}^{k} t'(R)$, where

$$t'(R) = \begin{cases} t(R) & \text{if } t(R) \leq m_t \\ m_t + W(1 + t(R) - m_t) & \text{otherwise.} \end{cases} \tag{8}$$

Here, W introduces a penalty cost for routes whose journey times exceed the maximum m_t. In our case we set W to m_t and include the addition of one in the formula to ensure that a route with $t(R) > m_t$ is always penalised more heavily than two routes with individual journey times of less than m_t.

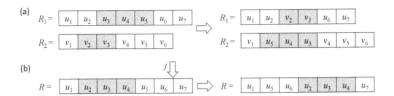

Fig. 2. Example application of (a) the Section Swap operator (here, the section u_3, u_4, u_5 has been inverted before insertion into R_2); and (b) the Extended Or-opt operator.

4.2   Local Search

As with many other VRP variants, our local search routine uses a combination of both inter- and intra-route neighbourhood operators. The following inter-route operators act on two routes $R_1, R_2 \in \mathcal{R}$. Without loss of generality, assume that $R_1 = (u_1, u_2, \ldots, u_{l_1})$ and $R_2 = (v_1, v_2, \ldots, v_{l_2})$.

Section Swap: Take two vertices in each route, u_{i_1}, u_{i_2} $(1 \leq i_1 \leq i_2 \leq l_1)$ and v_{j_1}, v_{j_2} $(1 \leq j_1 \leq j_2 \leq l_2)$, and use these to define the sections $u_{i_1}, \ldots, u_{i_2}$ and $v_{j_1}, \ldots, v_{j_2}$ within R_1 and R_2 respectively. Now swap the two sections, inverting either if this leads to a superior cost (see Fig. 2(a)).

Section Insert: Take a section in R_1, defined by u_{i_1} and u_{i_2} as above, together with an insertion point j $(1 \leq j \leq l_2 + 1)$ in R_2. Now remove the section from R_1 and insert it before vertex v_j in R_2, inverting the section if this leads to a better cost. If $j = l_2 + 1$, add the section to the end of R_2.

Note that these inter-route operators may result in too many students being assigned to R_1 or R_2, leading to a violation of Constraint (5). In our case, such moves are disallowed. Since multistops are permitted, it is also possible that they will result in a route containing a vertex $v \in V_1'$ more than once. Since each route must be a simple path, these need to be deleted. Assuming without loss of generality that a new route is to be produced by inserting the section $u_{i_1}, \ldots, u_{i_2}$ (possibly inverted) into a route $R = (v_1, v_2, \ldots, v_{l_2})$, and that some vertex in the section is already present in R, this is done by simply removing the relevant vertex from the section and reassigning its students to the other occurrence of the vertex in R.

Our three intra-route neighbourhood operators act on a single route $R = (u_1, u_2, \ldots, u_l) \in \mathcal{R}$. Their application does not affect the satisfaction of Constraints (3)–(5), nor do they introduce duplicate vertices into a route.

Swap: Take two vertices u_{i_1}, u_{i_2} $(1 \leq i_1 \leq i_2 \leq l)$ in R and swap their positions.
2-opt: Take two vertices u_{i_1}, u_{i_2} $(1 \leq i_1 \leq i_2 \leq l)$ and invert the section $u_{i_1}, \ldots, u_{i_2}$ within R.
Extended Or-opt: Take a section defined by u_{i_1} and u_{i_2} as above, together with an insertion point j outside of this section (i.e., $1 \leq j < i_1$ or $i_2 + 1 < j \leq l + 1$). Now remove the section and insert it before vertex u_j. If $j = l + 1$, then add

the section to the end of the route. Also, invert the section if this leads to a better cost (see Fig. 2(b)).

Note that, together, these five operators generalise a number of neighbourhood operators commonly featured in the literature. For example, our two inter-route operators include and extend the six outlined by Silva et al. [15] for the capacitated VRP. Similarly, they extend the basic VRP-based neighbourhood operators used with the bus routing problem considered in [14]. Our Extended Or-opt operator also generalises the more basic Or-opt, which only involves sections of up to three vertices [1].

Here, our local search procedure follows the steepest descent methodology: in each cycle all moves in all neighbourhoods are evaluated, and the move offering the largest reduction in cost is performed, breaking ties randomly. The process halts when no improving moves are identified. Note that the number of moves considered in each cycle is of $O(m^4)$, where $m = \sum_{i=1}^{k} |R_i|$ is the size of the incumbent solution. Though seemingly quite expensive, with appropriate bookkeeping the changes in cost caused by individual applications of these neighbourhood operators can always be calculated in constant time. Consequently, this growth rate was not found to be particularly restrictive.

4.3   Generating New Minimal Coverings via a Kick Operator

While our local search routine is able to alter and improve the cost of a solution, it does not alter the subset of bus stops being used V_1'. One way of doing this, as suggested in [14], would be to either swap a bus stop in a route with a currently unused bus stop, or simply remove a bus stop from a route altogether. However, besides not allowing the number of bus stops in V_1' to increase, this is unsuitable here because it fails to ensure the satisfaction of Constraint (4).

Given a minimal subset of bus stops V_1', our operator first removes a randomly chosen non-compulsory bus stop $v \in V_1'$, followed by x further non-compulsory bus stops, leaving a partial covering.[1] A different minimal covering is then constructed by selecting bus stops from the set $(V_1 - \{v\})$ using a randomised version of Chvatal's heuristic that, at each stage, arbitrarily selects any bus stop that will serve some currently unserved students, until all students are served. If necessary, randomly selected bus stops are then also removed until the covering is minimal.

Having produced a new minimal subset of bus stops $V_1'' \neq V_1'$, the current solution $\mathcal{R}$ needs to be repaired to reflect these changes. To do this, the closest bus stops in V_1'' to each address are first recalculated and bus stops from the set $(V_1' - V_1'')$ are deleted from routes in $\mathcal{R}$. Instances of multistops are also removed at this point so that each bus stop occurs at most once in $\mathcal{R}$. Randomly selected bus stops are then also removed from routes in $\mathcal{R}$ if their number of students exceeds the maximum capacity m_c. Finally bus stops in V_1'' not yet in $\mathcal{R}$ are

---

[1] In our case a value for x is selected randomly according to a binomial distribution $X \sim B(|V_1'|, 3/|V_1'|)$.

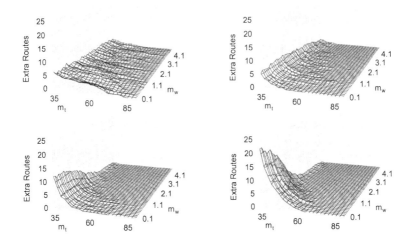

Fig. 3. Number of extra buses required by solutions for various values of m_t (in minutes) and m_w (in miles) for 25, 50, 100 and 250 bus stops respectively. Each point is the mean across five problem instances.

assigned to routes using the bin packing heuristic from Sect. 4.1. This results in a modified solution $\mathcal{R}$ obeying Constraints (3)–(5) as desired.

5   Experimentation

Our experiments consider the issues that affect the number of extra buses required in a solution compared to the lower bound of $\lceil (\sum_{i=1}^{n} s(v_i)) / m_c \rceil$. To do this, artificial problem instances were generated by placing a school at the centre of a circle with radius $r > m_e$. Bus stops were then randomly placed within this circle, followed by a set of addresses, ensuring that each address was at least m_e distance units from the school, but within m_w distance units of at least one bus stop. Distances between vertices are assumed to be Euclidean.

Figure 3 shows the effect of altering (a) the maximum walk distance m_w in our problem generator and (b) the maximum journey time m_t permitted by our algorithm, using 25, 50, 100 and 250 bus stops. In all instances we used a radius r of 15 miles and buses were assumed to travel at 30 mph—hence all bus stops are within 30 min of the school. The number of addresses was set to 400, with the number of students boarding each bus stop $s(v)$ selected randomly from the set $\{1, 2, 3, 4\}$, giving approximately 1,000 students per-instance. Finally, the maximum bus capacity m_c and minimum eligibility distance m_e were set to 70 and 3 miles respectively, with ten seconds of execution time permitted for each value of k. (The algorithm was written in C++ and executed on a 3.3 GHtz Windows 7 machine with 8 GB RAM.)

Figure 3 demonstrates that more routes (buses) are needed when both the maximum journey times m_t and the maximum walking distances m_w are low.

For low values for m_t this is quite natural: shorter journey limits imply the need for more routes in feasible solutions. On the other hand, for low values of m_w the instance generator clusters addresses tightly around bus stops; consequently, nearly all bus stops are compulsory, making the problem very similar to that described in Theorem 1. This means that any savings that could be achieved by only using a subset of the bus stops are not available, creating a need for additional routes. We also see that these effects increase for larger numbers of bus stops where, for low values of m_w in particular, more bus stops will need to be visited.

Considering multistops, we found that these occur more frequently when it is advantageous or necessary to assign large numbers of students to individual bus stops. This occurs for high values of m_w, where students are able to walk larger distances to bus stops (implying fewer bus stops in V_1'), or when the number of bus stops is small. From a bin packing perspective, more students per-stop implies larger items to pack into the bins, meaning that more of these items will need to be "split", resulting in a multistop.

As we might expect, the number of local optima visited by the algorithm within the ten second time limit (and therefore the number of kicks applied) is heavily influenced by the computational requirements of the local search routine, which is itself influenced by the size of a solution $\sum_{i=1}^{k} |R_i|$. To illustrate, for values of 25, 50, 100 and 250, these figures were seen to be approximately 250,000, 47,000, 4,000 and 60 respectively, suggesting that longer run times may be required for problem instances involving larger solutions.

6   Conclusions and Further Work

This paper has analysed a real-world school bus routing formulation that builds on previous models proposed in the literature by including bus stop selection, multistops, and dwell times. In doing so, relationships have been drawn with three well-known combinatorial optimisation problems.

Our experiments have demonstrated that our algorithm is often able to find solutions using the lower bound of $\lceil (\sum_{i=1}^{n} s(v_i)) / m_c \rceil$ routes. This is particularly so for instances where only a small proportion of bus stops need to be used, such as when the maximum walking distance of students is set quite high. Note, however, that in cases where all bus stops need to be used, our proposed kick operator has no effect, so it may be better to focus on extending the local search operator by, for example, including a tabu element.

As noted, the solution space in our current algorithm is restricted to bus stop subsets that correspond to minimal set coverings. However, according to Theorem 2, our use of multistops means that optimal solutions to a particular problem instance may not occur within this space. Future research will determine whether this restriction is beneficial, or whether it is preferable to use the larger space of all set coverings.

This paper has limited the empirical analysis to artificially generated problems; however, our research is ongoing and we are currently using this same method with large real-world problems generated using web mapping services. One feature of our current solutions to these problems is that, by ensuring the set of used bus stops corresponds to a minimal covering, large numbers of students are often assigned to a relatively small number of bus stops, rather than using more convenient bus stops that are closer to their home. We expect further improvements to the service might therefore be achieved by sometimes allowing additional bus stops to be used, though perhaps without increasing the number of routes unduly.

References

1. Babin, G., Deneault, S., Laporte, G.: Improvements to the Or-opt heuristic for the symmetric travelling salesman problem. J. Oper. Res. Soc. 58(3), 402–407 (2007)
2. Bertini, R., El-Geneidy, A.: Modeling transit trip time using archived bus dispatch system data. J. Transp. Eng. 130(1), 56–67 (2004)
3. Chvatal, V.: A greedy heuristic for the set-covering problem. Math. Oper. Res. 4(3), 233–235 (1979)
4. Dantzig, G., Ramser, J.: The truck dispatching problem. Manag. Sci. 60(1), 80–91 (1959)
5. Eksioglu, B., Volkan, V., Reisman, A.: The vehicle routing problem: a taxonomic review. Comput. Ind. Eng. 57(4), 1472–1483 (2009)
6. Garey, M., Johnson, D.: Computers and Intractability - A Guide to NP-Completeness, 1st edn. W. H. Freeman and Company, San Francisco (1979)
7. Laporte, G.: Fifty years of vehicle routing. Transp. Sci. 43, 408–416 (2009)
8. Letchford, A., Lysgaard, J., Eglese, R.: A branch-and-cut algorithm for the capacitated open vehicle routing problem. J. Oper. Res. Soc. 58, 1642–1651 (2007)
9. Lewis, R.: A Guide to Graph Colouring: Algorithms and Applications. Springer, Cham (2015). https://doi.org/10.1007/978-3-319-25730-3
10. Park, J., Kim, B.: The school bus routing problem: a review. Eur. J. Oper. Res. 202, 311–319 (2010)
11. Pillac, V., Gendreau, M., Guéret, C., Medagila, A.: A review of dynamic vehicle routing problems. Eur. J. Oper. Res. 225, 1–11 (2013)
12. Transit Cooperative Research Program: Transit Capacity and Quality of Service Manual, 3rd edn. (2013). ISBN 978-0-309-28344-1
13. Qin, H., Ming, W., Zhang, Z., Xie, Y., Lim, A.: A tabu search algorithm for the multi-period inspector scheduling problem. Comput. Oper. Res. 59, 78–93 (2015)
14. Schittekat, P., Kinable, J., Sörensen, K., Sevaux, M., Spieksma, F., Springael, J.: A metaheuristic for the school bus routing problem with bus stop selection. Eur. J. Oper. Res. 229, 518–528 (2013)
15. Silva, M., Subramanian, A., Ochi, L.S.: An interated local search heuristic for the split delivery vehicle routing problem. Comput. Oper. Res. 53, 234–239 (2015)
16. Wang, C., Zhirui, Y., Yuan, W., Yueru, X., Wei, W.: Modeling bus dwell time and time lost serving stop in China. J. Public Transp. 19(3), 55–77 (2016)

Heuristic, Branch-and-Bound Solver and Improved Space Reduction for the Median of Permutations Problem

Robin Milosz and Sylvie Hamel[✉]

DIRO, Université de Montréal, C.P. 6128 Succursale Centre-Ville,
Montréal, QC H3C 3J7, Canada
{robin.milosz,sylvie.hamel}@umontreal.ca

Abstract. Given a set $\mathcal{A} \subseteq \mathbb{S}_n$ of m permutations of $\{1, 2, \ldots, n\}$ and a distance function d, the **median** problem consists of finding a permutation π^* that is the "closest" of the m given permutations. Here, we study the problem under the Kendall-τ distance which counts the number of order disagreements between pairs of elements of permutations. In this article, we explore this NP-hard problem using three different approaches: a well parameterized heuristic, an improved space search reduction technique and a refined branch-and-bound solver.

1 Introduction

The problem of finding medians of a set of m permutations of $\{1, 2, \ldots, n\}$ under the Kendall-τ distance [13], often cited as the Kemeny Score Problem [12] consists of finding a permutation that agrees the most with the order of the m given permutations, *i.e.*, that minimizes the sum of order disagreements between pairs of elements of permutations. This problem has been proved to be NP-hard when $m \geq 4$, m even (first proved in [8], then corrected in [4]), but its complexity remains unknown for $m \geq 3$, m odd. A lot of work as been done in the last 15 years, either on deriving approximation algorithms [1,14,21] or fixed-parameter ones [3,11,20]. Other theoretical approaches aiming at reducing the search space for this problem have also been developed [2,5–7,19].

In this present work, we are interested in solving methods for the median of permutations problem, focusing on three different approaches. After introducing some basic definitions and notations in Sect. 2, we present our first approach in Sect. 3, an adaptation of the well known "Simulated Annealing" heuristic to our context. Second, in Sect. 4, we built ordering constraints for pairs of elements appearing in a median by merging previous approches [6,19] complemented by our simulated annealing heuristic, thus reducing significantly the search space for

This work is supported by a grant from the National Sciences and Engineering Research Council of Canada (NSERC) through an Individual Discovery Grant RGPIN-2016-04576 (Hamel) and by Fonds Nature et Technologies (FRQNT) through a Doctoral scholarship (Milosz).

L. Brankovic et al. (Eds.): IWOCA 2017, LNCS 10765, pp. 299–311, 2018.
https://doi.org/10.1007/978-3-319-78825-8_25

this median. Third, we present, in Sect. 5, an implementation of an exact solver: a branch-and-bound algorithm that is powered by the two previous approaches. Finally, Sect. 6 gives some thoughts on future works.

2   Median of Permutation: Definitions and Notations

A **permutation** π is a bijection of $[n] = \{1, 2 \ldots, n\}$ onto itself. The set of all permutations of $[n]$ is denoted $\mathbb{S}_n$. As usual we denote a permutation π of $[n]$ as $\pi = \pi_1 \pi_2 \ldots \pi_n$, and a segment of π by $\pi[i..j] = \pi_i \pi_{i+1} \ldots \pi_{j-1} \pi_j$. The cardinality of a set $\mathcal{S}$ will be denoted $\#\mathcal{S}$.

The **Kendall-τ distance**, denoted d_{KT}, counts the number of order disagreements between pairs of elements of two permutations and can be defined formally as follows: for permutations π and σ of $[n]$, we have that

$$d_{KT}(\pi, \sigma) = \#\{(i, j) | i < j \text{ and } [(\pi[i] < \pi[j] \text{ and } \sigma[i] > \sigma[j])$$
$$\text{or } (\pi[i] > \pi[j] \text{ and } \sigma[i] < \sigma[j])]\},$$

where $\pi[i]$ denotes the position of integer i in permutation π.

Given any set of permutations $\mathcal{A} \subseteq \mathbb{S}_n$ and a permutation $\pi \in \mathbb{S}_n$, we have $d_{KT}(\pi, \mathcal{A}) = \sum_{\sigma \in \mathcal{A}} d_{KT}(\pi, \sigma)$. The **problem of finding a median of $\mathcal{A}$ under the Kendall-τ distance** can be stated formally as follows: Given $\mathcal{A} \subseteq \mathbb{S}_n$, we want to find a permutation π^* of $\mathbb{S}_n$ such that $d_{KT}(\pi^*, \mathcal{A}) \leq d_{KT}(\pi, \mathcal{A}), \forall \pi \in \mathbb{S}_n$. Note that a set $\mathcal{A}$ can have more than one median. To keep track of the number of permutations in $\mathcal{A}$ that have a certain order between two elements, let us introduce the left/right distance matrices L and R.

Definition 1. *Let $L(\mathcal{A})$, be the* **left distance matrix** *of a set of m permutations $\mathcal{A} \subseteq \mathbb{S}_n$, where $L_{ij}(\mathcal{A})$ denotes the number of permutations of $\mathcal{A}$ having element i to the left of element j. Symmetrically, let $R(\mathcal{A})$, be the* **right distance matrix** *of $\mathcal{A}$, where $R_{ij}(\mathcal{A})$ denotes the number of permutations of $\mathcal{A}$ having element i to the right of element j. Obviously, $L_{ij}(\mathcal{A}) + R_{ij}(\mathcal{A}) = m$ and $L_{ij}(\mathcal{A}) = R_{ji}(\mathcal{A})$.*

We can calculate the distance between a permutation $\pi \in \mathbb{S}_n$ and a set of permutations $\mathcal{A} \subseteq \mathbb{S}_n$ using the right (or left) distance matrix as follow:

$$d_{KT}(\pi, \mathcal{A}) = \sum_{i=1}^{n} \sum_{\substack{j=1 \\ j \neq i \\ \pi[j] > \pi[i]}}^{n} R_{ij}(\mathcal{A}) = \sum_{i=1}^{n} \sum_{\substack{j=1 \\ j \neq i \\ \pi[j] > \pi[i]}}^{n} L_{ji}(\mathcal{A}).$$

3   A Heuristic Approach

Our first approach is based on the well known Simulated Annealing (SA) heuristic. This approach will give us an approximative solution for the median problem, an upper bound on the distance we are trying to minimize ($d_{KT}(\pi, \mathcal{A})$) and a direction for our branch-and-bound search.

3.1   Simulated Annealing

Simulated annealing (SA) is a probabilistic metaheuristic for locating a good approximation to the global optimum of a given function in a large search space [15]. It is an adaptation of the Metropolis-Hasting algorithm [16] and works best on large discrete space. For that reason, it is a good choice in our case where the search space is $\mathbb{S}_n$, the space of all permutations of $[n]$.

In a simulated annealing heuristic, each point s of the search space corresponds to a state of a physical system, and minimizing a function $f(s)$ corresponds to minimizing the internal energy of the system in state s. The goal is to bring the system, from a randomly chosen initial state, to a state with the minimum possible energy. At each step, the SA heuristic considers a neighbour s' of the current point s and (1) always moves to it if $f(s') < f(s)$ or (2) moves to it with a certain probability if $f(s') > f(s)$. This probability to do a "bad" move is almost 1 at the beginning of the heuristic when the system is heated but goes to 0 as the system is cooled. The heuristic stops either when the system reaches a good enough solution or when a certain number of moves has been made.

In our context, we are given a set of permutations $\mathcal{A} \subseteq \mathbb{S}_n$ for which we want to find a median. The function f we need to minimize, given the set $\mathcal{A}$, is $f(\pi) = d_{KT}(\pi, \mathcal{A})$, for $\pi \in \mathbb{S}_n$. We first choose randomly a starting point in our space i.e. a permutation $\pi \in \mathbb{S}_n$. This permutation is uniformly generated using the Fisher-Yates shuffle [9]. To find neighbours of a permutation π in our system, we consider circular moves defined as follows:

Definition 2. *Given $\pi \in \mathbb{S}_n$, we call **circular move** of a segment $\pi[i..j]$ of π, denoted $c[i, j](\pi)$, the cycling shifting of one position to the right, if $i < j$, of this segment inside the permutation π: $c[i, j](\pi) = \pi_1 \ldots \pi_{i-1} \pi_{\mathbf{j}} \pi_i \ldots \pi_{j-1} \pi_{j+1} \ldots \pi_n$ (if $i > j$, we shift to the left: $c[j, i](\pi) = \pi_1 \ldots \pi_{j-1} \pi_{j+1} \ldots \pi_i \pi_{\mathbf{j}} \pi_{i+1} \ldots \pi_n$).*

So, to choose a neighbour for our current permutation π, our SA heuristic first randomly generates two integers $i \neq j$, $1 \leq i, j \leq n$, and compute neighbour$(\pi) = c[i, j](\pi)$, if $i < j$ or neighbour$(\pi) = c[j, i](\pi)$, otherwise. To decide whether or not we move into state neighbour(π), we compute the difference of energy, noted ΔE, which in our case is the difference $d_{KT}(neighbour(\pi)), \mathcal{A}) - d_{KT}(\pi, \mathcal{A})$. If this difference is negative or null, the state neighbour(π) is closer to the median of $\mathcal{A}$ and we move to this state. If it is positive, we move to state neighbour(π) depending on the acceptance probability $e^{-\Delta E/T}$, where T is the temperature of the system. This process of going to neighbours states is repeated a fixed number of times during which the permutation with the lowest energy (distance to $\mathcal{A}$) are kept in a set. This set is returned at the end of the process. Algorithm 1, in Appendix A.I, gives the pseudo-code of our heuristic SA[1].

---

[1] Note that all appendices and the source code (Java) for testing can be found online at http://www-etud.iro.umontreal.ca/~miloszro/iwoca/iwoca.html.

3.2 Choosing the Parameters

The important parameters of a SA heuristic are: the initial temperature of the system, the cooling factor, the maximal number of movements, the function that propose a neighbouring alternative and the number of repetitions of SA. First, we choose the circular move (see Definition 2) as our neighbouring choosing function. This choice was made for two reasons. First, while looking at sets of medians for different sets of permutations $\mathcal{A} \subseteq \mathbb{S}_n$, we observe that most of the medians in a set could be obtained from one another by applying those circular moves. Second, we did try a lot of other moves (exchanging element $\pi[i]$ with one of its neighbour $\pi[i-1]$ or $\pi[i+1]$ or exchanging it with any other element $\pi[j]$, inverting the order of the elements of a block, etc.) that clearly did not converge as quickly as the circular moves.

For the rest of the SA parameters, note that for each instance of our problem, i.e. a set of permutations $\mathcal{A} \subseteq \mathbb{S}_n$, we have two important parameters: the number m of permutations in $\mathcal{A}$ and the size n of the permutations. This pair (m, n) is very relevant to describe the problem's difficulty. So, we were interested in tuning the SA parameters as function of n and m.

We made extensive testing to find out the optimal parameters for SA given the pair (m, n). More information on our choice of parameters are available in Appendix A.II. But here is an overview:

1. Initial solution: a random permutation generated by Fisher-Yates shuffle.
2. The cooling schedule is: $t_i = \alpha t_{i-1}$.
3. The initial temperature of the system set to: $t_0 \longleftarrow (0.25m + 4.0)n$
4. The cooling factor set to:

$$
\alpha = \begin{cases}
0.99 & \text{if } m = 3, 4 \\
0.95 & \text{if } n \leq 10 \\
0.99 & \text{if } 11 \leq n \leq 16 \\
0.999 & \text{if } 17 \leq n \leq 20 \\
0.9995 & \text{if } 21 \leq n \leq 24 \\
0.9998 & \text{otherwise}
\end{cases}
$$

5. The neighbour generating function: the circular move.
6. The number of allowed movements for a solution:

$$
nbMvts = \begin{cases}
0.6n^3 - 11n^2 + 127n & \text{if } m = 3 \\
0.9n^3 - 29n^2 + 435n - 1623 & \text{if } m = 4 \\
250 & \text{if } n \leq 7 \\
90n^2 - 1540n + 7000 & \text{if } 8 \leq n \leq 24 \\
35n^2 - 660n + 31000 & \text{if } 25 \leq n \leq 38 \\
80n^2 - 2300n + 27000 & \text{if } n > 38
\end{cases}
$$

7. The number of times to repeat the SA heuristic:

$$nbRuns = \begin{cases} \lceil 0.05n + 2 \rceil & \text{if } m = 3, 4 \\ \lceil 0.007nm + 3 \rceil & \text{if } m \text{ is odd} \\ \lceil 0.002mn + 3 \rceil & \text{if } m \text{ is even} \end{cases}$$

The initial temperature was set to accept (almost) all neighbours in the first iterations. The cooling constant and the number of movements where chosen to have a good compromise between the success probability (probability of achieving the optimal score) and the computing time for each (m, n). The number of SA runs was set such that the probability of overall success is above 96% for all instances $n \leq 38$ and $m \leq 50$. For greater m and n, we extrapolated the data to predict a similar SA behaviour.

As $m = 3, 4$ are unique cases, that do not seem to be affected by the cooling constant in our considered range and seem much easier to optimize, we treated them appart. For the rest, we found out that m odd problems are harder to solve than m even problems, and required more runs to reach the same success rate. Without surprise, as n and m are getting bigger, the problem is harder to solve.

4   Space Reduction Technique

In [18,19] we found theoretical properties of a set of permutations $\mathcal{A} \subseteq \mathbb{S}_n$ (called the **Major Order Theorems**) that can solve the relative order between some pairs of elements in median permutations of $\mathcal{A}$ thus reducing the search space. In this section we will show how to find additional constraints for the problem by merging this previous work on constraints with a lower bound idea of Conitzer *et al.* [6], giving us an even stronger lower bound, that can be then combined with the upper bound obtained by the simulated annealing method presented in Sect. 3. But first, let us quickly recalled our major order theorems in Sect. 4.1 and Conitzer *et al.* lower bound idea in Sect. 4.2.

4.1   Some Constraints

Given a set of permutations $\mathcal{A} \subseteq \mathbb{S}_n$, the major order theorems, presented in [18], solve the relative order between some pairs of elements in median permutations of $\mathcal{A}$. Thus, these theorems build a set of constraints that can be represented as a boolean matrix C, where $C_{ij} = 1$ if and only if we know that element i will precede element j (denoted $i \prec j$) in median permutations of $\mathcal{A}$. Note that this set of constraints reduces the search space for a median by cutting off all permutations breaking at least one constraint.

Here, we resume the ideas behind these theorems[2] but first, let us formally define the major order between pairs of elements.

---

[2] More detailed explanations and some examples for these Major Order Theorems can be found in Sect. 4 of [18].

Definition 3. *Let $\mathcal{A} \subseteq \mathbb{S}_n$ be a set of permutations. Let $L(\mathcal{A})$ and $R(\mathcal{A})$ be the left and right matrices of $\mathcal{A}$ as defined in Definition 1. Given two elements i and j, $1 \le i < j \le n$, we say that the **major order** between elements i and j is $i \prec j$ (resp. $j \prec i$) if $L_{ij}(\mathcal{A}) > R_{ij}(\mathcal{A})$ (resp. $R_{ij}(\mathcal{A}) > L_{ij}(\mathcal{A})$), the **minor order** is then $j \prec i$ (resp. $i \prec j$). We use δ_{ij} to denote the absolute value of the difference between the major and minor order of two elements i and j.*

The idea of the first major order theorem (**MOT1**) relies on the proximity of two elements in permutations of $\mathcal{A}$: if $i \prec j$ is the major order of elements i and j in permutations of $\mathcal{A}$, we can say that i will also be placed before j in all medians of $\mathcal{A}$ if the numbers of elements (including possible copies) between i and j in those permutations of $\mathcal{A}$ where $j \prec i$ is **less than** δ_{ij}. Intuitively, this multiset of elements between i and j, that we will called **interference multiset**, act as interference to the major order $i \prec j$ and so, if its cardinality is small enough, MOT1 gives a relative order for the pair (i, j) in medians of $\mathcal{A}$.

The second major order theorem (**MOT2**) built on the first one by reducing the cardinality of the interference multiset of a pair of elements i and j by removing from it any element that also appears in between i and j in permutations of $\mathcal{A}$ that follows the major order of this pair.

The idea of the third Major Order Theorem (**MOT3**) is to use previously found constraints (MOT1 and MOT2) to reduce even more the cardinality of the interference multiset of a pair of elements i and j, by removing from it all elements that cannot be in it. As an example, say that an element k is in the interference multiset of pair (i, j). This means that k appears in between i and j in at least one permutation of $\mathcal{A}$ where i and j are in their minor order. If we have already found using MOT1 or MOT2 that $C_{ki} = C_{kj} = 1$ or that $C_{ik} = C_{jk} = 1$ then k cannot be in between i and j and we can remove it from the interference multiset. This process is repeated until no new constraint is found.

Example 1. *The MOT3 is better illustrated with the following exemple: for $\mathcal{A} = \{78\underline{2}36\underline{1}54, 35\underline{1}786\underline{2}4, 5834\underline{1}2\underline{7}6\}$, the major order for 1 and 2 is $1 \prec 2$, $\delta_{12} = 1$ and we have $\{3, 6\}$ as the interference multiset. The 6 gets cancelled by the 6 between 1 and 2 in the second permutation (MOT2's way) as the 3 is eliminated by contraints $3 \prec 1$ and $3 \prec 2$ which were found by MOT1. Therefore the interference multiset is empty and $1 \prec 2$ is a valid constraint.*

In [19], we extended those major order theorems by considering the equality case (denoted MOTe) *i.e* the extended MOT theorems gives us the relative order of a pair of elements if the cardinality of the interference multiset for this pair is **less than or equal to** δ_{ij}. This case is more delicate as the method builds constraints only for a subset of the set of median permutations of $\mathcal{A}$, so care is to be taken to avoid possible contradicting constraints (it is possible that in one median of $\mathcal{A}$, $i \prec j$ and in another $j \prec i$; so using the MOTe theorems we will strictly find those medians of $\mathcal{A}$ satisfying one and only one of those "contradicting" constraint). However, the MOTe theorems have a much better efficiency than the MOT theorems, as you can see in Table 1.

4.2   A New Lower Bound

Given a set of permutations $\mathcal{A} \subseteq \mathbb{S}_n$, Davenport and Kalagnanam [7] propose a first intuitive lower bound on the median problem of $\mathcal{A}$: the sum, for all pairs of elements $i, j, 1 \leq i < j \leq n$, of the number of permutations in $\mathcal{A}$ having the pair (i, j) in its minor order. This bound corresponds to the possibility of ordering all pairs in a median of $\mathcal{A}$ with respect to their major order. If this ordering is possible without conflicting pairs, the bound is equal to the Kendall-τ distance of a median to $\mathcal{A}$. The bound can be easily compute by the following formula:

$$LowerBound_0 = \sum_{\substack{i<j \\ i,j \in [n]}} min\{R_{ij}, R_{ji}\}$$

Consider the directed weighted graph $G = (V, E)$, where each vertices $v \in V$ is a element of $[n]$ and where E is composed of two edges for each pair of vertices i, j: e_{ij} and e_{ji} with respected weights $w(e_{ij}) = R_{ji}(\mathcal{A})$ and $w(e_{ji}) = R_{ij}(\mathcal{A})$. The median problem can be reformulated as a minimum feedback arc set problem [10] which consist of finding the set of edges $E^* \subset E$ of minimal total weight $w(E^*) = \sum_{e \in E^*} w(e)$ to be removed from G to obtain a direct acyclic graph. Let G' be the graph obtained from G by "cancelling", for each pair of vertices, opposing ordering in permutations of $\mathcal{A}$: $w(e'_{ij}) = w(e_{ij}) - min\{R_{ij}, R_{ji}\}, \forall 1 \leq i < j \leq n$. Obviously $w(E^*) = w(E'^*) + LowerBound_0$.

Let $DC_{G'}$ be a set of disjoint directed cycles of G'. The previous lower bound can be augmented by adding for each cycle $c \in DC_{G'}$ the minimal weight of one of its edges:

$$LowerBound_1 = LowerBound_0 + \sum_{c \in DC_{G'}} min_{e \in c}\{w(e)\}.$$

In [6], Conitzer, Davenport and Kalagnanam push further the lower bound, proving that the cycles do not need to be disjoint. Let $JC_{G'}$ be any sequence $c_1, c_2, ..., c_l$ of G', that can share commun edges. Let $I(c, e)$ be an indicator function of the inclusion of an edge e to a cycle c, i.e. $I(c, e) = 1$ if $e \in c$ and 0 otherwise. If $v_i = min_{e \in c_i}\{w(e) - \sum_{j=1}^{i-1} I(c_j, e)v_j\}$, then we obtain the new following lower bound:

$$LowerBound_2 = LowerBound_0 + \sum_{i=1}^{l} v_i.$$

In practice, we can apply this lower bound method by iteratively searching for a cycle c in G', containing only non-zero weight edges, finding its minimal weight edge e, then subtracting $w(e)$ to the weight of all edges of c and adding it to the lower bound. This process is then repeating until no such cycle is left in G'.

The process of finding the strongest lower bound, i.e. the best sequence and choice of cycles, becomes a problem itself and can be resolved using linear programming. As we are interested here by an efficient pre-processing of the problem, we will use a restrained version of this previous lower bound that can be calculated quickly. Thus, only cycles of length 3 (3-cycles) will be considered.

Our contribution resides in taking advantage of a set of constraints (the one described in Sect. 4.1) while calculating $LowerBound_2$ as it provides additional information on the structure of the optimal solution. If $C_{ij} = 1$ then we know that the order of elements i and j in a median permutation of $\mathcal{A}$ will be $i \prec j$. In that case, we can add $w(e_{ji})$ to the lower bound (since all permutations with $j \prec i$ in $\mathcal{A}$ disagree with a median permutation) then set its value to $w(e_{ji}) = 0$ in G'.

At first glance, incorporating the constraints seems to be interesting but one will quickly observe that the constraints previously found by the MOT method are only of the type $C_{ij} = 1$ where $i \prec j$ is the major order, adding nothing to the lower bound because in the graph G', the associated minor order edge e'_{ji} has weight $w(e'_{ji}) = 0$, by construction.

Nevertheless, the superiority of calculating a lower bound with constraints will appear in Sect. 4.4, when combined with an upper bound.

For $\mathcal{A} \subseteq \mathbb{S}_n$, we will denote the lower bound with set of constraints C by $Lb_{\mathcal{A}}(C)$. If $C = \emptyset$, then $Lb_{\mathcal{A}}(\emptyset)$ becomes $LowerBound_2$ associated with 3-cycles, described above.

4.3   An Upper Bound

In Sect. 3, we detailed a simulated annealing heuristic that finds a approximative solution for our problem. The approximative solution is a valid permutation therefore its distance to $\mathcal{A}$ will be used as an upper bound. We will denote this upper bound by $Ub_{\mathcal{A}}$.

Naturally, if $Lb_{\mathcal{A}}(C) = Ub_{\mathcal{A}}$ for a particular instance of the problem and any set of valid constraints C, then the problem is solved: the median distance is $Ub_{\mathcal{A}}$ and the associated permutation to that upper bound will be a solution *i.e.* a median of $\mathcal{A}$.

4.4   Putting Everything Together

Given a set of permutations $\mathcal{A} \subseteq \mathbb{S}_n$, we can deduce a valid set of constraints C using the MOT methods described in Sect. 4.1 and then apply the technique described in Sect. 4.2 to obtain the lower bound $Lb_{\mathcal{A}}(C)$. Running the SA heuristic of Sect. 3 will get us the upper bound $Ub_{\mathcal{A}}$.

We can use these upper and lower bounds to search for new constraints simply by adding a new possible constraint and verifying if the lower bound has exceeded the upper bound with this new add on. To do so, let us choose a pair of elements (i, j) for which the ordering is still unknown in the median, *i.e.* for which $C_{ij} = 0$ and $C_{ji} = 0$. Now, let us suppose that $i \prec j$ in a median of $\mathcal{A}$ and let C' be the set of constraints C augmented with this new constraint $C_{ij} = 1$. If $Lb_{\mathcal{A}}(C') > Ub_{\mathcal{A}}$ then the added constraint $i \prec j$ is false, which means that $j \prec i$ in a median of $\mathcal{A}$ and we can add $C_{ji} = 1$ to our set of constraints C.

As finding a new constraint give an advantage to find others (as there is more knowledge about the structure of the optimal solution), we can redo the same process including unknown constraints that previously failed the test, repeating until no new constraint is found. We will called this new way of finding constraints the LUBC (lower-upper bounds and constraints) method.

Without surprise, as MOT3 method also benefits from new constraints, we propose a method that will iteratively alternate between MOT and LUBC until both are unable to find any additional constraint and call it MOT+LUBC. In Appendix B.I, Algorithms 2 and 3 give the pseudo-code of our MOT+LUBC method.

As seen before, the constraints found by the MOT method are only of the type $C_{ij} = 1$ where $i \prec j$ is the major order. An advantage of the LUBC is the possibility to find constraints of $\mathcal{A}$ that are of the minor type i.e. where $C_{ij} = 1$ and $i \prec j$ is the minor order ($R_{ij} > R_{ji}$). Recalling the construction of G', the weight of an edge associated with the major order is strictly positive and leads to a non-zero augmentation of the lower bound. When the LUBC methods finds any new valid constraint of the minor order type, the augmentation of the lower bound is advantageous to find further new contraints.

The great efficiency of this new method can be observed in Table 1, where we tested our different approaches on distributed random sets of m permutations of $[n]$, with $m \in [3; 50]$ and $n = 15, 20, 30$ or 45. On our instances, the gain of constraints is ranging from $+1\%$ to 41% passing from MOT to MOT+LUBC and from $+0.1\%$ to 36% passing from MOTe to MOTe+LUBC. We can note that all instances of $n \leq 20$ have a average resolution rate higher than 90% with the MOTe+LUBC method.

Tables 10 to 14 in Appendix B.II, gives more results for all of these methods. Appendix B.III discuss the time complexity of this new approach.

Table 1. Efficiency of the MOT, MOT+LUBC, MOTe, MOTe+LUBC approaches in terms of the proportion of ordering of pairs of elements solved, on sets of uniformly distributed random sets of m permutations, $m = 3$ and $m = 5x$, $1 \leq x \leq 10$, statistics generated over 80000 sets for $n = 15$, 50000 sets for $n = 20$, 10000 sets for $n = 30$ and 1000 sets for $n = 45$.

	$n = 15$				$n = 20$				$n = 30$				$n = 45$			
m	MOT	MOT+LUBC	MOTe	MOTe+LUBC	MOT	MOT+LUBC	MOTe	MOTe+LUBC	MOT	MOT+LUBC	MOTe	MOTe+LUBC	MOT	MOT+LUBC	MOTe	MOTe+LUBC
3	0.635	0.921	0.878	0.966	0.579	0.885	0.808	0.952	0.506	0.725	0.702	0.873	0.447	0.513	0.604	0.669
5	0.595	0.913	0.800	0.954	0.530	0.859	0.715	0.929	0.444	0.605	0.595	0.766	0.361	0.381	0.490	0.516
10	0.524	0.845	0.824	0.951	0.465	0.809	0.709	0.915	0.384	0.616	0.549	0.747	0.310	0.347	0.419	0.454
15	0.579	0.939	0.704	0.953	0.500	0.890	0.612	0.919	0.400	0.569	0.490	0.668	0.316	0.328	0.383	0.398
20	0.540	0.884	0.747	0.945	0.472	0.843	0.636	0.908	0.383	0.597	0.493	0.697	0.306	0.327	0.377	0.399
25	0.581	0.949	0.680	0.923	0.500	0.903	0.587	0.923	0.398	0.579	0.465	0.653	0.312	0.322	0.363	0.375
30	0.550	0.904	0.720	0.947	0.479	0.861	0.610	0.910	0.386	0.589	0.472	0.674	0.301	0.317	0.362	0.379
35	0.584	0.955	0.668	0.961	0.501	0.909	0.575	0.926	0.396	0.582	0.453	0.642	0.309	0.320	0.345	0.356
40	0.556	0.915	0.702	0.950	0.483	0.873	0.596	0.912	0.388	0.590	0.461	0.660	0.307	0.323	0.354	0.371
45	0.587	0.959	0.660	0.964	0.502	0.913	0.568	0.927	0.397	0.586	0.448	0.638	0.306	0.317	0.350	0.363
50	0.562	0.923	0.691	0.952	0.486	0.881	0.586	0.914	0.388	0.589	0.456	0.654	0.301	0.314	0.352	0.366

5   An Exact Approach

Our third approach is an exact branch-and-bound solver for the problem that combines the result of the simulated annealing method of Sect. 3 with the constraints obtained in Sect. 4.4 to avoid exploring not promising search sub-trees.

5.1   Branch-and-Bound

Our branch-and-bound algorithm simply constructs possible medians of $\mathcal{A} \subseteq \mathbb{S}_n$, *i.e.* permutations of $[n]$ by putting a new element to the right of the already known ones till no more element are available. Thus we explore a tree having the empty permutation at its root, permutations of any k elements of $[n]$ as nodes of level k and where the leaves are permutations of $\mathbb{S}_n$. Each node N will have a corresponding lower bound $Lb(N)$ representing the fact that for all permutations π derived from N, $d_{KT}(\pi, \mathcal{A}) \geq Lb(N)$.

If this lower bound $Lb(N)$ is higher than the current upper bound of the problem, then all the permutations derived from it will have a higher distance than one already found and node N with all its descendants can be omitted in the exploration. As the number of nodes/leaves is finite and the bounding method only cuts branches that do not represent possible medians, the BnB will always converge to the optimal solution.

More specifically, given our set $\mathcal{A} \subseteq \mathbb{S}_n$, we run the SA heuristic of Sect. 3 to have a first approximative median of $\mathcal{A}$, π_{approx}, and a first upper bound $Ub_\mathcal{A}$, which will always represents the score associated with the currently best solution found. The approximative solution will serve as guidance so that the first leaf that will be visited by the BnB will be π_{approx}. This guarantees us an efficient cutting of non-promising branches.

A node at level k is represented by a vector of size k, $x = [x_1, ..., x_k]$ which corresponds to a permutation in construction. Let S be the set of elements still to be placed *i.e* $S = [n] - \{x_1, x_2, ..., x_{k-1}, x_k\}$ and L a list which orders S. At the beginning of our BnB, $S = [n]$ and $L = \pi_{approx}$. The BnB will branch by choosing the next element from the list: $\ell_i \in L$ that will be placed at the immediate right of x_k. We are going to apply the bound and cuts $Bound_1$, Cut_1, Cut_2 and Cut_3 described in Sect. 5.2 below for each choice of ℓ_i. If it succeeds passing all the bounds test, $x_{k+1} := \ell_i$, and we go down to the new node $x' = [x_1, x_2, ..., x_{k-1}, x_k, x_{k+1}]$. If not, we try ℓ_{i+1} as a possible x_{k+1}. If we go down to a leaf, *i.e.* if $k + 1 = n$, its corresponding permutation π_{leaf} is compared to the best solution. If $d_{KT}(\pi_{leaf}, \mathcal{A}) = Ub_\mathcal{A}$, we add π_{leaf} to the current set of medians. If $d_{KT}(\pi_{leaf}, \mathcal{A}) < Ub_\mathcal{A}$, then it is our new upper bound and we change our set of medians so that it contain only this new optimal permutation π_{leaf}. The BnB backtracks after all possibilities of $\ell_i \in L$ for x_{k+1} had been explored or after investigating a leaf node.

5.2   Bounds and Cuts

Now, let us set everything that is needed to described our different bound and cuts. First, given a set of permutations $\mathcal{A} \subseteq \mathbb{S}_n$, we deduce a valid set of constraints $C(\mathcal{A})$ using the method MOTe+LUBC described in Sect. 4.4. We also pre-calculated, for each triplet of elements x, y and $z \in [n]$ the best ways to put them all together, one after the other, in a median of $\mathcal{A}$. All the other ways to order them consecutively in a permutation will be called a forbidden triplet, since a permutation containing this ordering cannot be a median of $\mathcal{A}$.

For each node $x = [x_1, ..., x_k]$, we can compute a distance $d(x, \mathcal{A})$ in the following way:

$$d(x, \mathcal{A}) = \underbrace{\sum_{i=1}^{|x|} \sum_{j=i+1}^{|x|} R_{x_i x_j}(\mathcal{A})}_{\substack{\text{contribution to the Kendall-}\tau \\ \text{distance for the elements} \\ \text{already placed}}} + \underbrace{\sum_{i=1}^{|x|} \sum_{j=1}^{|L|} R_{x_i l_j}(\mathcal{A})}_{\substack{\text{contribution obtained by the fact} \\ \text{that all elements of } x \text{ are to the} \\ \text{left of elements in } L \text{ (yet to be placed)}}}.$$

Finally, for each node x, let $b(x)$ be the boolean vector of length n representing which elements of $[n]$ are already placed in this node (i.e we have $b_i(x) = 1$, $1 \leq i \leq n$, if and only if $i = x_j$, $1 \leq j \leq k$). So if a node x' contained the same k elements of node x but in a different order, $b(x) = b(x')$. Our BnB will construct and update a set $TopScores$ of pairs $< b, v >$, where b is any boolean vector of length n and $v = \min d(x, \mathcal{A})$, for x already explored such that $b(x) = b$.

Now, if our current best solution is π_{best}, our current upper bound is $Ub_{\mathcal{A}} = d_{KT}(\pi_{best}, \mathcal{A})$, our current node is $x = [x_1, ..., x_k]$, L is an ordered list of the elements still to be placed, and we are studying $\ell_i \in L$ as a possible x_{k+1}, we have that:

- **Cut$_1$**: If (x_{k-1}, x_k, ℓ_i) is a forbidden triplet then x_{k+1} cannot be l_i and so it is rejected. (i.e we do not explore the subtree having node $x = [x_1, ..., x_k, \ell_i]$ as its root.)
- **Cut$_2$**: If there exist $\ell_j \in L$, $\ell_j \neq \ell_i$ such that $C(\mathcal{A})(\ell_j, \ell_i) = 1$ then we know that ℓ_i has to be to the right of ℓ_j in a median permutation of $\mathcal{A}$ and so x_{k+1} cannot be ℓ_i and is rejected.
- **Cut$_3$**: Let $x' = [x_1, \ldots, x_k, \ell_i]$. If $< b(x'), v > \in TopScores$ and $d(x', \mathcal{A}) > v$ then x_{k+1} cannot be ℓ_i and so it is rejected.
- **Bound$_1$**: Let a lower bound of a list L be $lb(L) = \sum_{i=1}^{|L|} \sum_{j=i+1}^{|L|} (min\{R_{l_i l_j}, R_{l_j l_i}\})$ $+tri(L)$, where $tri(L)$ is a simpler implementation of the lower bound using only 3-cycles described in Sect. 4.2. In this implementation, the 3-cycles are pre-calculated at the beginning, and the contribution of a cycle is added if and only if all three of its elements are in L. Let $x' = [x_1, \ldots, x_k, \ell_i]$ and let L' be the list L without ℓ_i. Let $Lb(x') = d(x', \mathcal{A}) + lb(L')$. If $Lb(x') > Ub_{\mathcal{A}}$ then x_{k+1} cannot be ℓ_i and so it is rejected.

Algorithm 4, in Appendix C, gives the pseudo-code of our BnB method. As a final note, our BnB can solve in reasonable time (a few seconds in average) any problem with $n \leq 38, m \leq 50$, since we kept all calculation in linear time at the node level for efficiency purpose. The case where $m = 4$ is the hardest case (and the most variable in execution time) for the BnB, opposite to SA, as the average number of medians of $\mathcal{A} \subseteq \mathbb{S}_n$ have been observed to be the biggest for all m.

In [17], Ali and Meilă made a thorough comparison of many solvers and heuristics, solving uniformly generated problems of size up to $n = 50, m = 100$. We did some quick testing of our BnB on similar problems (same n, m and uniformly generated sets) and we claim that it solves them in a comparable time, thus competing with the best solvers (BnB and Integer Linear Programming). More intensive testing will be done in the near future.

6   Conclusion

In this article, we studied the problem of finding the median of a set of m permutations $\mathcal{A} \subseteq \mathbb{S}_n$ under the Kendall-τ distance. This problem is known to be NP-hard for $m \geq 4$, m even. This work presents three different solving techniques for this problem; a well parameterized simulated annealing heuristic, a space reduction technique and an promising exact BnB solver.

Ideas for future works includes an extensive comparison with other exact solvers and heuristics, as well as testing on various synthetic and real life data sets. It would also be interesting to take into account the fact that the rankings considered are not always on the entire sets of elements involved. Furthermore, some ranking schemes often rank several elements in the same position, so rank ties are to be considered.

Acknowledgements. We would like to thanks our anonymous reviewers for their careful and inspiring comments. Be sure that the suggestions that were not included here, due to time and space constraints, will be integrate in the journal version of this article.

References

1. Ailon, N., Charikar, M., Newman, N.: Aggregating inconsistent information: ranking and clustering. J. ACM **55**(5), 1–27 (2008)
2. Betzler, N., Bredereck, R., Niedermeier, R.: Theoretical and empirical evaluation of data reduction for exact Kemeny Rank Aggregation. Auton. Agent. Multi-Agent Syst. **28**, 721–748 (2014)
3. Betzler, N., et al.: Average parameterization and partial kernelization for computing medians. J. Comput. Syst. Sci. **77**(4), 774–789 (2011)
4. Biedl, T., Brandenburg, F.J., Deng, X.: Crossings and permutations. In: Healy, P., Nikolov, N.S. (eds.) GD 2005. LNCS, vol. 3843, pp. 1–12. Springer, Heidelberg (2006). https://doi.org/10.1007/11618058_1
5. Blin, G., Crochemore, M., Hamel, S., Vialette, S.: Median of an odd number of permutations. Pure Math. Appl. **21**(2), 161–175 (2011)
6. Conitzer, V., Davenport, A., Kalagnanam, J.: Improved bounds for computing Kemeny rankings. In: Proceedings of the 21st Conference on Artificial Intelligence, AAAI 2006, vol. 1, pp. 620–626 (2006)

7. Davenport, A., Kalagnanam, J.: A computational study of the Kemeny rule for preference aggregation. In: Proceedings of the 19th National Conference on Artificial Intelligence, AAAI 2004, pp. 697–702 (2004)
8. Dwork, C., Kumar, R., Naor, M., Sivakumar, D.: Rank aggregation methods for the web. In: Proceedings of the 10th WWW, pp. 613–622 (2001)
9. Fisher, R.A., Yates, F.: Statistical Tables for Biological, Agricultural and Medical Research, 3rd edn, pp. 26–27. Oliver & Boyd, London (1948)
10. Karp, R.M.: Reducibility among combinatorial problems. In: Miller, R.E., Thatcher, J.W., Bohlinger, J.D. (eds.) Complexity of Computer Computations, pp. 85–103. Springer, Boston (1972). https://doi.org/10.1007/978-1-4684-2001-2_9
11. Karpinski, M., Schudy, W.: Faster algorithms for feedback arc set tournament, kemeny rank aggregation and betweenness tournament. In: Cheong, O., Chwa, K.-Y., Park, K. (eds.) ISAAC 2010. LNCS, vol. 6506, pp. 3–14. Springer, Heidelberg (2010). https://doi.org/10.1007/978-3-642-17517-6_3
12. Kemeny, J.: Mathematics without numbers. Daedalus **88**, 577–591 (1959)
13. Kendall, M.: A new measure of rank correlation. Biometrika **30**, 81–89 (1938)
14. Kenyon-Mathieu, C., Schudy, W.: How to rank with few errors. In: STOC 2007, pp. 95–103 (2007)
15. Kirkpatrick, S., Gelatt, C.D., Vecchi, M.P.: Optimization by simulated annealing. Science **220**(4598), 671–680 (1983)
16. Metropolis, N., Rosenbluth, A.W., Rosenbluth, M.N., Marshall, N., Teller, A.H., Teller, E.: Equation of state calculations by fast computing machines. J. Chem. Phys. **21**–**6**, 1087–1092 (1953)
17. Ali, A., Meilă, M.: Experiments with Kemeny ranking: what works when? Math. Soc. Sci. **64**, 28–40 (2012)
18. Milosz, R., Hamel, S.: Medians of permutations: building constraints. In: Govindarajan, S., Maheshwari, A. (eds.) CALDAM 2016. LNCS, vol. 9602, pp. 264–276. Springer, Cham (2016). https://doi.org/10.1007/978-3-319-29221-2_23
19. Milosz, R., Hamel, S.: Space reduction constraints for the median of permutations problem. J. Discret. Appl. Math. (submitted)
20. Nishimura, N., Simjour, N.: Parameterized enumeration of (locally-) optimal aggregations. In: Dehne, F., Solis-Oba, R., Sack, J.-R. (eds.) WADS 2013. LNCS, vol. 8037, pp. 512–523. Springer, Heidelberg (2013). https://doi.org/10.1007/978-3-642-40104-6_44
21. van Zuylen, A., Williamson, D.P.: Deterministic pivoting algorithms for constrained ranking and clustering problems. Math. Oper. Res. **34**(3), 594–620 (2009)

Efficient Lagrangian Heuristics for the Two-Stage Flow Shop with Job Dependent Buffer Requirements

Hanyu Gu, Julia Memar$^{(\boxtimes)}$, and Yakov Zinder

University of Technology Sydney, PO Box 123, Broadway, NSW 2007, Australia
{hanyu.gu,julia.memar,yakov.zinder}@uts.edu.au

Abstract. The paper is concerned with minimisation of total weighted completion time for the two-stage flow shop with a buffer. In contrast to the vast literature on this topic, the buffer requirement varies from job to job and a job occupies the buffer continuously from the start of its first operation till the completion of its second operation rather than only between operations. Such problems arise in supply chains requiring unloading and loading of minerals and in some multimedia systems. The problem is NP-hard and the straightforward integer programming approach is impossible even for modest problem sizes. The paper presents a Lagrangian relaxation based decomposition approach that allows to use for each problem, obtained by this decomposition, a very fast algorithm. Several Lagrangian heuristics are evaluated by means of computational experiments.

Keywords: Flow shop · Buffer · Total weighted completion time
Lagrangian relaxation

1 Introduction

Scheduling problems for the flow shops with buffers have been studied for several decades. Almost all existing publications are concerned with the buffers that restrict the number of jobs between operations (see, for example, [1,2]). Much less is known about the more general models where the buffer requirement varies from job to job, although such more general models better describe many practical situations [11].

In most existing publications a buffer restricts the number of jobs between operations, whereas many practical situations require models where a job occupies the buffer from the start of its processing and releases the buffer only at the completion of its last operation. Such situations arise, for example, in supply chains where change of the mode of transportation involves unloading and loading operations. Another example is sequencing media objects for auto-assembled presentations from digital libraries [5–8].

To the best of our knowledge, besides [5–8] only [4,9,10] studied the two-stage flow shop with a buffer where each job occupies the buffer for the entire

© Springer International Publishing AG, part of Springer Nature 2018
L. Brankovic et al. (Eds.): IWOCA 2017, LNCS 10765, pp. 312–324, 2018.
https://doi.org/10.1007/978-3-319-78825-8_26

duration of the job's processing and the buffer requirement varies from job to job. All these publications are concerned with the minimisation of the time required for completion of all jobs or/and due dates based objective functions. Our paper studies another commonly used objective function - the total weighted completion time.

As far as the optimisation methods are concerned, the previous publications suggested the branch-and-bound method [5,9], constructive heuristics [8], and various integer programming formulations and local search procedures [10], whereas our paper presents a Lagrangian relaxation based optimisation procedure - the approach that never has been used before for such flow shop problems.

The results of computational experiments below support the claim that the effectiveness of our optimisation procedure makes it suitable for various practical applications. It is important to stress that, although our method is based on an integer linear programming formulation, the optimisation procedure is designed in such a way that it does not need any mathematical programming software, can be easily implemented, and requires minimum computer memory, which is crucial advantage, for example, in the case of some multimedia applications.

1.1   Problem Description

The considered problem can be stated as follows. The set of n jobs $N = \{1, ..., n\}$ is to be processed by two machines - the first-stage machine and the second-stage machine. Each job i is to be processed on the first-stage machine during p_i^1 (the first operation of the job) and on the second-stage machine during p_i^2 (the second operation of the job). All processing times are integer. The second operation of a job can commence on the second-stage machine only after the completion of its first operation on the first-stage machine. Once an operation has started, it cannot be interrupted, i.e. no preemptions are allowed. Each machine can process at most one job at a time, and each job can be processed by at most one machine at a time. The processing of jobs commences at time $t = 0$.

In order to be processed, each job i requires $b(i)$ units of the buffer space. This buffer space is occupied by a job continuously from the start of its first operation till the completion of its second operation. At any point in time t, the total buffer requirement of all jobs that started their processing before or at t and have completion time of their second operation greater that t can not exceed B - the buffer capacity. As in [5–7,9,10], it is assumed that $b(i) = p_i^1$ for all $i \in N$. In other words, it is assumed that the buffer requirement of each job is determined by the duration of the first operation. This assumption is valid, for example, for the mentioned above multimedia application.

For each job i, let S_i^1 and S_i^2 be the starting times of the job's first and second operation, respectively. The goal is to construct a schedule with the smallest total weighted completion time $\sum_{i \in N} w_i C_i$, where w_i is a positive weight, characterising i, and $C_i = S_i^2 + p_i^2$ is the completion time of job i.

1.2 Outline of the Optimisation Procedure

Lagrangian relaxation (see, for example, [3]) is one of the most effective methods for solving difficult combinatorial optimisation problems. Section 2 presents an integer linear programming formulation of the considered scheduling problem and a Lagrangian relaxation of this formulation. This relaxation decomposes into subproblems, that can be solved separately. Section 2 presents a very fast recursive optimisation algorithm for these subproblems that also requires minimum computer memory. The jobs' starting times, determined by solving the Lagrangian relaxation, specify the order of jobs processing on each machine. Since the Lagrangian relaxation is obtained by dualizing some constraints of the integer linear programming formulation, in general, these orders do not give any feasible schedule. Section 3 presents algorithms for conversion of a pair of orders (one for each machine), determined by solving the Lagrangian relaxation, into a feasible schedule. In the literature on Lagrangian relaxation, such algorithms are referred to as Lagrangian heuristics. Section 4 presents the results of computational experiments. In order to find a good set of Lagrange multipliers for the Lagrangian relaxation, these experiments utilise the standard and most commonly used iterative procedure known as the subgradient method [3]. According to this method, at each iteration, the Lagrangian relaxation for the current set of Lagrangian multipliers is solved and a feasible schedule is constructed using one of the developed Lagrangian heuristics. The value of the objective function for this feasible schedule is compared with the best value of the objective function found so far and the smallest of these two is used as the current upper bound on the optimal value of the objective function in the calculation of a new set of Lagrangian multipliers. The best feasible schedule found in the course of these iterations is considered as the solution produced by the iterative procedure.

2 Lagrangian Relaxation

It is obvious that, in any optimal schedule, the completion time of a job can not exceed $T = \sum_{i \in N}(p_i^1 + p_i^2)$. For each $i \in N$, integer $0 \leq t < T$, $m \in \{1, 2\}$, let

$$x_{it}^m = \begin{cases} 1, & \text{if } S_i^m = t; \\ 0, & \text{otherwise.} \end{cases} \tag{1}$$

Then, similar to [10], the considered scheduling problem can be formulated as the following integer linear program:

$$\min \sum_{i=1}^{n} w_i \left(\sum_{t=1}^{T-1} t x_{it}^2 + p_i^2 \right) \tag{2}$$

subject to

$$\sum_{t=0}^{T-1} x_{it}^m = 1, \quad \text{for } 1 \le i \le n \text{ and } m \in \{1,2\} \tag{3}$$

$$\sum_{i=1}^{n} \sum_{\tau=\max\{0,t-p_i^m\}}^{t-1} x_{i\tau}^m \le 1, \quad \text{for } 1 \le t \le T \text{ and } m \in \{1,2\} \tag{4}$$

$$\sum_{t=1}^{T-1} t x_{it}^2 - \sum_{t=1}^{T-1} t x_{it}^1 \ge p_i^1, \quad \text{for } 1 \le i \le n \tag{5}$$

$$\sum_{\tau=0}^{t-1} \sum_{i=1}^{n} b(i) x_{i\tau}^1 - \sum_{\tau=0}^{t-p_i^2-1} \sum_{i=1}^{n} b(i) x_{i\tau}^2 \le B, \quad \text{for } 1 \le t \le T \tag{6}$$

$$x_{it}^m \in \{0,1\}, \quad \text{for } 1 \le i \le n, \ 0 \le t < T, \text{ and } m \in \{1,2\} \tag{7}$$

Dualizing (4) and (6) for chosen nonnegative Lagrange multipliers v_{tm} and u_t, where $1 \le t \le T$ and $m \in \{1,2\}$, gives the Lagrangian Relaxation

$$\min \sum_{i=1}^{n} w_i \left(\sum_{t=1}^{T-1} t x_{it}^2 + p_i^2 \right) + \sum_{t=1}^{T} \sum_{m=1}^{2} v_{tm} \left(\sum_{i=1}^{n} \sum_{\tau=\max\{0,t-p_i^m\}}^{t-1} x_{i\tau}^m - 1 \right)$$

$$+ \sum_{t=1}^{T} u_t \left(\sum_{\tau=0}^{t-1} \sum_{i=1}^{n} b(i) x_{i\tau}^1 - \sum_{\tau=0}^{t-p_i^2-1} \sum_{i=1}^{n} b(i) x_{i\tau}^2 - B \right)$$

subject to (3), (5) and (7).

Let v be the set of all v_{tm} and u be the set of all u_t. Let $LR(v,u)$ be the optimal value of the objective function of the Lagrangian Relaxation above. For each $i \in N$, let $Z_i(v,u)$ be the optimal value of the objective function of the integer linear program

$$\min w_i \sum_{t=1}^{T-1} t x_{it}^2 + \sum_{t=1}^{T} \sum_{m=1}^{2} v_{tm} \sum_{\tau=\max\{0,t-p_i^m\}}^{t-1} x_{i\tau}^m + b(i) \sum_{t=1}^{T} u_t \left(\sum_{\tau=0}^{t-1} x_{i\tau}^1 - \sum_{\tau=0}^{t-p_i^2-1} x_{i\tau}^2 \right) \tag{8}$$

subject to

$$\sum_{t=0}^{T-1} x_{it}^m = 1, \quad \text{for } m \in \{1,2\} \tag{9}$$

$$\sum_{t=1}^{T-1} t x_{it}^2 - \sum_{t=1}^{T-1} t x_{it}^1 \ge p_i^1 \tag{10}$$

$$x_{it}^m \in \{0,1\}, \quad \text{for } 0 \le t < T \text{ and } m \in \{1,2\} \tag{11}$$

It is easy to see that

$$LR(v, u) = \sum_{i=1}^{n} Z_i(v, u) + \sum_{i=1}^{n} w_i p_i^2 - \sum_{t=1}^{T} \sum_{m=1}^{2} v_{tm} - B \sum_{t=1}^{T} u_t \qquad (12)$$

Hence, for chosen Lagrange multipliers, the Lagrangian Relaxation can be solved by solving n separate integer linear programs (8)–(11). Each of these n problems can be solved very fast using the technique described below.

According to (9), exactly one x_{it}^1 and exactly one x_{it}^2 must be equal to 1 and all others must be zero. If $x_{ie}^1 = 1$ and $x_{ir}^2 = 1$, then the corresponding value of the objective function is

$$w_i r + \sum_{t=e+1}^{e+p_i^1} v_{t1} + \sum_{t=r+1}^{r+p_i^2} v_{t2} + b(i) \sum_{t=e+1}^{r+p_i^2} u_t.$$

Observe that, by virtue of (10), $e \leq r - p_i^1$ and consider function $f(r)$ defined for all $p_i^1 \leq r \leq T - p_i^2$ as follows

$$f(r) = \min_{0 \leq e \leq r - p_i^1} \left(\sum_{t=e+1}^{e+p_i^1} v_{t1} + b(i) \sum_{t=e+1}^{r+p_i^2} u_t \right). \qquad (13)$$

If $r > p_i^1$, then the straightforward algebraic transformations lead to

$$f(r) = \min \left[f(r-1) + b(i) u_{r+p_i^2}, \sum_{t=r-p_i^1+1}^{r} v_{t1} + b(i) \sum_{t=r-p_i^1+1}^{r+p_i^2} u_t \right] \qquad (14)$$

On the other hand, the direct substitution into (13) gives

$$f(p_i^1) = \sum_{t=1}^{p_i^1} v_{t1} + b(i) \sum_{t=1}^{p_i^1+p_i^2} u_t. \qquad (15)$$

Hence, using (14) and (15), all values $f(r)$ can be calculated recursively in $O(T)$ operations, and by virtue of

$$Z_i(v, u) = \min_{p_i^1 \leq r \leq T - p_i^2} \left(f(r) + w_i r + \sum_{t=r+1}^{r+p_i^2} v_{t2} \right),$$

the integer linear program (8)–(11) can be solved in $O(T)$ operations.

3   Lagrangian Heuristics

By virtue of (1), the optimal solution for the Lagrangian Relaxation specifies the order in which the jobs are to be processed on the first-stage machine and the order of job processing on the second-stage machine. It is convenient to refer to these orders as permutations π_1 and π_2 on $\{1, ..., n\}$. Unfortunately, a feasible schedule where the jobs are processed on the first-stage machine according to π_1 and on the second-stage machine according to π_2 may not exist. This section presents Lagrangian heuristics - algorithms that construct a feasible schedule based on a pair of permutations that was obtained by solving the Lagrangian Relaxation. The "no-wait" Lagrangian heuristic can violate the permutations, whereas the "wait" heuristic strictly follows these permutations. Of course, the latter Lagrangian heuristic is applicable only if there exists a feasible schedule with job processing according to the pair of permutations. The former approach is described in Subsect. 3.1. The latter approach, including a necessary and sufficient condition of the existence of a feasible schedule for the pair of permutations, is discussed in Subsect. 3.2.

3.1 No-Wait Algorithm

Let π_1 and π_2 be the permutations used for scheduling on the first and second-stage machines, correspondingly. We assume that $\pi_1 = \pi_2$, if we use only one of the two orders, defined by the starting times obtained during Lagrangian Relaxation stage of the algorithm. Otherwise π_1 and π_2 may be different. The algorithm uses Subroutine 1 for scheduling on the first-stage machine and Subroutine 2 - for scheduling on the second-stage machine. Denote by t_1 and t_2 the minimal starting time available for the unscheduled jobs on the first and the second-stage machine, correspondingly. Let In be the sum of buffer requirements of the jobs which are currently in buffer. Let n_1 and n_2 be the number of scheduled jobs on the first and the second-stage machine, correspondingly. Denote by cur_1 and cur_2 the number of the jobs scheduled during the current iteration of Subroutine 1 and Subroutine 2, correspondingly. Denote by pos_1 and pos_2 the position of the last job which was considered on the first and the second-stage machine, correspondingly. Set values of all these parameters to zero.

1. If all jobs are scheduled on both machines, stop. Otherwise go to next step.
2. If $n_1 > n$, go to step 5. Otherwise go to *Subroutine 1*.
3. If $cur_1 > 0$ after the last iteration of *Subroutine 1*, set $cur_1 = 0$, $pos_1 = 0$ and go to step 5. If $cur_1 = 0$, set $pos_1 = 0$, go to the next step.
4. If $t_1 < t_2$ determine the job j with the smallest $S_j^2 + p_j^2$, such that $S_j^2 + p_j^2 > t_1$; set $t_1 = S_j^2 + p_j^2$, set $In = In - b(j)$ and go to step 2. If $t_1 \geq t_2$, go to step 5.
5. If $n_2 > n$, go to step 1. Otherwise go to *Subroutine 2*.
6. If $cur_2 > 0$ after the last iteration of *Subroutine 2*, set $cur_2 = 0$, $pos_2 = 0$ and go to step 2. If $cur_2 = 0$, set $pos_2 = 0$, go to the next step.
7. If $t_2 < t_1$ determine the job j with the smallest $S_j^1 + p_j^1$, such that $S_j^1 + p_j^1 > t_2$; set $t_2 = S_j^1 + p_j^1$ and go to step 5. If $t_2 \geq t_1$, go to step 2.

Subroutine 1

1. Set $pos_1 = pos_1 + 1$. If $pos_1 = n + 1$, exit *Subroutine 1* and go to step 3 of the No-wait algorithm. Otherwise go to the next step.
2. Consider $i = \pi_1(pos_1)$. If S_i^1 has not been assigned yet and there is a place in the buffer for i, i.e $In + b(i) \le B$, then set $S_i^1 = t_1$, $In = In + b(i)$ and go to the next step. Otherwise go to step 1 of the subroutine.
3. A job j is released from the buffer only if it is completed on the second-stage machine by time t_1. If $t_2 - p_i^1 \le t_1 < t_2$, then set $In = In - b(j)$ for every job j such that $S_j^2 + p_j^2 - p_i^1 \le t_1 < S_j^2 + p_j^2$.
4. Set $t_1 = t_1 + p_i^1$, $cur_1 = cur_1 + 1$, $n_1 = n_1 + 1$ and go to step 1.

Subroutine 2

1. Set $pos_2 = pos_2 + 1$. If $pos_2 = n + 1$, exit *Subroutine 2* and go to step 6 of the No-wait algorithm. Otherwise go to the next step.
2. Consider $i = \pi_2(pos_2)$. If S_i^2 has not been assigned yet and i has been scheduled on the first-stage machine and $S_i^1 + p_i^1 \le t_2$, then set $S_i^2 = t_2$ and go to the next step. Otherwise go to step 1 of the subroutine.
3. If the job i is completed on the second-stage machine by t_1, then it is released from the buffer: if $S_i^2 + p_i^2 \le t_1$, set $In = In - b(i)$.
4. Set $t_2 = t_2 + p_i^2$, $cur_2 = cur_2 + 1$ and $n_2 = n_2 + 1$. Go to step 1 of the subroutine.

Lemma 1. *The No-wait algorithm constructs a feasible schedule.*

3.2 Wait Algorithm

Definition 1. *A pair of permutations (π_1, π_2) is feasible if there exists a schedule such that the jobs on the first-stage machine are processed in the order specified by π_1 and on the second-stage machine the jobs are processed in the order specified by π_2.*

Definition 2. *Job $i \in N$ is ordinary, if*

$$\sum_{1 \le u \le \pi_1^{-1}(i)} b(\pi_1(u)) > B. \tag{16}$$

Definition 3. *For every ordinary job $i \in N$ critical position $1 \le k_i \le n$ is defined as the smallest among all v for which the following inequality holds:*

$$\sum_{1 \le u \le \pi_1^{-1}(i)} b(\pi_1(u)) - \sum_{1 \le u \le v} b(\pi_2(u)) \le B. \tag{17}$$

Lemma 2. *If the pair of permutations (π_1, π_2) is feasible, then for each ordinary $i \in N$*

$$\pi_2^{-1}(i) > k_i. \tag{18}$$

Proof. Consider an arbitrary feasible schedule, in which the jobs are processed in the orders, specified by π_1 and π_2, and assume that $k_i > \pi_2^{-1}(i)$ for a job i. Consider all the jobs which are completed on the second-stage machine by the time $\tau = S_i^2 + p_i^2$. Since the considered schedule is feasible, $S_i^1 + p_i^1 \le S_i^2$, hence $\{j : \pi_1^{-1}(j) \le \pi_1^{-1}(i)\} \subseteq \{j : S_j^1 \le \tau\}$. At the same time $\{j : S_j^2 + p_j^2 \le \tau\} \subseteq \{j : \pi_2^{-1}(j) \le \pi_2^{-1}(i)\}$. For the feasible schedule the capacity of the buffer B should not be exceeded in any moment of time, for example τ:

$$\sum_{u:\, S_u^1 \le \tau} b(u) - \sum_{u:\, S_u^2 + p_u^2 \le \tau} b(u) \le B. \tag{19}$$

Since $k_i > \pi_2^{-1}(i)$, by virtue of Definition 3 and (19),

$$\sum_{u:\, S_u^1 \le \tau} b(u) - \sum_{u:\, S_u^2 + p_u^2 \le \tau} b(u) \ge \sum_{1 \le u \le \pi_1^{-1}(i)} b(\pi_1(u)) - \sum_{1 \le u \le \pi_2^{-1}(i)} b(\pi_2(u)) > B, \tag{20}$$

which contradicts (19). If $k_i = \pi_2^{-1}(i)$, then by Definition 3,

$$\sum_{1 \le u < \pi_1^{-1}(i)} b(\pi_1(u)) - \sum_{1 \le u \le \pi_2^{-1}(i)} b(\pi_2(u)) \le B - b(i), \tag{21}$$

which implies that job i has to be processed on the second-stage machine and leave the buffer before it can start on the first-stage machine, which contradicts to the assumption that the considered schedule is feasible. $\qquad\square$

Lemma 2 demonstrated that (18) is a necessary condition for feasibility of pair (π_1, π_2). To prove that (18) is also a sufficient feasibility condition we will need results of the following lemma.

Lemma 3. *If for the pair (π_1, π_2) the inequality (18) holds for each ordinary $i \in N$, then for any q such that $\pi_2^{-1}(q) \le k_i$, $\pi_1^{-1}(q) < \pi_1^{-1}(i)$.*

Proof. If q is not ordinary job and $\pi_1^{-1}(q) \ge \pi_1^{-1}(i)$, then

$$\sum_{1 \le u \le \pi_1^{-1}(i)} b(\pi_1(u)) \le \sum_{1 \le u \le \pi_1^{-1}(q)} b(\pi_1(u)) \le B, \tag{22}$$

which contradicts i being ordinary job. Assume that q is ordinary job and $\pi_1^{-1}(q) \ge \pi_1^{-1}(i)$. Then by virtue of (18), $k_q < \pi_2^{-1}(q) \le k_i$. Thus, by the assumption and taking into account Definition 3,

$$\sum_{1 \le u \le \pi_1^{-1}(q)} b(\pi_1(u)) - \sum_{1 \le u \le k_q} b(\pi_2(u)) \ge \sum_{1 \le u \le \pi_1^{-1}(i)} b(\pi_1(q)) - \sum_{1 \le u \le k_q} b(\pi_2(u)) > B, \tag{23}$$

which contradicts to the fact that k_q is the critical position for job q. $\qquad\square$

Wait algorithm. We assume that $\pi_1 = \pi_2$, if we use only one of the two orders, defined by the starting times obtained during Lagrangian Relaxation stage of the algorithm. Otherwise π_1 and π_2 may be different. The Wait algorithm employs *Repair routine* if (18) did not hold for (π_1, π_2) for all $i \in N$ at the start of the algorithm. The *Repair routine* checks whether or not $\pi_1^{-1}(j) < \pi_1^{-1}(i)$ for the considered job i and all jobs j such that $j = \pi_2(u)$ and $u \leq k_i$, and the routine changes π_2, if required. Denote by t_1 and t_2 the current minimal starting time available for unscheduled jobs on the first and the second-stage machine, correspondingly. Denote by pos_1 and pos_2 the current position in π_1 and π_2, correspondingly. Set $t_1 = t_2 = 0$ and $pos_1 = pos_2 = 1$.

1. If all jobs are scheduled on both machines, stop. Otherwise go to next step.
2. If $pos_1 > n$, go to step 5. Otherwise consider $i = \pi_1(pos_1)$. If i is not ordinary job or if i is ordinary job and S_q^2 has been assigned for $q = \pi_2(k_i)$ and $S_q^2 + p_q^2 \leq t_1$, go to the next step. Otherwise go to step 4.
3. Set $S_i^1 = t_1$ for $i = \pi_1(pos_1)$, $t_1 = t_1 + p_i^1$, $pos_1 = pos_1 + 1$. Go to step 5.
4. There are the following possibilities for the considered job $i = \pi_1(pos_1)$:
   4.1. If S_q^2 has been assigned for $q = \pi_2(k_i)$ but $S_q^2 + p_q^2 > t_1$, then set $t_1 = S_q^2 + p_q^2$, go to step 3 to schedule task i on the first-stage machine.
   4.2. If S_q^2 has not been assigned for $q = \pi_2(k_i)$ and (18) held for (π_1, π_2) for all $i \in N$ at the start of the algorithm, go to step 5.
   4.3. If S_q^2 has not been assigned for $q = \pi_2(k_i)$, (18) did not hold for (π_1, π_2) for all $i \in N$ at the start of the algorithm, go to *Repair routine*.
5. If $pos_2 > n$, go to step 1. Otherwise consider $j = \pi_2(pos_2)$. If S_j^1 has been assigned and $S_j^1 + p_j^1 \leq t_2$, go to the next step. Otherwise go to step 7.
6. Set $S_j^2 = t_2$ for $j = \pi_2(pos_2)$, $t_2 = t_2 + p_i^2$, $pos_2 = pos_2 + 1$. Go to step 2.
7. If job $j = \pi_2(pos_2)$ has been scheduled on the first-stage machine but $S_j^1 + p_j^1 > t_2$, set $t_2 = S_j^1 + p_j^1$ and go to step 6 to schedule job j on the second-stage machine. Otherwise go to step 2.

Repair Routine

1. If $k_i \geq \pi_2^{-1}(i)$, go to step 2 of the routine. Otherwise for $pos_2 \leq u \leq k_i$ and $j = \pi_2(u)$ check if $\pi_1^{-1}(j) < \pi_1^{-1}(i)$. If $\pi_1^{-1}(j) < \pi_1^{-1}(i)$ for all such j, go to step 5 of the algorithm. Otherwise go to the next step of the routine.
2. If $\displaystyle\sum_{1 \leq u \leq \pi_1^{-1}(i)} b(\pi_1(u)) - \sum_{1 \leq u < pos_2} b(\pi_2(u)) \leq B$, then set $k_i = pos_2 - 1$ and go to step 2 of the algorithm. Otherwise set $new_2 = pos_2$ and go to the next step of the routine.
3. Among jobs j such that $new_2 \leq \pi_2^{-1}(j) \leq n$ find the first job h, such that $\pi_1^{-1}(h) < \pi_1^{-1}(i)$. In permutation π_2 increase by one positions of all jobs j such that $new_2 \leq \pi_2^{-1}(j) < \pi_2^{-1}(h)$, set $\pi_2^{-1}(h) = new_2$, $new_2 = new_2 + 1$. If
$$\sum_{1 \leq u \leq \pi_1^{-1}(i)} b(\pi_1(u)) - \left(\sum_{1 \leq u \leq new_2} b(\pi_2(u)) \right) \leq B$$
then go to the next step of routine. Otherwise repeat the step 3 of the routine.
4. For $i = \pi_1(pos_1)$ set $k_i = \pi_2^{-1}(h)$ and go to the step 6 of the algorithm.

Lemma 4. *The Wait algorithm constructs a feasible schedule.*

Corollary 1. *If for the pair of permutations (π_1, π_2) the inequality (18) holds for each $i \in N$, then the pair is feasible.*

Proof. The statement of the corollary straightforwardly follows from Lemma 4.

4   Computational Experiments

The computational experiments aimed to compare the No-wait algorithm and Wait algorithm. The computational experiments were conducted by the second author on a personal computer with Intel Core $i5$ processor $CPU@1.70\,\mathrm{Ghz}$, using Ubuntu 14.04 LTS, with base memory 4096 MB. The algorithms were implemented using C programming language. The test instances were generated randomly with processing times chosen from the interval $[1, 10]$, and job's weights chosen from the interval $(0, 2]$. Two groups of 15 instances of 25 and 50 jobs were considered. The experiments were conducted for buffer sizes $B = b_{max}$ and $B = 2b_{max}$, where b_{max} is the maximum buffer requirement among all jobs of a set. The No-wait and Wait algorithms were compared in terms of initial value of the upper bound, the smallest upper bound produced by each algorithm, and how this value improved in comparison with the initial value. Denote by $NW1$, $NW2$ and $NW3$ the No-wait algorithm, with π_1 and π_2 defined by the starting times of jobs on the first, the second-stage machine or each of the machines, correspondingly; denote by $W1$, $W2$ and $W3$ the Wait algorithm with π_1 and π_2 defined by the starting times of jobs on the first, the second-stage machine or each of the machines, correspondingly. In all cases the starting times were obtained during Lagrangian Relaxation stage of the algorithm. For each job the parameter $W_i^0 = \frac{w_i}{(p_i^1 + p_i^2)}$ was calculated, and the list of jobs in non-increasing order of W_i^0 was constructed. To obtain an initial value of upper bound No-wait or Wait algorithm was employed with this order for both machines. In each heuristic the subgradient algorithm was run for 1000 iterations. The initial upper bounds for $NW1$, $NW2$ and $NW3$ have the same value which is provided in the second column of a table, denoted by INW. The initial upper bounds for $W1$, $W2$ and $W3$ have the same value which is provided in third column of a table, denoted by IW. The fourth column shows difference D between the initial bounds and calculated as $D = 1 - \frac{IW}{INW}$. The columns $5-10$ provide the smallest upper bound produced by the corresponding No-wait or Wait algorithms, the smallest value emphasised in bold.

   The columns 11–16 show the improvement of upper bound I, calculated as $I = 1 - \frac{NWi}{INW}$, for No-wait algorithms, and $I = 1 - \frac{Wi}{IW}$, for Wait algorithms, where $i = 1, 2, 3$. The results for 25 jobs sets and $B = b_{max}$ are presented in Table 1, for 25 jobs sets and $B = 2b_{max}$ - in Table 2; the results for 50 jobs sets and $B = b_{max}$ are presented in Table 3, for 50 jobs sets and $B = 2b_{max}$ - in Table 4.

Table 1. 25 jobs, $B = b_{max}$

Inst	Intial UB			Best UB						I, %					
	INW	IW	D, %	NW1	NW2	NW3	W1	W2	W3	NW1	NW2	NW3	W1	W2	W3
1	1942	1592	18	1672	1672	1569	**1423**	1455	1592	14	14	19	11	9	0
2	1994	1751	12	1707	1732	1738	**1584**	1599	1751	14	13	13	10	9	0
3	1524	1516	0	1394	1402	1403	1356	**1326**	1410	9	8	8	11	13	7
4	1875	1688	10	1573	1575	1581	**1506**	1553	1671	16	16	16	11	8	1
5	1731	1622	6	1472	1503	1475	**1411**	1411	1535	15	13	15	13	13	5
6	2275	2228	2	1999	2025	1981	1936	**1935**	1990	12	11	13	13	13	11
7	1897	1745	8	1645	1649	1640	**1630**	1661	1722	13	13	14	7	5	1
8	1603	1500	6	1481	1550	1478	**1277**	1291	1345	8	3	8	15	14	10
9	2192	1997	9	2055	1964	2055	1868	**1861**	1965	6	10	6	6	7	2
10	2401	2220	8	2173	2172	2063	**1984**	2023	2110	10	10	14	11	9	5
11	1542	1736	−13	**1483**	1506	1484	1535	1528	1575	4	2	4	12	12	9
12	1467	1499	−2	1382	1349	1364	**1285**	1305	1404	6	8	7	14	13	6
13	2045	1791	12	1708	1730	1708	1535	**1531**	1600	16	15	16	14	14	11
14	1876	1844	2	1797	1791	1787	1765	**1729**	1826	4	5	5	4	6	1
15	1838	1800	2	1721	1767	1736	**1714**	1728	1778	6	4	6	5	4	1

Table 2. 25 jobs, $B = 2b_{max}$

Inst	Intial UB			Best UB						I, %					
	INW	IW	D, %	NW1	NW2	NW3	W1	W2	W3	NW1	NW2	NW3	W1	W2	W3
1	1488	1210	19	1218	1303	1260	1151	**1118**	1210	18	12	15	5	8	0
2	1429	1401	2	1347	1306	1333	**1282**	1297	1370	6	9	7	8	7	2
3	1218	1173	4	1083	1086	1083	1088	**1081**	1150	11	11	11	7	8	2
4	1451	1409	3	1277	1404	1304	1249	**1245**	1332	12	3	10	11	12	5
5	1406	1270	10	1133	1132	1131	1108	**1102**	1212	19	20	20	13	13	5
6	1629	1618	1	1502	1498	1502	1495	**1488**	1578	8	8	8	8	8	2
7	1367	1273	7	1189	1189	1189	**1173**	1177	1268	13	13	13	8	8	0
8	1402	1293	8	1199	1223	1177	**1098**	1108	1121	14	13	16	15	14	13
9	1625	1474	9	1440	1442	1439	**1367**	1384	1474	11	11	11	7	6	0
10	1882	1713	9	1519	1532	1535	1486	**1459**	1653	19	19	18	13	15	3
11	1337	1340	0	1242	**1236**	1243	1253	1242	1340	7	8	7	6	7	0
12	1206	1133	6	1055	1062	1061	**1054**	1054	1133	13	12	12	7	7	0
13	1525	1335	12	1270	1292	1351	1184	**1175**	1335	17	15	11	11	12	0
14	1744	1414	19	1321	1376	1322	**1295**	1325	1383	24	21	24	8	6	2
15	1496	1418	5	1238	1249	1242	**1227**	1250	1302	17	17	17	13	12	8

It took 5–9.8 seconds to process a 25 jobs set, and it took 38–56 seconds to process a 50 jobs set. The overall time required for recursive procedure to calculate (12) constituted only 10% of processing time of a set. For small instances of 5 and 10 jobs all algorithms provided optimal/near optimal solutions for most instances, compared with the solutions obtained by CPLEX software. CPLEX failed to obtain a solution for a 25 jobs set within 10 hours limit.

Table 3. 50 jobs, $B = b_{max}$

Inst	Intial UB			Best UB						I, %					
	INW	IW	D, %	NW1	NW2	NW3	W1	W2	W3	NW1	NW2	NW3	W1	W2	W3
1	5789	5293	9	5152	5180	5336	**4944**	4974	5293	11	11	8	7	6	0
2	7277	6774	7	6432	6403	6427	**5749**	5791	6390	12	12	12	15	15	6
3	6176	5180	16	5577	5875	5540	4932	**4837**	5180	10	5	10	5	7	0
4	7218	5691	21	6027	5843	6050	**5217**	5426	5691	16	19	16	8	5	0
5	7821	8358	−7	7616	7524	7494	**7072**	7097	7795	3	4	4	15	15	7
6	5539	5778	−4	5367	5472	5430	**5237**	5296	5778	3	1	2	9	8	0
7	6207	5830	6	5929	5953	5545	**5199**	5249	5830	4	4	11	11	10	0
8	8053	7000	13	6839	7166	7135	**6289**	6345	7000	15	11	11	10	9	0
9	8896	9309	−5	8099	8389	8042	**7874**	7986	8739	9	6	10	15	14	6
10	6918	6198	10	6306	6506	6427	**5890**	5988	6198	9	6	7	5	3	0
11	5414	5171	4	4959	4815	4853	4655	**4640**	5171	8	11	10	10	10	0
12	6517	5725	12	5821	5975	5666	**5061**	5235	5725	11	8	13	12	9	0
13	7615	6535	14	7118	7173	7093	**6132**	6161	6535	7	6	7	6	6	0
14	7388	6980	6	6685	6870	6748	**6247**	6307	6980	10	7	9	11	10	0
15	8749	7571	13	7209	7508	7193	**6672**	6993	7571	18	14	18	12	8	0

Table 4. 50 jobs, $B = 2b_{max}$

Inst	Intial UB			Best UB						I, %					
	INW	IW	D, %	NW1	NW2	NW3	W1	W2	W3	NW1	NW2	NW3	W1	W2	W3
1	4124	4159	−1	3901	3925	3936	3911	**3853**	4159	5	5	5	6	7	0
2	6438	4907	24	4930	5351	5183	**4455**	4559	4907	23	17	19	9	7	0
3	4857	4175	14	4004	4249	4063	**3820**	3861	4175	18	13	16	9	8	0
4	5244	4273	19	4241	4248	4348	**4071**	4074	4273	19	19	17	5	5	0
5	6827	6131	10	5607	5623	5587	**5552**	5609	6131	18	18	18	9	9	0
6	4646	4323	7	4052	4146	4092	**3935**	3987	4323	13	11	12	9	8	0
7	5506	4590	17	4596	4747	4635	**4111**	4111	4590	17	14	16	10	10	0
8	5282	4946	6	4722	4922	4850	**4685**	4714	4946	11	7	8	5	5	0
9	7085	6308	11	6023	6303	6052	**5860**	5922	6308	15	11	15	7	6	0
10	5884	4793	19	4699	5120	4954	4674	**4667**	4793	20	13	16	2	3	0
11	4301	4141	4	**3722**	3787	3742	3780	3764	4141	13	12	13	9	9	0
12	5540	4416	20	4270	4405	4354	**4007**	4065	4416	23	20	21	9	8	0
13	6181	5168	16	5045	5577	5231	**4645**	4753	5168	18	10	15	10	8	0
14	6115	5141	16	4844	4922	4969	**4774**	4797	5141	21	20	19	7	7	0
15	6498	5321	18	5183	5187	5315	**5020**	5030	5321	20	20	18	6	5	0

The results demonstrate that Wait algorithms, which follow the order of jobs provided by Lagrangian Relaxation stage, produce feasible schedules with smaller value of the objective function, than No-wait algorithms. It appears that the order of jobs on the first-stage machine is more significant, as Wait algorithm $W1$, in which the jobs are scheduled in each machine exactly in the order, defined by starting times of jobs on the first-machine, provided the smallest upper bound for the most instances. Observe that Wait algorithm also provided smallest initial upper bound for most instances.

Further research will explore the applications to larger instances.

Authors are grateful to the anonymous reviewers for the useful comments and suggestions, which helped to improve the paper.

References

1. Brucker, P., Knust, S.: Complex Scheduling. GOR-Publications. Springer, Heidelberg (2012). https://doi.org/10.1007/978-3-642-23929-8
2. Emmons, H., Vairaktarakis, G.: Flow Shop Scheduling: Theoretical Results, Algorithms, and Applications. Springer, New York (2013). https://doi.org/10.1007/978-1-4614-5152-5
3. Fisher, M.L.: The Lagrangian relaxation method for solving integer programming problems. Manag. Sci. **50**(12), 1861–1871 (2004)
4. Fung, J., Zinder, Y.: Permutation schedules for a two-machine flow shop with storage. Oper. Res. Lett. **44**(2), 153–157 (2015)
5. Lin, F.-C., Hong, J.-S., Lin, B.M.T.: A two-machine flowshop problem with processing time-dependent buffer constraints - an application in multimedia presentations. Comput. Oper. Res. **36**(4), 1158–1175 (2009)
6. Lin, F.-C., Hong, J.-S., Lin, B.M.T.: Sequence optimization for media objects with due date constraints in multimedia presentations from digital libraries. Inf. Syst. **38**(1), 82–96 (2013)
7. Lin, F.-C., Lai, C.-Y., Hong, J.-S.: Minimize presentation lag by sequencing media objects for auto-assembled presentations from digital libraries. Data Knowl. Eng. **66**(3), 382–401 (2008)
8. Lin, F.-C., Lai, C.-Y., Hong, J.-S.: Heuristic algorithms for ordering media objects to reduce presentation lags in auto-assembled multimedia presentations from digital libraries. Electron. Libr. **27**(1), 134–148 (2009)
9. Kononov, A., Hong, J.-S., Kononova, P., Lin, F.-C.: Quantity-based buffer- constrained two-machine flowshop problem: active and passive prefetch models for multimedia applications. J. Sched. **15**(4), 487–497 (2012)
10. Kononova, P.A., Kochetov, Y.A.: The variable neibourhood search for two machine flowshop problem with passive prefetch. J. Appl. Ind. Math. **19**(5), 63–82 (2013)
11. Witt, A., Voß, S.: Simple heuristics for scheduling with limited intermediate storage. Comput. Oper. Res. **34**(8), 2293–2309 (2007)

Mixed Integer Programming

Linear Ordering Based MIP Formulations for the Vertex Separation or Pathwidth Problem

Sven Mallach$^{(\boxtimes)}$

Department of Computer Science, Universität zu Köln, Köln, Germany
mallach@informatik.uni-koeln.de

Abstract. We consider the task to compute the pathwidth of a graph which has been shown to be equivalent to the vertex separation problem. The latter is naturally modeled as a linear ordering problem w.r.t. the vertices of the graph. Mixed-integer programs proposed so far express linear orders using either position or set assignment variables. As we show, the lower bound on the pathwidth obtained when solving their linear programming relaxations is *zero* for *any* directed graph. We then present a new formulation based on conventional linear ordering variables and a slightly different perspective on the problem that sustains stronger lower bounds. An experimental evaluation of three mixed-integer programs, each representing one of the different modeling schemes, displays their potentials and limitations when used to solve the problem to optimality.

Keywords: Vertex separation · Pathwidth
Mixed integer programming

1 Introduction

A tree decomposition of an undirected graph $G = (V, E)$ is a collection of sets $X_i \subseteq V$, $i \in I$, along with a tree $T = (I, F)$ such that (a) $\bigcup_{i \in I} X_i = V$, (b) there is a set X_i, $i \in I$, with $\{v, w\} \subseteq X_i$ for each $\{v, w\} \in E$, and (c) for each $j \in V$, the tree-edges F connect all tree-vertices $i \in I$ where $j \in X_i$ [28]. The width of a tree decomposition is defined as $\max_{i \in I} |X_i| - 1$ and the treewidth $tw(G)$ of G is the minimum width among all its tree decompositions. Path decompositions and the pathwidth $pw(G)$ of an undirected graph G are defined analogously, just now requiring that T is actually a path [6]. It follows directly that $tw(G) \leq pw(G)$.

Both width parameters are of theoretical as well as of practical interest. While deciding whether an arbitrary graph has pathwidth at most k is itself $\mathcal{NP}$-complete [1, 25], many $\mathcal{NP}$-complete problems on graphs can be solved efficiently on instances of known constant bounded tree- respectively pathwidth. For example, Arnborg and Proskurowski gave linear time algorithms for the vertex cover, independent set, k-colorability, Hamiltonian circuit and further combinatorial problems in this case [2]. Typically, algorithms based on tree or

© Springer International Publishing AG, part of Springer Nature 2018
L. Brankovic et al. (Eds.): IWOCA 2017, LNCS 10765, pp. 327–340, 2018.
https://doi.org/10.1007/978-3-319-78825-8_27

path decompositions are fixed-parameter algorithms [10] or follow the dynamic programming [5] paradigm, i.e., their running times have constant factors that are exponential in the tree- or pathwidth. It is also known that for a graph with $|V| = n$, its pathwidth is at most $\mathcal{O}(\log n)$ times its treewidth [7] which can be used for the construction of approximation algorithms. In this paper, we deal with computing the exact pathwidth of a graph and exploit the fact that the (directed) pathwidth of a (directed) graph is equivalent to its (directed) vertex separation number [22,31]. Direct applications of the tree- and the pathwidth problems exist, among many others, in the context of VLSI design [14] and register allocation [4]. The directed vertex separation problem has, e.g., an application in optical communication networks [29]. For an overview of the several other (equivalence) relations with other graph-theoretical problems as well as applications, we refer to [6,7,15].

Exact approaches to compute the pathwidth have been studied rather rarely, especially compared to the treewidth. They can be roughly divided into enumerative [23,24,30], fixed-parameter [8,9,17,20] and combinatorial branch-and-bound algorithms [13,19], and finally mixed-integer programming (MIP) models [3,12,16,19,21,29]. While a combinatorial branch-and-bound method appears to be the currently fastest and most robust method for small and moderately sized graphs in practice, we aim at revealing the limitations and structural weaknesses of present MIP formulations and at improving their competitiveness. In this respect, the contribution of this paper is threefold.

First, we derive two MIP formulations that represent the current state-of-the-art. We then show that, for *any* given graph, the lower bound on its pathwidth obtained when solving their linear programming (LP) relaxations, i.e., the linear program that arises when neglecting any integrality requirements on the variables, is *zero*. Second, we propose a new MIP formulation whose LP relaxation yields stronger lower bounds and is still compact in size. It uses different variables than the previous ones and also exhibits a slightly different perspective on the problem. Finally, we evaluate the performance of the now three representatives when passed to a MIP solver in order to solve the problem to optimality.

The outline of this paper is as follows. Section 2 introduces the vertex separation problem formally. Section 3 summarizes related research about the vertex separation and pathwidth problems with emphasis on existing MIP and other practical solution methods. It provides the basis to derive representative MIP models reflecting the current state-of-the-art in Sect. 4. In Sect. 5, we present our novel MIP formulation and carry out experimental evaluations in Sect. 6.

2   The Vertex Separation Problem

Let $G = (V, A)$ be a directed graph (digraph) and let $\Pi(V)$ be the set of all permutations of the vertices V of G. For a given permutation or *linear order* $\pi \in \Pi(V)$, we denote with $\pi(v)$ the position of each $v \in V$ in π and consider the sets $L(\pi, v) = \{u \in V \mid \pi(u) \leq \pi(v)\}$ and $R(\pi, v) = \{w \in V \mid \pi(v) < \pi(w)\}$. They can be thought of as being generated by a cut $\delta(\pi, v)$ through the linear

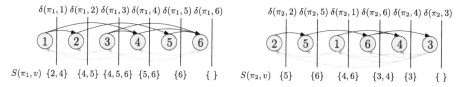

Fig. 1. A digraph drawn according to two different linear orders, $\pi_1 = \langle 1, 2, 3, 4, 5, 6 \rangle$ and $\pi_2 = \langle 2, 5, 1, 6, 4, 3 \rangle$, and illustrations of the associated cuts and separations.

order given by π that is carried out marginally close to the right of v. This is illustrated in Fig. 1 for an example graph. With each cut $\delta(\pi, v)$, we associate its corresponding *separation* $S(\pi, v) = \{ w \in R(\pi, v) \mid \exists u \in L(\pi, v) : (u, w) \in A \}$, i.e., informally, the subset of vertices in $R(\pi, v)$ that are 'hit' by arcs coming in from vertices in $L(\pi, v)$. For a fixed linear order $\pi \in \Pi(V)$, let $vs(\pi, G) = \max_{v \in V} |S(\pi, v)|$ be the corresponding maximum vertex separation. The *vertex separation problem* is to find a linear order $\pi^* \in \Pi(V)$, such that the maximum vertex separation is minimum, i.e., $vs(\pi^*, G) \leq vs(\pi, G)$ for all $\pi \in \Pi(V)$. The value $vs(\pi^*, G)$ is also referred to as the *vertex separation number* of G. For an undirected graph $G = (V, E)$, the vertex separation associated to a linear order $\pi \in \Pi(V)$ is $vs(\pi, G) = \max_{v \in V} |\{ w \in R(\pi, v) \mid \exists u \in L(\pi, v) : \{u, w\} \in E \}|$. Clearly, this value can be computed using a digraph-based method by replacing each edge $\{u, v\} \in E$ by two arcs $(u, v) \in A$ and $(v, u) \in A$.

As a remark, the previous definitions differ from other presentations in that L, R, S and the cuts are associated with a *vertex* rather than with a *position* in the linear order. While this makes no difference concerning the objective, it does in modeling, as we will see in the following sections.

3 Related Work

The amount of literature dealing with tree- and pathwidth as a theoretic concept is tremendous, so in the following, we restrict ourselves to research that deals with the exact solution of the pathwidth or vertex separation problem. As before, we consider undirected graphs $G = (V, E)$, and digraphs $G = (V, A)$, and throughout the paper, we have $n = |V|$ and, respectively, $m = |E|$ or $m = |A|$.

The proposal of combinatorial branch-and-bound algorithms started with the work of Solano and Pióro [29] in the context of wavelength-division multiplexing in optical communication networks. Coudert et al. later improved and extended their method by preprocessing and pruning techniques [13]. Comprehensive experimental studies in these articles show that it is currently the fastest and most robust method to tackle the pathwidth problem. Many instances with up to about 100 vertices can be solved routinely (still depending on the graph structure), and also some considerably larger ones. There is an implementation of their algorithm in SageMath[1] and also one of the dynamic programming algorithm presented in [8] (that is applicable to graphs with up to 31

---
[1] SageMath is an open-source mathematics library http://www.sagemath.org.

vertices only). A further combinatorial branch-and-bound scheme was proposed by Fraire Huacuja et al. [19]. On the more theoretical side, enumerative algorithms whose running times can be bounded by $\mathcal{O}(1.89^n)$ and $\mathcal{O}(1.9657^n)$ have respectively been given by Kitsunai et al. [23], and Suchan and Villanger [30], and exponential-time dynamic programming or fixed-parameter algorithms are presented in [8,9,17,20].

Among the existing MIP formulations, we distinguish two types of binary variables that have been used to model linear orders, namely *position assignment* and *set assignment variables*. Their formal definition will be given in Sect. 4. We are particularly interested in the *strength* of a formulation. A common strength measure is the lower bound on the objective (which is here the optimum vertex separation) obtained when solving a formulation's LP relaxation, i.e., the linear program that arises when neglecting any integrality constraints on the variables.

The first MIP formulation has been proposed by Solano and Pióro [29]. It is set assignment-based, has $3n^2 + 1 = \mathcal{O}(n^2)$ variables and $\mathcal{O}(nm)$ constraints. Computational experience with this model is hardly reported. However, a slightly adapted version has been implemented in SageMath and used for experiments, e.g., in [13,18]. As already noticed by Coudert in [12], its number of variables can be reduced to $2n^2 + 1$. Moreover, this reduction comes without losing strength, and it is easily seen that only n^2 of the variables need their integrality to be enforced explicitly as integrality of the others is then implied. The resulting formulation will be considered for experiments and stated formally as model MIP_S in Sect. 4. As proposed in [18], the number of constraints could be reduced to $\mathcal{O}(n^2)$ as well but the corresponding changes require to explicitly enforce integrality of all but one variable while there is no improvement concerning strength.

Duarte et al. [16] devised the first MIP formulation with position assignment variables (and a variable neighborhood search heuristic). It has $\mathcal{O}(n^4)$ variables and constraints. The quartic size stems from a straightforward linearization of quadratic terms. Using a more compact linearization and by creating some variables arc- instead of vertex-pair-based, the number of variables and constraints can be reduced to $\mathcal{O}(n^2m)$ and $\mathcal{O}(nm)$, respectively [19]. Another position assignment model by Gurski [21] involves $\mathcal{O}(n^6)$ variables which is impractical. However, as discussed by Coudert [12], it is possible to formulate a position assignment model with only $\mathcal{O}(n^2)$ variables and $\mathcal{O}(nm)$ constraints. Compared to the one by Duarte et al., its relaxation strength is as least as good whence we will consider it in our comparisons and state it as model MIP_P in the next section.

Finally, another integer program for undirected graphs has been presented by Biedl et al. [3]. Its main motivation was to show that the pathwidth problem can be formulated within a more general framework that assigns vertices and edges to grid coordinates. Due to a poor performance in their experiments, it was transformed into a satisfiability problem where graphs with $n + m < 45$ could almost always, but those where $n + m > 70$ could only rarely be solved.

4 Representative MIP Models

As announced in Sect. 3, we formulate two models, MIP_P and MIP_S, as the strongest and simultaneously most compact representatives for and with the same LP relaxation strength as the existing ones with either position or set assignment variables. We then show that the objective of these LP relaxations (and thus any model represented by them) is zero for any given digraph[2].

4.1 Position Assignments (MIP_P)

We start with the representative model MIP_P for position assignments. It models a linear order π in the same way as the models in [16, 19, 21] but introduces, as proposed by Coudert [12], fewer variables. These are the following:

- $a_{v,p} = 1$ if $\pi(v) = p$ and $a_{v,p} = 0$ otherwise (position assignment variables).
- $c_{v,p} = 1$ if $\pi(v) > p$ and there is a vertex u with $\pi(u) \leq p$ such that $(u, v) \in A$ (v 'counts' for the cut at position p), and $c_{v,p} = 0$ otherwise.
- Z: The objective variable that captures the vertex separation number.

The full model MIP_P is then:

$$
\begin{aligned}
\min\ & Z \\
\text{s.t.}\ & \sum_{p=1}^{n} a_{v,p} && = 1 && \text{for all } v \in V && (1) \\
& \sum_{v \in V} a_{v,p} && = 1 && \text{for all } 1 \leq p \leq n && (2) \\
& \sum_{q=1}^{p} (a_{u,q} - a_{v,q}) && \leq c_{v,p} && \text{for all } v \in V,\ (u,v) \in A,\ 1 \leq p \leq n && (3) \\
& \sum_{v \in V} c_{v,p} && \leq Z && \text{for all } 1 \leq p \leq n && (4) \\
& a_{v,p} && \in \{0,1\} && \text{for all } v \in V,\ 1 \leq p \leq n && (5) \\
& c_{v,p} && \in [0,1] && \text{for all } v \in V,\ 1 \leq p \leq n && (6) \\
& Z && \geq 0 && && (7)
\end{aligned}
$$

Equations (1) and (2) enforce a bijective mapping of vertices to positions. Inequalities (3) require $c_{v,p}$ to be equal to 1 whenever $\pi(v) > p$ and there is a vertex u with $\pi(u) \leq p$ and $(u, v) \in A$. Finally, the objective value Z is given by the maximum sum $\sum_{v \in V} c_{v,p}$ over all positions p (inequalities (4)). The number of variables is $2n^2 + 1$ and thus of order $\mathcal{O}(n^2)$, and the number of constraints is $3n + nm$ and thus of order $\mathcal{O}(nm)$.

4.2 Set Assignments (MIP_S)

The representative model MIP_S for the set-assignment approach arises from the initial one by Solano and Pióro [29] when successively applying the improvements made in the SageMath-implementation and those proposed by Coudert in [12].

---

[2] The model in [19] can be implemented such that its relaxation yields non-zero bounds for some graphs, but it remains of too excessive size and inferior to MIP_P in practice.

It constructs a linear order π of the vertices V by enforcing a collection of sets $\mathcal{W} = \{W_1, \ldots, W_n\}$ such that $W_p \subseteq V$ and $|W_p| = p$ for each $p \in \{1, \ldots, n\}$, and $W_p \subset W_q$ for any $p < q$. That is, W_1 specifies the vertex $v \in V$ with $\pi(v) = 1$, and in general, the rank $\pi(v)$ of vertex $v \in V$ is $\pi(v) = \min\{p \mid v \in W_p\}$.

Model MIP_S adopts the c- and objective variables from MIP_P but replaces the a-variables with variables $b_{v,p} = 1$ if $v \in W_p$ and $b_{v,p} = 0$ otherwise. Then the lines (1)–(3) and (5) of MIP_P are replaced by the following ones:

$$b_{v,p} - b_{v,p+1} \quad \leq 0 \qquad \text{for all } v \in V, \ 1 \leq p \leq n-1 \qquad (8)$$

$$\sum_{v \in V} b_{v,p} \quad = p \qquad \text{for all } 1 \leq p \leq n \qquad (9)$$

$$b_{u,p} - b_{v,p} \quad \leq c_{v,p} \qquad \text{for all } v \in V, \ (u,v) \in A, \ 1 \leq p \leq n \qquad (10)$$

$$b_{v,p} \quad \in \{0,1\} \qquad \text{for all } v \in V, \ 1 \leq p \leq n \qquad (11)$$

Inequalities (8) are *forwarding constraints* implementing the condition that $v \in W_q$ for any $q > p$ if $v \in W_p$. Equation (9) require the sets W_p to have their desired cardinalities. The inequalities (10) force $c_{v,p}$ to be equal to 1 whenever $v \notin W_p$ and there is a vertex $u \in W_p$ with $(u,v) \in A$. The number of variables is the same as in model MIP_P and thus again of order $\mathcal{O}(n^2)$. The number of constraints is $n(n-1) + n + nm + n$, i.e., a little larger but still of order $\mathcal{O}(nm)$.

4.3   Relaxation Strength of MIP_P and MIP_S

We denote the LP relaxations of MIP_P and MIP_S with LP_P and LP_S.

Theorem 1. *For any digraph $G = (V, A)$, there exists an optimum solution of LP_P and LP_S with objective value zero.*

Proof. We explicitly construct a feasible LP solution for both of the relaxations and any digraph $G = (V, A)$ with $|V| = n$ that has objective value zero.

In case of LP_P, set $a_{v,p} = \frac{1}{n}$ for each $v \in V$ and $1 \leq p \leq n$. It is easily verified that Eqs. (1) and (2) are satisfied. For any arc $(u,v) \in A$ and inequality of type (3), we obtain a left hand side of zero since $a_{u,q}$ equals $a_{v,q}$. Hence, each of these constraints imposes $c_{v,p} \geq 0$ which leads to $c_{v,p} = 0$ as we are minimizing.

Considering LP_S, set $b_{v,p} = \frac{p}{n}$ for each $v \in V$ and $1 \leq p \leq n$. This satisfies inequalities (8) strictly and Eqs. (9) exactly. Since inequalities (10) have only variables on their left hand sides that refer to the same position p, they again all evaluate to $c_{v,p} \geq 0$ such that an optimum solution has $Z = 0$. $\square$

5   A Novel Linear-Ordering Model (MIP_L)

While the vertex separation problem has often been explained in a linear ordering context, we are not aware of any promising mixed-integer program that is based on what is commonly known as *linear ordering variables* in the literature [27]. The only exception is [12] where such a formulation is indicated but then overcomplicated leading to unsatisfactory results.

Linear ordering variables $x_{v,w} \in \{0,1\}$ are defined for each pair of vertices $v, w \in V$ such that $v < w$ [27]. The variable $x_{v,w}$ is equal to 1 if and only if $\pi(v) < \pi(w)$, and hence equal to 0 if w precedes v. This will make it particularly easy to exploit the following properties within a corresponding new model MIP_L.

Lemma 1. *Let $v, w \in V$, $v \neq w$, and $(v, w) \in A$. Then $w \in S(\pi, v)$ if and only if $\pi(v) < \pi(w)$.*

Proof. Since v is the rightmost vertex of $L(\pi, v)$, w must be in $R(\pi, v)$ if it succeeds v. As we assume that $(v, w) \in A$, $w \in S(\pi, v)$ immediately follows in this case. Conversely, if $w \in S(\pi, v)$ then w must succeed v in π by definition. $\square$

Lemma 2. *Let $v, w \in V$, $v \neq w$, and $(v, w) \notin A$. Then $w \in S(\pi, v)$ if and only if $\pi(v) < \pi(w)$ and there is some $u \neq v, w$ such that $\pi(u) < \pi(v)$ and $(u, w) \in A$.*

Proof. First, suppose $w \in S(\pi, v)$. Then $w \in R(\pi, v)$ and hence $\pi(w) > \pi(v)$. Assume now that there is no such u as required. Then the only vertex in $L(\pi, v)$ that could cause $w \in S(\pi, v)$ is v itself. However, by assumption, $(v, w) \notin A$ which yields a contradiction. The converse direction is again immediate from the definition of $S(\pi, v)$. $\square$

Another central difference of MIP_L compared to MIP_P and MIP_S is that it defines cuts w.r.t. a *vertex* rather than a *position*. This leads to a replacement of the previous c-variables by new vertex-relational y-variables where $y_{v,w} = 1$ if $w \in S(\pi, v)$. We now formulate MIP_L entirely:

$$\min Z$$

$$\text{s.t. } x_{u,v} + x_{v,w} - x_{u,w} \geq 0 \qquad \text{for all } u, v, w \in V, u < v < w \qquad (12)$$

$$x_{u,v} + x_{v,w} - x_{u,w} \leq 1 \qquad \text{for all } u, v, w \in V, u < v < w \qquad (13)$$

$$x_{u,v} + x_{v,w} - 1 \quad \leq y_{v,w} \qquad \text{for all } u, v, w \in V, (v, w) \notin A, (u, w) \in A \quad (14)$$

$$\sum_{(v,w) \notin A} y_{v,w} + \sum_{(v,w) \in A} x_{v,w} \leq Z \qquad \text{for all } v \in V \qquad (15)$$

$$x_{v,w} \qquad \in \{0,1\} \quad \text{for all } v, w \in V, v < w \qquad (16)$$

$$y_{v,w} \qquad \in [0,1] \quad \text{for all } v, w \in V, v \neq w, (v, w) \notin A \qquad (17)$$

$$Z \qquad \geq 0$$

Integer valued linear ordering variables are in one-to-one correspondence with permutations of V if and only if they satisfy the so-called *3-dicycle-inequalities* (12) and (13) [27]. By Lemma 1, if $(v, w) \in A$, then the variables $y_{v,w}$ can be omitted and replaced by $x_{v,w}$ where necessary. If $(v, w) \notin A$, inequalities (14) implement Lemma 2, i.e., enforce $y_{v,w}$ to be 1 whenever there is a vertex u such that $\pi(u) < \pi(v)$, $(u, w) \in A$, and $\pi(w) > \pi(v)$. Finally, inequalities (15) let the objective function attain the desired value. Although not displayed to improve readability, in both inequalities (14) and (15), $x_{v,w}$ (and $x_{u,v}$) must be replaced by $1 - x_{w,v}$ $(1 - x_{v,u})$ if $v > w$ $(u > v)$ and we always refer to vertices $w \neq u, v$.

Model MIP_L has $\binom{n}{2} + n(n-1) - m + 1 = \mathcal{O}(n^2)$ variables. Variables $y_{v,w}$ (and hence also inequalities (14)) may further be omitted for vertices $w \in V$

that have no incoming arcs. The number of constraints is of order $\mathcal{O}(n^3)$ due to the three-dicycle inequalities. These are a natural candidate to be considered as cutting planes, i.e., they are omitted in the initial LP and then, as long as there exist inequalities that are violated by an LP solution, these are iteratively added and the LP is resolved. This increases the applicability to larger instances. In addition, every other inequality that is valid for the linear ordering problem (plenty exist, see e.g. [27]) is also valid for this formulation. When first neglecting the three-dicycle inequalities, the order of the remaining constraints is $\mathcal{O}(nm)$.

6   Experimental Evaluation

In the previous sections, we derived three different MIP models, MIP_P, MIP_S and MIP_L. Analogously to LP_P and LP_S, we denote the LP relaxation of MIP_L with LP_L. Our experiments shall give insights about the following questions:

(1) How good are the lower bounds on the pathwidth obtained with LP_L?
(2) How do LP_P, LP_S, and LP_L compare in terms of solution times?
(3) How do MIP_P, MIP_S and MIP_L perform with a sophisticated solver?

We compile a testbed of 147 bidirected graphs with at most 100 vertices from benchmark sets previously used for comparisons. First, we consider a set TWLIB consisting of 17 graphs from the TreewidthLIB. Second, we select six GRID and 20 TREE instances from the VSPLIB [16]. Then there are 20 graphs whose bidirected arcs represent the non-zero off-diagonal entries of instances from the Harwell-Boeing sparse matrix collection (denoted HB), and another set of 84 instances with only 16–24 vertices called SMALL that were introduced in [26].

All LPs and MIPs were solved using version 12.6 of CPLEX[3]. For maximum fairness and to reflect the effects of each formulation as purely as possible, all its internal cutting plane algorithms, heuristics, and presolve routines were disabled.

Each run was executed single-threadedly on a Debian Linux machine with an Intel Core i7-3770T processor running at 2.5 GHz and with 32 GB RAM.

6.1   Results

There is a table for each considered benchmark subset. Table 1 gives the results for the set HB, Table 2 for TWLIB, and Table 3 for the TREE and GRID instances. In case of the trees, we restricted the displayed results to four instances with indices 1, 6, and 11 as these are representative for all the graphs with indices 1–5, 6–10 and 11–15, respectively. Most (73 of 84) SMALL instances could be solved easily with all formulations whence the results for these are shown averaged in Table 4. Table 5 displays the instances that at least one formulation failed to solve to optimality within the time limit of ten minutes (wall clock time). Each of the Tables 1, 2, 3, 4 and 5, reports the following information: The instance

---
[3] IBM ILOG CPLEX Optimization Studio is a proprietary LP and MIP solver.

Table 1. Results for the HB instance test bed.

Instance	n	m	pw	LB	Time [s]			Time [s] or LB		
				LP_L	LP_L	LP_P	LP_S	MIP_L	MIP_P	MIP_S
ash85	85	438	8	3.4521	42.93	74.31	13.75	4.0000*	1.0000*	1.0000*
bcspwr01	39	92	4	1.5000	0.19	0.74	0.29	3.0000*	1.0000*	426.94
bcspwr02	49	118	3	1.7769	0.85	2.39	0.63	133.53	1.0000*	234.58
bcsstk01	48	352	13	4.7545	3.34	6.12	1.02	5.0000*	1.0000*	5.0000*
bcsstk02	66	4290	65	32.5000	5.85	389.99	28.52	32.5000*	0.0000*	2.0000*
can___24	24	136	5	3.0625	0.04	0.21	0.05	83.70	47.70	9.29
can___61	61	496	8	4.2394	3.51	22.77	2.52	5.0000*	1.0000*	3.0000*
can___62	62	156	4	1.8333	1.96	7.55	1.96	3.0000*	1.0000*	2.0000*
can___73	73	304	10	3.5556	17.60	27.88	4.01	4.0000*	1.0000*	4.0000*
can___96	96	672	13	4.0000	58.16	203.11	37.10	5.0000*	0.0000*	2.0000*
curtis54	54	248	6	2.9552	4.51	6.54	1.05	4.0000*	1.0000*	3.0000*
dwt___59	59	208	5	2.1667	2.23	7.82	1.48	3.0000*	1.0000*	4.0000*
dwt___66	66	254	2	1.9999	6.65	16.11	3.19	93.04	1.0000*	320.26
dwt___72	72	150	3	1.4473	4.44	13.66	3.24	115.68	1.0000*	208.56
dwt___87	87	454	8	3.8142	79.59	89.42	10.25	4.0000*	1.0000*	2.0000*
ibm32	32	180	10	3.9028	0.27	0.66	0.15	5.0000*	2.5227*	8.0000*
pores_1	30	206	7	4.1000	0.27	0.54	0.12	5.0000*	5.2381*	63.71
nos4	100	494	8	3.0000	51.18	161.99	37.92	4.0000*	1.0000*	4.0000*
steam3	80	848	7	5.4860	41.94	133.86	14.03	6.0000*	1.0000*	1.0000*
will57	57	254	5	2.8749	1.47	8.68	1.66	4.0000*	1.0000*	3.0000*

name, the number of vertices n, the number of directed arcs m and the pathwidth of the graph. Then, in column five, the lower bound obtained with LP_L is given. Columns 6–8 display the time required for solving the respective LP relaxations. The time specified for LP_L is the total time for iteratively solving LPs and adding three-dicycle-inequalities until a solution is found where none of these are violated. The last three columns give the results for solving MIP_L, MIP_P, and MIP_S. If an instance could be solved within the time limit by a particular formulation then the time needed is given in the respective table cell. Otherwise, the lower bound on the pathwidth at termination is displayed and marked with an asterisk. We now discuss the three posed research questions one-by-one.

How Good are the Lower Bounds on the Pathwidth Obtained with LP_L?

As could be expected, the lower bounds obtained depend on the graph structure. For complete digraphs on n vertices, it is half the true pathwidth (being $n - 1$) which can also be proven structurally. However, as the GRID instances show, the lower bound to pathwidth ratio can also be (made) arbitrarily bad (when increasing the number of vertices). When looking at the HB instances, the ratio is 30% for can___96, the best is 99% for dwt___66 and on average it is 50%. For

Table 2. Results for the TWLib instance test bed.

Instance	n	m	pw	LB	Time [s]			Time [s] or LB		
				LP_L	LP_L	LP_P	LP_S	MIP_L	MIP_P	MIP_S
barley	48	252	7	3.5850	1.66	4.39	0.80	4.0000*	2.0000*	4.3226*
david	87	812	13	7.2287	34.92	171.87	41.98	8.0000*	0.0000*	1.0000*
huck	74	602	10	5.6045	13.58	61.56	8.59	6.0000*	0.0000*	1.0000*
mainuk	48	396	7	4.7705	0.90	6.84	1.27	6.0000*	1.0000*	5.0000*
mildew	35	160	5	2.6246	0.18	0.74	0.20	4.0000*	1.0000*	176.65
myciel2	5	5	2	1.0000	0.00	0.00	0.00	0.00	0.01	0.00
myciel3	11	20	5	2.0833	0.00	0.00	0.00	1.75	1.10	0.34
myciel4	23	71	10	3.8580	0.06	0.15	0.04	5.2817*	7.2000*	147.44
myciel5	47	236	20	6.8750	2.18	8.02	1.20	8.0000*	1.0000*	6.0000*
queen5_5	25	320	18	6.6370	0.09	0.42	0.09	7.9205*	392.46	13.5714*
queen6_6	36	580	25	8.5000	0.85	3.65	0.54	8.9608*	1.0000*	8.0000*
queen7_7	49	952	35	10.4246	8.50	22.54	2.03	10.5145*	1.0000*	7.0000*
queen8_8	64	1456	45	12.2308	36.99	106.08	4.65	13.0000*	0.0000*	6.0000*
queen8_12	96	2736	65	15.5303	1790.72	1042.17	61.18	–	–	2.0000*
queen9_9	81	2112	58	14.1075	272.16	391.01	16.93	15.0000*	0.0000*	1.0000*
queen10_10	100	2940	72	16.0686	1348.58	1335.73	68.30	–	–	2.0000*
water	32	246	10	4.9286	0.36	0.93	0.21	5.7253*	1.0000*	9.0000*

Table 3. Results for the Tree and Grid instance test bed.

Instance	n	m	pw	LB	Time [s]			Time [s] or LB		
				LP_L	LP_L	LP_P	LP_S	MIP_L	MIP_P	MIP_S
tree_22_rot1	22	42	3	1.1250	0.01	0.04	0.04	32.14	1.9167*	12.81
tree_67_rot1	67	132	4	1.2500	1.32	8.14	2.49	2.0000*	1.0000*	2.0000*
tree_67_rot6	67	132	4	1.2656	1.09	7.80	2.43	2.0000*	1.0000*	2.0000*
tree_67_rot11	67	132	4	1.2812	1.51	7.68	2.49	2.0000*	1.0000*	2.0000*
grid_5	25	80	5	2.0000	0.04	0.16	0.04	78.16	552.65	12.60
grid_6	36	120	6	2.0000	0.22	0.72	0.22	4.0000*	2.3077*	57.28
grid_7	49	168	7	2.0000	0.61	3.17	0.61	2.2222*	1.0000*	277.32
grid_8	64	224	8	2.0000	1.21	13.13	2.19	2.0000*	1.0000*	4.7913*
grid_9	81	288	9	2.0000	4.08	42.08	9.29	3.0000*	1.0000*	2.0779*
grid_10	100	360	10	2.0000	20.08	125.20	23.88	2.0000*	1.0000*	2.0000*

the TWLib instances, the picture is similar but the quality of the lower bound drops considerably for the queen-graphs. For the Tree instances that are very similar to each other, the lower bounds obtained differ only marginally.

How Do LP_P, LP_S, and LP_L Compare in Terms of Solution Times?
Apart from a few exceptions, one can say that, across the instances considered, LP_S can be solved the fastest, followed by LP_L, while solving LP_P usually takes

Table 4. Averaged results for the SMALL instances solved by all formulations.

n	#	Avg. LB/pw [%]	Avg. time [s]			Avg. time [s]		
		LP_L	LP_L	LP_P	LP_S	MIP_L	MIP_P	MIP_S
16	4	51.09	0.0080	0.0220	0.0110	0.9440	27.2560	0.6160
17	10	57.11	0.0076	0.0216	0.0108	0.7984	33.8476	0.5592
18	10	57.30	0.0080	0.0260	0.0136	2.9644	46.6116	0.8476
19	10	54.28	0.0116	0.0348	0.0188	1.3064	84.8664	1.2656
20	9	53.55	0.0138	0.0431	0.0218	1.6124	95.1844	1.6596
21	7	57.83	0.0183	0.0514	0.0251	1.7971	62.2560	1.1634
22	6	54.77	0.0233	0.0680	0.0307	15.3127	184.4827	2.0993
23	8	55.58	0.0230	0.0750	0.0345	5.7990	150.7190	4.4000
24	9	56.20	0.0298	0.0987	0.0436	5.4227	231.2133	5.0178

Table 5. Results for the SMALL instances not solved by at least one formulation.

Instance	n	m	pw	LB	Time [s]			Time [s] or LB		
				LP_L	LP_L	LP_P	LP_S	MIP_L	MIP_P	MIP_S
p56_20_23	20	46	4	1.5000	0.01	0.04	0.02	5.83	3.0000*	7.88
p61_21_22	21	44	3	1.3308	0.02	0.04	0.02	0.89	2.0000*	3.86
p63_21_42	21	84	6	2.6939	0.02	0.07	0.02	207.52	5.0000*	10.48
p69_21_23	21	46	3	1.4000	0.01	0.05	0.02	0.94	2.0000*	6.65
p72_22_49	22	98	7	3.0625	0.03	0.12	0.04	5.2391*	235.09	3.03
p73_22_29	22	58	4	1.7500	0.01	0.06	0.02	17.54	3.0000*	13.26
p78_22_31	22	62	4	1.9028	0.02	0.06	0.02	10.61	3.0000*	2.03
p80_22_30	22	60	4	1.8644	0.02	0.09	0.02	8.68	2.3500*	2.52
p81_23_46	23	92	7	2.6429	0.04	0.11	0.03	4.7121*	5.8333*	14.27
p85_23_26	23	52	3	1.3854	0.01	0.08	0.03	3.54	1.6667*	15.31
p100_24_34	24	68	4	2.0000	0.02	0.10	0.04	10.28	3.0000*	40.27

considerably more time. There are some outliers where solving LP_L takes much longer, especially for queen_8_12 and queen_10_10 where even the time limit destined for the MIP experiment is exceeded. These extremes diminish when activating the presolve methods of CPLEX. Iterative addition of the three-dicycle inequalities is essential in order to achieve moderate solution times for LP_L. If all of them are added in advance, the solution times often degrade unsatisfactorily.

How Do MIP_P, MIP_S and MIP_L Perform with a Sophisticated Solver?

All models typically find an optimum solution quickly but often optimality cannot be proven within the time limit. Concerning the number of solved instances and derived lower bounds after ten minutes of computation time, MIP_P is clearly inferior to the other two models except for seven instances in total. MIP_S solves a few instances more than MIP_L within the time limit. On the other hand, for many unsolved instances, MIP_L provides a better lower bound than MIP_S when the timeout occurs. Further investigations revealed that the branch-and-bound scheme of CPLEX is able to improve the global lower bound much better when solving MIP_S than when solving MIP_L. More precisely, MIP_L starts with a non-zero bound but the bound then often stagnates early and solving the LPs takes more time compared to MIP_S which starts with a zero bound that can then often be improved steadily and the occurring subproblems can be enumerated quicker. So when raising the time limit, MIP_S solves or obtains the best bound for even slightly more instances although many remain unsolved even after 60 min.

6.2 Conclusion and Discussion

It became evident that the LP relaxation of MIP_L provides stronger lower bounds than those of MIP_S and MIP_P, and could still be solved relatively quickly in most of the cases. This advantage does however not necessarily translate into a better MIP solution performance. Contrarily, MIP_S could typically be solved the fastest (as also its relaxation) and could prove optimality for the highest number of instances within the time limit of ten minutes. MIP_P appeared to be inferior to both other models. However, none of them is yet competitive to the currently best combinatorial branch-and-bound methods to which many of the unsolved instances pose no challenge. Moreover, even the lower bounds obtained with MIP_L are often inferior to those achieved by combinatorial methods for treewidth (see, e.g., [11]). To become competitive, a prospective MIP formulation must exploit more structural knowledge to combine a good lower bound quality with effective search space reductions while retaining a fast relaxation performance.

References

1. Arnborg, S., Corneil, D.G., Proskurowski, A.: Complexity of finding embeddings in a k-tree. SIAM J. Algebr. Discret. Methods **8**(2), 277–284 (1987)
2. Arnborg, S., Proskurowski, A.: Linear time algorithms for NP-hard problems restricted to partial k-trees. Discret. Appl. Math. **23**(1), 11–24 (1989)
3. Biedl, T.C., Bläsius, T., Niedermann, B., Nöllenburg, M., Prutkin, R., Rutter, I.: Using ILP/SAT to determine pathwidth, visibility representations, and other grid-based graph drawings. CoRR, abs/1308.6778v2 (2015)
4. Bodlaender, H., Gustedt, J., Telle, J.A.: Linear-time register allocation for a fixed number of registers. In: Proceedings of 9th Annual ACM-SIAM Symposium on Discrete Algorithms, SODA 1998, Philadelphia, PA, USA, pp. 574–583. SIAM (1998)
5. Bodlaender, H.L.: Dynamic programming on graphs with bounded treewidth. In: Lepistö, T., Salomaa, A. (eds.) ICALP 1988. LNCS, vol. 317, pp. 105–118. Springer, Heidelberg (1988). https://doi.org/10.1007/3-540-19488-6_110
6. Bodlaender, H.L.: A tourist guide through treewidth. Acta Cybernetica **11**(1–2), 1–23 (1993)

7. Bodlaender, H.L.: A partial k-arboretum of graphs with bounded treewidth. Theoret. Comput. Sci. **209**(1–2), 1–45 (1998)
8. Bodlaender, H.L., Fomin, F.V., Koster, A.M., Kratsch, D., Thilikos, D.M.: A note on exact algorithms for vertex ordering problems on graphs. Theory Comput. Syst. **50**(3), 420–432 (2012)
9. Bodlaender, H.L., Kloks, T.: Efficient and constructive algorithms for the pathwidth and treewidth of graphs. J. Algorithms **21**(2), 358–402 (1996)
10. Bodlaender, H.L., Koster, A.M.C.A.: Combinatorial optimization on graphs of bounded treewidth. Comput. J. **51**(3), 255 (2007)
11. Bodlaender, H.L., Wolle, T., Koster, A.M.C.A.: Contraction and treewidth lower bounds. J. Graph Algorithms Appl. **10**(1), 5–49 (2006)
12. Coudert, D.: A note on integer linear programming formulations for linear ordering problems on graphs. Technical report hal-01271838, INRIA, February 2016
13. Coudert, D., Mazauric, D., Nisse, N.: Experimental evaluation of a branch-and-bound algorithm for computing pathwidth and directed pathwidth. J. Exp. Algorithmics **21**, 1.3:1–1.3:23 (2016)
14. Deo, N., Krishnamoorthy, M.S., Langston, M.A.: Exact and approximate solutions for the gate matrix layout problem. IEEE Trans. Comput. Aided Des. Integr. Circuits Syst. **6**(1), 79–84 (1987)
15. Díaz, J., Petit, J., Serna, M.: A survey of graph layout problems. ACM Comput. Surv. **34**(3), 313–356 (2002)
16. Duarte, A., Escudero, L.F., Martí, R., Mladenovic, N., Pantrigo, J.J., Sánchez-Oro, J.: Variable neighborhood search for the vertex separation problem. Comput. Oper. Res. **39**(12), 3247–3255 (2012)
17. Ellis, J.A., Sudborough, I.H., Turner, J.S.: The vertex separation and search number of a graph. Inf. Comput. **113**(1), 50–79 (1994)
18. Fraire-Huacuja, H.J., Castillo-García, N., López-Locés, M.C., Martínez Flores, J.A., Pazos R., R.A., González Barbosa, J.J., Carpio Valadez, J.M.: Integer linear programming formulation and exact algorithm for computing pathwidth. In: Melin, P., Castillo, O., Kacprzyk, J. (eds.) Nature-Inspired Design of Hybrid Intelligent Systems. SCI, vol. 667, pp. 673–686. Springer, Cham (2017). https://doi.org/10. 1007/978-3-319-47054-2_44
19. Fraire Huacuja, H.J., Castillo-García, N., Pazos Rangel, R.A., Martínez Flores, J.A., González Barbosa, J.J., Carpio Valadez, J.M.: Two new exact methods for the vertex separation problem. IJCOPI **6**(1), 31–41 (2015)
20. Fürer, M.: Faster computation of path-width. In: Mäkinen, V., Puglisi, S.J., Salmela, L. (eds.) IWOCA 2016. LNCS, vol. 9843, pp. 385–396. Springer, Cham (2016). https://doi.org/10.1007/978-3-319-44543-4_30
21. Gurski, F.: Linear programming formulations for computing graph layout parameters. Comput. J. **58**, 2921–2927 (2015)
22. Kinnersley, N.G.: The vertex separation number of a graph equals its path-width. Inf. Process. Lett. **42**(6), 345–350 (1992)
23. Kitsunai, K., Kobayashi, Y., Komuro, K., Tamaki, H., Tano, T.: Computing directed pathwidth in $O(1.89^n)$ time. In: Thilikos, D.M., Woeginger, G.J. (eds.) IPEC 2012. LNCS, vol. 7535, pp. 182–193. Springer, Heidelberg (2012). https:// doi.org/10.1007/978-3-642-33293-7_18
24. Kobayashi, Y., Komuro, K., Tamaki, H.: Search space reduction through commitments in pathwidth computation: an experimental study. In: Gudmundsson, J., Katajainen, J. (eds.) SEA 2014. LNCS, vol. 8504, pp. 388–399. Springer, Cham (2014). https://doi.org/10.1007/978-3-319-07959-2_33

25. Lengauer, T.: Black-white pebbles and graph separation. Acta Informatica **16**(4), 465–475 (1981)
26. Martí, R., Campos, V., Piñana, E.: A branch and bound algorithm for the matrix bandwidth minimization. Eur. J. Oper. Res. **186**(2), 513–528 (2008)
27. Martí, R., Reinelt, G.: The Linear Ordering Problem. Springer, Heidelberg (2011). https://doi.org/10.1007/978-3-642-16729-4
28. Robertson, N., Seymour, P.D.: Graph minors. II. Algorithmic aspects of tree-width. J. Algorithms **7**(3), 309–322 (1986)
29. Solano, F., Pióro, M.: Lightpath reconfiguration in WDM networks. IEEE/OSA J. Opt. Commun. Netw. **2**(12), 1010–1021 (2010)
30. Suchan, K., Villanger, Y.: Computing pathwidth faster than 2^n. In: Chen, J., Fomin, F.V. (eds.) IWPEC 2009. LNCS, vol. 5917, pp. 324–335. Springer, Heidelberg (2009). https://doi.org/10.1007/978-3-642-11269-0_27
31. Yang, B., Cao, Y.: Digraph searching, directed vertex separation and directed pathwidth. Discret. Appl. Math. **156**(10), 1822–1837 (2008)

Polynomial Algorithms

How to Answer a Small Batch of RMQs or LCA Queries in Practice

Mai Alzamel, Panagiotis Charalampopoulos$^{(\boxtimes)}$, Costas S. Iliopoulos, and Solon P. Pissis

Department of Informatics, King's College London, London, UK
{mai.alzamel,panagiotis.charalampopoulos,costas.iliopoulos, solon.pissis}@kcl.ac.uk

Abstract. In the Range Minimum Query (RMQ) problem, we are given an array A of n numbers and we are asked to answer queries of the following type: for indices i and j between 0 and $n-1$, query $\mathrm{RMQ}_A(i,j)$ returns the index of a minimum element in the subarray $A[i\mathinner{..}j]$. Answering a small batch of RMQs is a core computational task in many real-world applications, in particular due to the connection with the Lowest Common Ancestor (LCA) problem. With *small batch*, we mean that the number q of queries is $o(n)$ and we have them all at hand. It is therefore not relevant to build an $\Omega(n)$-sized data structure or spend $\Omega(n)$ time to build a more succinct one. It is well-known, among practitioners and elsewhere, that these data structures for online querying carry high constants in their pre-processing and querying time. We would thus like to answer this batch efficiently in practice. With *efficiently in practice*, we mean that we (ultimately) want to spend $n + \mathcal{O}(q)$ time and $\mathcal{O}(q)$ space. We write n to stress that the number of operations per entry of A should be a very small constant. Here we show how existing algorithms can be easily modified to satisfy these conditions. The presented experimental results highlight the practicality of this new scheme. The most significant improvement obtained is for answering a small batch of LCA queries. A library implementation of the presented algorithms is made available.

1 Introduction

In the Range Minimum Query (RMQ) problem, we are given an array A of n numbers and we are asked to answer queries of the following type: for indices i and j between 0 and $n-1$, query $\mathrm{RMQ}_A(i,j)$ returns the index of a minimum element in the subarray $A[i\mathinner{..}j]$.

The RMQ problem and the linearly equivalent Lowest Common Ancestor (LCA) problem [4] are very well-studied and several optimal algorithms exist to solve them. It was first shown by Harel and Tarjan [14] that a tree can be pre-processed in $\mathcal{O}(n)$ time so that LCA queries can be answered in $\mathcal{O}(1)$ time per query. A major breakthrough in practicable constant-time LCA-computation was made by Berkman and Vishkin [6]. Farach and Bender [4] further simplified

© Springer International Publishing AG, part of Springer Nature 2018
L. Brankovic et al. (Eds.): IWOCA 2017, LNCS 10765, pp. 343–355, 2018.
https://doi.org/10.1007/978-3-319-78825-8_28

this algorithm by showing that the RMQ problem is linearly equivalent to the LCA problem (shown also in [10]). The constants due to the reduction, however, remained quite large, making these algorithms impractical in most realistic cases. To this end, Fischer and Heun [9] presented yet another optimal, but also direct, algorithm for the RMQ problem. The same authors (but also others [15]) showed that due to large constants in the pre-processing and querying time implementations of this algorithm are often slower than implementations of the naive ones. Continuous efforts for engineering these solutions are being made [8].

In this article we try to address this problem, in particular when one wants to answer a relatively small batch of RMQs efficiently. This version of the problem is a core computational task in many real-world applications such as in *object inheritance* during static compilation of code [5] or in several *string matching* problems (see Sect. 5 for some). With *small batch*, we mean that the number q of the queries is $o(n)$ *and* we have them all at hand. It is therefore not relevant to build an $\Omega(n)$-sized data structure or spend $\Omega(n)$ time to build a more succinct one. It is well-known, among practitioners and elsewhere, that these data structures carry high constants in both their pre-processing and querying time. (Note that when $q = \Omega(n)$ one can use these data structures for this computation.) We would thus like to answer this batch efficiently in practice. With *efficiently in practice*, we mean that we (ultimately) want to spend $n + \mathcal{O}(q)$ time and $\mathcal{O}(q)$ space. We write n to stress that the number of operations per entry of A should be a very small constant; e.g. scan the array once or twice. In what follows, we show how existing algorithms can be easily modified to satisfy these conditions. Experimental results presented here highlight the practicality of this scheme. The most significant improvement obtained is for answering a small batch of LCA queries. The RMQ Batch problem can be defined as follows.

RMQ Batch
Input: An array A of size n of numbers and a list Q of q pairs of indices (i, j), $0 \leq i \leq j \leq n - 1$
Output: $\mathrm{RMQ}_A(i, j)$ for each $(i, j) \in Q$

The LCA Queries Batch problem can be defined as follows.

LCA Queries Batch
Input: A rooted tree T with n labelled nodes $0, 1, \ldots, n - 1$ and a list Q of q pairs of nodes (u, v)
Output: $\mathrm{LCA}_T(u, v)$ for each $(u, v) \in Q$

Our Computational Model. We assume the word-RAM model with word size $w = \Omega(\log n)$. For the RMQ Batch problem, we assume that we are given a rewritable array A of size n, each entry of which may be increased by n and still fit in a computer word. For the LCA Queries Batch problem, we assume that we are given (an $\mathcal{O}(n)$-sized representation of) a rewritable tree T allowing constant-time access to (at least) the nodes of T that are in some query in Q (see the representation in [12], for instance). All presented algorithms are deterministic.

2   Contracting the Input Array

Consider any two adjacent array entries $A[i]$ and $A[i + 1]$. Observe that if no query in Q starts or ends at i or at $i+1$ then, if $A[i] \neq A[i+1]$, $\max(A[i], A[i+1])$ will never be the answer to any of the queries in Q. Hence, the idea is that we want to contract array A, so that each block that does not contain the left or right endpoint of any query gets replaced by one element: its minimum. A similar idea, based on sorting the list Q, has been considered in the *External Memory* model [1] (see also [2]). In this section, we present a solution for our computational model, which avoids using $\Omega(n)$ space or time, but also avoids using $\Omega(\text{sort}(Q))$ time.

There are some technical details in order to update the queries for A into queries for the new array using only $\mathcal{O}(q)$ time and extra space. We first scan the array A once and find $\mu = \max_i A[i]$. We also create two auxiliary arrays $Z_0[0 \mathinner{\ldotp\ldotp} 2q - 1]$ and $Z_1[0 \mathinner{\ldotp\ldotp} 2q - 1]$. For each query $(i, j) \in Q$ we mark positions i (and j) in the array A as follows. If $A[i] \leq \mu$, then i has not been marked before. Let this be the k-th position, $k > 0$, that gets marked (we just store a counter for that). We store $A[i]$ in $Z_0[\mu + k \mod 2q]$ and replace the value that is stored in $A[i]$ by $\mu + k$. We also start a linked list at $Z_1[\mu + k \mod 2q]$, where we insert a pointer to query (i, j), so that we can update it later. If $A[i] > \mu$, then the position has already been marked; we just add a pointer to the respective query in the linked list starting at $Z_1[A[i] \mod 2q]$.

We then scan array A again and create a new array A_Q as follows: for each marked position j (i.e. $A[j] > \mu$), we copy the original value (i.e. $Z_0[A[j] \mod 2q]$) in A_Q, while each maximal block in A that does not contain a marked position is replaced by a single entry—its minimum. When we insert the original entry of a marked position j of A (i.e. $Z_0[A[j] \mod 2q]$) in A_Q at position p, we go through the linked list that is stored in $Z_1[A[j] \mod 2q]$, where we have stored pointers to all the queries of the form (i, j) or (j, k), and replace j by p in each of them. Thus, after we have scanned A, for each query $(i, j) \in Q$ on A, we will have stored the respective pair (i', j') on A_Q. Note that we need to scan array A only *once* if we know μ a priori (e.g. in LCP array [7]), or *twice* otherwise.

Example 1. Assume we are given array A and $Q = \{(4, 18), (0, 6), (6, 10)\}$.

A	17	22	38	4	5	8	2	8	9	21	0	12	8	7	13	3	6	14	1	36	0	4

Then A_Q is as follows.

A_Q	17	4	5	8	2	8	0	3	1	0

While creating A_Q, we also store in an auxiliary array the function $f : \{0, 1, \ldots, |A_Q| - 1\} \to \{0, 1, \ldots, n-1\}$ between positions of A_Q and the respective original positions in A.

Now notice that A_Q and the auxiliary arrays are all of size $\mathcal{O}(q)$ since in the worst case we mark $2q$ distinct elements of A and contract $2q + 1$ blocks that do

not contain a marked position. (We can actually throw away everything before
the first marked position and everything after the last marked position and get
$4q - 1$ instead.) The whole procedure takes $n + \mathcal{O}(q)$ time and $\mathcal{O}(q)$ space. Note
that if $\mathrm{RMQ}_{A_Q}(i', j') = \ell$ then $\mathrm{RMQ}_A(i, j) = f(\ell)$.

We can finally retrieve the original input array if required by replacing $A[f(j)]$
by $A_Q[j]$ for every j in the domain of f in $\mathcal{O}(q)$ time.

3   Small RMQ Batch

3.1   An $n + \mathcal{O}(q \log q)$-time and $\mathcal{O}(q)$-space Algorithm

The algorithm presented in this section is a modification of the *Sparse Table*
algorithm by Bender and Farach-Colton [4] applied on array A_Q; we denote it
by ST-RMQ. The modification is based on the fact that **(i)** we do not want
to consume $\Omega(q \log q)$ extra space to answer the q queries; and **(ii)** we do not
want to necessarily do all the pre-processing work of the algorithm in [4], which
is designed to answer any of the $\Theta(q^2)$ possible queries online. We denote this
modified algorithm by ST-RMQ$_{\mathrm{CON}}$ and formalise it below.

ST-RMQ$_{\mathrm{CON}}(A, Q)$

```
 1   A_Q ← CONTRACT(A, Q)
 2   Store function f; store (i', j') for every (i, j) ∈ Q
 3   for each (i, j) ∈ Q do
 4       if i = j then
 5           REPORT((i, i), i)
 6       else Add (i, j) in bucket B_⌊log(j'−i')⌋
 7   t ← max{r|B_r ≠ ∅} + 1
 8   for m = 0 to |A_Q| − 1 do
 9       D[m] ← (A_Q[m], m)
10   for k = 0 to t − 1 do
11       for each (i, j) ∈ B_k do
12           (a, p) ← min(D[i'], D[j' − 2^k + 1])
13           REPORT((i, j), f(p))
14       for m = 0 to |A_Q| − 1 do
15           if m + 2^k ≤ |A_Q| − 1 then
16               D[m] ← min(D[m], D[m + 2^k])
```

The idea is to first put each $(i, j) \in Q$ with $i \neq j$ in a bucket B_k based on
the k for which $2^k \leq j' - i' < 2^{k+1}$—we can have at most $\lceil \log(|A_Q| - 1) \rceil$ such
buckets. In this process, if we find queries of the form $(i, i) \in Q$, we answer them
on the spot. We can do this in $\mathcal{O}(q)$ time.

We then create an array D of size $|A_Q|$ where we will store 2-tuples (a, p).
In Step k, $D[m]$ will store the minimum value across $A_Q[m .. m + 2^k - 1]$, as
well as the position p, $m \leq p < m + 2^k$ where it occurs. We initialise it as
$D[m] = (A_Q[m], m)$ and we will then update it by utilising the *doubling technique*.
At Step 0 we answer all (trivial) queries that are stored in B_0; they are of the form

$(i, i + 1)$ and the answer can be found by looking at $\min(D[i'], D[i' + 1])$—note that we compare elements of D lexicographically. When we are done with B_0 we have to update D by setting $D[m] = \min(D[m], D[m + 2^0])$ for all $m < |A_Q| - 1$.

Generally, in Step k, we answer the queries of B_k as follows. For query (i, j), we find the answer by obtaining $\min(D[i'], D[j' - 2^k + 1]) = (a, p)$. We then return $f(p)$. The point is that $\{i', \ldots, i' + 2^k - 1\} \cup \{j' - 2^k + 1, \ldots, j'\} = \{i', \ldots, j'\}$. When we are done with B_k we set $D[m] = \min(D[m], D[m + 2^k])$ if $m + 2^k \leq |A_Q| - 1$.

We do this until we have gone through all t non-empty buckets (i.e. $t = \max\{r | B_r \neq \emptyset\} + 1$). Updating D takes $\mathcal{O}(q)$ time in each step, and we need in total $\mathcal{O}(q)$ time for the queries. We thus need $\mathcal{O}(qt)$ time for this part of the algorithm. Since $t = \max\{\lfloor \log(j' - i') \rfloor | (f(i'), f(j')) \in Q)\} = \mathcal{O}(\log q)$, this time is $\mathcal{O}(q \log q)$. The overall time complexity of the algorithm is thus $n + \mathcal{O}(q \log q)$. Notably, the space required is only $\mathcal{O}(q)$ as we overwrite D in each step.

3.2 $n + \mathcal{O}(q)$-time and $\mathcal{O}(q)$-space Algorithms

Offline-Based Algorithm. Given an array A of n numbers its *Cartesian tree* is defined as follows. The root of the Cartesian tree is $A[i] = \min\{A[0], \ldots, A[n-1]\}$, its left subtree is computed recursively on $A[0], \ldots, A[i-1]$ and its right subtree on $A[i + 1], \ldots, A[n - 1]$. An LCA instance can be obtained from an RMQ instance on an array A by letting T be the Cartesian tree of A that can be constructed in $\mathcal{O}(n)$ time [10]. It is easy to see that $\text{RMQ}_A(i, j)$ in A translates to $\text{LCA}_T(A[i], A[j])$ in T. The first step of this algorithm is to create array A_Q in $n + \mathcal{O}(q)$ time similarly to algorithm ST-RMQ$_{\text{CON}}$. The second step is to construct the Cartesian tree T_Q of A_Q in $\mathcal{O}(q)$ time and extra space. Finally, we apply the offline algorithm by Gabow and Tarjan [11] to answer q LCA$_{T_Q}$ queries in $\mathcal{O}(q)$ time and extra space. This takes overall $n + \mathcal{O}(q)$ time and $\mathcal{O}(q)$ extra space. We denote this algorithm by OFF-RMQ$_{\text{CON}}$. We denote by OFF-RMQ the same algorithm applied on array A.

Online-Based Algorithm. The first step of this algorithm is to create array A_Q in $n + \mathcal{O}(q)$ time similarly to algorithm ST-RMQ$_{\text{CON}}$. We can then apply the algorithm by Fischer and Heun [9] on array A_Q to obtain overall an $n + \mathcal{O}(q)$-time and $\mathcal{O}(q)$-space algorithm. We denote this algorithm by ON-RMQ$_{\text{CON}}$. We denote by ON-RMQ the same algorithm applied on array A.

Note that in the case when $q = \Omega(n)$, i.e. the batch is not so small, we can choose to apply algorithm OFF-RMQ or algorithm ON-RMQ on array A directly thus obtaining an algorithm that always works in $n + \mathcal{O}(q)$ time and $\mathcal{O}(\min\{n, q\})$ extra space. We therefore obtain the following result asymptotically.

Theorem 1. *The RMQ Batch problem can be solved in $n + \mathcal{O}(q)$ time and $\mathcal{O}(\min\{n, q\})$ extra space.*

4 Small LCA Queries Batch

In the LCA problem, we are given a rooted tree T having n labelled nodes and we are asked to answer queries of the following type: for nodes u and v, query $\text{LCA}_T(u, v)$ returns the node furthest from the root that is an ancestor of both u and v. There exists a time-optimal algorithm by Gabow and Tarjan [11] to answer a batch Q of q LCA queries in $\mathcal{O}(n + q)$ time and $\mathcal{O}(n)$ extra space. We denote this algorithm by OFF-LCA. In this section, we present a simple but *non-trivial* algorithm for improving this, for $q = o(n)$, to $n + \mathcal{O}(q)$ time and $\mathcal{O}(q)$ extra space.

It is well-known (see [4] for the details) that an RMQ instance A can be obtained from an LCA instance on a tree T by writing down the depths of the nodes visited during an *Euler tour* of T. That is, A is obtained by listing all node-visitations in a depth-first search (DFS) traversal of T starting from the root. The LCA of two nodes translates to an RMQ (where we compare nodes based on their level) between the first occurrences of these nodes in A.

We proceed largely as in Sect. 2. For each query $(u, v) \in Q$, we mark nodes u (and v) in T as follows. If $u < n$ then u has not been marked before. Let this be the k-th node, $k > 0$, that gets marked (we just store a counter for that). We also create two arrays $Z_0[0\mathinner{\ldotp\ldotp}2q-1]$ and $Z_1[0\mathinner{\ldotp\ldotp}2q-1]$. We store u in $Z_0[n-1+k \bmod 2q]$ and replace u by $n-1+k$. We also start a linked list at $Z_1[n-1+k \bmod 2q]$, where we insert a pointer to query (u, v), so that we can update it later. If $u > n - 1$, the node has already been marked, and we just add a pointer to the respective query in the linked list starting at $Z_1[u \bmod 2q]$.

We then do a single DFS traversal on T and create two new arrays E_Q and L_Q as follows. When a marked node v (i.e. $v > n - 1$) is visited for the *first time*, we write down in E_Q its original value (i.e. $Z_0[v \bmod 2q]$), while for each maximal sequence of visited nodes that are not marked we write down a single entry—the one with the *minimum tree level*. At the same time, we store in $L_Q[v]$ the level of the node added in $E_Q[v]$. While creating E_Q, we also store in an auxiliary array the function $f : \{0, 1, \ldots, |E_Q| - 1\} \to \{0, 1, \ldots, n - 1\}$ between positions of E_Q and the respective node labels in T.

When we insert the original entry of a marked node u of T (i.e. $Z_0[u \bmod 2q]$) in E_Q at position p, we go through the linked list that is stored in $Z_1[u \bmod 2q]$, where we have stored pointers to all the queries of the form (u, v) or (w, u), and replace u by p in each of these queries. Thus, after we have finished the traversal on T, for each LCA query $(u, v) \in Q$ on T, we will have stored the respective RMQ pair (u', v') on L_Q; where u' (resp. v') corresponds to the *first occurrence* of node u (resp. v) in the traversal. Thus we traverse T only *once*.

Now notice that E_Q and the auxiliary arrays are all of size $\mathcal{O}(q)$ since in the worst case we mark $2q$ distinct nodes of T and contract $2q + 1$ sequences of visited nodes that do not contain a marked node. (We can actually throw away everything before the first marked node and everything after the last marked node and get $4q - 1$ instead.) The whole procedure takes $n + \mathcal{O}(q)$ time and $\mathcal{O}(q)$ space. We are now in a position to apply algorithm ON-RMQ on L_Q to obtain the final bound. To answer the queries, note that if $\text{RMQ}_{L_Q}(u', v') = \ell$ then

$\mathrm{LCA}_T(u, v) = E_Q[\ell]$. We denote this algorithm by ON-LCA$_{\mathsf{CON}}$. Alternatively, we can apply algorithm ST-RMQ on L_Q to solve this problem in $n + \mathcal{O}(q \log q)$ and $\mathcal{O}(q)$ extra space; we denote this algorithm by ST-LCA$_{\mathsf{CON}}$.

We can finally retrieve the original input tree if required by replacing node $f(v)$ by $E_Q[v]$ for every v in the domain of f in $\mathcal{O}(q)$ time.

Note that in the case when $q = \Omega(n)$, i.e. the batch is not so small, we can choose to apply algorithm OFF-LCA on tree T directly, thus obtaining an algorithm that always works in $n + \mathcal{O}(q)$ time and $\mathcal{O}(\min\{n, q\})$ extra space. We therefore obtain the following result asymptotically.

Theorem 2. *The LCA Queries Batch problem can be solved in $n + \mathcal{O}(q)$ time and $\mathcal{O}(\min\{n, q\})$ extra space.*

5   Applications

We consider the well-known application of answering q LCA queries on the suffix tree of a string. The *suffix tree* $T(S)$ of a non-empty string S of length n is a compact trie representing all suffixes of S (see [7], for details). The nodes of the trie which become nodes of the suffix tree are called *explicit* nodes, while the other nodes are called *implicit*. Each edge of the suffix tree can be viewed as an upward maximal path of implicit nodes starting with an explicit node. Moreover, each node belongs to a unique path of that kind. Then, each node of the trie can be represented in the suffix tree by the edge it belongs to and an index within the corresponding path. The *path-label* of a node v is the concatenation of the edge labels along the path from the root to v. The nodes whose path-label corresponds to a suffix of S are called *terminal*. Given two terminal nodes u and v in $T(S)$, representing suffixes $S[i \,..\, n - 1]$ and $S[j \,..\, n - 1]$, the *string depth* of node $\mathrm{LCA}_{T(S)}(u, v)$ corresponds to the *length* of their longest common prefix, also known as their longest common extension (LCE) [15].

In many textbook solutions for classical string matching problems (e.g. maximal palindromic factors, approximate string matching with k-mismatches, approximate string matching with k-differences, online string search with the suffix array, etc.) we have that $q = \Omega(n)$ and/or the queries have to be answered *online*. In other algorithms, however, q can be *much smaller* on average (in practice) and the queries can be answered *offline*. We describe here a few such solutions. The common idea, as in many fast average-case algorithms, is to minimise the number of queries by *filtering out* queries that can never lead to a valid solution.

Text Indexing. Suppose we are given the suffix tree $T(S)$ of a text S of length n and we are asked to create the suffix links for the internal nodes. This may be necessary if the construction algorithm does not compute suffix links (e.g. construction via suffix array) but they are needed for an application of interest [18]. The *suffix link* of a node v with path-label αy is a pointer to the node path-labelled y, where $\alpha \in \Sigma$ is a single letter and y is a string. The suffix link of v

exists if v is a non-root internal node of T. The suffix links can be computed as follows. The first step is to mark each internal node v of the suffix tree with a pair of leaves (i, j) such that leaves labelled i and j are in subtrees rooted at different children of v. This can be done by a DFS traversal of the tree. (Note that if an internal node v has only one child then it must be terminal; assume that it represents the suffix $S[t .. n - 1]$. We thus create a suffix link to the node representing $S[t + 1 .. n - 1]$.) Given an internal node v marked with (i, j), note that $v = \mathrm{LCA}_{T(S)}(i, j)$, and let αy be its path-label. To create the suffix link from v, node u with path-label y can be obtained by the query $\mathrm{LCA}_{T(S)}(i + 1, j + 1)$. We can create a batch of LCA queries consisting of all such pairs. Note that in randomly generated texts, the number of internal nodes of $T(S)$ is $\mathcal{O}(n/h)$ on average, where h is the alphabet's entropy [21]; thus the standard $\Theta(n)$-time and $\Theta(n)$-space solution to this problem, building the LCA data structure over $T(S)$ [4], is not satisfactory.

Finding Frequent Gapped Factors in Texts. We are given a text S of length n, and positive integers ℓ_1, ℓ_2, d, and $k > 1$. The problem is to find all couples (u, v), such that string uwv, for *any* string w (known as *gap* or *spacer*), $|w| = d$, occurs in S at least k times, $|u| = \ell_1$, $|v| = \ell_2$ [16,20]. The first step is to build $T(S)$. We then locate all subtrees rooted at an explicit node with string depth at least ℓ_1 and whose parent has string depth less than ℓ_1, corresponding to factors u repeated in S. From these subtrees, we only consider the ones with at least k terminal nodes. Note that if k is large enough, we may have only a few such subtrees. For each subtree with $k' \geq k$ terminal nodes, representing suffixes $S[i_1 .. n - 1], S[i_2 .. n - 1], \ldots, S[i_{k'} .. n - 1]$, we create a batch of LCA queries between all pairs $(i_j + \ell_1 + d, i_{j'} + \ell_1 + d)$ and report occurrences when LCA queries extend pairwise matches to length at least ℓ_2 for a set of at least k such suffixes. (This algorithm can be easily generalised for any number of gaps.)

Pattern Matching on Weighted Sequences. A *weighted sequence* specifies the probability of occurrence of each letter of the alphabet for every position. A weighted sequence thus represents many different strings, each with the probability of occurrence equal to the product of probabilities of its letters at subsequent positions of the weighted sequence. The problem is to find all occurrences of a (standard) pattern P of length m with probability at least $1/z$ in a weighted sequence S of length n [17]. The first step is to construct the heavy string of S, denoted by $H(S)$, by assigning to $H(S)[i]$ the most probable letter of $S[i]$ (resolving ties arbitrarily). The second step is to build $T(P\$H(S))$, $\$ \notin \Sigma$. We can then compute the first mismatch between P and every substring of $H(S)$. Note that the number of positions in S where two or more letters occur with probability at least $1/z$ can be small, and so we consider only these positions to cause a legitimate mismatch between P and a factor of $H(S)$. We then use $\mathcal{O}(\log z)$ batches of LCA queries per such starting position to extend a match to length at least m. This is because P cannot match a weighted sequence S with probability $1/z$ if more than $\lfloor \log z \rfloor$ mismatches occur between P and $H(S)$ [17].

Pattern Matching with Don't Care Letters. We are given a pattern P of length m, with $m - k$ letters from alphabet Σ and k occurrences of a don't care letter (matching itself and any letter from Σ), and a text S of length n. The problem is to find all occurrences of P in S [19]. The first step is to build $T(P'\$S)$, $\$ \notin \Sigma$, where P' is the string obtained from P by replacing don't care letters with a letter $\# \notin \Sigma$. We then locate the subtree rooted at the highest explicit node corresponding to the longest factor f of P' without $\#$'s. We also locate, in the same subtree, all V terminal nodes corresponding to starting positions of f in S. Note that if f is long enough, we may have only a few such nodes. Since we know where the don't care letters occur in P, we can create a batch of kV LCA queries. An occurrence is then reported when LCA queries extend a match to length at least m. (This algorithm can be easily generalised for any number of patterns.)

Circular String Matching. We are given a pattern P of length m and a text S of length n. The problem is to find all occurrences of P or any of its cyclic shifts in S [3]. The first step is to build $T(PP\$P^R P^R \# S\% S^R)$, where $\$, \#, \% \notin \Sigma$, and X^R denotes the reverse image of string X. We then conceptually split P in two fragments of lengths $\lceil m/2 \rceil$ and $\lfloor m/2 \rfloor$. Any cyclic shift of P contains as a factor at least one of the two fragments. We thus locate the two subtrees rooted at the highest explicit nodes corresponding to the fragments. We also locate in the same subtrees all V terminal nodes corresponding to starting positions of the fragments in S. Note that if m is long enough, we may have only a few such nodes. We create a batch of at most $2V$ LCA queries in order to extend to the left and to the right and report occurrences when LCA queries extend a match to length at least m. (This algorithm can be easily generalised for any number of patterns.)

6   Experimental Results

We have implemented algorithms ST-RMQ$_{\text{CON}}$, OFF-RMQ$_{\text{CON}}$, and ON-RMQ$_{\text{CON}}$ in the C++ programming language. We have also implemented the same algorithms applied on the original array A, denoted by ST-RMQ, OFF-RMQ, and ON-RMQ, respectively; as well as the brute-force algorithm for answering RMQs in the two corresponding flavours, denoted by BF-RMQ$_{\text{CON}}$ and BF-RMQ. For the implementation of ON-RMQ$_{\text{CON}}$ and ON-RMQ we used the sdsl-lite library [13]. If an algorithm requires $f(n, q)$ time and $g(n, q)$ extra space, we say that the algorithm has complexity $<f(n, q), g(n, q)>$. Table 1 summarises the implemented algorithms. The following experiments were conducted on a Desktop PC using one core of Intel Core i5-4690 CPU at 3.50 GHz and 16GB of RAM. All programs were compiled with g++ version 5.4.0 at optimisation level 3 (-O3).

Experiment I. We generated random (uniform distribution) input arrays of $n = 1,000,000$ and $n = 100,000,000$ entries (integers), and random (uniform distribution) lists of queries of sizes varying from $\sqrt{n}$ to $128\sqrt{n}$, doubling each

Table 1. Time and space complexities of algorithms for answering RMQs offline.

Non-contracted		Contracted	
ST-RMQ	$<\mathcal{O}(n\log n + q), \mathcal{O}(n\log n)>$	ST-RMQ$_{CON}$	$<n + \mathcal{O}(q\log q), \mathcal{O}(q)>$
ON-RMQ	$<\mathcal{O}(n + q), \mathcal{O}(n)>$	ON-RMQ$_{CON}$	$<n + \mathcal{O}(q), \mathcal{O}(q)>$
OFF-RMQ	$<\mathcal{O}(n + q), \mathcal{O}(n)>$	OFF-RMQ$_{CON}$	$<n + \mathcal{O}(q), \mathcal{O}(q)>$
BF-RMQ	$<\mathcal{O}(qn), \mathcal{O}(1)>$	BF-RMQ$_{CON}$	$<n + \mathcal{O}(q^2), \mathcal{O}(q)>$

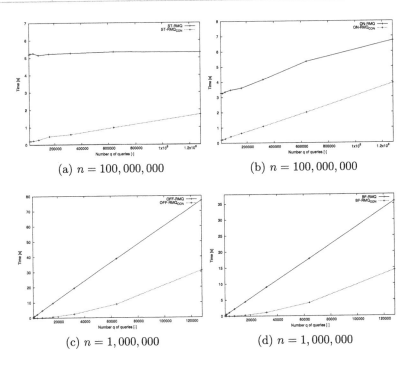

(a) $n = 100,000,000$ (b) $n = 100,000,000$

(c) $n = 1,000,000$ (d) $n = 1,000,000$

Fig. 1. Impact of the proposed scheme on the RMQ algorithms of Table 1.

time. We compared the runtime of the implementations of the algorithms in Table 1 on these inputs; in particular, for each algorithm, we compared the standard implementation against the one with the contracted array. We used the large array, $n = 100,000,000$, for ST-RMQ and ON-RMQ because they are significantly faster and the small one, $n = 1,000,000$, for OFF-RMQ and BF-RMQ. The results plotted in Fig. 1 show that the proposed scheme of contracting the input array improves the performance for all implementations substantially.

Experiment II. We generated random input arrays of $n = 1,000,000,000$ entries, and random lists of queries of sizes varying from $\sqrt{n}$ to $128\sqrt{n}$, doubling each time. We then compared the runtime of ON-RMQ$_{CON}$ and ST-RMQ$_{CON}$ on these inputs. The results are plotted in Fig. 2a. We observe that ST-RMQ$_{CON}$ becomes two times faster than ON-RMQ$_{CON}$ as q grows. Notably, it was not possible to

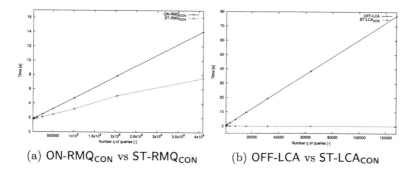

(a) ON-RMQ$_{CON}$ vs ST-RMQ$_{CON}$ (b) OFF-LCA vs ST-LCA$_{CON}$

Fig. 2. Elapsed-time comparison of ON-RMQ$_{CON}$ and ST-RMQ$_{CON}$ algorithms for $n = 1,000,000,000$ (left); and of OFF-LCA and ST-LCA$_{CON}$ algorithms for $n = 1,000,000$ (right).

run this experiment with ON-RMQ, which implements a *succinct* data structure for answering RMQs, due to insufficient amount of main memory.

Experiment III. In addition, we have implemented algorithms ST-LCA$_{CON}$ and OFF-LCA for answering LCA queries. We first generated a random input array of $n = 1,000,000$ entries and used this array to compute its Cartesian tree. Next we generated random lists of LCA queries of sizes varying from $\sqrt{n}$ to $128\sqrt{n}$, doubling each time. We then compared the runtime of OFF-LCA and ST-LCA$_{CON}$ on these inputs. The results plotted in Fig. 2b show that the implementation of ST-LCA$_{CON}$ is more than two orders of magnitude faster than the implementation of OFF-LCA, highlighting the impact of the proposed scheme on LCA queries.

7   Final Remarks

In this article, we presented a new family of algorithms for answering a small batch of RMQs or LCA queries in practice. The main purpose was to show that if the number q of queries is small with respect to n and we have them all at hand existing algorithms for RMQs and LCA queries can be easily modified to perform in $n + \mathcal{O}(q)$ time and $\mathcal{O}(q)$ extra space. The presented experimental results indeed show that with this new scheme significant practical improvements can be obtained; in particular, for answering a small batch of LCA queries.

Specifically, algorithms ST-RMQ$_{CON}$ and ST-LCA$_{CON}$, our modifications to the *Sparse Table* algorithm whose main catch is $\Theta(n \log n)$ space [4], seem to be the best way to answer in practice a small batch of RMQs and LCA queries, respectively. A library implementation of ST-RMQ$_{CON}$ is available at https://github.com/solonas13/rmqo under the GNU General Public License.

References

1. Afshani, P., Sitchinava, N.: I/O-efficient range minima queries. In: Ravi, R., Gørtz, I.L. (eds.) SWAT 2014. LNCS, vol. 8503, pp. 1–12. Springer, Cham (2014). https://doi.org/10.1007/978-3-319-08404-6_1
2. Arge, L., Fischer, J., Sanders, P., Sitchinava, N.: On (dynamic) range minimum queries in external memory. In: Dehne, F., Solis-Oba, R., Sack, J.-R. (eds.) WADS 2013. LNCS, vol. 8037, pp. 37–48. Springer, Heidelberg (2013). https://doi.org/10.1007/978-3-642-40104-6_4
3. Athar, T., Barton, C., Bland, W., Gao, J., Iliopoulos, C.S., Liu, C., Pissis, S.P.: Fast circular dictionary-matching algorithm. Math. Struct. Comput. Sci. **27**(2), 143–156 (2017)
4. Bender, M.A., Farach-Colton, M.: The LCA problem revisited. In: Gonnet, G.H., Viola, A. (eds.) LATIN 2000. LNCS, vol. 1776, pp. 88–94. Springer, Heidelberg (2000). https://doi.org/10.1007/10719839_9
5. Bender, M.A., Farach-Colton, M., Pemmasani, G., Skiena, S., Sumazin, P.: Lowest common ancestors in trees and directed acyclic graphs. J. Algorithms **57**(2), 75–94 (2005)
6. Berkman, O., Vishkin, U.: Recursive star-tree parallel data structure. SIAM J. Comput. **22**(2), 221–242 (1993)
7. Crochemore, M., Hancart, C., Lecroq, T.: Algorithms on Strings. Cambridge University Press, Cambridge (2007)
8. Ferrada, H., Navarro, G.: Improved range minimum queries. J. Discret. Algorithms **43**, 72–80 (2016)
9. Fischer, J., Heun, V.: Theoretical and practical improvements on the RMQ-problem, with applications to LCA and LCE. In: Lewenstein, M., Valiente, G. (eds.) CPM 2006. LNCS, vol. 4009, pp. 36–48. Springer, Heidelberg (2006). https://doi.org/10.1007/11780441_5
10. Gabow, H.N., Bentley, J.L., Tarjan, R.E.: Scaling and related techniques for geometry problems. In: STOC 1984, pp. 135–143. ACM (1984)
11. Gabow, H.N., Tarjan, R.E.: A linear-time algorithm for a special case of disjoint set union. J. Comput. Syst. Sci. **30**(2), 209–221 (1985)
12. Geary, R.F., Rahman, N., Raman, R., Raman, V.: A simple optimal representation for balanced parentheses. Theoret. Comput. Sci. **368**(3), 231–246 (2006)
13. Gog, S., Beller, T., Moffat, A., Petri, M.: From theory to practice: plug and play with succinct data structures. In: Gudmundsson, J., Katajainen, J. (eds.) SEA 2014. LNCS, vol. 8504, pp. 326–337. Springer, Cham (2014). https://doi.org/10.1007/978-3-319-07959-2_28
14. Harel, D., Tarjan, R.E.: Fast algorithms for finding nearest common ancestors. SIAM J. Comput. **13**(2), 338–355 (1984)
15. Ilie, L., Navarro, G., Tinta, L.: The longest common extension problem revisited and applications to approximate string searching. J. Discret. Algorithms **8**(4), 418–428 (2010)
16. Iliopoulos, C., Mchugh, J., Peterlongo, P., Pisanti, N., Rytter, W., Sagot, M.-F.: A first approach to finding common motifs with gaps. Int. J. Found. Comput. Sci. **16**(6), 1145–1155 (2005)
17. Kociumaka, T., Pissis, S.P., Radoszewski, J.: Pattern matching and consensus problems on weighted sequences and profiles. In: ISAAC 2016. LIPIcs, vol. 64, pp. 46:1–46:12. Schloss Dagstuhl-Leibniz-Zentrum fuer Informatik (2016)

18. Mäkinen, V., Belazzougui, D., Cunial, F., Tomescu, A.I.: Genome-Scale Algorithm Design: Biological Sequence Analysis in the Era of High-Throughput Sequencing. Cambridge University Press, Cambridge (2015)
19. Pinter, R.Y.: Efficient string matching with don't-care patterns. In: Apostolico, A., Galil, Z. (eds.) Combinatorial Algorithms on Words. NATO ASI Series, vol. F12, pp. 11–29. Springer, Heidelberg (1985). https://doi.org/10.1007/978-3-642-82456-2_2
20. Pissis, S.P.: MoTeX-II: structured motif extraction from large-scale datasets. BMC Bioinform. **15**, 235 (2014)
21. Régnier, M., Jacquet, P.: New results on the size of tries. IEEE Trans. Inf. Theory **35**(1), 203–205 (1989)

Computing Asymmetric Median Tree of Two Trees via Better Bipartite Matching Algorithm

Ramesh Rajaby[1,2] and Wing-Kin Sung[2,3(✉)]

[1] NUS Graduate School for Integrative Sciences and Engineering,
National University of Singapore, 28 Medical Drive, Singapore 117456, Singapore
e0011356@u.nus.edu
[2] School of Computing, National University of Singapore,
13 Computing Drive, Singapore 117417, Singapore
ksung@comp.nus.edu.sg
[3] Genome Institute of Singapore, 60 Biopolis Street, Genome,
Singapore 138672, Singapore

Abstract. Maximum bipartite matching is a fundamental problem in computer science with many applications. The HopcroftKarp algorithm can find a maximum bipartite matching of a bipartite graph G in $O(\sqrt{n}m)$ time where n and m are the number of nodes and edges, respectively, in the bipartite graph G. However, when G is dense (i.e., $m = O(n^2)$), the Hopcroft–Karp algorithm runs in $O(n^{2.5})$ time.

In this paper, we consider a special case where the bipartite graph G is formed as a union of ℓ complete bipartite graphs. In such case, even when G has $O(n^2)$ edges, we show that a maximum bipartite graph can be found in $O(\sqrt{n}(n + \ell) \log n)$ time.

We also describe how to apply our solution to compute the asymmetric median tree of two phylogenetic trees. We improve the running time from $O(n^{2.5})$ to $O(n^{1.5} \log^3 n)$.

A phylogenetic tree is a rooted, unordered, leaf-labeled tree in which every internal node has at least two children and all leaves have different labels, which represent the taxa. Using different data-sources, we may obtain different phylogenetic trees for the same set of taxa, and it is thus important to find a consensus of these different phylogenetic trees. This problem was first proposed by Adam [1]. After that, many different consensus tree definitions have been proposed. They include Adam's tree [1], strict consensus tree [12], loose consensus tree [2], majority-rule consensus tree [10], asymmetric median tree problem [11], greedy consensus tree problem [3,6], R* consensus tree problem [3], etc. Please refer to the surveys in [3], Chap. 30 in [5], and Chap. 8.4 in [13] for more details about different consensus trees and their advantages and disadvantages.

Here, we would like to focus on the asymmetric median tree problem [11], which is introduced by Phillips and Warnow. Informally, it aims to find a tree T that maximizes the number of common clusters with the input trees. (Section 1.1

© Springer International Publishing AG, part of Springer Nature 2018
L. Brankovic et al. (Eds.): IWOCA 2017, LNCS 10765, pp. 356–367, 2018.
https://doi.org/10.1007/978-3-319-78825-8_29

formally defines this problem.) The problem is NP-hard in general, but fortunately Phillips and Warnow have shown that an asymmetric median tree for two trees can be found in polynomial time. Basically, the asymmetric median tree problem for two trees can be transformed into a problem of maximum matching in a bipartite graph, where the bipartite graph has $O(n)$ nodes and $O(n^2)$ edges. Then, by the Hopcroft-Karp algorithm [7], the problem can be solved in $O(n^{2.5})$ time.

In this paper, we aim to improve the time complexity for computing the asymmetric median tree for two trees. Our basic technique is an improved algorithm for finding a maximum matching for a special type of bipartite graph. Although the transformed bipartite graph has $O(n^2)$ edges, we observe that it can be described using ℓ complete bipartite graphs where $\ell = O(n \log^2 n)$. (Section 1.2 gives a formal definition of this special type of bipartite graph.) We show that, for this special type of bipartite graph, we can compute the maximum matching using $O(\sqrt{n}(n+\ell) \log n)$ time. Hence, we show that the asymmetric median tree problem for two trees can be solved in $O(\sqrt{n}(n \log^2 n) \log n) = O(n^{1.5} \log^3 n)$ time.

The organization of this paper is as follows. First, we define the asymmetric median tree problem and the maximum bipartite matching problem. Second, we give the $O(\sqrt{n}(n+\ell) \log n)$-time algorithm for computing the maximum matching in a ℓ-interval-pair bipartite graph. Finally, we describe the $O(n^{1.5} \log^3 n)$-time algorithm.

1   Problem Definition

This section formally defines the asymmetric median tree problem and the maximum bipartite matching problem.

1.1   Asymmetric Median Tree Problem

For any phylogenetic tree T, the set of all internal nodes in T is denoted by $V(T)$ and the set of all leaf labels in T by $\Lambda(T)$. For any $u \in V(T)$, T^u denotes the subtree of T rooted at the node u. The leaf-label set $\Lambda(T^u)$ is called a cluster of T. We denote $C(T) = \{\Lambda(T^u) \mid u \in V(T)\}$ be the set of all clusters in T.

Given any two trees T and T' with the same leaf-label set. The similarity of T and T' is defined as $sim(T, T') = |C(T) \cap C(T')|$.

Now, we can define the asymmetric median tree problem. The input consists of k phylogenetic trees $T_1, T_2, \ldots, T_k$, where every tree T_i is leaf-labeled by the same set of taxa L (assume $|L| = n$). The problem aims to find a tree T that maximizes $\sum_{i=1}^{k} sim(T, T_i)$. For example, for the trees T_1, T_2 and T in Fig. 1, we have $sim(T, T_1) = 4$ and $sim(T, T_2) = 2$. In fact, T is the tree that maximizes $sim(T, T_1) + sim(T, T_2)$. Hence, T is an asymmetric median tree of T_1 and T_2.

Phillips and Warnow [11] showed that the asymmetric median tree problem is NP-hard when $k \geq 3$. When $k = 2$, they gave an $O(n^{2.5})$-time algorithm to compute an asymmetric median tree of T_1 and T_2. Below, we improve the running time of this algorithm.

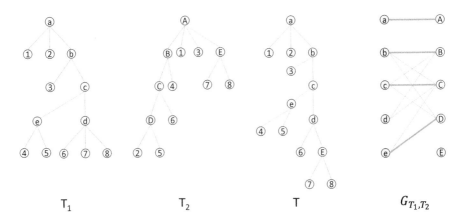

Fig. 1. T_1 and T_2 are two phylogenetic trees leaf-labeled by $\{1,2,3,4,5,6,7,8\}$. T is an asymmetric median tree for T_1 and T_2. G_{T_1,T_2} is the incompatible graph for T_1 and T_2. It is the union of three complete bipartite subgraphs: $\{b,c\} \times \{B,C,D\}$, $\{d\} \times \{B,C\}$ and $\{e\} \times \{C,D\}$. The maximum bipartite matching of G_{T_1,T_2} consists of three edges $\{(b,B),(c,D),(e,D)\}$. The maximum independent set of G_{T_1,T_2} is $\{a,b,c,d,e,A,E\}$.

1.2   Maximum Bipartite Matching Problem

Maximum bipartite matching problem is one of the fundamental problems in computer science. It finds applications in a number of problems including MAST [9], subtree isomorphism, etc.

Given a bipartite graph $G = (U,V,E)$, where $E \subseteq U \times V$, the maximum bipartite matching problem aims to find the maximum subset M of E such that the edges in M do not share endpoints.

This problem can be solved by the Hopcroft-Karp algorithm in $O(\sqrt{n}m)$ time, where $n = |U| + |V|$ and $m = |E|$.

However, when G is dense (i.e. $m = O(n^2)$), the Hopcroft-Karp algorithm runs in $O(n^{2.5})$ time, which is inefficient.

Here, we ask if G can be represented differently so that we can solve the bipartite matching more efficiently. In particular, we study the following representation. We fix the ordering of the nodes in U and V, say, $U = \{u_1, \ldots, u_{n_1}\}$ and $V = \{v_1, \ldots, v_{n_2}\}$ where $n = n_1 + n_2$. For any $1 \le i \le j \le n_1$, we denote $\{u_i, \ldots, u_j\}$ be an interval of U. Similarly, for any $1 \le i \le j \le n_2$, we denote $\{v_i, \ldots, v_j\}$ be an interval of V. For any interval I of U and any interval J of V, the complete bipartite graph $I \times J$ is called an interval pair. Suppose the edges in G can be partitioned into ℓ interval pairs $\{I_k \times J_k \mid k = 1, \ldots, \ell, I_k$ is an interval of U and J_k is an interval of $V\}$. Then, we call G a ℓ-interval-pair bipartite graph. (Note that, for any bipartite graph G with m edges, G is a ℓ-interval-pair bipartite graph with $\ell \le m$.)

For example, the bipartite graph G_{T_1,T_2} in Fig. 1 has 10 edges. Moreover, G_{T_1,T_2} can be represented as the union of 3 interval pairs: $\{\{b,c\} \times$

$\{B,C,D\},\{d\}\times\{B,C\},\{e\}\times\{C,D\}\}$. Hence, G_{T_1,T_2} is a 3-interval-pair bipartite graph.

Below, we show that if a bipartite graph G is a ℓ-interval-pair bipartite graph, the maximum bipartite matching of G can be computed in $O(\sqrt{n}(n+\ell)\log n)$ time.

2   Computing Maximum Matching in a ℓ-Interval-Pair Bipartite Graph

Consider two arrays of nodes $U[1..n_1]$ and $V[1..n_2]$ where $n = n_1 + n_2$. Consider a set of interval pairs $\mathcal{I} = \{(I_1, J_1), \ldots, (I_\ell, J_\ell)\}$. The ℓ-interval-pair bipartite graph $G_\mathcal{I}$ is a bipartite graph whose edge set is the union of $\{I_i \times J_i \mid i = 1, \ldots, \ell\}$.

We aim to compute the maximum bipartite matching of $G_\mathcal{I}$. Here, we describe an algorithm which runs in $O(\sqrt{n}(n+\ell)\log n)$ time.

We need some preliminary lemmas.

Lemma 1. *We can maintain a bit array $A[1..n]$ that supports the following 3 operations:*

- *Initialize the bit array such that $A[i] = 0$ for $1 \leq i \leq n$. This operation takes $O(n)$ time.*
- *$flip(A[i])$: This operation sets $A[i] = 1 - A[i]$. It takes $O(\log n)$ time.*
- *$zero(A, i..j)$: This operation reports $\{k \mid A[k] = 0, i \leq k \leq j\}$. It takes $O(\log n + occ)$ time where occ is the size of $\{k \mid A[k] = 0, i \leq k \leq j\}$.*

Proof. We just maintain the set $\{i \mid A[i] = 0\}$ using a balanced binary search tree.   □

Lemma 2. *We can maintain a set of intervals $\{I_1, \ldots, I_\ell\}$ supporting the following operations:*

- *(1) Insert/deletion an interval in $O(\log \ell)$ time.*
- *(2) Find all intervals overlap with a point q in $O(\log \ell + occ)$ time, where occ is the number of intervals overlapping with q.*

Proof. We use interval trees.   □

We have some definitions first. Given a matching M, an alternating path is a path starting from an unmatched node whose edges alternate between match and unmatch edges. An augmenting path is an alternating path that starts and ends at an unmatched node. Note that if we flip the match and unmatch edges in an augmenting path, the size of the matching is increased by 1.

We run a modified version of the Hopcroft–Karp algorithm. Initially, we set the matching M to be an empty matching. Then, the algorithm iterates two phases to improve the matching M. The two phases are as follows.

- **Phase 1**: Given the current matching M, let d be the length of the shortest augmenting path. Let $W^k = \{w \in U \cup V \mid$ the shortest alternating path starting from an unmatched node in U to w is of length $k\}$. This phase computes $W^0, W^1, \ldots, W^d$.
- **Phase 2**: From $W^0, W^1, \ldots, W^d$, Phase 2 finds a maximal set of vertex-disjoint shortest augmenting paths of length d and flips the match and unmatch edges, obtaining an improved matching M.

These two phases will be iterated until there is no augmenting path (i.e., we cannot further improve the matching M). Hopcroft and Karp showed that this algorithm iterates the two phases at most $O(\sqrt{n})$ times. Below, we show that both phases 1 and 2 take $O((\ell+n) \log n)$ time. In summary, we have the following theorem.

Theorem 1. *Given a ℓ-interval-pair bipartite graph G with n nodes. The maximum matching of G can be found in $O(\sqrt{n}(\ell + n) \log n)$ time.*

2.1   Phase 1

We first discuss the first phase. The input is a matching M. Let d be the length of the shortest augmenting path. This step aims to find W^k for $k = 1, 2, \ldots, d$.

Let $\overline{U}$ and $\overline{V}$ be all nodes in U and V, respectively, which are unmatched by M; more precisely, $\overline{U} = \{u \in U \mid (u, v) \notin M\}$ and $\overline{V} = \{v \in V \mid (u, v) \notin M\}$. By definition, $W^0 = \overline{U}$.

Lemma 3. $W^0, W^1, \ldots, W^d$ *satisfy the following properties:*

1. $W^k \subseteq U$ *if k is even while $W^k \subseteq V$ if k is odd.*
2. W^i *and W^j are pairwise disjoint for all $0 \leq i, j \leq d$.*

Proof. For (1), since the alternating path starts from unmatched node in U, we have $W^0 \subseteq U$. Then, since it is a bipartite graph, the next nodes are from V; hence, $W^1 \subseteq V$. Subsequently, the nodes are alternating between U and V.

For (2), by contrary, assume w appears in W^i and W^j where $i < j$. $w \in W^i$ implies the shortest alternating path from some node in $\overline{U}$ to w is of length i. Then, w cannot appear in W^j.   □

We create a bit array $b_V[1..n_2]$ and initialize $b_V[j] = 0$ for all j. We also create an interval tree $\mathcal{I}$ and insert $\{I_h \mid (I_h, J_h) \in \mathcal{L}\}$ into it. By definition, $W^0 = \overline{U}$; then, for odd $d = 1, 3, 5, \ldots$, the algorithm iteratively identifies W^d and W^{d+1} using three steps.

First, let $Q_U = \{I_h \in \mathcal{I} \mid I_h$ properly overlaps with $W^{d-1}\}$. All intervals in Q_U are removed from $\mathcal{I}$.

Second, W^d is set to be $\bigcup_{I_h \in Q_U} zero(b_V, J_h)$, which is the set of unmatched nodes in V which are reachable from W^{d-1} (recall that (I_h, J_h) is an interval pair, i.e., $I_h \times J_h$ is a complete bipartite graph). We set $b_V[j] = 1$ for all $j \in \bigcup_{I_h \in Q_U} zero(b_V, J_h)$. If W^d overlaps with $\overline{V}$, d is the length of the shortest augmenting path. Then, we report $(W^0, W^1, \ldots, W^d)$.

Third, in this step, $W^d \subseteq V$ and all nodes in W^d are matched. W^{d+1} is set to be $\bigcup_{j \in W^d} \{i \mid (U[i], V[j]) \in M\}$.

Figure 2 details the algorithm for Phase 1. The following lemma shows that Phase 1 runs in $O((\ell + n) \log n)$ time.

Lemma 4. *Phase 1 runs in $O((\ell + n) \log n)$ time.*

Proof. Refer to Fig. 2. Steps 1, 2 and 4 take $O(n)$ time. Step 3 takes $O(\ell \log \ell)$ time. Then, in the for loop, every interval will be identified and deleted at most once from $\mathcal{I}$. Hence, the time for processing the intervals in $\mathcal{I}$ is $O(\ell \log \ell)$. As $W^0, \ldots, W^d$ are all pairwise disjoint, the time for processing the nodes is $O(n \log n)$.

In total, the running time is $O(n \log n + \ell \log \ell)$. Since $\ell = O(n^2)$, the lemma follows. $\square$

Algorithm FindShortAugmentingPath$(M, U[1..n_1], V[1..n_2], \mathcal{L})$

Require: M is a matching of a bipartite graph subgraph of $U \times V$, which is the union of $\{I_h \times J_h \mid (I_h, J_h) \in \mathcal{L}\}$.

Ensure: $(V_0, V_1, \ldots, V_d)$ where d is the length of the shortest augmenting path

1: Let $\overline{U} = \{u \in U \mid (u, v) \notin M\}$ and $\overline{V} = \{v \in V \mid (u, v) \notin M\}$;
2: Initialize a bit array $b_V[1..n_2]$ and set $b_V[j] = 0$;
3: Build the interval tree $\mathcal{I}$ for $\{I_h \mid (I_h, J_h) \in \mathcal{L}\}$;
4: Set $W^0 = \overline{U}$;
5: **for** $d = 1$ to n step 2 **do**
6:     By the interval tree $\mathcal{I}$, set Q_U be the set of I_h in $\mathcal{I}$ such that I_h overlaps with W^{d-1};
7:     Remove Q_U from $\mathcal{I}$;
8:     Set $W^d = \bigcup_{I_h \in Q_U} zero(b_V, J_h)$;
9:     Perform $flip(b_V[j])$ for $j \in W^d$;
10:    **if** $W^d \cap \overline{V} \neq \emptyset$ **then**
11:        Set $W^d = W^d \cap \overline{V}$;
12:        Return $(W^0, W^1, \ldots, W^d)$;
13:    **end if**
14:    Set $W^{d+1} = \bigcup_{j \in W^d} \{i \mid (U[i], V[j]) \in M\}$;
15: **end for**
16: Return nil;

Fig. 2. FindShortAugmentingPath$(M, U, V, \mathcal{L})$ reports $V_0, \ldots, V_d$ where d is the length of the shortest augment path.

2.2   Phase 2

Let M be the current mathing in $G_{\mathcal{L}}$. Let d be the length of the shortest augmenting path (note that d is odd). Phase 1 identified $W^0, \ldots, W^d$. In Phase 2, we aim to identify a maximal set of shortest augmenting paths in $G_{\mathcal{L}}$. Let $\mathcal{J}$ be the interval tree for $\{J_h \mid (I_h, J_h) \in \mathcal{L}\}$. Let $b_U[1..n_1]$ be a bit vector fully initialized to zeroes.

Given $(W^0, W^1, \ldots, W^d)$, we can perform depth-first search to identify a maximal set of vertex disjoint paths from $(W^0, \ldots, W^d)$ where each path $(w_0, w_1, \ldots, w_d)$ is an augmenting path such that $w_p \in W^p$.

We first define the recursive formula $DFS\_Path()$. For any w in W^k, for any $k \leq d$, we define DFS_Path(w, k) be an alternating path $(w_0, w_1, \ldots, w_k)$ if exists, where $w_x \in W^x$ for $x = 0, 1, \ldots, k$ and $w = w_k$; otherwise, DFS_Path$(w, k) = nil$. The following lemma states a recursive formula for finding such an alternating path.

Lemma 5. *For any w in W^k, for any $k \leq d$, we have $DFS\_Path(w, k)$*

$$= \begin{cases} (w) & \text{if } k = 0 \\ (p, w) & \text{if } k \text{ is even, } (w, v) \in M \text{ and } p = DFS\_Path(v, k-1) \neq nil \\ (p', w) & \text{if } k \text{ is odd, } u \in I_h - (W^k \cup \ldots \cup W^d), \ w \in J_h, \ (I_h, J_h) \in \mathcal{L} \text{ and} \\ & p' = DFS\_Path(u, k-1) \neq nil \\ nil & \text{otherwise} \end{cases}$$

Proof. When $k = 0$, we report the alternating path of length 0 starting from $w \in W^0$. By definition, we report (w).

When k is even, the alternating path is $(w_0, \ldots, w_{k-1}, w_k = w)$ where (w_{k-1}, w) is a match edge in M. Suppose the matching edge in M ending at w is (v, w). Then, the alternating path is $(w_0, \ldots, w_{k-1} = v, w_k = w)$. Note that $(w_0, \ldots, w_{k-1} = v) = DFS\_Path(v, k-1)$.

When k is odd, the alternating path is $(w_0, \ldots, w_{k-1} = u, w_k = w)$ where (u, w) is an unmatch edge. This means that there exists $(I_h, J_h) \in \mathcal{L}$ such that $u \in I_h$ and $w \in J_h$. We also require u does not belong to $W^k \cup \ldots \cup W^d$ since we need to ensure the alternating path is the shortest path starting from w. Hence, we require $u \in I_h - (W^k \cup \ldots \cup W^d)$. This is accomplished by keeping track of the visited nodes in the bit array b_U. Furthermore, we also require the alternating path $DFS\_Path(v, k-1) = p' = (w_0, \ldots, w_{k-1} = u)$ exists, then (p', w) is an alternating path of length k ends at w. The lemma follows. $\qquad \square$

By the above lemma, we can select a node $v \in W^d$ and run the recursive algorithm $DFS\_Path(v, d)$ to find an augmenting path. During the execution of the recursive algorithm, we mark all nodes visited. Figure 3 details the recursive algorithm. To find another augmenting path, we find another unmarked node $v \in W^d$ and rerun the recursive algorithm $DFS\_Path(v, d)$. This process is repeated until we obtain a maximal set of augmenting paths.

The complexity analysis is similar to Lemma 4. Note that each interval in $\mathcal{J}$ is identified and removed once, and we make $O(n)$ queries to $\mathcal{J}$ since we visit each node at most once, which gives us $O((\ell + n) \log \ell)$ time in total. Furthermore, we make $O(\ell)$ queries to b_U. Therefore the running time of this step is $O((\ell + n) \log n)$ time.

Algorithm DFS(w, k)

Require: w is a node in W^k

Ensure: report an alternating path $(w_0, w_1, \ldots, w_k = w)$, where $w_j \in W^j$, that has
  not been reported before.

1: **if** $k = 0$ **then**
2:    return (w);
3: **else if** k is even **then**
4:    Let v be the node in W^{k-1} such that $(w, v) \in M$;
5:    $p = DFS(v, k - 1)$;
6:    If $p \neq nil$, then return (p, w);
7: **else**
8:    **for** every J_h in $\mathcal{J}$ that overlaps with w **do**
9:       **for** every $u \in zero(b_U, I_h)$ such that $u \in W^{k-1}$ **do**
10:         Mark $b_U[u] = 1$;
11:         $p = DFS(u, k - 1)$;
12:         If $p \neq nil$, then return (p, w);
13:      **end for**
14:      Remove J_h from $\mathcal{J}$;
15:   **end for**
16:   Return nil;
17: **end if**

Fig. 3. By DFS(u, d), we report a path $(w_0, w_1, \ldots, w_d)$ where $w_j \in W^j$ which are disjoint from all paths reported previously.

3 Compute the Asymmetric Median Tree of Two Trees by Maximum Bipartite Matching

Phillips and Warnow [11] proposed a method to compute the asymmetric median tree of two trees. Below, we describe their method.

We first need some definitions. For any clusters $C_1 \in C(T_1)$ and $C_2 \in C(T_2)$, C_1 and C_2 are compatible if either (1) $C_1 \subseteq C_2$, (2) $C_2 \subseteq C_1$ or (3) $C_1 \cap C_2 = \emptyset$; otherwise, C_1 and C_2 are not compatible. We denote G_{T_1, T_2} as a conflict graph if G_{T_1, T_2} is a bipartite subgraph of $C(T_1) \times C(T_2)$ and the edge set of G_{T_1, T_2} is $\{(C_1, C_2) \in C(T_1) \times C(T_2) \mid C_1$ and C_2 are not compatible$\}$.

Phillips and Warnow proved the following lemma.

Lemma 6. *Consider two trees T_1 and T_2 Let $\mathcal{C}$ be the maximum independent set of G_{T_1, T_2}. Let T be the asymmetric median tree of T_1 and T_2. We have $C(T) = \mathcal{C}$.*

As an illustration, consider the example in Fig. 1. T is the asymmetric median tree of T_1 and T_2. $C(T)$ is the clusters for $\{a, b, c, d, e, E\}$. The maximum independent set of G_{T_1, T_2} is $\{a, A, b, c, d, e, E\}$. Since a and A correspond to the same cluster, the maximum independent set of G_{T_1, T_2} equals $C(T)$.

By Lemma 6, the asymmetric median tree of T_1 and T_2 can be computed as follows. The method first constructs the conflict graph G_{T_1, T_2}. Then, it computes the maximum matching M of G_{T_1,T_2}. By Konig's theorem, we can obtain a maximum independent set $\mathcal{C}$ of G_{T_1,T_2} from M. For $i = 1, 2$, let T_i' be the tree containing the clusters $C(T_i) \cap \mathcal{C}$. Both T_1' and T_2' can be constructed in $O(n)$ time. By the merge operation of [8], we can merge T_1' and T_2' in $O(n)$ time. Figure 4 shows the pseudocode of the algorithm. The following lemma gives the running time of the algorithm.

Lemma 7. *Let T_1 and T_2 be two trees leaf-labeled by L, where $|L| = n$. Let $t_{construct}$ be the time required to construct G_{T_1,T_2}. Let $t_{matching}$ be the time required to compute the maximum matching of G_{T_1,T_2}. The running time of the algorithm FindAsymmetricMedianTree(T_1, T_2) is $O(n + t_{construct} + t_{matching})$.*

Proof. Apart from the steps for constructing G_{T_1,T_2} and finding maximum matching of G_{T_1,T_2}, all other steps run in $O(n)$ time. The lemma follows. $\qquad\square$

As G_{T_1,T_2} contains $O(n^2)$ edges, G_{T_1,T_2} can be constructed in $O(n^2)$ time. Using Hopcroft–Karp algorithm, the maximum matching of G_{T_1,T_2} can be computed in $O(\sqrt{n}n^2) = O(n^{2.5})$ time. Hence, Phillips and Warnow [11] proposed an $O(n^{2.5})$-time algorithm to compute the asymmetric median tree of two trees.

Algorithm FindAsymmetricMedianTree(T_1, T_2)

Require: Two phylogenetic trees T_1 and T_2 leaf-labeled by L.
Ensure: An asymmetric median tree T of T_1 and T_2
1: Build G_{T_1,T_2};
2: Compute the maximum matching M of G_{T_1,T_2}.
3: By Konig's theorem, we can obtain a maximum independent set $\mathcal{C}$ of G_{T_1,T_2} from M using linear time;
4: Let T_1' be the tree formed by remove all clusters not in $\mathcal{C}$ from T_1;
5: Let T_2' be the tree formed by remove all clusters not in $\mathcal{C}$ from T_2;
6: Set $T = merge(T_1', T_2')$;
7: Return T;

Fig. 4. The algorithm FindAsymmetricMedianTree computes the asymmetric median tree of T_1 and T_2.

Below, we show that G_{T_1,T_2} is an $O(n \log^2 n)$-interval-pair bipartite graph, that is, G_{T_1,T_2} is formed by the union of $O(n \log^2 n)$ interval pairs. Together with the maximum matching algorithm in Sect. 2, we have the following theorem.

Theorem 2. *An asymmetric median tree of T_1 and T_2 can be found in $O(n^{1.5} \log^3 n)$ time.*

Proof. By Lemma 10, G_{T_1,T_2} is an $O(n \log^2 n)$-interval-pair bipartite graph. Lemma 13 shows that these $O(n \log^2 n)$ interval pairs can be found using $O(n \log^3 n)$ time.

Then, by Theorem 1, the maximum matching of G_{T_1,T_2} can be found in $O(\sqrt{n}(n \log^2 n) \log n) = O(n^{1.5} \log^3 n)$ time. The theorem follows. $\qquad\square$

3.1 G_{T_1,T_2} is the Union of $O(n \log^2 n)$ Interval Pairs

This section shows that the edges in G_{T_1,T_2} can be partitioned into a few interval pairs.

First, we need to decompose the nodes in $V(T_i)$, for $i = 1, 2$, into a set of paths. We use core path decomposition [4].

For every node $w \in V(T_i)$, denote w' to be its core child if w' is a child of u and $|\Lambda(T_i^{w'})|$ has the biggest size among all the children of w. Call (w, w') the core edge. Denote $P(T_i)$ to be the set of core paths formed by core edges. Note that $P(T_i)$ forms a partition of all nodes in T_i. For each core path $P_i \in P(T_i)$, denote $r(P_i)$ be the most ancestral node of P_i. For example, in Fig. 1, T_1 has two core paths: (a, b, c, d) and (e). T_2 has two core paths: (A, B, C, D) and (E).

For every core path $P_1 \in P(T_1)$ and $P_2 \in P(T_2)$, we denote G_{P_1,P_2} to be $\{(T_1^u, T_2^v) \in G_{T_1,T_2} \mid u \in P_1, v \in P_2\}$. In other words, G_{P_1,P_2} is a bipartite graph containing edges (T_1^u, T_2^v) denoting incompatibility between T_1^u and T_2^v, where $u \in P_1$ and $v \in P_2$. Hence, G_{T_1,T_2} is the bipartite graph formed by the union of edges of G_{P_1,P_2} among all core paths $P_1 \in P(T_1)$ and $P_2 \in P(T_2)$.

The following lemma gives the upper bound of the number of edges in G_{P_1,P_2}.

Lemma 8. *Consider any two paths $P_1 = (u_1, u_2, \ldots, u_p)$ in T_1 and $P_2 = (v_1, v_2, \ldots, v_q)$ in T_2, where u_1 and v_1 are the most ancestral internal nodes. G_{P_1,P_2} is a ℓ-interval-pair bipartite graph where $\ell = O(|\Lambda(T_1^{u_1}) \cap \Lambda(T_2^{v_1})|)$.*

Proof. (Sketch) For every $u \in P_1$, we define $\alpha(u)$ to be the largest index such that $\Lambda(T_1^u) \subseteq \Lambda(T_2^{v_{\alpha(u)}})$, while $\alpha(u) = 0$ if no such node exists in P. We define $\beta(u)$ to be the smallest index such that either $\Lambda(T_2^{v_{\beta(u)}}) \subseteq \Lambda(T_1^u)$ or $\Lambda(T_1^u) \cap \Lambda(T_2^{v_{\beta(u)}}) = \emptyset$, while $\beta(u) = q + 1$ if no such node exists in P. By definition, we have $\alpha(u) \leq \beta(u)$, and $\alpha(u) = \beta(u)$ if and only if $\Lambda(T_1^u) = \Lambda(T_2^{v_{\alpha(u)}})$.

For every u_i in P_1 such that $\alpha(u_i) < \beta(u_i)$, we can show that $G_{(u_i),P_2}$ is a 1-interval-pair bipartite graph, and the interval pair is $\{u_i\} \times \{v_{\alpha(u_i)+1}, \ldots, v_{\beta(u_i)-1}\}$.

Let $i_1, \ldots, i_k$ be indexes such that $(\Gamma(T_1^{u_{i_r}}) - \Gamma(T_1^{u_{i_r+1}})) \cap \Gamma(T_2^{v_1}) \neq \emptyset$. Note that $k \leq |\Gamma(T_1^{u_1}) \cap \Gamma(T_2^{v_1})|$.

Let $I_0 = [1..i_1-1]$, $I_k = [i_k+1..p]$ and $I_r = [i_r+1..i_{r+1}-1]$ for $r = 1, \ldots, k-1$. For every I_r, we can show that $\alpha(u_i) = \alpha(u_{i'})$ and $\beta(u_i) = \beta(u_{i'})$ for every $i, i' \in I_r$.

The intervals in $\{I_r \mid r = 0, \ldots, k\} \cup \{\{i_r\} \mid r = 1, \ldots, k\}$ partition $[1..p]$ into $2k + 1$ intervals. Since $k \leq |\Gamma(T_1^{u_1}) \cap \Gamma(T_2^{v_1})|$. The lemma follows. □

Lemma 9. $\sum_{P_1 \in P(T_1), P_2 \in P(T_2)} |\Lambda(T_1^{r(P_1)}) \cap \Lambda(T_2^{r(P_2)})| = O(n \log^2 n)$.

Proof. By the property of the core path decomposition, a leaf label i can only appear in $O(\log n)$ core paths in a tree T. Hence each leaf label will appear in the intersection of at most $O(\log^2 n)$ pairs of core paths. Since there are n leaf labels, the lemma follows. □

Lemma 10. G_{T_1,T_2} is the union of $O(n \log^2 n)$ interval pairs.

Proof. By Lemma 8, G_{P_1,P_2} is a $|\Lambda(T_1^{r(P_1)}) \cap \Lambda(T_2^{r(P_2)})|$-interval pair bipartite graph. Recall that G_{T_1,T_2} equals $\bigcup_{P_1 \in P(T_1), P_2 \in P(T_2)} G_{P_1,P_2}$. Hence, the total number of interval pairs in G_{T_1,T_2} equals $\sum_{P_1 \in P(T_1), P_2 \in P(T_2)} |\Lambda(T_1^{r(P_1)}) \cap \Lambda(T_2^{r(P_2)})|$. By Lemma 9, the lemma follows.   □

Algorithm BuildGraph(T_1, T_2)
Require: T_1 and T_2 are two trees leaf-labeled by L
Ensure: A set $\mathcal{L}$ of ℓ interval pairs such that $G_{T_1,T_2} = G_{\mathcal{L}}$
1: Let $U[1..n_1]$ be the concatenation of the core paths of T_1 where n_1 is the number of internal nodes in T_1;
2: Let $V[1..n_2]$ be the concatenation of the core paths of T_2 where n_1 is the number of internal nodes in T_2;
3: $\mathcal{L} = \emptyset$;
4: Using Lemma 11, find the set $\mathcal{E} = \{(P_1, P_2) \mid P_1 \in P(T_1), P_2 \in P(T_2), \Lambda(T_1^{r(P_1)}) \cap \Lambda(T_2^{r(P_2)}) \neq \emptyset\}$. Furthermore, we also store the set $\Lambda(T_1^{r(P_1)}) \cap \Lambda(T_2^{r(P_2)})$ for every $(P_1, P_2) \in \mathcal{E}$;
5: **for** every core path pairs $(P_1, P_2) \in \mathcal{E}$ **do**
6:    Compute the set S of interval pairs for G_{P_1,P_2} using Lemma 12;
7:    Set $\mathcal{L} = \mathcal{L} \cup S$;
8: **end for**
9: Return $\mathcal{L}$;

Fig. 5. BuildGraph(T_1, T_2) constructs the bipartite graph G_{T_1,T_2}, which is represented as a set of interval pairs $\mathcal{L}$.

Denote the set $\mathcal{E} = \{(P_1, P_2) \mid P_1 \in P(T_1), P_2 \in P(T_2), \Lambda(T_1^{r(P_1)}) \cap \Lambda(T_2^{r(P_2)}) \neq \emptyset\}$.

Lemma 11. *$\mathcal{E}$ can be constructed using $O(n \log^2 n)$ time. Furthermore, we also store the set $\Lambda(T_1^{r(P_1)}) \cap \Lambda(T_2^{r(P_2)})$ for every $(P_1, P_2) \in \mathcal{E}$.*

Proof. For $i = 1, 2$, for each leaf $x \in L$, observe that the path between x and the root crosses at most $\log n$ core paths in T_i. Denoted $\mathcal{D}_i(x) = \{P \in P(T_1) \mid x \in \Lambda(T_i^{r(P)})\}$ be such a set of core paths. Using $O(n \log n)$ time, we can compute $\mathcal{D}_1(x)$ and $\mathcal{D}_2(x)$ for all $x \in L$.

Then, we obtain a set of $O(n \log^2 n)$ tuples, which is $\bigcup_{x \in L} \mathcal{D}_1(x) \times \mathcal{D}_2(x) \times \{x\}$. By stable sorting, we can obtain the set $\mathcal{E} = \{(P_1, P_2) \mid P_1 \in P(T_1), P_2 \in P(T_2), \Lambda(T_1^{r(P_1)}) \cap \Lambda(T_2^{r(P_2)}) \neq \emptyset\}$. Furthermore, for each $(P_1, P_2) \in \mathcal{E}$, we can compute the set $\Lambda(T_1^{r(P_1)}) \cap \Lambda(T_2^{r(P_2)})$ for every $(P_1, P_2) \in \mathcal{E}$. The total time complexity is $O(n \log^2 n)$.   □

Lemma 12. *For any two paths $P_1 = (u_1, u_2, \ldots, u_p)$ in T_1 and $P_2 = (v_1, v_2, \ldots, v_q)$ in T_2, given the set $\Lambda(T_1^{r(P_1)}) \cap \Lambda(T_2^{r(P_2)})$, we can construct G_{P_1,P_2} using $O(|\Lambda(T_1^{r(P_1)}) \cap \Lambda(T_2^{r(P_2)})|)$ time.*

Proof. To be shown in the journal version. □

Lemma 13. *The algorithm in Fig. 5 can build the conflict graph G_{T_1,T_2} using $O(n \log^3 n)$ time.*

Proof. Lemma 11 constructs the set $\mathcal{E}$ using $O(n \log^2 n)$ time. For every heavy paths $P_1 \in P(T_1)$ and $P_2 \in P(T_2)$, Lemma 12 constructs the interval pairs for G_{P_1,P_2} using $O(|\Lambda(T_1^{r(P_1)}) \cap \Lambda(T_2^{r(P_2)})| \log n)$ time. The total time for constructing G_{T_1,T_2} equals

$$\sum_{P_1 \in P(T_1), P_2 \in P(T_2)} |\Lambda(T_1^{r(P_1)}) \cap \Lambda(T_2^{r(P_2)})| \log n,$$

which equals $O(n \log^3 n)$ time. The lemma follows. □

References

1. Adams III, E.N.: Consensus techniques and the comparison of taxonomic trees. Syst. Biol. **21**(4), 390–397 (1972)
2. Bremer, K.: Combinable component consensus. Cladistics **6**(4), 369–372 (1990)
3. Bryant, D.: A classification of consensus methods for phylogenetics. In: Janowitz, M.F., Lapointe, F.-J., McMorris, F.R., Mirkin, B., Roberts, F.S. (eds.) Bioconsensus. DIMACS Series in Discrete Mathematics and Theoretical Computer Science, vol. 61, pp. 163–184. American Mathematical Society (2003)
4. Cole, R., Farach-Colton, M., Hariharan, R., Przytycka, T., Thorup, M.: An o(nlog n) algorithm for the maximum agreement subtree problem for binary trees. SIAM J. Comput. **30**(5), 1385–1404 (2000)
5. Felsenstein, J.: Inferring Phylogenies. Sinauer Associates Inc., Sunderland (2004)
6. Felsenstein, J.: PHYLIP, version 3.6. Software package, Department of Genome Sciences, University of Washington, Seattle, U.S.A. (2005)
7. Hopcroft, J.E., Karp, R.M.: An $n^{5/2}$ algorithm for maximum matchings in bipartite graphs. SIAM J. Comput. **2**(4), 225–231 (1973)
8. Jansson, J., Shen, C., Sung, W.-K.: Improved algorithms for constructing consensus trees. J. ACM **63**(3), 1–24 (2016)
9. Kao, M.-Y., Lam, T.W., Sung, W.-K., Ting, H.-F.: Cavity matchings, label compressions, and unrooted evolutionary trees. SIAM J. Comput. **30**(2), 602–624 (2000)
10. Margush, T., McMorris, F.R.: Consensus n-trees. Bull. Math. Biol. **43**(2), 239–244 (1981)
11. Phillips, C., Warnow, T.J.: The asymmetric median tree—a new model for building consensus trees. Discrete Appl. Math. **71**(1–3), 311–335 (1996)
12. Sokal, R.R., Rohlf, F.J.: Taxonomic congruence in the Leptopodomorpha re-examined. Syst. Zool. **30**(3), 309–325 (1981)
13. Sung, W.-K.: Algorithms in Bioinformatics: A Practical Introduction. Chapman & Hall/CRC, Boca Raton (2010)

Privacy

Privacy-Preserving and Co-utile
Distributed Social Credit

Josep Domingo-Ferrer$^{(\boxtimes)}$ (iD)

UNESCO Chair in Data Privacy, Department of Computer Science and Mathematics,
Universitat Rovira i Virgili, Av. Països Catalans 26, Tarragona 43007, Catalonia
josep.domingo@urv.cat

Abstract. Reputation is a powerful incentive for agents to abide by the
prescribed rules of an interaction. In computer science, reputation can be
phrased as being an artificial incentive that can turn into self-enforcing
protocols that would not be such otherwise. Quite recently, China has
announced a national reputation system that will be launched in the
future under the name of social credit system. However, to be gener-
alizable without damaging the privacy of citizens/agents, a reputation
system must be decentralized and privacy-preserving. We present a peer-
to-peer fully distributed reputation protocol in which the anonymity of
both the scoring and the scored agents is maintained. At the same time,
the reputation protocol itself is co-utile, that is, the rational option for
all agents is to honestly fulfill their part in the protocol.

Keywords: Protocols · Reputation · P2P · Self-enforcement
Co-utility · Privacy

1 Introduction

Ensuring that agents will honestly follow their roles in an interaction has always
been a thorny issue. In the presence of a common legal framework, the agents
may have an incentive not to deviate from the legally established procedures.
However, in an open environment such as the information society, common legal
frameworks are often lacking. In this case, the prescribed rules of interaction,
called a *protocol* in the computer science jargon, must be designed to that they
deter deviations and are thus *self-enforcing*.

Reputation is a powerful incentive for agents to adhere to the prescribed interaction rules, both in the information society and in society at large. The Chinese
government has realized this and they have recently announced a national reputation system called Social Credit System [1]. However, unless properly implemented, such a system has the potential of becoming a mass surveillance system.
Technical details of the Chinese reputation system are not yet known, although
it may be partly inspired on Alibaba's Sesame Credit [12]. Anyway, since no
standard social credit system exists yet, it makes a lot of sense to investigate

© Springer International Publishing AG, part of Springer Nature 2018
L. Brankovic et al. (Eds.): IWOCA 2017, LNCS 10765, pp. 371–382, 2018.
https://doi.org/10.1007/978-3-319-78825-8_30

what properties a generalized reputation system should satisfy in order to be socially acceptable.

Privacy preservation and decentralization, or even better distributedness, stand out as obvious desiderata, and they are intertwined:

- Privacy should be preserved as much as possible for those agents whose reputation is computed: the legitimate interest of society is to know how well a certain agent has performed in the past, but this does not require knowing the detail of all the transactions the agent has been involved in. Whereas in a centralized reputation system, there is no way around the central authority learning the behavior in all transactions, a distributed system may be more privacy-preserving. After all, reputation is an aggregate metric of behavior, not a history of the behavior of an agent in each and every past transaction.
- Privacy should also be preserved for those agents participating in the computation of the reputation of other agents. Whereas in a centralized system it does not make sense to request privacy for the central authority computing reputations, in a decentralized system reputation may be computed by peers. Unless the privacy of those peers is protected, they might be subject to bribery, extortion or retaliation aimed at altering the reputations they compute.

1.1 Contribution and Plan of This Paper

In this paper, we describe a peer-to-peer reputation system that preserves privacy both for the agents involved in computing the reputations and for the agents whose reputation is computed. This makes it a good candidate to implement a generalized social credit system.

Section 2 recalls the basics of co-utility, which is a property related to a protocol being self-enforcing and beneficial for the participants (a distributed reputation protocol must be co-utile for peers to rationally collaborate in the protocol). Section 3 states the requirements of a co-utile privacy-preserving reputation system. In Sect. 4 we recall the EigenTrust distributed reputation system, which meets many of our requirements. In Sect. 5, we describe a co-utile and weakly anonymous extension of EigenTrust. Then, in Sect. 6, we present a novel further extension, that also offers enhanced privacy to agents, both when their reputation is computed and when they help computing the reputations of other agents. Conclusions and future work directions are gathered in Sect. 7.

2 Co-utility

When setting forth for decentralized systems, peer-to-peer (P2P) architectures are the most appealing ones, because they empower individual agents. The challenge in P2P is how to ensure that peers will collaborate as expected. *Co-utility* [3,4] is a specially attractive form of self-enforcing collaboration between agents. Co-utile protocols are those in which helping other (rational) agents increase their utilities is also the best way to increase one's own utility.

We can formalize the definition of co-utility in game-theoretic terms. A *game* is an abstraction of a scenario in which a set of rational agents can take decisions [7,10]. We will focus on sequential perfect-information games, consisting of several rounds so that the agent choosing an action in a certain round knows the actions chosen in previous rounds (by others and herself). A game of this class can be represented in the so-called *extensive form*, that is a tree in which: (i) nodes are the points where decisions are made, (ii) each node is labeled with the name of the agent making the decision at that node, (iii) edges going out from a node represent the available actions that can be chosen at the node, and (iv) each of the terminal nodes (leaves) of the tree is labeled with the tuple of payoffs that agents obtain when the node is reached. Now we can give game-theoretic definitions of *protocol, self-enforcing protocol* and *co-utility*.

Definition 1 (Protocol). *Given a perfect-information game G in extensive form represented as a tree, a protocol is either a path from the root to a leaf or a subtree from the root to several leaves. In the latter case, alternative edges are labeled with probabilities of being chosen.*

Definition 2 (Self-enforcing protocol). *A protocol P on a game G is self-enforcing if no agent can increase her utility by deviating from P, provided that the other agents stick to P. Equivalently, at each successive node of the protocol path, sticking to the next action prescribed by the protocol (taking the next edge in the path) is an equilibrium of the remaining subgame of G (the subtree rooted at the current node). More technically, P on G is self-enforcing if and only if P is a subgame perfect equilibrium of G.*

Definition 3 (Co-utility). *A self-enforcing protocol P on a game G is co-utile if it is Pareto optimal and the utility derived by each participating agent is strictly greater than the utility the agent derives from not participating.*

A co-utile protocol P is Pareto-optimal in the sense that there is no alternative protocol P' giving greater utilities to all agents and strictly greater utility to at least one agent. See [3] for more details on co-utility.

3 Requirements of a Co-utile and Distributed Reputation Protocol

Many reputation mechanisms have been proposed in the literature for peer-to-peer communities, either centralized or distributed (see the surveys [5,9]). Reputation is a very versatile artificial utility that can be used both as a reward and as a penalty. In fact, the reputation incentive can be added to a non-co-utile protocol to render it co-utile: it is a matter of rewarding with reputation increases those actions within the protocol and penalizing any deviation with reputation decreases. Yet, *the reputation calculation and management protocol itself should be naturally co-utile*, without requiring any additional incentives.

Otherwise, computing reputations would not be rationally sustainable and would not serve its purpose of inducing co-utility or self-enforcement in other P2P protocols.

Specifically, we want a reputation protocol with the following features, which should make it amenable to co-utility, privacy-preserving and hence socially acceptable as a generalized reputation protocol:

- *Decentralization.* Reputations should be computed and enforced by the peers themselves rather than by a central authority. A central authority knows everything on everyone (no privacy) and is a single point of failure.
- *Privacy protection for those agents computing reputations and those whose reputation is computed.* See justification in Sect. 1 above. Surprisingly, most reputation schemes in the literature provide very little privacy or no privacy at all [11].
- *Low overhead.* The computational cost (bandwidth, storage, calculation) of computing reputations should be low (linear or quasi-linear). Otherwise, the negative utility of those costs might dominate the benefits brought by reputation.
- *Proper management of new agents.* Newcomers should not enjoy any reputation advantage. Otherwise, malicious peers may be motivated to create new anonymous identifiers after abusing the system in order to regain the advantages of a good reputation.
- *Attack tolerance.* A number of attacks may be orchestrated in order to subvert the reputation system. Since we assume rational players, we must make the cost of such attacks unattractively high. Following the classification in [5], we must avoid the following attacks: *self-promotion* (agents falsely increasing their reputations), *whitewashing* (creating new clean identities to get rid of bad reputations), *slandering* (falsely lowering other agents' reputations in order for one's own reputation to become comparatively higher) and *denial of service* (agents blocking the calculation and dissemination of reputation values).

After examining a number of decentralized reputation mechanisms [5], we selected EigenTrust [6] as a starting point to obtain a co-utile distributed reputation protocol. EigenTrust offers most of the desirable features identified above: distributed reputation calculation, low overhead and robustness to attacks. This reputation scheme is designed to filter out inauthentic content in peer-to-peer file sharing networks. Its basic idea is to calculate a global reputation for each agent based on aggregating the local opinions of the peers that have interacted with the agent. If we represent the local opinions by a matrix whose component (i, j) contains the opinion of agent i on agent j, the distributed calculation mechanism computes global reputation values that approximate the left principal eigenvector of this matrix. For unstructured networks, other solutions can be used, such as gossip-based reputation distribution protocols [13].

4 The EigenTrust Distributed Reputation System

In EigenTrust, a local reputation is assigned by an agent to another agent with whom the former has directly interacted, as a function of her opinion on the latter's behavior in the interaction. Then, the global reputation of an agent is computed in a distributed way, by means of a protocol whereby the agents share their local reputation values. The original EigenTrust system was designed for P2P file sharing, and the file receiver's opinion on a file download is just binary: satisfactory or not satisfactory. Since we want to be able to compute reputations based on opinions that have several categories or are even continuous, in [2] we extended EigenTrust to compute reputations based on non-binary opinions.

4.1 Computing Local Reputations

The opinion of an agent $\mathcal{P}_i$ on another agent $\mathcal{P}_j$ with whom $\mathcal{P}_i$ has directly interacted is the reputation s_{ij} of $\mathcal{P}_j$ local to $\mathcal{P}_i$. We define this value as the aggregation of payoffs (either positive or negative) that $\mathcal{P}_i$ has obtained from the set of transactions (Y_{ij}) performed with $\mathcal{P}_j$:

$$s_{ij} = \sum_{y_{ij} \in Y_{ij}} payoff_i(y_{ij}).$$

Payoffs may be binary (positive/negative opinion), discrete (an opinion from a discrete scale) or continuous (for example, the cost incurred in the transaction, in terms of bandwidth, time, etc.).

4.2 Computing Global Reputations

We review how EigenTrust computes global reputations from local reputations. First, in order to properly aggregate the local reputation values computed by peers, a normalized version c_{ij} of every local reputation s_{ij} assigned by peer $\mathcal{P}_i$ to any other peer $\mathcal{P}_j$ is computed as:

$$c_{ij} = \frac{\max(s_{ij}, 0)}{\sum_j \max(s_{ij}, 0)}.$$

The normalized local reputations lie between 0 and 1 and the sum of all normalized local reputations awarded by $\mathcal{P}_i$ to other peers is 1. In other words, each agent has a reputation budget of only 1 that she has to split among her peers proportionally to her positive experiences. In this way, all agents have the same influence on the global reputation. In particular, there is no dominance by the agents with more experiences and peers cannot collude by assigning arbitrarily high values to good peers. Regarding the truncation of negative values to 0, on the one hand it prevents $\mathcal{P}_i$ from assigning arbitrarily low values to other peers (thus neutralizing slandering), and on the other hand it sets newcomers on an equal footing with peers with whom $\mathcal{P}_i$ had a negative experience (thus neutralizing whitewashing).

The next step is to disseminate normalized local reputations and have them aggregated through the network peers by leveraging transitive reputation. Any agent $\mathcal{P}_i$ can compute $\hat{t}_{ik}^{(0)}$, an approximation of the reputation of a potentially unknown peer $\mathcal{P}_k$, by asking the peers with whom $\mathcal{P}_i$ has interacted ($\mathcal{P}_j$) for their local reputation w.r.t. $\mathcal{P}_k$, that is c_{jk}. Since $\mathcal{P}_i$ has already computed the local normalized reputation w.r.t. $\mathcal{P}_j$, that is c_{ij}, $\mathcal{P}_i$ can compute a local estimate of the reputation t_{ik} of $\mathcal{P}_k$ by using c_{ij} to weight $\mathcal{P}_j$'s local reputation; specifically, $\hat{t}_{ik}^{(0)} = \sum_j c_{ij} c_{jk}$. Thanks to the local normalization, $\hat{t}_{ik}^{(0)}$ takes values between 0 and 1. Observe that if we call $\mathbf{c}_i = (c_{i1}, \ldots, c_{in})^T$ and $\mathbf{C} = [c_{ij}]$, then $\hat{\mathbf{t}}_i^{(0)} = \mathbf{C}^T \mathbf{c}_i$, where $\hat{\mathbf{t}}_i^{(0)} = (\hat{t}_{i1}^{(0)}, \ldots, \hat{t}_{in}^{(0)})^T$. If every agent $\mathcal{P}_i$ computes $\hat{\mathbf{t}}_i^{(0)}$, in the next iteration $\mathcal{P}_i$ can compute $\hat{\mathbf{t}}_i^{(1)} = \mathbf{C}^T(\mathbf{C}^T \mathbf{c}_i)$. After m iterations, $\mathcal{P}_i$ will compute $\hat{\mathbf{t}}_i^{(m-1)} = (\mathbf{C}^T)^m \mathbf{c}_i$. Under the assumptions that $\mathbf{C}$ is irreducible and aperiodic [6], in an ideal and static setting the succession of reputation vectors computed by any peer will converge to the same vector for every peer, which we call $\mathbf{t} = (t_1, \ldots, t_n)^T$ and is the left principal eigenvector of $\mathbf{C}$. The j-th component of $\mathbf{t}$ represents the global reputation of the system on agent $\mathcal{P}_j$, for $j = 1, \ldots, n$.

Unfortunately, computing the global reputations by the above method is not efficient, because it takes too much communication. This and other issues are fixed in the proposals discussed in the next sections.

5 Co-utile and Weakly Anonymous Computation of Global Reputations

In the same EigenTrust paper [6], the authors gave a secure version of the protocol to compute global reputations, in which not every peer contributed to computing the global reputation on every other peer. In [2], we extended that version to make it co-utile and ensure some level of anonymity, and we summarize this extension here.

In our extension, each agent $\mathcal{P}_i$ has an initial global reputation $t_i^{(0)}$ (based on previous experiences or assigned by default) and M score managers that will update her reputation value. Given a pseudonym ID_i of agent $\mathcal{P}_i$ (the pseudonym is $\mathcal{P}_i$'s identifier in the P2P network), her score managers are defined by a distributed hash table (DHT), which maps ID_i to M score managers whose pseudonyms are closest (according to an agreed distance), respectively, to values $h_0(ID_i), \ldots, h_{M-1}(ID_i)$, where $h_0, h_1, \ldots, h_{M-1}$ are hash functions. The use of pseudonyms guarantees some level of anonymity (a weak level, as argued in the next section) and the use of hash functions prevents anyone from choosing a particular pseudonym as her score manager.

With the above arrangement, on average every agent is the score manager for M agents, so the work is balanced. Let D_i be the set of daughters of $\mathcal{P}_i$, that is, the set of agents for whom $\mathcal{P}_i$ is a score manager. During the computation of the global reputation, each $\mathcal{P}_i$ learns, for each $\mathcal{P}_d \in D_i$, the set A_d of agents that directly interacted with $\mathcal{P}_d$ (to provide or receive help) and receives the normalized local reputations c_{jd} on $\mathcal{P}_d$ from each $\mathcal{P}_j \in A_d$. The terms c_{ji} for $j \notin A_i$ are zero. Then $\mathcal{P}_i$ engages in an iterative refinement, for $k = 0, 1, 2, \ldots$:

$$t_d^{(k+1)} = c_{1d}t_1^{(k)} + c_{2d}t_2^{(k)} + \ldots + c_{nd}t_n^{(k)}. \tag{1}$$

Note that, in Expression (1), the weight attached to each normalized local reputation c_{jd} on $\mathcal{P}_d$ received from $\mathcal{P}_j$ is the current global reputation $t_j^{(k)}$ of $\mathcal{P}_j$. The refinement ends when the global reputation for $\mathcal{P}_d$ changes less than a small value $\epsilon > 0$ from k to $k + 1$. The resulting global reputations t_d are kept for a period until they are recomputed. The length of the reputation update period is a parameter of the system.

With this arrangement, not only the entire computation is mediated by the score managers, but also the dissemination of global reputations. For any $\mathcal{P}_i$, her global reputation t_i can be obtained from her score managers.

Briefly speaking, *this protocol to compute global reputations is co-utile because it encourages the agents to collaborate*. The reason is that the impact of their opinions on the computation of the global reputations increases when they are active (that is, when they are members of as many sets A_* as possible). See [2] for a detailed co-utility analysis.

6 Co-utile and Privacy-Preserving Computation of Global Reputations

The protocol in Sect. 5 provides only weak anonymity, because it uses a single pseudonym for each agent. Indeed, if all the interactions of an agent $\mathcal{P}_i$ are carried out under the same pseudonym, in the end the agent can be identified from its graph of interactions alone. This is well known in graph anonymization, where it is clear that just replacing the node labels with pseudonyms is not a sufficient anonymization (see, *e.g.* [8]).

A possible way to improve the privacy of agents is to split the identity of each agent into as many pseudonyms as the agent wishes. We consider two cases: multiple pseudonyms with independent reputations and multiple linkable pseudonyms.

6.1 Multiple Pseudonyms with Independent Reputations

This case makes sense if it is socially tolerable that an agent leads several parallel lives each with its own independent reputation. Splitting into parallel lives is not necessarily unfair because:

- If an agent behaves very well under a pseudonym and earns a high reputation with that pseudonym, she will only enjoy the benefits of high reputation under that pseudonym.
- On the other hand, if the agent behaves very poorly under another pseudonym, she will suffer the effects of a bad reputation only under that pseudonym.

In practice, this scenario amounts to splitting an agent into several surrogate agents. So instead of agent $\mathcal{P}_i$ having a single pseudonym ID_i, she will be allowed to have several pseudonyms $ID_i^1, ID_i^2, \ldots$, etc. Otherwise, the scheme will work exactly as described in Sect. 5.

6.2   Multiple Linkable Pseudonyms

In this case, the agent can lead several independent lives under different pseudonyms. However, at any point she can choose to link some of her pseudonyms and thereby merge their corresponding reputations. This makes sense *if the motivation of the agent when using several pseudonyms is to improve her privacy, but she would like to enjoy the same (or a similar) reputation no matter which among her pseudonyms she is using.* To allow this, the following protocols can be followed.

We assume $\mathcal{P}_i$ wants to create k pseudonyms that are linkable to each other at a later time. Protocol 1 creates multiple linkable pseudonyms.

Protocol 1 (Linkable pseudonym creation)

1. *$\mathcal{P}_i$ chooses strings ID_i^1, ID_i^2, ..., ID_i^k.*
2. *$\mathcal{P}_i$ computes*

$$H(ID_i^1 || H_i^1, K_i^1) = R_i^1$$
$$H(ID_i^2 || H_i^2, K_i^2) = R_i^2 \qquad (2)$$
$$\vdots \quad \vdots \quad \vdots$$
$$H(ID_i^k || H_i^k, K_i^k) = R_i^k$$

where $||$ is the concatenation operator, $H(\cdot, \cdot)$ is a secure keyed hash function,

$$H_i^j = H'(ID_i^1 || \ldots || ID_i^{j-1} || ID_i^{j+1} || \ldots || ID_i^k), \quad j = 1, \ldots, k$$

with $H'(\cdot)$ being a secure hash function, and K_i^1, ..., K_i^k are random keys. The values H_i^j and the keys K_i^j are known only to $\mathcal{P}_i$ for $j = 1, \ldots, k$.
3. *$\mathcal{P}_i$ can now operate under $ID_i^1 || R_i^1$, $ID_i^2 || R_i^2$, ..., $ID_i^k || R_i^k$ as her pseudonyms.*

Linkable pseudonyms are longer than normal pseudonyms, because they are appended the hash image as a suffix. Although their random-looking suffix makes them distinguishable from normal pseudonyms, this should not be a problem as long as a sufficient number of agents is using linkable pseudonyms. However, if a single agent (or very few agents) used them, the linkable pseudonyms could be easily linked by anyone, and the anonymity gain of the agent's having several pseudonyms would be canceled.

If, at a certain point of time, $\mathcal{P}_i$ wants to link some of her linkable pseudonyms, say without loss of generality, $ID_i^1 || R_i^1$, $ID_i^2 || R_i^2$, ..., $ID_i^{k'} || R_i^{k'}$, for $k' \leq k$, she can do so using the following protocol.

Protocol 2 (Pseudonym linkage)

1. $\mathcal{P}_i$ sends $ID_i^1||R_i^1, \ldots, ID_i^{k'}||R_i^{k'}, K_i^1, \ldots, K_i^{k'}, H_i^1, \ldots, H_i^{k'}$ to the score managers corresponding to the k' pseudonyms to be linked, that is, to the M score managers whose pseudonyms are closest to $h_0(ID_i^1||R_i^1), \ldots,$ $h_{M-1}(ID_i^1||R_i^1)$, and to the M score managers whose pseudonyms are closest to $h_0(ID_i^2||R_i^2), \ldots, h_{M-1}(ID_i^2||R_i^2)$, and so on, up to the M score managers whose pseudonyms are closest to $h_0(ID_i^{k'}||R_i^{k'}), \ldots, h_{M-1}(ID_i^{k'}||R_i^{k'})$.
2. The score managers check that the k' pseudonyms are currently marked as unlinked. Any pseudonyms that have been linked in previous instances of the protocol are discarded. Without loss of generality, assume that the first k'' pseudonyms are unlinked, with $k'' \leq k'$.
3. The score managers verify whether the following equations hold:

$$H(ID_i^1||H_i^1, K_i^1) \overset{?}{=} R_i^1$$
$$H(ID_i^2||H_i^2, K_i^2) \overset{?}{=} R_i^2 \qquad\qquad (3)$$
$$\vdots \ \ \vdots \ \ \vdots$$
$$H(ID_i^{k''}||H_i^{k''}, K_i^{k''}) \overset{?}{=} R_i^{k''}$$

where the j-th equation is verified by the M score managers managing $ID_i^j||R_i^j$, for $j = 1, \ldots, k''$.
4. If all checks in Expression (3) hold, the score managers of $ID_i^j||R_i^j$, for $j = 1, \ldots, k''$, will consider the k'' pseudonyms as linked from now on. If only a subset of at least two equations holds, then the score managers will consider as linked only the subset of pseudonyms corresponding to that subset. If only one equation holds, no pseudonyms will be linked. The implications of having a set of linked pseudonyms are:
   (a) The newly linked pseudonyms are marked by the corresponding score managers as "linked".
   (b) Each time the global reputation of one of the linked pseudonyms is updated:
      i. The score managers of that pseudonym notify the pseudonym's updated reputation to the score managers of the other linked pseudonyms.
      ii. The score managers of all linked pseudonyms recompute the reputation of the linked pseudonyms as the aggregation of the reputations individually earned by those pseudonyms (using any agreed aggregation operator, like for example the mean).
      iii. The score managers replace the reputations of the linked pseudonyms with the newly computed aggregated reputation. A small random perturbation can be added to the aggregated reputation of each pseudonym, to avoid that exactly equal reputations leak to everyone that pseudonyms are linked.

An interesting point of Protocol 2 is that, when $\mathcal{P}_i$ wants to link only a strict subset of her pseudonyms ($k' < k$), the rest of her pseudonyms are not disclosed by the protocol as belonging to $\mathcal{P}_i$.

Note 4 (On the co-utility of multiple linkable pseudonyms). The extension to multiple linkable pseudonyms presented in this section is in the interest of agent $\mathcal{P}_i$. Since Protocol 1 involves only $\mathcal{P}_i$, it does not endanger the co-utility of the protocol previous to the extension (justified in [2]). On the other hand, Protocol 2 involves $\mathcal{P}_i$ and the score managers. The motivation of $\mathcal{P}_i$ to adhere to the protocol is clear. Regarding the score managers, they also can be assumed to follow the protocol in their own interest, because the availability of accurate reputations is good for all peers (the score managers do not deviate for the same reasons they did not deviate from the protocol previous to the extension, see the co-utility analysis in [2]).

Note 5 (On disagreements between score managers). We have argued in the previous note that rationally selfish score managers will adhere to Protocol 2. However, some score managers might be not only selfish, but interested in malicious deviation (*e.g.* they may be offered money to alter a certain agent's reputation). Yet, consistently with the EigenTrust security model, we assume that a majority among the M score managers assigned to each agent are honest (even if rationally selfish), so that the reputation value reported by the majority can be assumed to be correct. If small random perturbations are added to differentiate the aggregate reputation reported for each linked pseudonym (as suggested at the end of Protocol 2), only those reputation differences beyond the perturbation range will be regarded as disagreements.

Note 6 (Confidentiality of pseudonym linkage). When several pseudonyms are linked, the linkage becomes known to the score managers of the linked pseudonyms, but, unless those managers tell other peers, no one else needs to know about the linkage. Hence, linked pseudonyms are better in terms of privacy than a single pseudonym replacing all of them.

Proposition 7 (Security of pseudonym linkage). *Only the agent who created a pseudonym can link it to other pseudonyms. Given a pseudonym created by an agent $\mathcal{P}_i$, no other agent $\mathcal{P}_j$ can create and link a pseudonym to a pseudonym of $\mathcal{P}_i$ without the latter's consent.*

Proof. Non-linkable pseudonyms do not need to be considered, because they lack the hash images R needed for the score managers to verify the linkage.

For a linkable pseudonym $ID_i^j\|R_i^j$ to be linked to other pseudonyms, the score managers of the pseudonym need to be provided with K_i^j and H_i^j, so that they can verify the j-th equation in Expressions (3). But, if $ID_i^j\|R_i^j$ is still unlinked, K_i^j is only known to the agent $\mathcal{P}_i$ who created the pseudonym; this is ensured by the security of the keyed hash function employed in Expressions (2). On the other hand, if $ID_i^j\|R_i^j$ is already linked, by design of Protocol 2 no one can link it again.

Regarding the second statement, agent $\mathcal{P}_l$ can fabricate a pseudonym $ID_l^{j'}\|R_l^{j'}$ with the aim of linking it to a pseudonym $ID_i^j\|R_i^j$ created by agent $\mathcal{P}_i$. In order for the two pseudonyms to be considered as linked, the score managers for $ID_l^{j'}\|R_l^{j'}$ should receive $K_l^{j'}$ and $H_l^{j'}$ such that $H(ID_l^{j'}\|H_l^{j'}, K_l^{j'}) =$

$R_l^{j'}$, and the score managers for $ID_i^j||R_i^j$ should receive K_i^j and H_i^j such that $H(ID_i^j||H_i^j, K_i^j) = R_i^j$. Now, $\mathcal{P}_l$ can choose any $H_l^{j'}$ and $K_l^{j'}$ and take $R_l^{j'} := H(ID_l^{j'}||H_l^{j'}, K_l^{j'})$. However, regarding $ID_i^j||R_i^j$, only $\mathcal{P}_i$ knows K_i^j if the pseudonym is unlinked; if it was already linked, it cannot be linked again. □

6.3 Generalization: Pseudonyms Allowing Multiple Linkages

In the protocols described in Sect. 6.2, a linkable pseudonym can only be linked once. This is because there is a single secret key K_i^j for a pseudonym ID_i^j: after the agent $\mathcal{P}_i$ owning the pseudonym discloses K_i^j to perform the linkage in Protocol 2, the owner would no longer be the only one able to link the pseudonym if further linkages were allowed.

A way to overcome this problem is to create linkable pseudonyms with several secret keys. For example, to allow up to ℓ linkages, $\mathcal{P}_i$ could create and disseminate a pseudonym as: $ID_i^j||R_i^{j,1}, \ldots, R_i^{j,\ell}$ where

$$H(ID_i^j||H_i^j, K_i^{j,1}) = R_i^{j,1} \tag{4}$$

$$H(ID_i^j||H_i^j, K_i^{j,2}) = R_i^{j,2} \tag{5}$$

$$\vdots \quad \vdots \quad \vdots$$

$$H(ID_i^j||H_i^j, K_i^{j,\ell}) = R_i^{j,\ell}.$$

In this way, there are ℓ secret keys. Up to ℓ linkages of the pseudonym can be performed by using a slightly generalized version of Protocol 2, where in the first linkage $\mathcal{P}_i$ would reveal $K_i^{j,1}$ that satifies Eq. (4), in the second linkage $\mathcal{P}_i$ would reveal $K^{j,2}$ that satisfies Eq. (5), and so on. The score managers for the pseudonym would maintain a counter with the number of times the pseudonym has been linked, and they would not accept the same key more than once.

7 Conclusions and Future Work

We have presented a peer-to-peer fully distributed reputation system that is privacy-preserving, in that it allows peers to use any number of pseudonyms and link some of them if they want to enjoy the same (or a similar) reputation under several of their pseudonyms. A reputation system of this kind is a better candidate than a centralized reputation system for generalized use in a social credit system.

Future research will be devoted to improving the management of pseudonyms. Specifically, we will investigate solutions that improve the confidentiality of pseudonym linkage. Also, allowing multiple linkages of a pseudonym without expanding its length deserves further work. Another interesting direction is to create revocable pseudonyms that can be attributed to a certain agent with the help of a trusted third party.

Acknowledgments and Disclaimer. Partial support to this work has been received from the Templeton World Charity Foundation (grant TWCF0095/AB60 "CO-UTILITY"), ARC (grant DP160100913), the European Commission (projects H2020-644024 "CLARUS" and H2020-700540 "CANVAS"), the Government of Catalonia (ICREA Acadèmia Prize) and the Spanish Government (projects TIN2014-57364-C2-1-R "SmartGlacis" and TIN 2015-70054-REDC). The author holds the UNESCO Chair in Data Privacy, but the views in this paper are the author's own and are not necessarily shared by UNESCO.

References

1. Creemers, R.: China Copyright and Media, 15 March 2018. https://chinacopyrightandmedia.wordpress.com/about/
2. Domingo-Ferrer, J., Farràs, O., Martínez, S., Sánchez, D., Soria-Comas, J.: Self-enforcing protocols via co-utile reputation management. Inf. Sci. **367–368**, 159–175 (2016)
3. Domingo-Ferrer, J., Martínez, S., Sánchez, D., Soria-Comas, J.: Co-utility: self-enforcing protocols for the mutual benefit of participants. Eng. Appl. Artif. Intell. **59**, 148–158 (2017)
4. Domingo-Ferrer, J., Sánchez, D., Soria-Comas, J.: Co-utility - self-enforcing collaborative protocols with mutual help. Prog. Artif. Intell. **5**(2), 105–110 (2016)
5. Hoffman, K., Zage, D., Nita-Rotaru, C.: A survey of attack and defense techniques for reputation systems. ACM Comput. Surv. **42**(1) (2009). Article no. 1
6. Kamvar, S.D., Schlosser, M.T., Garcia-Molina, H.: The EigenTrust algorithm for reputation management in P2P networks. In: Proceedings of the 12th International Conference on World Wide Web, pp. 640–651. ACM (2003)
7. Leyton-Brown, K., Shoham, Y.: Essentials of Game Theory: A Concise Multidisciplinary Introduction. Morgan and Claypool, San Rafael (2008)
8. Liu, K., Terzi, E.: Towards identity anonymization on graphs. In: Proceedings of the 2008 ACM SIGMOD International Conference on Management of Data - SIGMOD 2008, pp. 93–106. ACM (2008)
9. Marti, S., Garcia-Molina, H.: Taxonomy of trust: categorizing P2P reputation systems. Comput. Netw. **50**(4), 472–484 (2006)
10. Osborne, M., Rubinstein, A.: A Course in Game Theory. MIT Press, Cambridge (1994)
11. Singh, A., Liu, L.: TrustMe: anonymous management of trust relationships in decentralized P2P systems. In: Proceedings of the Third International Conference on Peer-to-Peer Computing (P2P 2003), pp. 142–149 (2003)
12. Hsu, S.: China's new social credit system. The Diplomat, 10 May 2015. http://thediplomat.com/2015/05/chinas-new-social-credit-system/
13. Zhou, R., Hwang, K., Cai, M.: GossipTrust for fast reputation aggregation in peer-to-peer networks. IEEE Trans. Knowl. Data Eng. **20**(9), 1282–1295 (2008)

Combinatorial Algorithms and Methods for Security of Statistical Databases Related to the Work of Mirka Miller

Andrei Kelarev[1]([✉]) [iD], Jennifer Seberry[2], Leanne Rylands[3] [iD], and Xun Yi[1] [iD]

[1] School of Science, RMIT University,
GPO Box 2476, Melbourne, VIC 3001, Australia
andrei.kelarev@gmail.com, xun.yi@rmit.edu.au
[2] School of Computing and Information Technology, University of Wollongong,
Northfields Avenue, Wollongong, NSW 2522, Australia
jennie@uow.edu.au
[3] School of Computing, Engineering and Mathematics,
Western Sydney University, Locked Bay 1797, Penrith, NSW 2751, Australia
l.rylands@westernsydney.edu.au

Abstract. This article gives a survey of combinatorial algorithms and methods for database security related to the work of Mirka Miller. The main contributions of Mirka Miller and coauthors to the security of statistical databases include the introduction of Static Audit Expert and theorems determining time complexity of its combinatorial algorithms, a polynomial time algorithm for deciding whether the maximum possible usability can be achieved in statistical database with a special class of answerable statistics, NP-completeness of similar problems concerning several other types of databases, sharp upper bounds on the number of compromise-free queries in certain categories of statistical databases, and analogous results on applications of Static Audit Expert for the prevention of relative compromise.

Keywords: Combinatorial algorithms · NP-completeness
Privacy in data mining · Database security · Time complexity
Sharp upper bounds

1 Introduction

This article surveys combinatorial algorithms and methods for maintaining the security of statistical databases. We include concise statements of the main theorems and results related to the work of Mirka Miller. For background information and preliminaries, the readers are referred to [6,17,18,42,48]. An excellent overview of various concepts used in statistical disclosure control with detailed explanations and examples illustrating the major notions is given in [4].

Statistical databases are databases in which only statistical types of queries are allowed. They store records with data on individuals (companies, organizations, etc.) and can output statistics concerning subsets of individuals providing

© Springer International Publishing AG, part of Springer Nature 2018
L. Brankovic et al. (Eds.): IWOCA 2017, LNCS 10765, pp. 383–394, 2018.
https://doi.org/10.1007/978-3-319-78825-8_31

aggregated information on groups of records in the database, while protecting confidential data of individuals from disclosure. Users can pose statistical queries, which are either answered (precisely or approximately), or rejected by a control mechanism to ensure the privacy of confidential data of individuals. Statistical databases are very important for numerous practical application. For example, answers to statistical queries can help medical researchers to evaluate the effectiveness of medications or certain lifestyle changes for the treatment or prevention of various conditions.

It is usually possible to deduce confidential information by comparing the results of several different queries. The security problem for statistical databases is to develop control mechanisms that will prevent direct or indirect disclosure of confidential data by the release of statistics as answers to statistical queries.

2   Classical Compromise

If the value of a protected attribute of an individual record can be derived, then the database is said to have been (positively) *compromised*. It is shown in [33] how supplementary knowledge available from other sources can be exploited to obtain values of a confidential attribute. The following types of supplementary knowledge are defined in [33]. *Supplementary knowledge of type I* is knowledge of the values of attributes which uniquely identify a particular record or a particular subset of records in a database. *Supplementary knowledge of type II* is knowledge of the value of a confidential attribute for a particular individual.

Let us denote the numerical attributes contained in a statistical database by $A_0, A_1, \ldots, A_m$. Without loss of generality we may assume that the users can submit queries on statistics concerning the attribute A_0 and the values of attributes $A_1, \ldots, A_m$ are used to select subsets of records for these queries. Then A_0 is called a *quantitative attribute* and $A_1, \ldots, A_m$ are called *characteristic attributes* for such queries. The set of records chosen for a query by specifying conditions on the characteristic attributes is called the *query set*. Denote by n the number of records stored in the database. Let $x_1, x_2, \ldots, x_n$ be the (protected) values of the quantitative attribute in these records.

A *SUM query* is a sum of the form $a_1 x_1 + \cdots + a_n x_n$, where $a_i = 1$ if the i-th record belongs to the query set, and $a_i = 0$ otherwise. For SUM queries, it is enough to consider 1-dimensional statistical databases, or databases with only one quantitative attribute. An arbitrary set of SUM queries in a multi-dimensional statistical database can be represented as a disjoint union of SUM queries corresponding to different quantitative attributes, and each of these subsets can be viewed as a set of SUM queries of the corresponding 1-dimensional database. A set of SUM queries can be recorded as a system of linear equations of the form

$$MX = V, \tag{1}$$

where $X = (x_1, \ldots, x_n)$ and V is the vector with the values returned by the SUM queries corresponding to the rows of the matrix M. Each query corresponds to a row of the matrix M. It is enough to store only linearly independent queries in the matrix M, since if several queries are known, then all their linear combinations are known too. The standard elementary row and column operations used to simplify systems of linear equations (1) result in a new system with rows corresponding to new queries with the outcomes equal to the corresponding values in the column V again. Therefore, we can assume that (1) has been simplified and stores a so-called *normalized query basis matrix*, so that $M = M_k$, where

$$M_k = (I_k | M'_k) \qquad (2)$$

and I_k is a $(k \times k)$ identity matrix. Then the matrix M is said to be in a *normalized* form. The row vectors of M_k form a basis of the space of all queries with outcomes which are known, since they all can be derived using linear combinations of query vectors.

Audit Expert is a system using a normalized basis matrix to store all queries answered so far (cf. [15]). When a new query is added, Audit Expert adds it to the matrix and then reduces it to a normalized basis form again.

Theorem 1 ([15]). *The time complexity of the combinatorial algorithm dynamically processing the query matrix of the Audit Expert and maintaining it in a normalized form for a set of k consecutive queries is $O(k^2)$. The statistical database is compromised if and only if the normalized query basis matrix M_k has a row with exactly one nonzero entry.*

The paper [38] suggested using a Static Audit Expert, where the query basis matrix is fixed by the system (possibly the database administrator) in advance. A user's query is then allowed to be answered if it belongs to the vector space spanned by the rows of the matrix.

Theorem 2 ([38]). *The time complexity of the combinatorial algorithm for processing each new query by a Static Audit Expert with a predesigned query matrix in a normalized basis form is $O(k)$.*

This shows that Static Audit Expert is substantially more efficient than the dynamic Audit Expert. The maximum number of answerable queries, for databases where all SUM queries are posable, was determined in [35], where a combinatorial algorithm for constructing these sets of queries was also given.

Theorem 3 ([35]). (i) *In a 1-dimensional database with n real entries, the maximum number of SUM queries answerable without a compromise is equal to $\binom{n}{\lfloor n/2 \rfloor}$.*
(ii) ([24]) *The maximum is achieved if and only if the set of all entries is partitioned into two parts of size $\lfloor n/2 \rfloor$ and $\lceil n/2 \rceil$, and each allowed query set has equal numbers of elements from both parts.*

The *usability* of a statistical database is defined as the ratio of the maximum number of valid statistics that can be disclosed without a database compromise to the total number of valid statistics in the database.

If a confidential statistic based on one record has been revealed, then the term 1-compromise is used. The problem of preventing a compromise (1-compromise) can also be called the problem of preserving anonymity (1-anonymity).

Theorem 4 ([8,23]). *In a statistical database of size n where all statistics are valid, the usability for 1-compromise is equal to $\binom{n}{\lfloor n/2 \rfloor}$.*

Theorem 5 ([11]). *In a statistical database where all statistics are valid and a fixed set of statistics are confidential and should not be disclosed, it is an NP-complete problem to decide whether the usability $\binom{n}{\lfloor n/2 \rfloor}$ can be achieved.*

Theorem 6 ([10]). *There exists a polynomial time algorithm to decide whether the usability $\binom{n}{\lfloor n/2 \rfloor}$ can be achieved in a statistical database where each statistic is based on at most two records, or each record appears in at most two statistics. It is an NP-complete problem to answer this question for statistical databases, where each statistic is based on exactly four records or each record appears in at most three statistics.*

Range queries are a special case of SUM queries. A *range query* is a sum of the form $a_1 x_1 + \cdots + a_n x_n$, where $x_1, \ldots, x_n$ are values of the quantitative attribute A_0 in all records of the query set, and the query set is not arbitrary, but is selected using a range defined by inequalities as follows. Let $b_1, \ldots, b_m$ and $c_1, \ldots, c_m$ be real numbers such that $b_1 \le c_1, \ldots, b_m \le c_m$. A *query set* of a range query is a set of all records $(r_0, r_1, \ldots, r_m)$ of the database such that the following inequalities hold: $b_1 \le r_1 \le c_1, \ldots, b_m \le r_m \le c_m$. The value of the range query is the sum of the values of the quantitative attribute A_0 in all records in the query set.

The paper [9] presents several new results concerning the usability of statistical databases for general SUM, COUNT and MEAN queries as well as for the corresponding range queries, and combinatorial algorithms for constructing such sets of queries. In certain special cases the authors derive the usability of m-dimensional statistical databases for all $m \ge 1$.

The paper [3] is devoted to special sets of queries, where each record in a database is contained in at most two queries. Sets of queries of this sort are called *queries of type α*. For a set Q of queries, a graph $G = G(Q)$ is associated with Q. The vertices of $G(Q)$ correspond to the queries in Q and edges of $G(Q)$ correspond to records of the database. The authors of [3] introduce the notion of the *L-core* of the graph G. This concept makes it possible to formulate necessary and sufficient conditions for the set Q to be compromise free. The paper [3] shows how to determine the *L-core* from the eigenvalues of the graph, and proposes an algorithm for computing the *L-core* directly from the graph.

Theorem 7 ([3]). *Let Q be a query set of type α for a statistical database and let $G = G(Q)$ be the graph associated with Q. Then Q is compromise-free if and only if G coincides with its L-core.*

Several articles investigated range queries, where the values of the quantitative attribute A_0 are confidential and should not be compromised. For $i = 1, \ldots, k$, denote by d_i the number of distinct values of the characteristic attribute A_i in the database. The main result of the paper [27] shows that the largest set of all range queries, which does not lead to a compromise, is uniquely determined and coincides with the set of all range queries with an even number of records.

Theorem 8 ([27]). *Let D be a k-dimensional database of size $d_1 \times \cdots \times d_k$. Then the usability of D is equal to $1 - \frac{1}{2^k} \prod_{i=1}^{k} f(d_i)$, where $f(x) = (x+2)/(x+1)$ for x even, $f(x) = (x+1)/x$ for x odd.*

It follows that the usability of the database always belongs to the segment

$$\left[1 - \frac{1}{2^k} \prod_{i=1}^{k} \frac{d_i + 2}{d_i + 1}, 1 - \frac{1}{2^k} \prod_{i=1}^{k} \frac{d_i + 1}{d_i}\right]. \tag{3}$$

In [5], a formula is given for the usability of range queries in a 1-dimensional database that is allowed to contain many indistinguishable copies of some records.

3 Relative Compromise

A new type of compromise, which does not involve the disclosure of exact values, was introduced in [37]. Namely, a set S of records in a statistical database is said to be *relatively compromised* with respect to a field F if the relative order of magnitude of the F-values of the records in S becomes known [37]. It is shown in [37] that even when the exact confidential information remains protected, relative compromise may still be possible. Possible consequences of relative compromise are studied too. By applying block designs for the design of queries, it is shown in [37] that a relative compromise can be achieved even if the overlap of any two query sets is restricted not to exceed one element. In the case of SUM queries of fixed query size, the paper [37] used block designs to derive a number of conditions for the relative compromise to occur.

Theorem 9 ([38]). *Let D be a 1-dimensional database with n records, where SUM queries are allowed, and let M_k be the normalized query basis matrix of the Audit Expert. Then there is a relative compromise if and only if at least one of the following conditions is satisfied.*

(i) *There exists a row of the normalized query basis matrix containing exactly one nonzero element.*
(ii) *A row of the normalized query basis matrix contains exactly two nonzero entries which sum to zero.*
(iii) *There exist two rows $i \neq j$ of the normalized query basis matrix $M_k = (I_k | M_k')$ such that the rows M_i' and M_j' are identical.*

The paper [38] classifies various types of compromise and extends the mechanism of Audit Expert to exclude relative compromises for SUM queries. The paper [36] used Audit Expert to determine the maximal number of answerable SUM queries preventing a relative compromise.

Theorem 10 ([36]). *Let D be a 1-dimensional database with n records, where SUM queries are allowed. The maximum number of SUM queries preventing relative compromised can be achieved by using a Static Audit Expert with the normalized query basis matrix $M_{n-1} = (I_{n-1}|M'_{n-1})$, where the transpose of M'_{n-1} is equal to*

$$(-[n/2], \ldots, -3, -2, 1, 2, 3, \ldots, [(n+1)/2]).$$

It follows from the results of [47] that the exact value of the maximum number of SUM queries in Theorem 10 coincides with the middle coefficient of the polynomial

$$(1+x)(1+x^2)\cdots(1+x^{[n/2]})(1+x)(1+x^2)\cdots(1+x^{[(n+1)/2]})$$

if the order of the polynomial is even, and it coincides with both of the two middle coefficients if the order of this polynomial is odd.

4   Group Compromise or k-compromise

Statistics revealing information about a subset of k or fewer individuals may also need to be protected, because supplementary information often allows an attacker to derive private data about an individual from such statistics. The disclosure of a statistic based on k or fewer records in the database is called a k-compromise. The prevention of a k-compromise can also be called the preservation of k-anonymity.

It was shown in [24] that the usability of k-compromise in a statistical database with n records is in $O(n^{-1-k/2})$. Denote by $G(n, k)$ the maximum number of SUM queries, which can prevent k-compromise in the 1-dimensional database with n records. For any positive integers n and k, the next theorem determines the value $G(n, k)$ up to a constant factor less than $1/2$.

Theorem 11 ([1]). *If $n/2 \le k < n$, then*

$$\frac{k+1}{n}\binom{n}{k+1} < G(n, k) \le \binom{n}{k+1}.$$

If $2 \le k < n/2$ and n is odd, then

$$\frac{n+1}{n-1}\binom{n-1}{\frac{n-3}{2}} \le G(n, k) < 2\binom{n-1}{\frac{n-3}{2}}.$$

If $2 \le k < n/2$ and n is even, then

$$\frac{n+2}{2n-2}\binom{n}{\frac{n-2}{2}} \le G(n, k) < \binom{n}{\frac{n-2}{2}}.$$

Further, denote by $G(n, m, k)$ (resp., $G(n, \le m, k)$) the maximum number of SUM queries in the database, where each of the sums contains m (resp., at most m) summands, and k-compromise is prevented.

Theorem 12 ([16]). *Let $m \leq n$ be positive integers, and let $t = \lfloor n/m \rfloor$. Then the following conditions are satisfied.*

(i) *If $m \ll n$, then*

$$G(n, m, 1) = t \binom{n-t}{m-1}.$$

(ii) *If $n \to \infty$, then*

$$G(n, \leq m, 1) = t \binom{n-t}{m-1}(1 + o(1)).$$

Theorem 13 ([1]). *Let $k < m \leq n$ be positive integers, and let*

$$t \in \{\lfloor n/m \rfloor, \lfloor (n+1)/m \rfloor\}.$$

Then the following equality holds:

$$G(n, m, k) = t \binom{n-t}{m-1}.$$

Furthermore, the optimal set of SUM queries involving k summands corresponds to the set of $(0,1)$-solutions of weight m to the linear equation

$$(m-1)x_1 + \cdots + (m-1)x_t - x_{t+1} - \cdots - x_n = 0. \tag{4}$$

If $\lfloor n/m \rfloor = \lfloor (n+1)/m \rfloor$, then the optimal set of SUM queries is unique up to permutation of the elements. If, however, $\lfloor n/m \rfloor \neq \lfloor (n+1)/m \rfloor$, then there exist precisely two (up to permutation of the elements) optimal sets of SUM queries determined by the linear equation (4) corresponding to $t = \lfloor n/m \rfloor$ or $t = \lfloor (n+1)/m \rfloor$, respectively.

Corollary 1 ([1]). *Let $k < m$ be positive integers. If $n \to \infty$, then*

$$G(n, \leq m, k) = t \binom{n-t}{m-1}(1 + o(1)).$$

Theorem 14 ([1]). *Let k, m and n be positive integers satisfying one of the following conditions:*

(i) *$m \geq 8$, $n \geq m^2$,*
(ii) *$4 \leq m \leq 7$ and $n \geq cm^2$ for a positive constant c,*

and let $t = \lfloor n/m \rfloor$. Then the following equality holds:

$$G(n, \leq m, k) = t \binom{n-t}{m-1}.$$

Conjecture 1 ([1]). *If $n \geq 3$, then*

$$G(n, 2) = \sum_{i=0}^{t} \binom{t}{i} \binom{n-t}{2i},$$

where $t = \lfloor n/3 \rfloor$.

In [12] a new result is obtained for the problem of maximizing the number of disclosed range queries preventing k-compromise, where k is odd, in the case of a 1-dimensional statistical database.

Theorem 15 ([12]). *Let D be a 1-dimensional database with n records, where range queries are allowed, and let $k = 2\ell - 1$, where $\ell > 1$ is a positive integer. Then the maximum number of elements in a k-compromise free set of range queries in D is equal to $\lfloor n/2\ell \rfloor (\lfloor n/\ell \rfloor - \lfloor n/2\ell \rfloor)$.*

The paper [13] determines the maximum number of sum totals that can be disclosed without leading to a 2-compromise in a 1-dimensional database for range queries. The following theorem was proved in [13] for a 1-dimensional statistical database of size n, where n is odd, or n is even and is greater than 52. For all other values of n, these formulas were proved in [32].

Theorem 16 ([13,32]). *Let D be a 1-dimensional database with n records, where range queries are allowed. Then the maximum number $\mu_2(n)$ of elements in a 2-compromise free set of range queries in D is equal to*

$$\mu_2(n) = \begin{cases} (n+1)^2/16 & \text{for odd } n \geq 1; \\ n^2/16 & \text{for even } n \neq 12. \end{cases} \tag{5}$$

The paper [45] uses graphs to represent trust levels in informational relations among entities for the purposes of treating the requirements of access to confidential data for maintaining privacy and security.

The paper [44] introduces a Hippocratic security method for managing a collection of statistical databases by a virtual community at several institutions following a collection of management rules.

An attacker can often gain insight into confidential records stored in a statistical database using additional available information about the types of attributes stored in the database (called *working knowledge*), general restrictions on the values of the attributes in the real world (called *supplementary knowledge*), or additional restrictions on the values of attributes caused by various legal systems (called *legal knowledge*). The paper [39] proposed to use knowledge based systems capturing working knowledge, supplemental knowledge and legal knowledge to regulate access to statistical databases for the prevention of compromise.

5   Generalizations and Other Related Results

A new method for maintaining the integrity of data in publicly accessible databases is developed in [26]. The method is based on the recent development of pseudo-random function families and sibling intractable function families.

A practical method for maintaining anonymous and verifiable databases with public data held in separate databases is introduced in [25]. It prevents unauthorized users from collecting and collating private data concerning individuals from these separate databases. The method is based on the use of smartcards and the improved Leighton-Micali protocol for the distribution of keys and can be extended to mobile computing environments.

The security problems and possible mechanisms for the prevention of compromises are discussed in [34] with particular attention devoted to medical databases, where confidentiality is paramount. The paper concludes with a proposal for a security subsystem to be incorporated in a database management system. Applications of value added networks in managing the security of information stored in statistical databases in the health informatics sector are discussed in [40,41].

It is explained in [20] that the multidimensional matrix model of statistical databases and the multidimensional cubes of On-Line Analytical Processing (OLAP) are essentially the same. The paper investigates the application of decision trees to mining information from statistical databases and studies robust noise addition methods to ensure the preservation of privacy. Methods for preserving privacy and enabling k-means clustering are proposed in [31,43].

A novel noise addition framework for a statistical database containing several numerical attributes and a single categorical attribute is studied in [28]. Data perturbation techniques for the prevention of disclosure of confidential values are studied in [29] in order to handle categorical attributes without a natural order of their values. A novel approach towards clustering of such categorical values is proposed in [29] and is used to perturb data. It applies horizontal partitioning and clusters the values of a given categorical attribute rather than the records of the datasets. An experimental study was performed to compare the resulting perturbation system DETECTIVE in its effectiveness with another system called CACTUS [29].

Notice that k-anonymity is a broad concept applicable in various settings. For example, in [14] it is studied for recommendation systems. It is shown in [19] that permutation is the most essential principle underlying any anonymization of microdata that involves the utility and privacy guarantees. Any anonymization for microdata can be regarded as a permutation combined with the possible addition of a small amount of noise. This lead to a new natural privacy model called (d, v, f)-permuted privacy. It incorporates subject-verifiability, i.e., the ability of every subject supplying original data to verify privacy.

The paper [46] explains how sum labellings of graphs can be used for representing the access structure of a secret sharing scheme. Another privacy-preserving framework using novel noise addition techniques is investigated in [30]. It uses noise addition to categorical values as well, so that attributes of all types are protected. An experimental study of the practical system VICUS incorporating noise addition for categorical attributes is carried out in [22].

Statistical disclosure control is also discussed in [21], where a strategic dependency model of a statistical data warehouse system is proposed and an associated model of trust is explored.

The problem of evaluating and comparing privacy provided by various techniques is tackled in [2], where a novel entropy based security measure is proposed. It can be applied to any generalization, restriction or data modification technique for preserving privacy of statistical databases. This measure is used in [2] in an empirical study evaluating and comparing the methods of query restriction, sampling and noise addition.

A new method for achieving k-anonymity of network graph data prior to its release is considered in [7] for privacy protection. The method is based on randomizing the location of the triangles in the graph. It is shown that this new method preserves the main structural characteristics of the graph, which can provide valuable information for the study of the graph, while preserving k-anonymity.

Acknowledgements. The authors are grateful to three reviewers for comments and corrections that have helped to improve this paper. This work has been supported by Discovery grant DP160100913 from Australian Research Council.

References

1. Ahlswede, R., Aydinian, H.: On security of statistical databases. SIAM J. Discrete Math. **25**, 1778–1791 (2011)
2. Alfalayleh, M., Brankovic, L.: Quantifying privacy: a novel entropy-based measure of disclosure risk. In: Kratochvíl, J., Miller, M., Froncek, D. (eds.) IWOCA 2014. LNCS, vol. 8986, pp. 24–36. Springer, Cham (2015). https://doi.org/10.1007/978-3-319-19315-1_3
3. Brankovic, L., Cvetković, D.: The eigenspace of the eigenvalue -2 in generalized line graphs and a problem in security of statistical databases. Univ. Beograd. Publ. Elektrotehn. Fak. Ser. Mat. **14**, 37–48 (2003)
4. Brankovic, L., Giggins, H.: Statistical database security. In: Petković, M., Jonker, W. (eds.) Security, Privacy, and Trust in Modern Data Management. DCSA, pp. 167–181. Springer, Heidelberg (2007). https://doi.org/10.1007/978-3-540-69861-6_12
5. Brankovic, L., Horak, P., Miller, M.: An optimization problem in statistical database security. SIAM J. Discrete Math. **13**(3), 346–353 (2000)
6. Brankovic, L., Islam, M.Z., Giggins, H.: Privacy-preserving data mining. In: Petković, M., Jonker, W. (eds.) Security, Privacy, and Trust in Modern Data Management. DCSA, pp. 151–165. Springer, Heidelberg (2007). https://doi.org/10.1007/978-3-540-69861-6_11
7. Brankovic, L., Lopez, M., Miller, M., Sebe, F.: Triangle randomization for social network data anonymization. Ars Math. Contemp. **7**, 461–477 (2014)
8. Brankovic, L., Miller, M.: An application of combinatorics to the security of statistical databases. Austral. Math. Soc. Gaz. **22**(4), 173–177 (1995)
9. Brankovic, L., Miller, M., Horak, P., Wrightson, G.: Usability of compromise-free statistical databases. In: Proceedings of the International Working Conference on Scientific and Statistical Database Management, Melbourne, Australia, 29–30 January, pp. 144–154 (1997)
10. Brankovic, L., Miller, M., Širáň, J.: Graphs, 0–1 matrices, and usability of statistical databases. Congr. Numer. **120**, 169–182 (1996)
11. Brankovic, L., Miller, M., Širáň, J.: Towards a practical auditing method for the prevention of statistical database compromise. In: Proceedings of the Seventh Australasian Database Conference, Melbourne, Australia, 29–30 January, pp. 177–184 (1996)

12. Brankovic, L., Miller, M., Širáň, J.: On range query dsability of statistical databases. Int. J. Comput. Math. **79**(12), 1265–1271 (2002)
13. Branković, L., Širáň, J.: 2-compromise usability in 1-dimensional statistical databases. In: Ibarra, O.H., Zhang, L. (eds.) COCOON 2002. LNCS, vol. 2387, pp. 448–455. Springer, Heidelberg (2002). https://doi.org/10.1007/3-540-45655-4_48
14. Casino, F., Domingo-Ferrer, J., Patsakis, C., Puig, D., Solanas, A.: A k-anonymous approach to privacy preserving collaborative filtering. J. Comput. Syst. Sci. **81**, 1000–1011 (2015)
15. Chin, F.Y., Ozsoyoglu, G.: Auditing and inference control in statistical databases. IEEE Trans. Software Eng. **8**(6), 574–582 (1982)
16. Demetrovics, J., Katona, G.O.H., Miklós, D.: On the security of individual data. In: Seipel, D., Turull-Torres, J.M. (eds.) FoIKS 2004. LNCS, vol. 2942, pp. 49–58. Springer, Heidelberg (2004). https://doi.org/10.1007/978-3-540-24627-5_5
17. Domingo-Ferrer, J.: Inference Control in Statistical Databases, vol. 2316, 1st edn. Springer, Berlin (2002). https://doi.org/10.1007/3-540-47804-3
18. Domingo-Ferrer, J.: A survey of inference control methods for privacy-preserving data mining. In: Aggarwal, C.C., Yu, P.S. (eds.) Privacy-Preserving Data Mining Models and Algorithms. ADBS, vol. 34, pp. 53–80. Springer, Boston (2008). https://doi.org/10.1007/978-0-387-70992-5_3
19. Domingo-Ferrer, J., Muralidhar, K.: New directions in anonymization: permutation paradigm, verifiability by subjects and intruders, transparency to users. Inf. Sci. **337–338**, 11–24 (2016)
20. Estivill-Castro, V., Brankovic, L.: Data swapping: balancing privacy against precision in mining for logic rules. In: Mohania, M., Tjoa, A.M. (eds.) DaWaK 1999. LNCS, vol. 1676, pp. 389–398. Springer, Heidelberg (1999). https://doi.org/10.1007/3-540-48298-9_41
21. Giggins, H., Brankovic, L.: Statistical disclosure control: to trust or not to trust. In: Proceedings of the International Symposium on Computer Science and its Applications, pp. 108–113. IEEE Computer Society (2008)
22. Giggins, H., Brankovic, L.: VICUS - a noise addition technique for categorical data. In: Proceedings of the Tenth Australasian Data Mining Conference, AusDM 2012, Conferences in Research and Practice in Information Technology (CRPIT), vol. 134, pp. 139–148 (2012)
23. Griggs, J.R.: Concentrating subset sums at k points. Bull. Inst. Comb. Appl. **20**, 65–74 (1997)
24. Griggs, J.R.: Database security and the distribution of subset sums in R^m. In: Graph Theory and Combinatorial Biology. Bolyai Society Mathematical Studies, vol. 7, pp. 223–252 (1997)
25. Hardjono, T., Seberry, J.: Applications of smartcards for anonymous and verifiable databases. Comput. Secur. **14**, 465–472 (1995)
26. Hardjono, T., Zheng, Y., Seberry, J.: Database authentication revisited. Comput. Secur. **13**, 573–580 (1994)
27. Horak, P., Brankovic, L., Miller, M.: A combinatorial problem in database security. Discrete Appl. Math. **91**(1–3), 119–126 (1999)
28. Islam, M.Z., Brankovic, L.: A framework for privacy preserving classification in data mining. In: Proceedings of the 2nd Workshop on Australasian Information Security, Data Mining and Web Intelligence, and Software Internationalisation, vol. 32, pp. 163–168 (2004)

29. Islam, M.Z., Brankovic, L.: DETECTIVE: a decision tree based categorical value clustering and perturbation technique for preserving privacy in data mining. In: Proceedings of the 3rd IEEE International Conference on Industrial Informatics, INDIN 2005, pp. 701–708 (2005)

30. Islam, M.Z., Brankovic, L.: Privacy preserving data mining: a noise addition framework using a novel clustering technique. Knowl.-Based Syst. **24**, 1214–1223 (2011)

31. Liu, D., Bertino, E., Yi, X.: Privacy of outsourced k-means clustering. In: Proceedings of the 9th ACM Symposium on Information, Computer and Communications Security, ASIA CCS 2014, pp. 123–133 (2014)

32. Mathieson, L., King, T., Brankovic, L.: 2-compromise: usability in 1-dimensional statistical database. Research Gate (2008). https://www.researchgate.net/publication/228973056

33. Miller, M.: A model of statistical databse compromise incorporating supplementary knowledge. In: Databases in the 1990's, pp. 258–267 (1991)

34. Miller, M., Cooper, J.: Security considerations for present and future medical databases. Int. J. Med. Inform. **41**, 39–46 (1996)

35. Miller, M., Roberts, I., Simpson, J.: Application of symmetric chains to an optimization problem in the security of statistical databases. Bull. Inst. Comb. Appl. **2**, 47–58 (1991)

36. Miller, M., Roberts, I., Simpson, J.: Prevention of relative compromise in statistical databases using audit expert. Bull. Inst. Comb. Appl. **10**, 51–62 (1994)

37. Miller, M., Seberry, J.: Relative compromise of statistical databases. Aust. Comput. J. **21**(2), 56–61 (1989)

38. Miller, M., Seberry, J.: Audit expert and statistical database security. In: Databases in the 1990's, pp. 149–174 (1991)

39. Mishra, V., Stranieri, A., Miller, M., Ryan, J.: Knowledge based regulation of statistical databases. WSEAS Trans. Inf. Sci. Appl. **3**(2), 239–244 (2006)

40. Pacheco, F., Cooper, J., Bomba, D., Morris, S., Miller, M., Brankovic, L.: Education issues in health informatics. Inform. Healthc. **4**, 101–105 (1995)

41. Pacheco, F., Cooper, J., Bomba, D., Morris, S., Miller, M., Brankovic, L.: Value added networks (VANs) and their benefit to a health information system. Inform. Healthc. **4**, 141–144 (1995)

42. Pieprzyk, J., Hardjono, T., Seberry, J.: Fundamentals of Computer Security. Springer, Berlin (2003). https://doi.org/10.1007/978-3-662-07324-7

43. Rao, F.Y., Samanthula, B., Bertino, E., Yi, X., Liu, D.: Privacy-preserving and outsourced multi-user k-means clustering. In: Proceedings of the IEEE Conference on Collaboration and Internet Computing, CIC 2015, pp. 80–89 (2015)

44. Skinner, G., Chang, E., McMahon, M., Aisbett, J., Miller, M.: Shield privacy Hippocratic security method for virtual community. In: IECON Proceedings of the Industrial Electronics Conference, pp. 472–479 (2004)

45. Skinner, G., Miller, M.: Managing privacy, trust, security, and context relationships using weighted graph representations. WSEAS Trans. Inf. Sci. Appl. **3**(2), 283–290 (2006)

46. Slamet, S., Sugeng, K.A., Miller, M.: Sum graph based access structure in a secret sharing scheme. J. Prime Res. Math. **2**, 113–119 (2006)

47. Stanley, R.P.: Weyl groups, the hard Lefshetz theorem, and the Sperner property. SIAM J. Algebr. Discrete Meth. **1**, 168–184 (1980)

48. Yi, X., Paulet, R., Bertino, E.: Private Information Retrieval. Morgan and Claypool, San Rafael (2013)

String Algorithms

Shortest Unique Palindromic Substring Queries in Optimal Time

Yuto Nakashima[1,2(✉)], Hiroe Inoue[1], Takuya Mieno[1], Shunsuke Inenaga[1], Hideo Bannai[1], and Masayuki Takeda[1]

[1] Department of Informatics, Kyushu University, Fukuoka, Japan
{yuto.nakashima,hiroe.inoue,inenaga,bannai,takeda}@inf.kyushu-u.ac.jp
[2] Japan Society for the Promotion of Science (JSPS), Tokyo, Japan

Abstract. A palindrome is a string that reads the same forward and backward. A palindromic substring P of a string S is called a shortest unique palindromic substring ($SUPS$) for an interval $[s, t]$ in S, if P occurs exactly once in S, this occurrence of P contains interval $[s, t]$, and every palindromic substring of S which contains interval $[s, t]$ and is shorter than P occurs at least twice in S. The $SUPS$ problem is, given a string S, to preprocess S so that for any subsequent query interval $[s, t]$ all the $SUPS$s for interval $[s, t]$ can be answered quickly. We present an optimal solution to this problem. Namely, we show how to preprocess a given string S of length n in $O(n)$ time and space so that all $SUPS$s for any subsequent query interval can be answered in $O(\alpha + 1)$ time, where α is the number of outputs.

1 Introduction

A substring $S[i..j]$ of a string S is called a *shortest unique substring* (SUS) for a position p if $S[i..j]$ is the shortest substring s.t. $S[i..j]$ is unique in S (i.e., $S[i..j]$ occurs exactly once in S), and $[i..j]$ contains p (i.e., $i \leq p \leq j$). Recently, Pei et al. [13] proposed the *point SUS problem*, preprocessing a given string S of length n so that we can return a SUS for any given query position efficiently. This problem was considered for some applications in bioinformatics, e.g., polymerase chain reaction (PCR) primer design in molecular biology. Pei et al. [13] proposed an algorithm which returns a SUS for any given position in constant time after $O(n^2)$-time preprocessing. After that, Tsuruta et al. [15] and Ileri et al. [9] independently showed optimal $O(n)$-time preprocessing and constant query time algorithms. They also showed optimal $O(n)$-time preprocessing and $O(k)$ query time algorithms which return all SUSs for any given position where k is the number of outputs. Moreover, Hon et al. [6] proposed an in-place algorithm which returns a SUS. A more general problem called *interval SUS problem*, where a query is an interval, was considered by Hu et al. [7]. They proposed an optimal $O(n)$-time preprocessing and $O(k)$ query time algorithm which returns all SUSs containing a given query interval. Most recently, Mieno et al. [12] proposed an efficient algorithm for interval SUS problem when the input string is represented by *run-length encoding*.

© Springer International Publishing AG, part of Springer Nature 2018
L. Brankovic et al. (Eds.): IWOCA 2017, LNCS 10765, pp. 397–408, 2018.
https://doi.org/10.1007/978-3-319-78825-8_32

In this paper, we consider a new variant of interval SUS problems concerning palindromes. A substring $S[i..j]$ is called a palindromic substring of S if $S[i..j]$ and the reversed string of $S[i..j]$ is the same string. The study of combinatorial properties and structures on palindromes is still an important and well studied topic in stringology [1,3–5,8,14]. Droubay et al. [3] showed a string of length n can contain at most $n + 1$ distinct palindromes. Moreover, Groult et al. [5] proposed a linear time algorithm for computing all distinct palindromes in a string.

Our new problem can be described as follows. A substring $S[i..j]$ of a string S is called a *shortest unique palindromic substring* (*SUPS*) for an interval $[s, t]$ if $S[i..j]$ is the shortest substring s.t. $S[i..j]$ is unique in S, $[i..j]$ contains $[s, t]$, and $S[i..j]$ is a palindromic substring. The *interval SUPS problem* is to preprocess a given string S of length n so that we can return all *SUPS*s for any query interval efficiently. For this problem, we propose an optimal $O(n)$-time preprocessing and $O(\alpha + 1)$-time query algorithm, where α is the number of outputs. Potential applications of our algorithm are in bioinformatics; It is known that the presence of particular (e.g., unique) palindromic sequences can affect immunostimulatory activities of oligonucleotides [10,16]. The size and the number of palindromes also influence the activity. Since any unique palindromic sequence can be obtained easily from a shorter unique palindromic sequences, we can focus on the shortest unique palindromic substrings.

The contents of our paper are as follows. In Sect. 2, we state some definitions and properties on strings. In Sect. 3, we explain properties on $SUPS$ and our query algorithm. In Sect. 4, we show the main part of the preprocessing phase of our algorithm. Finally, we conclude.

2   Preliminaries

2.1   Strings

Let Σ be an integer *alphabet*. An element of Σ^* is called a *string*. The length of a string S is denoted by $|S|$. The empty string ε is a string of length 0, namely, $|\varepsilon| = 0$. Let Σ^+ be the set of non-empty strings, i.e., $\Sigma^+ = \Sigma^* - \{\varepsilon\}$. For a string $S = xyz$, x, y and z are called a *prefix*, *substring*, and *suffix* of S, respectively. A prefix x and a suffix z of S are respectively called a *proper prefix* and *proper suffix* of S, if $x \neq S$ and $z \neq S$. The i-th character of a string S is denoted by $S[i]$, where $1 \leq i \leq |S|$. For a string S and two integers $1 \leq i \leq j \leq |S|$, let $S[i..j]$ denote the substring of S that begins at position i and ends at position j. For convenience, let $S[i..j] = \varepsilon$ when $i > j$.

2.2   Palindromes

Let S^R denote the reversed string of S, that is, $S^R = S[|S|] \cdots S[1]$. A string S is called a palindrome if $S = S^R$. Let $P \subset \Sigma^*$ be the set of palindromes. A substring $S[i..j]$ of S is said to be a palindromic substring of S, if $S[i..j] \in P$.

The center of a palindromic substring $S[i..j]$ of S is $\frac{i+j}{2}$. Thus a string S of length $n \geq 1$ has $2n - 1$ centers $(1, 1.5, \ldots, n - 0.5, n)$. The following lemma can be easily obtained by the definition of palindromes.

Lemma 1. *Let S be a palindrome. For any integers i, j s.t. $1 \leq i \leq j \leq |S|$, $S[|S| - j + 1..|S| - i + 1] = S[i..j]^R$ holds.*

2.3 *MUPS*s, *SUPS*s and Our Problem

For any non-empty strings S and w, let $occ_S(w)$ denote the set of occurrences of w in S, namely, $occ_S(w) = \{i \mid 1 \leq i \leq |S| - |w| + 1, w = S[i..i + |w| - 1]\}$. A substring w of a string S is called a *unique substring* (resp. a *repeat*) of S if $|occ_S(w)| = 1$ (resp. $|occ_S(w)| \geq 2$). In the sequel, we will identify each unique substring w of S with its corresponding (unique) interval $[i, j]$ in S such that $w = S[i..j]$. A substring $S[i..j]$ is said to be *unique palindromic substring* if $S[i..j]$ is a unique substring in S and a palindromic substring. We will say that an interval $[i_1, j_1]$ contains an interval $[i_2, j_2]$ if $i_1 \leq i_2 \leq j_2 \leq j_1$ holds. The following notation is useful in our algorithm.

Definition 1 (Minimal Unique Palindromic Substring (*MUPS*)). *A string $S[i..j]$ is a MUPS in S if $S[i..j]$ satisfies all the following conditions;*

- *$S[i..j]$ is a unique palindromic substring in S,*
- *$S[i + 1..j - 1]$ is a repeat in S or $1 \leq |S[i..j]| \leq 2$.*

Let $\mathcal{M}_S$ denote the set of intervals of all *MUPS*s in S and let $mups_i = [b_i, e_i]$ denote the i-th *MUPS* in $\mathcal{M}_S$ where $1 \leq i \leq m$ and m is the number of *MUPS*s in S. We assume that *MUPS*s in $\mathcal{M}_S$ are sorted in increasing order of beginning positions. For convenience, we define $mups_0 = [-1, -1]$, $mups_{m+1} = [n+1, n+1]$.

Example 1 (MUPS). For $S = \texttt{acbaaabcbcbcbaab}$, $\mathcal{M}_S = \{[4, 6], [8, 12], [13, 16]\}$ (see also Fig. 1).

Definition 2 (Shortest Unique Palindromic Substring (*SUPS*)). *A string $S[i..j]$ is a SUPS for an interval $[s, t]$ in S if $S[i..j]$ satisfies all the following conditions;*

- *$S[i..j]$ is a unique palindromic substring in S,*
- *$[i, j]$ contains $[s, t]$,*
- *no unique palindromic substring $S[i'..j']$ containing $[s, t]$ with $j' - i' < j - i$ exists.*

Example 2 (SUPS). Let $S = \texttt{acbaaabcbcbcbaab}$. *SUPS* for interval $[6, 7]$ is the $S[3..7] = \texttt{baaab}$. *SUPS* for interval $[7, 8]$ are $S[2..8] = \texttt{cbaaabc}$ and $S[7..13] = \texttt{bcbcbcb}$. *SUPS* for interval $[4, 13]$ does not exist. (see also Fig. 1).

In this paper, we tackle the following problem.

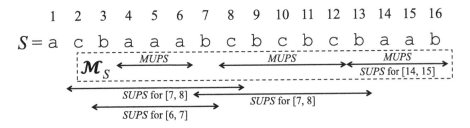

Fig. 1. This figure shows all *MUPS*s for S = acbaaabcbcbcbaab and some *SUPS* described in Example 2.

Problem 1 (SUPS problem).

– **Preprocess**: String S of length n.
– **Query**: An interval $[s, t](1 \leq s \leq t \leq n)$.
– **Return**: All the *SUPS*s for interval $[s, t]$.

2.4   Computation Model

Our model of computation is the word RAM: We shall assume that the computer word size is at least $\lceil \log_2 n \rceil$, and hence, standard operations on values representing lengths and positions of strings can be manipulated in constant time. Space complexities will be determined by the number of computer words (not bits).

3   Solution to the *SUPS* Problem

In this section, we show how to compute all *SUPS*s for any query interval $[s, t]$.

3.1   Properties on *SUPS* and *MUPS*

In our algorithm, we compute *SUPS*s by using *MUPS*s. Firstly, we show the following lemma. Lemma 2 states that *MUPS*s cannot nest in each other.

Lemma 2. *For any pair of distinct MUPSs, one cannot contain the other.*

Proof. Consider two *MUPS*s u, v such that u contains v. If u and v have the same center, then u is not a *MUPS*. On the other hand, if u and v have a different center, we have from Lemma 1 and that v is a palindromic substring, v occurs in u at least twice. This contradicts that v is unique.

From this lemma, we can see that no pair of distinct *MUPS*s begin nor end at the same position. This fact implies that the number of *MUPS*s is at most n for any string of length n. The following lemma states a characterization of *SUPS*s by *MUPS*s.

Lemma 3. *For any SUPS $S[i..j]$ for some interval, there exists exactly one MUPS that is contained in $[i, j]$. Furthermore, the MUPS has the same center as $S[i..j]$.*

Proof. Let $S[i..j]$ be a *SUPS* for some interval. $S[i..j]$ contains a *MUPS* $S[x_1..y_1]$ of the same center, i.e., $\frac{i+j}{2} = \frac{x_1+y_1}{2}$, s.t. $j - i \geq y_1 - x_1$. Suppose that there exists another *MUPS* $S[x_2..y_2]$ contained in $[i, j]$. From Lemma 2, $S[x_1..y_1]$ and $S[x_2..y_2]$ do not have the same center. On the other hand, if $S[x_1..y_1]$ and $S[x_2..y_2]$ have different centers, then $S[x_2..y_2]$ occurs at least two times in $S[i..j]$ by Lemma 1, since $S[x_2..y_2] = S[x_2..y_2]^R$. This contradicts that $S[x_2..y_2]$ is a *MUPS*.

From the above lemma, any *SUPS* contains exactly one *MUPS* which has the same center (see also Fig. 1). Below, we will describe the relationship between a query interval $[s, t]$ and the *MUPS* contained in a *SUPS* for $[s, t]$. Before explaining this, we define the following notations.

- $\mathcal{M}([s, t])$: the set of *MUPSs* containing $[s, t]$.
- $predMUPS[t] = i$ s.t. $i = \max\{k \mid e_k \leq t\}$.
- $succMUPS[s] = i$ s.t. $i = \min\{k \mid s \leq b_k\}$.

In other words, $mups_{predMUPS[t]}$ is the rightmost *MUPS* which ends before position $t+1$, and $mups_{succMUPS[s]}$ is the leftmost *MUPS* which begins after position $s - 1$.

Lemma 4. *Let $S[i..j]$ be a SUPS for an interval $[s, t]$. Then, the unique MUPS $S[x..y]$ contained in $[i, j]$ is in $\{predMUPS[t]\} \cup \mathcal{M}([s, t]) \cup \{succMUPS[s]\}$.*

Proof. Assume to the contrary that there exists a *SUPS* $S[i..j]$ that contains a *MUPS* $S[x..y] \notin \{predMUPS[t]\} \cup \mathcal{M}([s, t]) \cup \{succMUPS[s]\}$. Since $S[x..y] \notin \mathcal{M}([s, t])$, $[x, y]$ does not contain $[s, t]$. Thus, there can be the following two cases:

- If $y < t$, there must exist *MUPS* $[x', y']$ s.t. $y < y' \leq t$, since $S[x, y] \neq predMUPS[t]$. By Lemma 2, $x < x'$. Thus $i \leq x < x' \leq y' \leq t \leq j$ holds. However, this contradicts Lemma 3.
- If $s < x$, there must exist *MUPS* $[x', y']$ s.t. $s \leq x' < x$, since $S[x, y] \neq succMUPS[s]$. By Lemma 2, $y' \leq y$. Thus $i \leq s \leq x' \leq y' \leq y \leq j$ holds, However, this contradicts Lemma 3.

Therefore the lemma holds.

Next, we want to explain how *SUPSs* are related to *MUPSs*. It is easy to see that there may not be a *SUPS* for some query interval. We first show a case where there are no *SUPSs* for a given query. The following corollary is obtained from Lemma 3.

Corollary 1. *Let $S[x_1..y_1]$ and $S[x_2..y_2]$ be MUPSs contained in a query interval $[s, t]$. There is no SUPS for an interval $[s, t]$.*

From this corollary, a *SUPS* for an interval $[s, t]$ can exist if the number of *MUPS*s contained in $[s, t]$ is less than or equal to 1. The following two lemmas show what the *SUPS* for $[s, t]$ is, when $[s, t]$ contains only one *MUPS*, and when $[s, t]$ does not contain any *MUPS*s.

Lemma 5. *Let $S[x..y]$ be the only MUPS contained in the query interval $[s, t]$. If $S[x - z, y + z]$ is a palindromic substring where $z = \max\{x - s, t - y\}$, then $S[x - z, y + z]$ is the SUPS for $[s, t]$. Otherwise, there is no SUPS for $[s, t]$.*

Proof. Assume that there exists a *SUPS* u for $[s, t]$ which has the same center with a *MUPS* other than $S[x..y]$. By the definition of *SUPS*, u should contain $[s, t]$. Since $[s, t]$ contains $[x, y]$, u contains two *MUPS*s, a contradiction. Thus, there can be no *SUPS* s.t. the center is not $\frac{x+y}{2}$. It is clear that $S[x - z, y + z]$ is a unique palindromic substring if $S[x - z, y + z]$ is a palindromic substring where $z = \max\{x - s, t - y\}$. Therefore the lemma holds.

Lemma 6. *Let $[s, t]$ be the query interval. Then SUPSs for $[s, t]$ are the shortest of the following candidates.*

1. *$S[x..y]$ s.t. $[x, y] \in \mathcal{M}([s, t])$,*
2. *$S[x - t + y..t]$ s.t. $[x, y] = predMUPS([s, t])$, if it is a palindromic substring,*
3. *$S[s..y + x - s]$ s.t. $[x, y] = succMUPS([s, t])$, if it is a palindromic substring.*

Proof. It is clear that $S[x..y]$ is a unique palindromic substring containing $[s, t]$ if $[x, y] \in \mathcal{M}([s, t])$ exists. It is also clear that if $[x, y] = predMUPS([s, t])$ or $[x, y] = succMUPS([s, t])$, then $S[x - t + y..t]$ or $[s..y + x - s]$, respectively, are unique palindromic substrings, if they are palindromic substrings. By Lemma 4, we do not need to consider palindromic substrings which have the same center as *MUPS*s other than the candidates considered above. Thus the shortest of the candidates is *SUPS* for $[s, t]$ (see also Fig. 2).

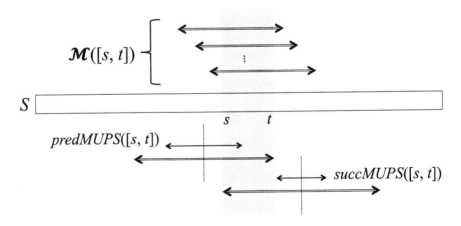

Fig. 2. Double arrows represent the candidates of *SUPS* for $[s, t]$. The shortest of the candidates is *SUPS* for $[s, t]$.

From the above arguments, the number of *MUPS*s is useful to compute *SUPS*s for a query interval. The following lemma shows how to compute the number of *MUPS*s contained in a given interval.

Lemma 7. *For any interval* $[s,t]$,

- *if* $succMUPS[s] > predMUPS[t]$, $[s,t]$ *contains no MUPS*,
- *if* $succMUPS[s] = predMUPS[t]$, $[s,t]$ *contains only one MUPS*, $mups_{succMUPS[s]} = mups_{predMUPS[t]}$, *and*
- *if* $succMUPS[s] < predMUPS[t]$, $[s,t]$ *contains at least two MUPSs*.

Proof. – Let $j = succMUPS[s] > predMUPS[t] = i$. Then $b_i < s \le b_j$ and $e_i \le t < e_j$ hold, and thus neither of $mups_i$ and $mups_j$ are contained in $[s,t]$. If we assume that $[s,t]$ contains a *MUPS* $mups_k$ for some k, it should be that $i < k < j$, $b_i < s \le b_k < b_j$. However, this contradicts that $j = succMUPS[s]$ (see also the top in Fig. 3).

- Let $succMUPS[s] = predMUPS[t] = i$. Since $succMUPS[s] = i$, b_{i-1} should be less than s, and b_i at least s. Since $predMUPS[t] = i$, e_{i+1} should be larger than t, and e_i at most t. Thus $[s,t]$ only contains $mups_i$ (see also the middle in Fig. 3).
- Let $i = succMUPS[s] < predMUPS[t] = j$. Then $s \le b_i < b_j$ and $e_i < e_j \le t$ hold, which implies $s \le b_i \le e_i < t$ and $s < b_j \le e_j \le t$. Thus, both $mups_i$ and $mups_j$ are contained in $[s,t]$ (see also the bottom in Fig. 3).

3.2 Tools

Here, we show some tools for our algorithm.

Lemma 8 (e.g., [14]). *For any interval* $[i,j]$ *in* S *of length* n, *we can check whether* $S[i..j]$ *is a palindromic substring or not in* $O(n)$ *preprocessing time and constant query time with* $O(n)$ *space.*

Manacher's algorithm [11] can compute all maximal palindromic substrings in linear time. If we have the array of radiuses of maximal palindromic substrings for all $2n - 1$ centers, we can check whether a given substring $S[i..j]$ is a palindromic substring or not in constant time.

Range Minimum Queries (RmQ). Let A be an integer array of size n. A *range minimum query* $RmQ_A(i,j)$ returns the index of a minimum element in the subarray $A[i,j]$ for given a query interval $[i,j](1 \le i \le j \le n)$, i.e., it returns one of $\arg\min_{i \le k \le j}\{A[k]\}$. It is well-known (see e.g., [2]) that after an $O(n)$-time preprocessing over the input array A, $RmQ_A(i,j)$ can be answered in $O(1)$ time for any query interval $[i,j]$, using $O(n)$ space.

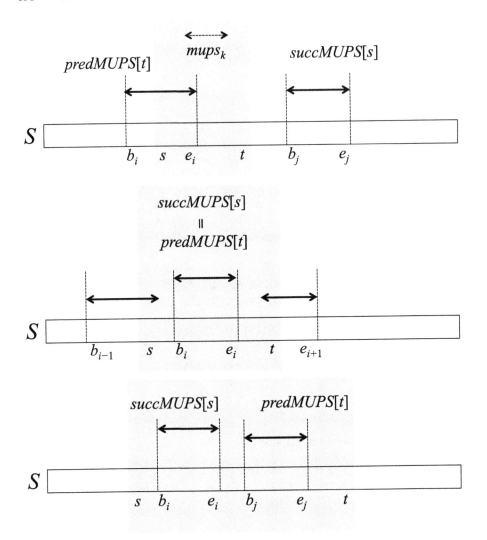

Fig. 3. Illustrations for proof of Lemma 7.

3.3 Algorithm

Due to the arguments in Sect. 3.1, if we can compute *predMUPS*, the shortest *MUPS*s in $\mathcal{M}([s,t])$ and *succMUPS* for a query interval $[s,t]$, then, we can compute *SUPS*s for $[s,t]$. Below, we will describe our solution to the *SUPS* problem.

Preprocessing Phase. First, we compute $\mathcal{M}_S$ for a given string S of length n in increasing order of beginning positions. We show, in the next section, that this can be done in $O(n)$ time and space. After computing $\mathcal{M}_S$, we compute the arrays $predMUPS$ and $succMUPS$. It is easy to see that we can also compute these arrays in $O(n)$ time by using $\mathcal{M}_S$. In the query phase, we are required to compute the shortest $MUPS$s that contain the query interval $[s,t]$. To do so efficiently, we prepare the following array. Let $Mlen$ be an array of length $m = |\mathcal{M}_S|$, and the i-th entry $Mlen[i]$ holds the length of $mups_i$, i.e., $Mlen[i] = |mups_i| = e_i - b_i + 1$. We also preprocess $Mlen$ for RmQ queries. This can be done in $O(m)$ time and space as noted in Sect. 3.2. Thus, since $m = O(n)$, the total preprocessing is $O(n)$ time and space.

Query Phase. First, we compute how many $MUPS$s are contained in a query interval $[s,t]$ by using Lemma 7, which we denote by num. This can be done in $O(1)$ time given arrays $predMUPS$ and $succMUPS$.

- If $num = 0$, let $mups_i = predMUPS([s,t])$ and $mups_j = succMUPS([s,t])$, i.e., $i = predMUPS[s]$ and $j = succMUPS[t]$. We check whether $S[b_i-t+e_i..t]$ and $S[s..e_j + b_j - s]$ are palindromic substrings or not. If so, then they are candidates of $SUPS$s for $[s,t]$ by Lemma 6. Let q be the length of the shortest candidates which can be found in the above. Second, we compute the shortest $MUPS$ in $\mathcal{M}([s,t])$, if their lengths are at most q. In other words, we compute the smallest values in $Mlen[i+1..j-1]$, if they are at most q. We can compute all such $MUPS$s in linear time w.r.t. the number of such $MUPS$s by using RmQ queries on $Mlen[i + 1, j - 1]$; if $k = RmQ_{Mlen}(i + 1, j - 1)$ and $Mlen[k] \leq q$, then we consider the range $Mlen[i+1..k-1]$ and $Mlen[k+1, j-1]$ and recurse. Otherwise, we stop the recursion. Finally, we return the shortest candidates as $SUPS$.
- If $num = 1$, let $mups_i$ be the $MUPS$ contained in $[s,t]$. First, we check whether $S[b_i - z, e_i + z]$ is a palindromic substring or not by using Lemma 8 where $z = \max\{b_i - s, t - e_i\}$. If so, then return $[b_i - z, e_i + z]$, otherwise $SUPS$ for $[s,t]$ does not exist.
- If $num \geq 2$, then, from Corollary 1, $SUPS$ for $[s,t]$ does not exist.

Therefore, we obtain the following.

Theorem 1. *After constructing an $O(n)$-space data structure of a given string of length n in $O(n)$ time, we can compute all $SUPS$s for a given query interval $[s,t]$ in $O(\alpha + 1)$ time where α is the number of outputs.*

4   Computing $MUPS$s

In this section, we show how to compute $\mathcal{M}_S$ in $O(n)$ time and space. Let DP_S be the set of *distinct palindromic substrings* in S, and $strM_S = \{S[i,j] \mid [i,j] \in \mathcal{M}_S\}$. Our idea of computing $\mathcal{M}_S$ is based on the following lemma.

Lemma 9. $strM_S \subseteq DP_S$.

Proof. It is clear that any string in $strM_S$ is a palindromic substring of S.

An algorithm for computing all distinct palindromic substrings in string in linear time and space was proposed by Groult et al. [5]. We show a linear time and space algorithm which computes $\mathcal{M}_S$ by modifying Groult et al.'s algorithm.

4.1 Tools

We show some tools for computing $\mathcal{M}_S$ below.

- **Longest previous factor array (LPF).** We denote the longest previous factor array of S by LPF_S. The i-th entry $(1 \leq i \leq n)$ is the length of the longest prefix of $S[i..n]$ which occurs at a position less than i.
- **Inverse suffix array (ISA).** We denote the inverse suffix array of S by ISA_S. The i-th entry $(1 \leq i \leq n)$ is the lexicographic order of $S[i..n]$ in all suffixes of S.
- **Longest common prefix array (LCP).** We denote the longest common prefix array of S by LCP_S. The i-th entry $(2 \leq i \leq n)$ is the length of the longest common prefix of the lexicographically i-th suffix of S and the $(i-1)$-th suffix of S.

4.2 Computing Distinct Palindromes

Here, we show a summary of Groult et al.'s algorithm. The following lemma states the main idea.

Lemma 10 ([3]). *The number of distinct palindromic substrings in S is equal to the number of prefixes of S s.t. its longest palindromic suffix is unique in the prefix.*

Since counting suffixes that uniquely occur in a prefix implies that only the leftmost occurrences of substrings, and thus distinct substrings are counted, their algorithm finds all the distinct palindromic substrings by:

- computing the longest palindromic suffix of each prefix of S, and
- checking whether each longest palindromic suffix occurs uniquely in the prefix or not.

They first propose an algorithm which computes all the longest palindromic suffixes in linear time. They then check, in constant time, the uniqueness of the occurrence in the prefix by using the LPF array, thus computing DP_S in linear time and space.

4.3 Computing All $MUPS$s

Finally, we show how to modify Groult et al.'s algorithm. As mentioned, they compute the leftmost occurrence of each distinct palindromic substring. We call such a palindromic substring, the leftmost palindromic substring. It is clear that if a leftmost palindromic substring w is unique in S and is a minimal palindromic substring, then w is a $MUPS$. Thus, we add operations to check the uniqueness and minimality of each leftmost palindromic substring. We can do these operations by using ISA and LCP array.

Let $S[i..j]$ be a leftmost palindromic substring in S. First, we check whether $S[i..j]$ is unique or not in S. If $ISA[i] = k$, $S[i..n]$ is the lexicographically k-th suffix of S. $S[i..j]$ is unique in S iff $LCP[k] < j - i + 1$ and $LCP[k + 1] < j - i + 1$. Thus we can check whether $S[i..j]$ is unique or not in constant time. Finally, we check whether $S[i..j]$ is a minimal palindromic substring or not. By definition, $S[i..j]$ is minimal palindromic substring if $j - i + 1 \leq 2$, i.e., $S[i..j]$ has no shorter unique palindromic substring. If $j - i + 1 > 2$, then we check whether $S[i + 1..j - 1]$ is unique or not by using ISA and LCP in a similar way. Thus we can also check whether $S[i..j]$ is minimal or not in constant time. By the above arguments, we can compute all $MUPSs$ in linear time and space.

5 Conclusions

We consider a new problem called the shortest unique palindromic substring problem. We proposed an optimal linear time preprocessing algorithm so that all $SUPSs$ for any given query interval can be answered in linear time w.r.t. the number of outputs. The key idea was to use palindromic properties in order to obtain a characterization of $SUPS$, more precisely, that a palindromic substring cannot contain a unique palindromic substring with a different center.

References

1. Bannai, H., Gagie, T., Inenaga, S., Kärkkäinen, J., Kempa, D., Piątkowski, M., Puglisi, S.J., Sugimoto, S.: Diverse palindromic factorization is NP-complete. In: Potapov, I. (ed.) DLT 2015. LNCS, vol. 9168, pp. 85–96. Springer, Cham (2015). https://doi.org/10.1007/978-3-319-21500-6_6
2. Bender, M.A., Farach-Colton, M.: The LCA problem revisited. In: Gonnet, G.H., Viola, A. (eds.) LATIN 2000. LNCS, vol. 1776, pp. 88–94. Springer, Heidelberg (2000). https://doi.org/10.1007/10719839_9
3. Droubay, X., Justin, J., Pirillo, G.: Episturmian words and some constructions of de Luca and Rauzy. Theor. Comput. Sci. **255**(1–2), 539–553 (2001)
4. Fici, G., Gagie, T., Kärkkäinen, J., Kempa, D.: A subquadratic algorithm for minimum palindromic factorization. J. Discrete Algorithms **28**, 41–48 (2014)
5. Groult, R., Prieur, É., Richomme, G.: Counting distinct palindromes in a word in linear time. Inf. Process. Lett. **110**(20), 908–912 (2010)
6. Hon, W.-K., Thankachan, S.V., Xu, B.: An in-place framework for exact and approximate shortest unique substring queries. In: Elbassioni, K., Makino, K. (eds.) ISAAC 2015. LNCS, vol. 9472, pp. 755–767. Springer, Heidelberg (2015). https://doi.org/10.1007/978-3-662-48971-0_63
7. Hu, X., Pei, J., Tao, Y.: Shortest unique queries on strings. In: Moura, E., Crochemore, M. (eds.) SPIRE 2014. LNCS, vol. 8799, pp. 161–172. Springer, Cham (2014). https://doi.org/10.1007/978-3-319-11918-2_16
8. I, T., Sugimoto, S., Inenaga, S., Bannai, H., Takeda, M.: Computing palindromic factorizations and palindromic covers on-line. In: Kulikov, A.S., Kuznetsov, S.O., Pevzner, P. (eds.) CPM 2014. LNCS, vol. 8486, pp. 150–161. Springer, Cham (2014). https://doi.org/10.1007/978-3-319-07566-2_16
9. İleri, A.M., Külekci, M.O., Xu, B.: Shortest unique substring query revisited. In: Kulikov, A.S., Kuznetsov, S.O., Pevzner, P. (eds.) CPM 2014. LNCS, vol. 8486, pp. 172–181. Springer, Cham (2014). https://doi.org/10.1007/978-3-319-07566-2_18

10. Kuramoto, E., Yano, O., Kimura, Y., Baba, M., Makino, T., Yamamoto, S., Yamamoto, T., Kataoka, T., Tokunaga, T.: Oligonucleotide sequences required for natural killer cell activation. Jpn. J. Cancer Res. **83**(11), 1128–1131 (1992)
11. Manacher, G.: A new linear-time "on-line" algorithm for finding the smallest initial palindrome of a string. J. ACM **22**, 346–351 (1975)
12. Mieno, T., Inenaga, S., Bannai, H., Takeda, M.: Shortest unique substring queries on run-length encoded strings. In: Proceedings of MFCS 2016, pp. 69:1–69:11 (2016)
13. Pei, J., Wu, W.C.H., Yeh, M.Y.: On shortest unique substring queries. In: Proceedings of ICDE 2013, pp. 937–948 (2013)
14. Rubinchik, M., Shur, A.M.: EERTREE: an efficient data structure for processing palindromes in strings. In: Lipták, Z., Smyth, W.F. (eds.) IWOCA 2015. LNCS, vol. 9538, pp. 321–333. Springer, Cham (2016). https://doi.org/10.1007/978-3-319-29516-9_27
15. Tsuruta, K., Inenaga, S., Bannai, H., Takeda, M.: Shortest unique substrings queries in optimal time. In: Geffert, V., Preneel, B., Rovan, B., Štuller, J., Tjoa, A.M. (eds.) SOFSEM 2014. LNCS, vol. 8327, pp. 503–513. Springer, Cham (2014). https://doi.org/10.1007/978-3-319-04298-5_44
16. Yamamoto, S., Yamamoto, T., Kataoka, T., Kuramoto, E., Yano, O., Tokunaga, T.: Unique palindromic sequences in synthetic oligonucleotides are required to induce IFN [correction of INF] and augment IFN-mediated [correction of INF] natural killer activity. J. Immunol. **148**(12), 4072–4076 (1992)

A Faster Implementation of Online
Run-Length Burrows-Wheeler Transform

Tatsuya Ohno, Yoshimasa Takabatake, Tomohiro I, and Hiroshi Sakamoto[⊠]

Kyushu Institute of Technology, 680-4 Kawazu, Iizuka, Fukuoka 820-8502, Japan
{t_ohno,takabatake}@donald.ai.kyutech.ac.jp,
{tomohiro,hiroshi}@ai.kyutech.ac.jp

Abstract. Run-length encoding Burrows-Wheeler Transformed strings, resulting in *Run-Length BWT (RLBWT)*, is a powerful tool for processing highly repetitive strings. We propose a new algorithm for online RLBWT working in run-compressed space, which runs in $O(n \lg r)$ time and $O(r \lg n)$ bits of space, where n is the length of input string S received so far and r is the number of runs in the BWT of the reversed S. We improve the state-of-the-art algorithm for online RLBWT in terms of empirical construction time. Adopting the dynamic list for maintaining a total order, we can replace rank queries in a dynamic wavelet tree on a run-length compressed string by the direct comparison of labels in a dynamic list. The empirical result for various benchmarks show the efficiency of our algorithm, especially for highly repetitive strings.

1 Introduction

1.1 Motivation

The *Burrows-Wheeler Transform (BWT)* [8] is one of the most successful and elegant technique for lossless compression. When a string contains several frequent substrings, the transformed string would have several *runs*, i.e., maximal repeat of a symbol. Then, such a BWT string is easily compressed by run-length compression. We refer to the run-length compressed string as the *Run-Length BWT (RLBWT)* of the original string. Because of the definition of BWT, the number r of runs in the RLBWT is closely related to the easiness of compression of the original string. In fact, r can be (up to) exponentially smaller than the text length, and several studies [4,12,18,19] showed that r is available for a measure of repetitiveness.

After the invention of BWT, various applications have been proposed for string processing [7,9,10]. The most notable one would be the BWT based self-index, called FM index [10], which allows us to search patterns efficiently while storing text in the entropy-based compressed space. However, the traditional entropy-based compression is not enough to process highly repetitive strings because it does not capture the compressibility in terms of repetitiveness. Therefore several authors have studied "repetitive-aware" self-indexes based on

© Springer International Publishing AG, part of Springer Nature 2018
L. Brankovic et al. (Eds.): IWOCA 2017, LNCS 10765, pp. 409–419, 2018.
https://doi.org/10.1007/978-3-319-78825-8_33

RLBWT [4,12,18,19]. In particular, a self-index in [4] works in space proportional to the sizes of the RLBWT and LZ77 [20], another powerful compressor that can capture repetitiveness.

When it comes to constructing the RLBWT, a major concern is to reduce the working space depending on the repetitiveness of a given text. Namely, the problem is to construct the RLBWT *online in run-length compressed space*. It has been suggested in [12] that we can solve the problem using a dynamic data structure supporting rank queries on run-length encoded strings. An implementation appears very recently in [1,17], proving its merit in space reduction. However the throughput is considerably sacrificed probably due to its use of dynamic succinct data structure. To ameliorate the time consumption, we present a novel algorithm for online RLBWT and show experimentally that our implementation runs faster with reasonable increase of memory consumption. Since Policriti and Prezza [16] recently proposed algorithms to compute LZ77 factorization in compressed space via RLBWT, online RLBWT becomes more and more important, and therefore, practical time-space tradeoffs are worth exploring.

1.2 Our Contribution

Given an input string $S = S[1]S[2]\cdots S[n]$ of length n in online manner, the algorithm described in [16] constructs the RLBWT of the reversed string $S^R = S[n]\cdots S[2]S[1]$ in $O(r\lg n)$ bits of space and $O(n\lg r)$ time, where r is the number of runs appearing in the BWT of S^R. When a new input symbol c is appended, whereas the BWT of Sc requires (in the worst case) sorting all the suffixes again, the BWT of $(Sc)^R$ requires just inserting c into the BWT of S^R, and the insert position can be efficiently computed by rank operations on the BWT of S^R. Hence a dynamic data structure on a run-length compressed string supporting rank operations allows to construct the RLBWT online. However, the algorithm of [16] internally uses rank operations on dynamic wavelet trees, which is considerably slow in practice.

In order to get a faster implementation, we replace the work carried out on dynamic wavelet trees by a comparison of integers using the dynamic maintenance of a total order. Here, the *Order-Maintenance Problem* is to maintain a total order of elements subject to $insert(X, Y)$: insert a new element Y immediately after X in the total order, $delete(X)$: remove X from the total order, and $order(X, Y)$: determine whether $X > Y$ in the total order. Bender et al. [5] proposed a simple algorithm for this problem to allow $O(1)$ amortized insertion and deletion time and $O(1)$ worst-case query time. Adopting this technique, we develop a novel data structure for computing the insert position of c in the current BWT by a comparison of integers, instead of heavy rank operations on dynamic wavelet trees.

Compared to the baseline [16], we significantly improve the throughput of RLBWT with reasonable increase of memory consumption. Although there is a tradeoff between memory consumption and throughput performance, as shown in the experimental results, the working space of our algorithm is still sufficiently smaller than the input size, especially for highly repetitive strings.

2   Preliminaries

Let Σ be an ordered *alphabet*. An element of Σ^* is called a *string*. The length of a string S is denoted by $|S|$. The empty string ε is the string of length 0, namely, $|\varepsilon| = 0$. For a string $S = XYZ$, strings X, Y, and Z are called a *prefix*, *substring*, and *suffix* of S, respectively. For $1 \le i \le |S|$, the ith character of a string S is denoted by $S[i]$. For $1 \le i \le j \le |S|$, let $S[i..j] = S[i] \cdots S[j]$, i.e., $S[i..j]$ is the substring of S starting at position i and ending at position j in S. For convenience, let $S[i..j] = \varepsilon$ if $j < i$.

In the *run-length encoding (RLE)* of a string S, a maximal run c^e (for some $c \in \Sigma$ and $e \in \mathcal{N}$) of a single character in S is encoded by a pair (c, e), where we refer to c and respectively e as the *head* and *exponent* of the run. Since each run is encoded in $O(1)$ words (under Word RAM model with word size $\Omega(\lg |S|)$), we refer to the number of runs as the size of the RLE. For example, $S = \texttt{aaaabbcccacc} = \texttt{a}^4\texttt{b}^2\texttt{c}^3\texttt{a}^1\texttt{c}^2$ is encoded as $(\texttt{a}, 4), (\texttt{b}, 2), (\texttt{c}, 3), (\texttt{a}, 1), (\texttt{c}, 2)$, and the size of the RLE is five.

For any string S and any $c \in \Sigma$, let $\mathsf{occ}_c(S)$ denote the number of occurrences of c in S. Also, let $\mathsf{occ}_{<c}(S)$ denote the number of occurrences of any character smaller than c in S, i.e., $\mathsf{occ}_{<c}(S) = \sum_{c' < c} \mathsf{occ}_{c'}(S)$. For any $c \in \Sigma$ and position i $(1 \le i \le |S|)$, $\mathsf{rank}_c(S, i)$ denotes the number of occurrences of c in $S[1..i]$, i.e., $\mathsf{rank}_c(S, i) = \mathsf{occ}_c(S[1..i])$. For any $c \in \Sigma$ and i $(1 \le i \le \mathsf{occ}_c(S))$, $\mathsf{select}_c(S, i)$ denotes the position of the ith c in S, i.e., $\mathsf{select}_c(S, i) = \min\{j \mid \mathsf{rank}_c(S, j) = i\}$. Also we let $\mathsf{access}(S, i)$ denote the query to ask for $S[i]$. We will consider data structures to answer $\mathsf{occ}_{<c}$, rank, select, and access without having S explicitly.

2.1   BWT

Here we define the BWT of a string $S \in \Sigma^+$, denoted by BWT_S. For convenience, we assume that S ends with a terminator $\$ \in \Sigma$ whose lexicographic order is smaller than any character in $S[1..|S| - 1]$. BWT_S is obtained by sorting all non-empty suffixes of S lexicographically and putting the immediately preceding character of each suffix (or $\$$ if there is no preceding character) in the order.

For the online construction of BWT, it is convenient to consider "prepending" (rather than appending) a character to S because it does not change the lexicographic order among existing suffixes.[1] Namely, for some $c \in \Sigma$, we consider updating BWT_S to BWT_{cS} efficiently. The task is to replace the unique occurrence of $\$$ in BWT_S with c, and insert $\$$ into appropriate position. Since replacing can be easily done if we keep track of the current position of $\$$, the main task is to find the new position of $\$$ to insert, which can be done with a standard operation on BWT as follows: Let i be the position of $\$$ in BWT_S, then the new position is computed by $\mathsf{rank}_c(\mathsf{BWT}_S, i) + \mathsf{occ}_{<c}(S) + 1$ because the new suffix cS is the $(\mathsf{rank}_c(\mathsf{BWT}_S, i) + 1)$th lexicographically smallest suffix among those starting with c, and there are $\mathsf{occ}_{<c}(S)$ suffixes starting with

---

[1] Or appending a character but constructing BWT for reversed string.

some c' ($< c$). Thus, BWT can be constructed online using a data structure that supports rank, $\mathrm{occ}_{<c}$, and insert queries.

Let RLBWT_S denote the run-length encoding of BWT_S. In Sect. 3, we study data structures that supports rank_c, $\mathrm{occ}_{<c}$ and insert queries on run-length encoded strings, which can be directly used to construct RLBWT_S online in $O(|S| \lg r)$ time and $O(r \lg |S|)$ bits of space, where r is the size of RLE of BWT_S.

2.2   Searchable Partial Sums with Indels

We use a data structure for the *searchable partial sums with indels (SPSI)* problem as a tool. The SPSI data structure T ought to maintain a dynamic sequence $Z[1..m]$ of non-negative integers (called *weights*) to support the following queries as well as insertion/deletion of weights:

- $\mathsf{T}.\mathsf{sum}(k)$: Return the partial sum $\sum_{j=1}^{k} Z[j]$.
- $\mathsf{T}.\mathsf{search}(i)$: For an integer i ($1 \leq i \leq \mathsf{T}.\mathsf{sum}(m)$), return the minimum index k such that $\mathsf{T}.\mathsf{sum}(k) \geq i$.
- $\mathsf{T}.\mathsf{update}(k, \delta)$: For a (possibly negative) integer δ with $Z[k] + \delta \geq 0$, update $Z[k]$ to $Z[k] + \delta$.

We employ a simple implementation of T based on a B+tree whose kth leaf corresponds to $Z[k]$.[2] Let B (≥ 3) be the parameter of B+trees that represents the arity of an internal node. Namely the number of children of each internal node ranges from $B/2$ to B (unless m is too small), and thus, the height of the tree is $O(\log_B m)$. An internal node has two integer arrays LA and WA of length B such that $LA[j]$ (resp. $WA[j]$) stores the sum of #leaves (resp. weights) under the subtrees of up to the jth child of the node.

Using these arrays, we can easily answer $\mathsf{T}.\mathsf{sum}$ and $\mathsf{T}.\mathsf{search}$ queries in $O(\log_B m)$ time while traversing the tree from the root to a leaf: For example, $\mathsf{T}.\mathsf{sum}(k)$ can be computed by traversing to the kth leaf (navigated by LA) while summing up the weights of the subtrees existing to the left of the traversed path by WA. It is the same for $\mathsf{T}.\mathsf{search}(i)$ (except switching the roles of LA and WA). For $\mathsf{T}.\mathsf{update}(k, \delta)$ query, we only have to update LA and WA of the nodes in the path from the root to the kth leaf, which takes $O(B \log_B m)$ time. Also, indels can be done in $O(B \log_B m)$ time with standard split/merge operations of B+trees.

Naively the space usage is $O(m \lg M)$ bits, where M is the sum of all weights. Here we consider improving this to $O(m \lg(M/m))$ bits. Let us call an internal node whose children are leaves a *bottom node*, for which we introduce new arity parameter B_L, differentiated from B for the other internal nodes. For a bottom node, we discard LA, WA and the pointers to the leaves. Instead we let it store the weights of its children in a space efficient way. For example, using gamma encoding, the total space usage for the bottom nodes becomes $O(\sum_{j=1}^{m} \lg Z[j]) = O(m \lg(M/m))$ bits. The other (upper) part of T uses $O(m \lg M/B_L)$ bits, which

---

[2] More sophisticated solutions can be found in [6,11,15], but none of them has been implemented to the best of our knowledge.

can be controlled by B_L. The queries can be supported in $O(B_L + B \log_B m/B_L)$ time. Hence, setting $B = O(1)$ and $B_L = \Theta(\lg m)$, we get the next lemma.

Lemma 1. *For a dynamic sequence of weights, there is a SPSI data structure of $O(m \lg(M/m))$ bits supporting queries in $O(\lg m)$ time, where m is the current length of the sequence and M is the sum of weights.*

3 Dynamic Rank/Select Data Structures on Run-Length Encoded Strings

In this section, we study dynamic rank/select data structures working on run-length encoded strings. Note that select and delete queries are not needed for online RLBWT algorithms, but we also provide them as they may find other applications. Throughout this section, we let X denote the current string with $n = |X|$, RLE size r, and containing σ distinct characters. We consider the following update queries as well as rank_c, select_c, access and $\mathsf{occ}_{<c}$ queries on X:

- $\mathsf{insert}(X, i, c^e)$: For a position i $(1 \le i \le n+1)$, $c \in \Sigma$ and $e \in \mathcal{N}$, insert c^e between $X[i-1]$ and $X[i]$, i.e., $X \leftarrow X[1..i-1]c^e X[i..n]$.
- $\mathsf{delete}(X, i, e)$: For a position i $(1 \le i \le n-e+1)$ such that $X[i..i+e-1] \in c^e$ for some $c \in \Sigma$, delete $X[i..i+e-1]$, i.e., $X \leftarrow X[1..i-1]X[i+e..n]$.

Theorem 1. *There is a data structure that occupies $O(r \lg n)$ bits of space and supports rank_c, select_c, access, $\mathsf{occ}_{<c}$, insert and delete in $O(\lg r)$ time.*

We will describe two data structures holding the complexities of Theorem 1 in theory but likely exhibiting different time-space tradeoffs in practice. In Subsect. 3.1, we show an existing data structure. On the basis of this data structure, in Subsect. 3.2, we present our new data structure to get a faster implementation.

We note that the problem to support $\mathsf{occ}_{<c}$ in $O(\sigma \lg n)$ bits of space and $O(\lg \sigma)$ time is somewhat standard. For instance, we can think about the SPSI data structure of Lemma 1 storing $\mathsf{occ}_c(X)$'s in increasing order of c. It is easy to modify the data structure so that, for a given c, we can traverse from the root to the leaf corresponding to the predecessor of c, where we mean by the predecessor of c the largest character c' that is smaller than c and appears in X. Then $\mathsf{occ}_{<c}$ queries can be supported in a similar way to sum queries using WA. Thus in the following subsections, we focus on the other queries.

3.1 Existing Data Structure

Here we review the data structure described in [16] with implementation available in [1,17].[3] In theory it satisfies Theorem 1 though its actual implementation has the time complexity of $O(\lg \sigma \lg r)$ slower than $O(\lg r)$.

Let $(c_1, e_1), (c_2, e_2), \ldots, (c_r, e_r)$ be the RLE of X. The data structure consists of three components (see also Fig. 1 for the first two):

---

[3] The basic idea of the algorithm originates from the work of RLFM+ index in [12].

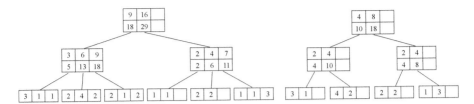

Fig. 1. For $X = \mathrm{a^3b^1a^1c^2a^4b^2a^2c^1a^2b^1c^1a^2c^2a^1b^1a^3}$, examples of T_{all} (left) and T_a (right) with $B = B_L = 3$ are shown. Note that the other components of the data structure (T_b, T_c and H) are omitted here. T_{all} holds the sequence $[3,1,1,2,4,2,2,1,2,1,1,2,2,1,1,3]$ of the exponents in its leaves, and T_a holds the sequence $[3,1,4,2,2,2,1,3]$ of the exponents of a's runs in its leaves. For a node having two rows, the first row represents LA and the second WA.

1. T_{all}: SPSI data structure for the sequence $e_1e_2\cdots e_r$ of all exponents.
2. T_c (for every $c \in \Sigma$): SPSI data structure for the sequence of the exponents of c's run.
3. H: Dynamic rank/select data structure for the head string $H = c_1c_2\cdots c_r$. There is a data structure (e.g., see [13,14]), with which H can be implemented in $r\lg\sigma + o(r\lg\sigma) + O(\sigma\lg r)$ bits while supporting queries in $O(\lg r)$ time. (However, the actual implementation of [1,17] employs a simpler algorithm based on wavelet trees that has $O(\lg\sigma\lg r)$ query time.)

Note that for every run c^e there are two copies of its exponent, one in T_{all} and the other in T_c. Since $\sigma \leq r \leq n$ holds, the data structure (excluding $\mathsf{occ}_{<c}$ data structure) uses $r\lg\sigma + o(r\lg\sigma) + O(r\lg(n/r) + \sigma\lg r) = O(r\lg n)$ bits.

Let us demonstrate how to support $\mathsf{rank}_c(X,i)$. Firstly by computing $k \leftarrow \mathsf{T}_{all}.\mathsf{search}(i)$ we can find that $X[i]$ is in the kth run. Next by computing $k_c \leftarrow \mathsf{H.rank}_c(H,k)$ we notice that, up to the kth run, there are k_c runs with head c. Here we can check if the head of the kth run is c, and compute the number e of c's in the kth run appearing after $X[i]$. Finally, $\mathsf{T}_c.\mathsf{sum}(k_c) - e$ tells the answer of $\mathsf{rank}_c(X,i)$. It is easy to see that each step can be done in $O(\lg r)$ time.

Note that H plays an important role to bridge two trees T_{all} and T_c by converting the indexes k and k_c. The update queries also use this mechanism: We first locate the update position in T_{all}, then find the update position in T_c by bridging two trees with H. After locating the positions, the updates can be done in each dynamic data structure. $\mathsf{select}_c(X,i)$ can be answered by first locating ith c in T_c, finding the corresponding position in T_{all} with $\mathsf{H.select}_c$, then computing the partial sum up to the position in T_{all}. Finally, $\mathsf{access}(X,i)$ is answered by $\mathsf{H.access}(H, \mathsf{T}_{all}.\mathsf{search}(i))$.

3.2 New Data Structure

Now we present our new data structure satisfying Theorem 1. We share some of the basic concepts with the data structure described in Sect. 3.1. For example, our data structure also uses the idea of answering queries by going back and

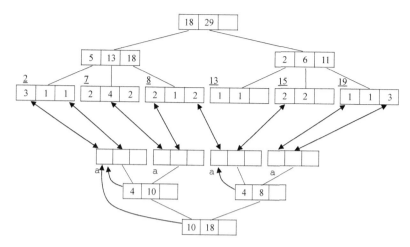

Fig. 2. For $X = \mathsf{a}^3\mathsf{b}^1\mathsf{a}^1\mathsf{c}^2\mathsf{a}^4\mathsf{b}^2\mathsf{a}^2\mathsf{c}^1\mathsf{a}^2\mathsf{b}^1\mathsf{c}^1\mathsf{a}^2\mathsf{c}^2\mathsf{a}^1\mathsf{b}^1\mathsf{a}^3$ (same as the one in Fig. 1), examples of modified T_{all} (up) and T_a (down) with $B = B_L = 3$ are shown, where T_a is illustrated upside down. Note that the data structure related to T_b and T_c (e.g., pointers of Change1 to them) are omitted here. Each pair of leaves corresponding to the same run is connected by bidirectional pointers (Change1). Each internal node of T_a has pointer to its leftmost leaf (Change2). The character a is stored in each bottom node of T_a (Change3). Each bottom node of T_{all} stores a label (underlined number) that is monotonically increasing from left to right (Change4). LAs and weights in the leaves of T_a are discarded (Change5).

forth between T_{all} and T_c. However we do not use H to bridge the two trees. Succinct data structures (like H) are attractive if the space limitation is critical, but otherwise the suffered slow-down might be intolerable. Therefore we design a fast algorithm that bridges the two trees in a more direct manner, while taking care not to blow up the space too much.

In order to do without H, we make some changes to T_{all} and T_c (see also Fig. 2):

1. We maintain bidirectional pointers connecting every pair of leaves representing the same run in T_{all} and T_c (recall that every run with head c has exactly one corresponding leaf in each of T_{all} and T_c).
2. For every internal node of T_c, we store a pointer to the leftmost leaf in the subtree rooted at the node.
3. For every bottom node of T_c, we store the character c.
4. For every bottom node of T_{all}, we store a label (a positive integer) such that the labels of bottom nodes are monotonically increasing from left to right. Since bottom nodes are inserted/deleted dynamically, we utilize the algorithm [5] for the order-maintenance problem to maintain the labels.
5. Minor implementation notes: Every LA can be discarded as our data structure does not use the navigation of indexes. Also, we can quit storing the leaf-level weights in T_c as it can be retrieved using the pointer to the corresponding leaf in T_{all}.

Space Analysis. In the change list, Changes1–4 increase the space usage while Change5 reduces. It is easy to see that the increase fits in $O(r \lg n)$ bits. More precisely, since Changes2–4 are made to internal nodes, the increase by these changes is $O(r \lg n / B_L)$ bits, which is more or less controllable by B_L (recall that B_L is arity parameter for bottom nodes, and we have $O(r \lg n / B_L) = O(r \lg(n/r))$ by setting $B_L = \Theta(\lg r)$ for Lemma 1). On the other hand, Change1 is made to leaves and takes $2r \lg r$ bits of space. Thus, the total space usage of the data structure (excluding $\mathsf{occ}_{<c}$ data structure) is $2r \lg r + O(r \lg(n/r)) = O(r \lg n)$ bits.

By this analysis, it is expected that $2r \lg r$ becomes a leading term when the ratio n/r is small, i.e., compressibility in terms of RLE is not high. It should be compared to $r \lg \sigma + o(r \lg \sigma) + O(r \lg(n/r) + \sigma \lg r)$ bits used by the data structure of Sect. 3.1, in which $r \lg r$ term does not exist. Hence, the smaller the ratio n/r, the larger the gap between the two data structures in space usage will be. On the other hand, when the $r \lg(n/r)$ term is leading, i.e., r is sufficiently smaller than n, the increase by the $r \lg r$ term would be relatively moderate.

Answering Queries. We show how to answer queries on our data structure. All queries are supported in $O(\lg r)$ time.

$\mathsf{access}(X, i)$: We first traverse from the root of T_{all} to the run containing $X[i]$ (navigated by WA), jump to the corresponding leaf of T_c by pointer of Change1, then read the character stored in the bottom node of T_c due to Change3.

$\mathsf{select}_c(X, i)$: We first traverse from the root of T_c to the run containing ith c (navigated by WA). At the same time, we can compute the rank i' of ith c within the run. Next we jump to the corresponding leaf in T_{all} by pointer of Change1, then compute the sum of characters appearing strictly before the leaf while going up the tree. The answer to $\mathsf{select}_c(X, i)$ is the sum plus i'.

$\mathsf{rank}_c(X, i)$: Recalling the essence of the algorithm described in Sect. 3.1, we can answer $\mathsf{rank}_c(X, i)$ if we locate the leaf of T_c representing the rightmost c's run that starts at or before position i. In order to locate such leaf v, we first traverse from the root of T_{all} to the run containing $X[i]$ (navigated by WA). If we are lucky, we may find a c's run in the bottom node containing $X[i]$, in which case we can easily get v or the successor of v by using the pointer of Change1 outgoing from the c's run. Otherwise, we search for v traversing T_c from the root navigated by labels of Change4. Let t be the label of the bottom node containing $X[i]$. Then, it holds that v is the rightmost leaf pointing to a node of T_{all} with label smaller than t. Since the order of labels is maintained, we can use t as a key for binary search, i.e., we notice that an internal node u (and its succeeding siblings) cannot contain v if the leftmost leaf in the subtree rooted at u points to a node of T_{all} with label greater than t. Using the pointer of Change2 to jump to the leftmost leaf, we can conduct each comparison in $O(1)$ time, and thus, we can find v in $O(\lg r)$ time.

Update queries: The main task is to locate the update positions both in T_{all} and T_c, and this is exactly what we did in rank_c query—locating the run containing $X[i]$ and v. After locating the update positions, the update can be done in

Table 1. Computation time in seconds and working space in mega bytes to construct the RLBWT of r runs from each dataset of size $|S|$ using the proposed method (ours) and the previous method (PP).

Dataset	$|S|$ (MB)	r	Time (sec)		Space (MB)	
			ours	PP	ours	PP
fib41	255.503	42	27	552	0.004	0.067
rs.13	206.706	76	16	623	0.005	0.068
tm29	256.000	82	24	802	0.005	0.068
dblp.xml.00001.1	100.000	172, 195	37	2, 060	2.428	1.307
dblp.xml.00001.2	100.000	175, 278	37	2, 070	2.446	1.322
dblp.xml.0001.1	100.000	240, 376	40	2, 100	4.381	1.586
dblp.xml.0001.2	100.000	269, 690	40	2, 105	4.565	1.730
dna.001.1	100.000	1, 717, 162	58	1, 667	35.966	5.729
english.001.2	100.000	1, 436, 696	58	2, 153	20.680	6.166
proteins.001.1	100.000	1, 278, 264	58	1, 839	19.790	5.133
sources.001.2	100.000	1, 211, 104	49	2, 141	19.673	5.721
cere	439.917	11, 575, 582	534	7, 597	186.073	43.341
coreutils	195.772	4, 732, 794	128	4, 479	81.642	22.301
einstein.de.txt	88.461	99, 833	30	1, 807	2.083	1.106
einstein.en.txt	445.963	286, 697	182	9, 293	4.836	2.296
Escherichia_Coli	107.469	15, 045, 277	154	2, 047	316.184	36.655
influenza	147.637	3, 018, 824	91	2, 501	72.730	12.386
kernel	246.011	2, 780, 095	146	5, 333	41.758	12.510
para	409.380	15, 635, 177	547	7, 364	329.901	52.005
world_leaders	44.792	583, 396	17	857	9.335	2.891
boost	1024.000	63, 710	320	20, 327	1.161	0.904
samtools	1024.000	562, 326	440	21, 375	9.734	3.595
sdsl	1024.000	758, 657	419	21, 014	17.760	4.803

$O(\lg r)$ time in each tree. When the update operation invokes insertion/deletion of a bottom node of T_{all}, we maintain labels of Change4 using the algorithm of [5]. We note that the algorithm of [5] takes $O(\lg r)$ amortized time per "indel of bottom node", and hence, takes $O(1)$ amortized time per "indel of leaf" (recall that $B_L = \Theta(\lg r)$, and one indel of bottom node needs $\Theta(\lg r)$ indels of leaves). In addition, the algorithm is quite simple and efficiently implementable without any data structure than labels themselves.

4   Experiments

We implemented in C++ the online RLBWT construction algorithm based on our new rank/select data structure described in Sect. 3.2 (the source code is available at [2]). We evaluate the performance of our method comparing with the state-of-the-art implementation [1] (we refer to it as PP taking the authors' initials of [16]) of the algorithm based on the data structure described in Sect. 3.1. We tested on highly repetitive datasets in repcorpus[4], well-known corpus in this field, and some larger datasets created from git repositories. For the latter, we use the script [3] to create 1024 MB texts (obtained by concatenating source files from the latest revisions of a given repository, and truncated to be 1024 MB) from the repositories for boost[5], samtools[6] and sdsl-lite[7] (all accessed at 2017-03-27). The programs were compiled using g++6.3.0 with -Ofast -march=native option. The experiments were conducted on a 6core Xeon E5-1650V3 (3.5 GHz) machine with 32 GB memory running Linux CentOS7.

Table 1 shows the comparison of the two methods on construction time and working space. The result shows that our method significantly improves the construction time of PP as we intended. Especially for dumpfiles of Wikipedia articles (einstein.de.txt and einstein.en.txt), our method ran 60 times faster than PP. Our method also shows good performance for the 1024 MB texts from git repositories. On the other hand, the working space is increased (except the artificial datasets, which are extremely compressible) by 1.3–8.7 times. Especially for less compressible datasets in terms of RLBWT like Escherichia_Coli, the space usage tends to be worse as predicted by space analysis in Sect. 3.2. Still for most of the other datasets the working space of our method keeps way below the input size.

5   Conclusion

We have proposed an improvement of online construction of RLBWT [1,17], intended to speed up the construction time. We significantly improved the throughput of original RLBWT with reasonable increase of memory consumption for the benchmarks from various domain. By applying our new algorithm to the algorithm of computing LZ77 factorization in compressed space using RLBWT [16], we would immediately improve the throughput of [16]. As LZ77 plays a central role in many problems on string processing, engineering/optimizing implementation for compressed LZ77 computation is important future work.

Acknowledgments. This work was supported by JST CREST (Grant Number JPMJCR1402), and KAKENHI (Grant Numbers 17H01791 and 16K16009).

---

[4] See http://pizzachili.dcc.uchile.cl/repcorpus/statistics.pdf for statistics of the datasets.

[5] https://github.com/boostorg/boost.

[6] https://github.com/samtools/samtools.

[7] https://github.com/simongog/sdsl-lite.

References

1. DYNAMIC: Dynamic succinct/compressed data structures library. https://github.com/xxsds/DYNAMIC
2. Online RLBWT. https://github.com/itomomoti/OnlineRLBWT
3. Get-git-revisions: Get all revisions of a git repository. https://github.com/nicolaprezza/get-git-revisions
4. Belazzougui, D., Cunial, F., Gagie, T., Prezza, N., Raffinot, M.: Composite repetition-aware data structures. In: Cicalese, F., Porat, E., Vaccaro, U. (eds.) CPM 2015. LNCS, vol. 9133, pp. 26–39. Springer, Cham (2015). https://doi.org/10.1007/978-3-319-19929-0_3
5. Bender, M.A., Cole, R., Demaine, E.D., Farach-Colton, M., Zito, J.: Two simplified algorithms for maintaining order in a list. In: Möhring, R., Raman, R. (eds.) ESA 2002. LNCS, vol. 2461, pp. 152–164. Springer, Heidelberg (2002). https://doi.org/10.1007/3-540-45749-6_17
6. Bille, P., Cording, P.H., Gørtz, I.L., Skjoldjensen, F.R., Vildhøj, H.W., Vind, S.: Dynamic relative compression, dynamic partial sums, and substring concatenation. In: ISAAC, pp. 18:1–18:13 (2016)
7. Bowe, A., Onodera, T., Sadakane, K., Shibuya, T.: Succinct de Bruijn graphs. In: Raphael, B., Tang, J. (eds.) WABI 2012. LNCS, vol. 7534, pp. 225–235. Springer, Heidelberg (2012). https://doi.org/10.1007/978-3-642-33122-0_18
8. Burrows, M., Wheeler, D.J.: A block-sorting lossless data compression algorithm. Technical report, HP Labs (1994)
9. Ferragina, P., Luccio, F., Manzini, G., Muthukrishnan, S.: Structuring labeled trees for optimal succinctness, and beyond. In: FOCS, pp. 184–196 (2005)
10. Ferragina, P., Manzini, G.: Opportunistic data structures with applications. In: FOCS, pp. 390–398 (2000)
11. Hon, W., Sadakane, K., Sung, W.: Succinct data structures for searchable partial sums with optimal worst-case performance. Theor. Comput. Sci. 412(39), 5176–5186 (2011)
12. Mäkinen, V., Navarro, G., Sirén, J., Välimäki, N.: Storage and retrieval of highly repetitive sequence collections. J. Comput. Biol. 17(3), 281–308 (2010)
13. Munro, J.I., Nekrich, Y.: Compressed data structures for dynamic sequences. In: Bansal, N., Finocchi, I. (eds.) ESA 2015. LNCS, vol. 9294, pp. 891–902. Springer, Heidelberg (2015). https://doi.org/10.1007/978-3-662-48350-3_74
14. Navarro, G., Nekrich, Y.: Optimal dynamic sequence representations. SIAM J. Comput. 43(5), 1781–1806 (2014)
15. Navarro, G., Sadakane, K.: Fully functional static and dynamic succinct trees. ACM Trans. Algorithms 10(3), 16 (2014)
16. Policriti, A., Prezza, N.: Computing LZ77 in run-compressed space. In: DCC, pp. 23–32 (2016)
17. Prezza, N.: A framework of dynamic data structures for string processing. In: SEA (2017 to appear)
18. Sirén, J.: Compressed Full-Text Indexes for Highly Repetitive Collections. Ph.D. thesis, University of Helsinki (2012)
19. Sirén, J., Välimäki, N., Mäkinen, V., Navarro, G.: Run-length compressed indexes are superior for highly repetitive sequence collections. In: Amir, A., Turpin, A., Moffat, A. (eds.) SPIRE 2008. LNCS, vol. 5280, pp. 164–175. Springer, Heidelberg (2008). https://doi.org/10.1007/978-3-540-89097-3_17
20. Ziv, J., Lempel, A.: A universal algorithm for sequential data compression. IEEE Trans. Inf. Theory IT 23(3), 337–349 (1977)

Computing Abelian String Regularities Based on RLE

Shiho Sugimoto[1]($\boxtimes$), Naoki Noda[2], Shunsuke Inenaga[1], Hideo Bannai[1], and Masayuki Takeda[1]

[1] Department of Informatics, Kyushu University, Fukuoka, Japan
{shiho.sugimoto,inenaga,bannai,takeda}@inf.kyushu-u.ac.jp
[2] Department of Physics, Kyushu University, Fukuoka, Japan

Abstract. Two strings x and y are said to be Abelian equivalent if x is a permutation of y, or vice versa. If a string z satisfies $z = xy$ with x and y being Abelian equivalent, then z is said to be an *Abelian square*. If a string w can be factorized into a sequence $v_1, \ldots, v_s$ of strings such that $v_1, \ldots, v_{s-1}$ are all Abelian equivalent and v_s is a substring of a permutation of v_1, then w is said to have a *regular Abelian period* (p, t) where $p = |v_1|$ and $t = |v_s|$. If a substring $w_1[i..i+\ell-1]$ of a string w_1 and a substring $w_2[j..j + \ell - 1]$ of another string w_2 are Abelian equivalent, then the substrings are said to be a common Abelian factor of w_1 and w_2 and if the length ℓ is the maximum of such then the substrings are said to be a *longest common Abelian factor* of w_1 and w_2. We propose efficient algorithms which compute these Abelian regularities using the *run length encoding (RLE)* of strings. For a given string w of length n whose RLE is of size m, we propose algorithms which compute all Abelian squares occurring in w in $O(mn)$ time, and all regular Abelian periods of w in $O(mn)$ time. For two given strings w_1 and w_2 of total length n and of total RLE size m, we propose an algorithm which computes all longest common Abelian factors in $O(m^2 n)$ time.

1 Introduction

Two strings s_1 and s_2 are said to be *Abelian equivalent* if s_1 is a permutation of s_2, or vice versa. For instance, strings *ababaac* and *caaabba* are Abelian equivalent. Since the seminal paper by Erdős [7] published in 1961, the study of Abelian equivalence on strings has attracted much attention, both in word combinatorics and string algorithmics.

1.1 Our Problems and Previous Results

In this paper, we are interested in the following algorithmic problems related to Abelian regularities of strings.

© Springer International Publishing AG, part of Springer Nature 2018
L. Brankovic et al. (Eds.): IWOCA 2017, LNCS 10765, pp. 420–431, 2018.
https://doi.org/10.1007/978-3-319-78825-8_34

1. Compute *Abelian squares* in a given string.
2. Compute *regular Abelian periods* of a given string.
3. Compute *longest common Abelian factors* of two given strings.

Cummings and Smyth [6] proposed an $O(n^2)$-time algorithm to solve Problem 1, where n is the length of the given string. Crochemore et al. [5] proposed an alternative $O(n^2)$-time solution to the same problem. Recently, Kociumaka et al. [12] showed how to compute all Abelian squares in $O(s + \frac{n^2}{\log^2 n})$ time, where s is the number of outputs.

Related to Problem 2, various kinds of Abelian periods of strings have been considered: An integer p is said to be a *full Abelian period* of a string w iff there is a decomposition $u_1, \ldots, u_z$ of w such that $|u_i| = p$ for all $1 \leq i \leq z$ and $u_1, \ldots, u_z$ are all Abelian equivalent. A pair (p, t) of integers is said to be a *regular Abelian period* (or simply an *Abelian period*) of a string w iff there is a decomposition $v_1, \ldots, v_s$ of w such that p is a full Abelian period of $v_1 \cdots v_{s-1}$, $|v_i| = p$ for all $1 \leq i \leq s - 1$, and v_s is a permutation of a substring of v_1 (and hence $t \leq p$). A triple (h, p, t) of integers is said to be a *weak Abelian period* of a string w iff there is a decomposition $y_1, \ldots, y_r$ of w such that (p, t) is an Abelian period of $y_2 \cdots y_r$, $|y_1| = h$, $|y_i| = p$ for all $2 \leq i \leq r - 1$, $|y_r| = t$, and y_1 is a permutation of a substring of y_2 (and hence $h \leq p$).

The study on Abelian periodicity of strings was initiated by Constantinescu and Ilie [4]. Fici et al. [9] gave an $O(n \log \log n)$-time algorithm to compute all full Abelian periods. Later, Kociumaka et al. [11] showed an optimal $O(n)$-time algorithm to compute all full Abelian periods. Fici et al. [9] also showed an $O(n^2)$-time algorithm to compute all regular Abelian periods for a given string of length n. Kociumaka et al. [11] also developed an algorithm which finds all regular Abelian periods in $O(n(\log \log n + \log \sigma))$ time, where σ is the alphabet size. Fici et al. [8] proposed an algorithm which computes all weak Abelian periods in $O(\sigma n^2)$ time, and later Crochemore et al. [5] proposed an improved $O(n^2)$-time algorithm to compute all weak Abelian periods. Kociumaka et al. [12] showed how to compute all *shortest* weak Abelian periods in $O(n^2/\sqrt{\log n})$ time.

Consider two strings w_1 and w_2. A pair (s_1, s_2) of a substring s_1 of w_1 and a substring s_2 of w_2 is said to be a *common Abelian factor* of w_1 and w_2, iff s_1 and s_2 are Abelian equivalent. Alatabbi et al. [1] proposed an $O(\sigma n^2)$-time and $O(\sigma n)$-space algorithm to solve Problem 3 of computing all *longest common Abelian factors* (*LCAFs*) of two given strings of total length n. Later, Grabowski [10] showed an algorithm which finds all LCAFs in $O(\sigma n^2)$ time with $O(n)$ space. He also presented an $O((\frac{\sigma}{k} + \log \sigma)n^2 \log n)$-time $O(kn)$-space algorithm for a parameter $k \leq \frac{\sigma}{\log \sigma}$. Recently, Badkobeh et al. [3] proposed an $O(n \log^2 n \log^* n)$-time $O(n \log^2 n)$-space algorithm for finding all LCAFs.

1.2 Our Contribution

In this paper, we show that we can accelerate computation of Abelian regularities of strings via *run length encoding* (*RLE*) of strings. Namely, if m is the size of the RLE of a given string w of length n, we show that:

(1) All Abelian squares in w can be computed in $O(mn)$ time.
(2) All regular Abelian periods of w can be computed in $O(mn)$ time.

Since $m \leq n$ always holds, solution (1) is at least as efficient as the $O(n^2)$-time solutions by Cummings and Smyth [6] and by Crochemore et al. [5], and can be much faster when the input string w is highly compressible by RLE. Amir et al. [2] proposed an $O(\sigma(m^2 + n))$-time algorithm to compute all Abelian squares using RLEs. Our $O(mn)$-time solution is faster than theirs when $\frac{\sigma m^2}{m-\sigma} = \omega(n)$. Solution (2) is faster than the $O(n(\log \log n + \log \sigma))$-time solution by Kociumaka et al. [11] for highly RLE-compressible strings with $\log \log n = \omega(m)^1$.

Also, if m is the total size of the RLEs of two given strings w_1 and w_2 of total length n, we show that:

(3) All longest common Abelian factors of w_1 and w_2 can be computed in $O(m^2 n)$ time.

Our solution (3) is faster than the $O(\sigma n^2)$-time solution by Grabowski [10] when $\sigma n = \omega(m^2)$, and is faster than the fastest variant of the other solution by Grabowski [10] (choosing $k = \frac{\sigma}{\log \sigma}$) when $\sqrt{n \log n \log \sigma} = \omega(m)$. Also, solution (3) is faster than the $O(n \log^2 n \log^* n)$-time solution by Badkobeh et al. [3] when $\log n \sqrt{\log^* n} = \omega(m)$.

The time bounds of our algorithms are all deterministic. Proofs omitted due to lack of space can be found in a full version of this paper [13].

2 Preliminaries

Let $\Sigma = \{c_1, \ldots, c_\sigma\}$ be an ordered alphabet of size σ. An element of Σ^* is called a string. For any string w, $|w|$ denotes the length of w. The empty string is denoted by ε. Let $\Sigma^+ = \Sigma^* - \{\varepsilon\}$. For any $1 \leq i \leq |w|$, $w[i]$ denotes the i-th symbol of w. For a string $w = xyz$, strings x, y, and z are called a *prefix*, *substring*, and *suffix* of w, respectively. The substring of w that begins at position i and ends at position j is denoted by $w[i..j]$ for $1 \leq i \leq j \leq |w|$. For convenience, let $w[i..j] = \varepsilon$ for $i > j$.

For any string $w \in \Sigma^*$, its *Parikh vector* $\mathcal{P}_w$ is an array of length σ such that for any $1 \leq i \leq |\Sigma|$, $\mathcal{P}_w[i]$ is the number of occurrences of each character $c_i \in \Sigma$ in w. For example, for string $w = \mathsf{abaab}$ over alphabet $\Sigma = \{\mathsf{a}, \mathsf{b}\}$, $\mathcal{P}_w = \langle 3, 2 \rangle$. We say that strings x and y are *Abelian equivalent* if $\mathcal{P}_x = \mathcal{P}_y$. Note that $\mathcal{P}_x = \mathcal{P}_y$ iff x and y are permutations of each other. When x is a substring of a permutation of y, then we write $\mathcal{P}_x \subseteq \mathcal{P}_y$. For any Parikh vectors P and Q, let $diff(P, Q) = |\{i \mid P[i] \neq Q[i], 1 \leq i \leq \sigma\}|$.

A string w of length $2k > 0$ is called an *Abelian square* if it is a concatenation of two Abelian equivalent strings of length k each, i.e., $\mathcal{P}_{w[1..k]} = \mathcal{P}_{w[k+1..2k]}$. A string w is said to have a *regular Abelian period* (p, t) if w can be factorized into a sequence $v_1, \ldots, v_s$ of substrings such that $p = |v_1| = \cdots = |v_{s-1}|$, $|v_s| = t$,

---

1 Since we can w.l.o.g. assume that $\sigma \leq m$, the $\log \sigma$ term is negligible here.

$\mathcal{P}_{v_i} = \mathcal{P}_{v_1}$ for all $2 \leq i < s$, and $\mathcal{P}_{v_s} \subseteq \mathcal{P}_{v_1}$. For any strings $w_1, w_2 \in \Sigma^*$, if a substring $w_1[i..i + \ell - 1]$ of w_1 and a substring $w_2[j..j + \ell - 1]$ of w_2 are Abelian equivalent, then the pair of substrings is said to be a *common Abelian factor* of w_1 and w_2. When the length ℓ is the maximum of such then the pair of substrings is said to be a *longest common Abelian factor* of w_1 and w_2.

The *run length encoding* (*RLE*) of string w of length n, denoted $RLE(w)$, is a compact representation of w which encodes each maximal character run $w[i..i + p - 1]$ by a^p, if (1) $w[j] = a$ for all $i \leq j \leq i + p - 1$, (2) $w[i - 1] \neq w[i]$ or $i = 1$, and (3) $w[i+p-1] \neq w[i+p]$ or $i+p-1 = n$. E.g., $RLE(\mathsf{aabbbbcccaaa\$}) = \mathsf{a^2b^4c^3a^3\$^1}$. The *size* of $RLE(w) = a_1^{p_1} \cdots a_m^{p_m}$ is the number m of maximal character runs in w, and each $a_i^{p_i}$ is called an *RLE factor* of $RLE(w)$. Notice that $m \leq n$ always holds. Also, since at most m distinct characters can appear in w, in what follows we will assume that $\sigma \leq m$. Even if the underlying alphabet is large, we can sort the characters appearing in w in $O(m \log m)$ time and use this ordering in Parikh vectors. Since all of our algorithms will require at least $O(mn)$ time, this $O(m \log m)$-time preprocessing is negligible.

For any $1 \leq i \leq j \leq n$, let $RLE(w)[i..j] = a_i^{p_i} \cdots a_j^{p_j}$. For convenience let $RLE(w)[i..j] = \varepsilon$ for $i > j$. For $RLE(w) = a_1^{p_1} \cdots a_m^{p_m}$, let $RLE\_Bound(w) = \{1 + \sum_{i=1}^{k} p_k \mid 1 \leq k < m\} \cup \{1, n\}$. For any $1 \leq i \leq n$, let $succ(i) = \min\{j \in RLE\_Bound(w) \mid j > i\}$. Namely, $succ(i)$ is the smallest position in w that is greater than i and is either the beginning position of an RLE factor in w or the last position n in w.

3   Computing Regular Abelian Periods Using RLEs

In this section, we propose an algorithm which computes all regular Abelian periods of a given string.

Theorem 1. *Given a string w of length n over an alphabet of size σ, we can compute all regular Abelian periods of w in $O(mn)$ time and $O(n)$ working space, where m is the size of $RLE(w)$.*

Proof. Our algorithm is very simple. We use a single window for each length $d = 1, \ldots, \lfloor \frac{n}{2} \rfloor$. For an arbitrarily fixed d, consider a decomposition $v_1, \ldots, v_s$ of w such that $v_i = w[(i-1)d+1..id]$ for $1 \leq i \leq \lfloor \frac{n}{d} \rfloor$ and $v_s = w[n-(n \bmod d)+1..n]$. Each v_i is called a *block*, and each block of length d is called a *complete block*.

There are two cases to consider.

Case (a): If w is a unary string (i.e., $RLE(w) = a^n$ for some $a \in \Sigma$). In this case, $(d, (n \bmod d))$ is a regular Abelian period of w for any d. Also, note that this is the only case where $(d, (n \bmod d))$ can be a regular Abelian period of any string of length n with $RLE(v_i) = a^d$ for some complete block v_i. Clearly, it takes a total of $O(n)$ time and $O(1)$ space in this case.

Case (b): If w contains at least two distinct characters, then observe that a complete block v_i is fully contained in a single RLE factor iff $succ(1 + (i - 1)d) = succ(id)$. Let S be an array of length n such that $S[j] = succ(j)$ for each $1 \leq j \leq n$. We precompute this array S in $O(n)$ time by a simple left-to-right

1	2	3	4	5	6	7	8	9	10	11	12	13	14	15	16	17
a	a	b	b	a	a	a	b	a	b	a	a	a	a	b	b	a

Fig. 1. $(3, 2)$ is a regular Abelian period of string $w = aabbaaababaaaabbaa$ since $\mathcal{P}_{w[1..3]} = \mathcal{P}_{w[4..6]} = \mathcal{P}_{w[7..9]} = \mathcal{P}_{w[10..12]} = \mathcal{P}_{w[13..15]} \supset \mathcal{P}_{w[16..17]}$.

scan over w. Using the precomputed array S, we can check in $O(m)$ time if there exists a complete block v_i satisfying $succ(1 + (i - 1)d) = succ(id)$; we process each complete block v_i in increasing order of i (from left to right), and stop as soon as we find the first complete block v_i with $succ(1 + (i - 1)d) = succ(id)$. If there exists such a complete block, then we can immediately determine that $(d, (n \bmod d))$ is not a regular Abelian period (recall also Case (a) above.)

Assume every complete block v_i overlaps at least two RLE factors. For each v_i, let $m_i \geq 2$ be the number of RLE factors of $RLE(w)$ that v_i overlaps (i.e., m_i is the size of $RLE(v_i)$). We can compute $\mathcal{P}_{v_i}$ in $O(m_i)$ time from $RLE(v_i)$, using the exponents of the elements of $RLE(v_i)$. We can compare $\mathcal{P}_{v_i}$ and $\mathcal{P}_{v_{i-1}}$ in $O(m_i)$ time, since there can be at most m_i distinct characters in v_i and hence it is enough to check the m_i entries of the Parikh vectors. Since there are $\lfloor \frac{n}{d} \rfloor$ complete blocks and each complete block overlaps more than one RLE factor, we have $\lfloor \frac{n}{d} \rfloor \leq \frac{1}{2} \sum_{i=1}^{s-1} m_i$. Moreover, since each RLE factor is counted by a unique m_i or by a unique pair of m_{i-1} and m_i for some i, we have $\sum_{i=1}^{s} m_i \leq 2m$. Overall, it takes $O(\sigma + \frac{n}{d} + \sum_{i=1}^{s} m_i) = O(m)$ time to test if $(d, (n \bmod d))$ is a regular Abelian period of w. Consequently, it takes $O(mn)$ total time to compute all regular Abelian periods of w for all d's in this case. Since we use the array S of length n and we maintain two Parikh vectors of the two adjacent v_{i-1} and v_i for each i, the space requirement is $O(\sigma + n) = O(n)$. □

For example, let $w = aabbaaababaaaabbaa$ and $d = 3$. See also Fig. 1 for illustration. We have $RLE(w) = a^2 b^2 a^3 b^1 a^1 b^1 a^4 b^2 a^1$. Then, we compute $\mathcal{P}_{v_1} = \langle 2, 1 \rangle$ from $RLE(v_1) = a^2 b^1$, $\mathcal{P}_{v_2} = \langle 2, 1 \rangle$ from $RLE(v_2) = b^1 a^2$, $\mathcal{P}_{v_3} = \langle 2, 1 \rangle$ from $RLE(v_3) = a^1 b^1 a^1$, $\mathcal{P}_{v_4} = \langle 2, 1 \rangle$ from $RLE(v_4) = b^1 a^2$, $\mathcal{P}_{v_5} = \langle 2, 1 \rangle$ from $RLE(v_5) = a^2 b^1$, and $\mathcal{P}_{v_6} = \langle 1, 1 \rangle$ from $RLE(v_6) = b^1 a^1$. Since $\mathcal{P}_{v_i} = \mathcal{P}_{v_1}$ for $1 \leq i \leq 5$ and $\mathcal{P}_{v_6} \subset \mathcal{P}_{v_1}$, $(3, 2)$ is a regular Abelian period of the string w.

4   Computing Abelian Squares Using RLEs

In this section, we describe our algorithm to compute all Abelian squares occurring in a given string w of length n. Our algorithm is based on the algorithm of Cummings and Smyth [6] which computes all Abelian squares in w in $O(n^2)$ time. We will improve the running time to $O(mn)$, where m is the size of $RLE(w)$.

4.1   Cummings and Smyth's $O(n^2)$-Time Algorithm

We recall the $O(n^2)$-time algorithm proposed by Cummings and Smyth [6]. To compute Abelian squares in a given string w, their algorithm aligns two adjacent sliding windows of length d each, for every $1 \leq d \leq \lfloor \frac{n}{2} \rfloor$.

Consider an arbitrary fixed d. For each position $1 \leq i \leq n - 2d + 1$ in w, let L_i and R_i denote the left and right windows aligned at position i. Namely, $L_i = w[i..i+d-1]$ and $R_i = w[i+d..i+2d-1]$. At the beginning, the algorithm computes $\mathcal{P}_{L_1}$ and $\mathcal{P}_{R_1}$ for position 1 in w. It takes $O(d)$ time to compute these Parikh vectors and $O(\sigma)$ time to compute $\mathit{diff}(\mathcal{P}_{L_1}, \mathcal{P}_{R_1})$. Assume $\mathcal{P}_{L_i}$, $\mathcal{P}_{R_i}$, and $\mathit{diff}(\mathcal{P}_{L_i}, \mathcal{P}_{R_i})$ have been computed for position $i \geq 1$, and $\mathcal{P}_{L_{i+1}}$, $\mathcal{P}_{R_{i+1}}$, and $\mathit{diff}(\mathcal{P}_{L_{i+1}}, \mathcal{P}_{R_{i+1}})$ is to be computed for the next position $i + 1$. A key observation is that given $\mathcal{P}_{L_i}$, then $\mathcal{P}_{L_{i+1}}$ for the left window L_{i+1} for the next position $i + 1$ can be easily computed in $O(1)$ time, since at most two entries of the Parikh vector can change. The same applies to $\mathcal{P}_{R_i}$ and $\mathcal{P}_{R_{i+1}}$. Also, given $\mathit{diff}(\mathcal{P}_{L_i}, \mathcal{P}_{R_i})$ for the two adjacent windows L_i and R_i for position i, then it takes $O(1)$ time to determine whether or not $\mathit{diff}(\mathcal{P}_{L_{i+1}}, \mathcal{P}_{R_{i+1}}) = 0$ for the two adjacent windows L_{i+1} and R_{i+1} for the next position $i + 1$. Hence, for each d, it takes $O(n)$ time to find all Abelian squares of length $2d$, and thus it takes a total of $O(n^2)$ time for all $1 \leq d \leq \lfloor \frac{n}{2} \rfloor$.

4.2   Our $O(mn)$-Time Algorithm

We propose an algorithm which computes all Abelian squares in a given string w of length n in $O(mn)$ time, where m is the size of $RLE(w)$.

Our algorithm will output consecutive Abelian squares $w[i..i+2d-1]$, $w[i+1..i+2d]$, ..., $w[j..j+2d-1]$ of length $2d$ each as a triple $\langle i, j, d \rangle$. A single Abelian square $w[i..i+2d-1]$ of length $2d$ will be represented by $\langle i, i, d \rangle$.

For any position i in w, let $beg(L_i)$ and $end(L_i)$ respectively denote the beginning and ending positions of the left window L_i, and let $beg(R_i)$ and $end(R_i)$ respectively denote the beginning and ending positions of the right window R_i. Namely, $beg(L_i) = i$, $end(L_i) = i + d - 1$, $beg(R_i) = i + d$, and $end(R_i) = i+2d-1$. Cummings and Smyth's algorithm described above increases each of $beg(L_i)$, $end(L_i)$, $beg(R_i)$, and $end(R_i)$ one by one, and tests all positions $i = 1, \ldots, n - 2d + 1$ in w. Hence their algorithm takes $O(n)$ time for each window size d.

In what follows, we show that it is indeed enough to check only $O(m)$ positions in w for each window size d. The outline of our algorithm is as follows. As Cummings and Smyth's algorithm, we use two adjacent windows of size d, and slide the windows. However, unlike Cummings and Smyth's algorithm where the windows are shifted by one position, in our algorithm the windows can be shifted by more than one position. The positions that are not skipped and are explicitly examined will be characterized by the RLE of w, and the equivalence of the Parikh vectors of the two adjacent windows for the skipped positions can easily be checked by simple arithmetics.

Now we describe our algorithm in detail. First, we compute $RLE(w)$ and let m be its size. Consider an arbitrarily fixed window length $d \geq 1$.

Initial Step for Position 1. Initially, we compute $\mathcal{P}_{L_1}$ and $\mathcal{P}_{R_1}$ for position 1. We can compute these Parikh vectors in $O(m)$ time and $O(\sigma)$ space using the same method as in the algorithm of Theorem 1 in Sect. 3.

Steps for Positions Larger Than 1. For each position $i \geq 1$ in a given string w, let $D_1^i = succ(beg(L_i)) - beg(L_i)$, $D_2^i = succ(beg(R_i)) - beg(R_i)$, and $D_3^i = succ(end(R_i) + 1) - end(R_i) - 1$. The *break point* for each position i, denoted bp(i), is defined by $i + \min\{D_1^i, D_2^i, D_3^i\}$. Assume the left window is aligned at position i in w. Then, we jump to the break point bp(i) directly from i. In other words, the two windows L_i and R_i are directly shifted to $L_{\mathrm{bp}(i)}$ and $R_{\mathrm{bp}(i)}$, respectively.

It depends on the value of $diff(\mathcal{P}_{L_i}, \mathcal{P}_{R_i})$ whether there can be an Abelian square between positions i and bp(i). Note that $diff(\mathcal{P}_{L_i}, \mathcal{P}_{R_i}) \neq 1$. Below, we characterize the other cases in detail.

Lemma 1. *Assume $diff(\mathcal{P}_{L_i}, \mathcal{P}_{R_i}) = 0$. Then, for any $i < j \leq$ bp(i), j is the beginning position of an Abelian square of length $2d$ iff $w[beg(L_i)] = w[beg(R_i)] = w[end(R_i) + 1]$.*

Lemma 2. *Assume $diff(\mathcal{P}_{L_i}, \mathcal{P}_{R_i}) = 2$. Let c_p be the unique character which occurs more in the left window L_i than in the right window R_i, and c_q be the unique character which occurs more in the right window R_i than in the left window L_i. Let $x = \mathcal{P}_{L_i}[p] - \mathcal{P}_{R_i}[p] = \mathcal{P}_{R_i}[q] - \mathcal{P}_{L_i}[q] > 0$, and assume $x \leq \min\{D_1^i, D_2^i, D_3^i\}$. Then, $i + x$ is the beginning position of an Abelian square of length $2d$ iff $w[beg(L_i)] = c_p$, $w[beg(R_i)] = c_q = w[end(R_i) + 1]$. Also, this is the only Abelian square of length $2d$ beginning at positions between i and* bp(i).

Lemma 3. *Assume $diff(\mathcal{P}_{L_i}, \mathcal{P}_{R_i}) = 2$. Let c_p be the unique character which occurs more in the left window L_i than in the right window R_i, and c_q be the unique character which occurs more in the right window R_i than in the left window L_i. Let $x = \mathcal{P}_{L_i}[p] - \mathcal{P}_{R_i}[p] = \mathcal{P}_{R_i}[q] - \mathcal{P}_{L_i}[q] > 0$, and assume $\frac{x}{2} \leq \min\{D_1^i, D_2^i, D_3^i\}$. Then, $i + \frac{x}{2}$ is the beginning position of an Abelian square of length $2d$ iff $w[beg(L_i)] = c_p = w[end(R_i) + 1]$, $w[beg(R_i)] = c_q$. Also, this is the only Abelian square of length $2d$ beginning at positions between i and* bp(i).

Lemma 4. *Assume $diff(\mathcal{P}_{L_i}, \mathcal{P}_{R_i}) = 3$. Let $c_p = w[beg(L_i)]$, $c_{p'} = w[end(R_i) + 1]$, and $c_q = w[beg(R_i)]$. Then, $i + x$ with $i < i + x \leq$ bp(i) is the beginning position of an Abelian square of length $2d$ iff $0 < x = \mathcal{P}_{L_i}[p] - \mathcal{P}_{R_i}[p] = \mathcal{P}_{L_i}[p'] - \mathcal{P}_{R_i}[p'] = \frac{\mathcal{P}_{R_i}[q] - \mathcal{P}_{L_i}[q]}{2} \leq \min\{D_1^i, D_2^i, D_3^i\}$. Also, this is the only Abelian square of length $2d$ beginning at positions between i and* bp(i).

Lemma 5. *Assume $diff(\mathcal{P}_{L_i}, \mathcal{P}_{R_i}) \geq 4$. Then, there exists no Abelian square of length $2d$ beginning at any position j with $i < j \leq$ bp(i).*

Main Result. We are ready to show the main result of this section.

Theorem 2. *Given a string w of the length n over an alphabet of size σ, we can compute all Abelian squares in w in $O(mn)$ time and $O(n)$ working space, where m is the size of $RLE(w)$.*

Proof. Consider an arbitrarily fixed window length d. As was explained, it takes $O(m)$ time to compute $\mathcal{P}_{L_1}$, $\mathcal{P}_{R_1}$, and $\textit{diff}(\mathcal{P}_{L_1}, \mathcal{P}_{R_1})$ for the initial position 1. Suppose that the two windows are aligned at some position $i \geq 1$. Then, our algorithm computes Abelian squares starting at positions between i and $\text{bp}(i)$ using one of Lemmas 1, 2, 3, 4, and 5, depending on the value of $\textit{diff}(\mathcal{P}_{L_1}, \mathcal{P}_{R_i})$. In each case, all Abelian squares of length $2d$ starting at positions between i and $\text{bp}(i)$ can be computed in $O(1)$ time by simple arithmetics. Then, the left and right windows L_i and R_i are shifted to $L_{\text{bp}(i)}$ and $R_{\text{bp}(i)}$, respectively. Using the array S as in Theorem 1, we can compute $\text{bp}(i)$ in $O(1)$ time for a given position i in w.

Let us analyze the number of times the windows are shifted for each d. Since $\text{bp}(i) = i + \min\{D_1^i, D_2^i, D_3^i\}$, for each position p there can be at most three distinct positions i, j, k such that $p = \text{bp}(i) = \text{bp}(j) = \text{bp}(k)$. Thus, for each d we shift the two adjacent windows at most $3m$ times.

Overall, our algorithm runs in $O(mn)$ time for all window lengths $d = 1, \ldots, \lfloor n/2 \rfloor$. The space requirement is $O(n)$ since we need to maintain the Parikh vectors of the two sliding windows and the array S.   □

Example on how our algorithm computes all Abelian squares using RLEs can be found in Appendix of this paper [13].

5 Computing Longest Common Abelian Factors Using RLEs

In this section, we introduce our RLE-based algorithm which computes longest common Abelian factors of two given strings w_1 and w_2. Formally, we solve the following problem. Let $n = \min\{|w_1|, |w_2|\}$. Given two strings w_1 and w_2, compute the length $l = \max\{d \in [1, n] \mid 1 \leq \exists i \leq |w_1|, 1 \leq \exists k \leq |w_2|$ s.t. $\mathcal{P}_{w_1[i..i+d-1]} = \mathcal{P}_{w_2[k..k+d-1]}\}$ of the longest common Abelian factor(s) of w_1 and w_2, together with a pair (i, k) of positions on w_1 and w_2 such that $\mathcal{P}_{w_1[i..i+l-1]} = \mathcal{P}_{w_2[k..k+l-1]}$.

Our algorithm uses an idea from Alatabbi et al.'s algorithm [1]. For each window size d, their algorithm computes the Parikh vectors of all substrings of w_1 and w_2 of length d in $O(\sigma n)$ time, using two windows of length d each. Then they sort the Parikh vectors in $O(\sigma n)$ time, and output the largest d for which common Parikh vectors exist for w_1 and w_2, together with the lists of respective occurrences of longest common Abelian factors. The total time requirement is clearly $O(\sigma n^2)$.

Our algorithm is different from Alatabbi et al.'s algorithm in that (1) we use RLEs of strings w_1 and w_2 and (2) we avoid to sort the Parikh vectors. As in the previous sections, for a given window length d ($1 \leq n$), we shift two windows of length d over both $RLE(w_1)$ and $RLE(w_2)$, and stops when we reach a break point of $RLE(w_1)$ or $RLE(w_2)$. We then check if there is a common Abelian factor in the ranges of w_1 and w_2 we are looking at.

Since we use a single window for each of the input strings w_1 and w_2, we need to modify the definition of the break points. Let U_i and V_k be the sliding

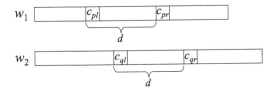

Fig. 2. Conceptual drawing of c_{p_l}, c_{p_r}, c_{q_r}, and c_{q_l}.

windows for w_1 and w_2 that are aligned at position i of w_1 and at position k of w_2, respectively. For each position $i \geq 1$ in w_1, let $\mathrm{bp}_1(i) = i + \min\{D_1^i, D_2^i\}$, where $D_1^i = succ(beg(U_i)) - i$ and $D_2^i = succ(end(U_i)) - i$. For each position $k \geq 1$ in w_2, $\mathrm{bp}_2(k)$ is defined analogously. Let $p_l = beg(U_i)$, $p_r = end(U_i) + 1$, $q_l = beg(V_k)$ and $q_r = end(V_k) + 1$.

Consider an arbitrarily fixed window length d. Assume that we have just shifted the window on w_1 from position i (i.e., $U_i = w_1[i..i + d - 1]$) to the break point $\mathrm{bp}_1(i)$ (i.e., $U_{\mathrm{bp}_1(i)} = w_1[\mathrm{bp}_1(i)..\mathrm{bp}_1(i) + d - 1]$). Let $c_{p_l} = w_1[i]$ and $c_{p_r} = w_1[i + d]$ (see also Fig. 2).

For characters c_{p_l} and c_{p_r}, we consider the minimum and maximum numbers of their occurrences during the slide from position i to $\mathrm{bp}_1(i)$. Let $min(p_l) = \mathcal{P}_{w_1[\mathrm{bp}_1(i)..\mathrm{bp}_1(i)+d-1]}[p_l]$, $max(p_l) = \mathcal{P}_{w_1[i..i+d-1]}[p_l]$, $min(p_r) = \mathcal{P}_{w_1[i..i+d-1]}[p_r]$ and $max(p_r) = \mathcal{P}_{w_1[\mathrm{bp}_1(i)..\mathrm{bp}_1(i)+d-1]}[p_r]$. We will use these values to determine if there is a common Abelian factor of length d for w_1 and w_2.

Also, assume that we have just shifted the window on w_2 from position k (i.e., $V_k = w_2[k..k + d - 1]$) to the break point $\mathrm{bp}_2(k)$ (i.e., $V_{\mathrm{bp}_2(k)} = w_2[\mathrm{bp}_2(k)..\mathrm{bp}_2(k) + d - 1]$). Let $c_{q_l} = w_2[k]$ and $c_{q_r} = w_2[k + d]$ (see also Fig. 2). For characters c_{q_l} and c_{q_r}, we also consider the minimum and maximum numbers of occurrences of of these characters during the slide from position k to $\mathrm{bp}_2(k)$. Let $min(q_l) = \mathcal{P}_{w_2[\mathrm{bp}_2(k)..\mathrm{bp}_2(k)+d-1]}[q_l]$, $max(q_l) = \mathcal{P}_{w_2[k..k+d-1]}[q_l]$, $min(q_r) = \mathcal{P}_{w_2[k..k+d-1]}[q_r]$ and $max(q_r) = \mathcal{P}_{w_2[\mathrm{bp}_2(k)..\mathrm{bp}_2(k)+d-1]}[q_r]$.

Let m be the total size of $RLE(w_1)$ and $RLE(w_2)$, and l be the length of longest common Abelian factors of w_1 and w_2. Our algorithm computes an $O(m^2)$-size representation of every pair (i, k) of positions for which $(w_1[i..i+l-1], w_2[k..k + l - 1])$ is a longest common Abelian factor of w_1 and w_2.

In the lemmas which follow, we assume that $\mathcal{P}_{w_1[i..i+d-1]}[v] = \mathcal{P}_{w_2[k..k+d-1]}[v]$ for any $v \in \{1, .., \sigma\} \setminus \{p_l, p_r, q_l, q_r\}$. This is because, if this condition is not satisfied, then there cannot be an Abelian common factor of length d for positions between i to $\mathrm{bp}_1(i)$ in w_1 and position between k to $\mathrm{bp}_2(k)$ in w_2.

Lemma 6. *Assume $c_{p_l} = c_{p_r}$ and $c_{q_l} = c_{q_r}$. Then, for any pair of positions $i \leq i' \leq \mathrm{bp}_1(i)$ and $k \leq k' \leq \mathrm{bp}_2(k)$, $(w_1[i'..i' + d - 1], w_2[k'..k' + d - 1])$ is an Abelian common factor of length d iff $\mathcal{P}_{w_1[i..i+d-1]} = \mathcal{P}_{w_2[k..k+d-1]}$.*

Proof. Since $c_{p_l} = c_{p_r}$ and $c_{q_l} = c_{q_r}$, the Parikh vectors of the sliding windows do not change during the slides from i to $\mathrm{bp}_1(i)$ and from k to $\mathrm{bp}_2(k)$. Thus the lemma holds. $\square$

Lemma 7. *Assume $c_{p_l} = c_{q_l} \neq c_{p_r} = c_{q_r}$. There is a common Abelian common factor $(w_1[i+x..i+x+d-1], w_2[k+y..k+y+d-1])$ of length d iff $0 \leq x \leq \mathrm{bp}_1(i) - i$, $0 \leq y \leq \mathrm{bp}_2(k) - k$ and $x - y = max(p_l) - max(q_l) = min(q_r) - min(p_r)$.*

Lemma 8. *Assume $c_{p_r} \neq c_{p_l} = c_{q_l} \neq c_{q_r}$ and $c_{p_r} \neq c_{q_r}$. There is a common Abelian factor $(w_1[i+x..i+x+d-1], w_2[k+y..k+y+d-1])$ of length diff $\mathrm{bp}_1(i) - i \geq x = \mathcal{P}_{w_2[k..k+d-1]}[p_r] - min(p_r) \geq 0$, $\mathrm{bp}_2(k) - k \geq y = \mathcal{P}_{w_1[i..i+d-1]}[q_r] - min(q_r) \geq 0$ and $\mathcal{P}_{w_1[i..i+d-1]}[p_l] - x = \mathcal{P}_{w_2[k..k+d-1]}[q_l] - y$.*

Lemma 9. *Assume $c_{p_l} \neq c_{p_r} = c_{q_r} \neq c_{q_l}$ and $c_{p_l} \neq c_{q_l}$. There is a common Abelian factor $(w_1[i+x..i+x+d-1], w_2[k+y..k+y+d-1])$ of length d iff $x = max(p_l) - \mathcal{P}_{w_2[k..k+d-1]}[p_l] \geq 0$, $y = max(q_l) - \mathcal{P}_{w_1[i..i+d-1]}[q_l] \geq 0$ and $\mathcal{P}_{w_1[i..i+d-1]}[p_r] + x = \mathcal{P}_{w_2[k..k+d-1]}[q_r] + y$.*

Lemma 10. *Assume $c_{p_l} = c_{q_r} \neq c_{p_r} = c_{q_l}$. There is a common Abelian factor $(w_1[i+x..i+x+d-1], w_2[k+y..k+y+d-1])$ of length d iff $x + y = min(p_r) - max(q_l) = max(q_l) - min(p_r)$, $0 \leq x \leq \mathrm{bp}_1(i) - i$ and $0 \leq y \leq \mathrm{bp}_2(k) - k$.*

Lemma 11. *Assume $c_{p_l}, c_{p_r}, c_{q_l}$ and c_{q_r} are mutually distinct. There is a common Abelian factor $(w_1[i+x..i+x+d-1], w_2[k+y..k+y+d-1])$ of length d iff $0 \leq x = max(p_l) - \mathcal{P}_{w_2[k..k+d-1]}[p_l] = \mathcal{P}_{w_2[k..k+d-1]}[p_r] - min(p_r) \leq \mathrm{bp}_1(i) - i$ and $0 \leq y = max(q_l) - \mathcal{P}_{w_1[i..i+d-1]}[q_l] = \mathcal{P}_{w_1[i..i+d-1]}[q_r] - min(q_r) \leq \mathrm{bp}_2(k) - k$.*

Lemma 12. *Assume $c_{q_l} \neq c_{p_l} = c_{p_r} \neq c_{q_r}$ and $c_{q_l} \neq c_{q_r}$. There is a common Abelian factor $(w_1[i+x..i+x+d-1], w_2[k+y..k+y+d-1])$ of length d iff $0 \leq x \leq \mathrm{bp}_1(i) - i$, $0 \leq y = max(q_l) - \mathcal{P}_{w_1[i..i+d-1]}[q_l] = \mathcal{P}_{w_1[i..i+d-1]}[q_r] - min(q_r) \leq \mathrm{bp}_2(k) - k$ and $\mathcal{P}_{w_1[i..i+d-1]}[p_l] = \mathcal{P}_{w_2[k..k+d-1]}[p_l]$.*

Lemma 13. *Assume $c_{p_l} \neq c_{q_l} = c_{q_r} \neq c_{p_r}$ and $c_{p_l} \neq c_{p_r}$. There is a common Abelian factor $(w_1[i+x..i+x+d-1], w_2[k+y..k+y+d-1])$ of length d iff $0 \leq y \leq \mathrm{bp}_2(k) - k$ and $x = max(p_l) - \mathcal{P}_{w_2[k..k+d-1]}[p_l] = \mathcal{P}_{w_2[k..k+d-1]}[p_r] - min(p_r) \geq 0$.*

Lemma 14. *Assume $c_{p_r} \neq c_{p_l} = c_{q_r} \neq c_{q_l}$ and $c_{p_r} \neq c_{q_l}$. There is a common Abelian factor $(w_1[i+x..i+x+d-1], w_2[k+y..k+y+d-1])$ of length d iff $0 \leq x = \mathcal{P}_{w_2[k..k+d-1]}[p_r] - min(p_r) \leq \mathrm{bp}_1(i) - i$, $0 \leq y = max(q_l) - \mathcal{P}_{w_1[i..i+d-1]}[q_l] \leq \mathrm{bp}_2(k) - k$ and $x + y = \mathcal{P}_{w_1[i..i+d-1]}[p_l] - \mathcal{P}_{w_2[k..k+d-1]}[q_r] = max(p_l) - min(q_r)$.*

Lemma 15. *Assume $c_{p_l} \neq c_{q_l} = c_{p_r} \neq c_{q_r}$ and $c_{p_l} \neq c_{q_r}$. There is a common Abelian factor $(w_1[i+x..i+x+d-1], w_2[k+y..k+y+d-1])$ of length d iff $0 \leq x = max(p_l) - \mathcal{P}_{w_2[k..k+d-1]}[p_l] \leq \mathrm{bp}_1(i) - i$, $0 \leq y = \mathcal{P}_{w_1[i..i+d-1]}[q_r] - min(q_r) \leq \mathrm{bp}_2(k) - k$ and $x + y = \mathcal{P}_{w_2[k..k+d-1]}[q_l] - \mathcal{P}_{w_1[i..i+d-1]}[p_r] = max(q_l) - min(p_r)$.*

Theorem 3. *Given two strings w_1 and w_2, we can compute an $O(m^2)$-size representation of all longest common Abelian factors of w_1 and w_2 in $O(m^2 n)$ time with $O(\sigma)$ working space, where m and n are the total size of the RLEs and the total length of w_1 and w_2, respectively.*

Proof. The correctness follows from Lemmas 6–15.

Let m_1, m_2 be the sizes of $RLE(w_1)$ and $RLE(w_2)$, respectively. Let $n_{\min} = \min\{|w_1|, |w_2|\}$. For each fixed window size d, the window for w_1 shifts over w_1 $O(m_1)$ times. For each shift of the window for w_1, the window for w_2 shifts over w_2 $O(m_2)$ times. Thus, we have $O(m_1 \cdot m_2 \cdot n_{\min})$ total shifts. Since all the conditions in Lemmas 6–15 can be tested in $O(1)$ time each by simple arithmetic, the total time complexity is $O(m_1 m_2 n_{\min} + n)$, where the n term denotes the cost to compute $RLE(w_1)$ and $RLE(w_2)$. Thus, it is clearly bounded by $O(m^2 n)$. Next, we focus on the output size. Let l be the length of the longest common Abelian factors of w_1 and w_2. Using Lemmas 7–15, for each pair of the shifts of the two windows we can compute an $O(1)$-size representation of the longest common Abelian factors found. Since there are $O(m_1 \cdot m_2)$ shifts for window length l, the output size is bounded by $O(m^2)$. The working space is $O(\sigma)$, since we only need to maintain two Parikh vectors for the two sliding windows. □

Examples. We show an example of how our algorithm computes a common Abelian factor of length 4 for two input strings $w_1 = aaaaacbbbcc$ and $w_2 = cccaaccbbbb$.

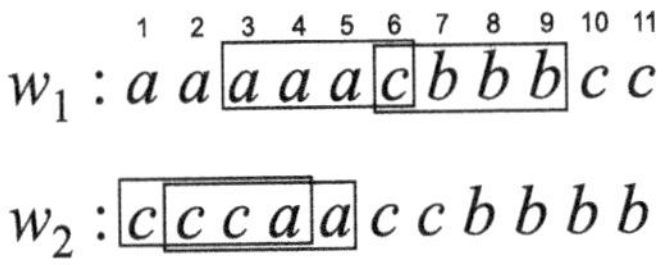

Fig. 3. Showing two sliding windows of length $d = 4$, where $i = 3$, $\mathrm{bp}_1(i) = 6$, $k = 1$, $\mathrm{bp}_2(k) = 2$, $c_{p_l} = a$, $c_{p_r} = b$, $c_{q_l} = c$, $c_{q_r} = a$.

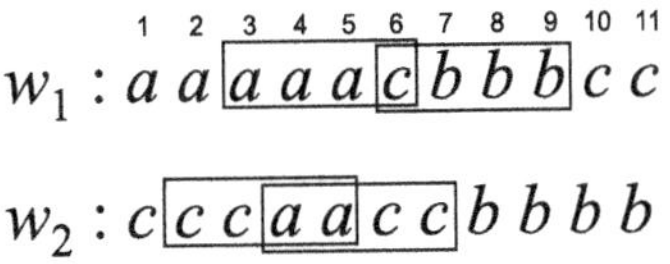

Fig. 4. Showing two sliding windows of length $d = 4$, where $i = 3$, $\mathrm{bp}_1(i) = 6$, $k = 2$, $\mathrm{bp}_2(k) = 4$, $c_{p_l} = a$, $c_{p_r} = b$, $c_{q_l} = c$, $c_{q_r} = c$.

Suppose that the window for w_1 is now aligned at position 3 of w_1 ($U_3 = w_1[3..6]$). We then shift it to position $\mathrm{bp}_1(3) = 6$ ($U_6 = w_1[6..9]$). For this shift of the window on w_1, we test $O(m_2)$ shifts of the window over the second string w_2, as follows. We begin with position 1 of the other string w_2 ($V_1 = w_2[1..4]$), and shift the window to position $\mathrm{bp}_2(1) = 2$. See also Fig. 3. It follows from Lemma 14 that there is no common Abelian factor during these slides. Next, the window for w_2 is shifted from position 2 to position $\mathrm{bp}_2(2) = 4$ ($V_4 = w_2[4..7]$). See also Fig. 4. It follows from Lemma 13 that there is no common Abelian factor during the slides.Next, the window for w_2 is shifted from position 4 to position $\mathrm{bp}_2(4) = 6$ ($V_6 = w_2[6..9]$). See also Fig. 5. Since the numbers of occurrences of c on w_1 and w_2 are different and c is not equal to a or b, there is no common Abelian factor during the slides. Next, the window for w_2 is shifted from position 6 to position $\mathrm{bp}_2(6) = 8$. See Fig. 5. It follows from Lemma 9 that there is a common Abelian factor ($w_1[6..9], w_2[7..10]$) of length $d = 4$ (Fig. 6).

$$w_1 : a\ a\ \boxed{a\ a\ a\ \boxed{c}\ b\ b\ b}\ c\ c$$

$$w_2 : c\ c\ c\ \boxed{a\ a\ \boxed{c\ c}\ b\ b}\ b\ b$$

Fig. 5. Showing two sliding windows of length $d = 4$, where $i = 3$, $\mathrm{bp}_1(i) = 3$, $k = 4$, $\mathrm{bp}_2(k) = 6$, $c_{p_l} = a$, $c_{p_r} = b$, $c_{q_l} = a$, $c_{q_r} = b$.

$$w_1 : a\ a\ \boxed{a\ a\ a\ \boxed{c}\ b\ b\ b}\ c\ c$$

$$w_2 : c\ c\ c\ a\ a\ \boxed{c\ c\ \boxed{b\ b}\ b\ b}$$

Fig. 6. Showing two sliding windows of length $d = 4$, where $i = 3$, $\mathrm{bp}_1(i) = 3$, $k = 6$, $\mathrm{bp}_2(k) = 3$, $c_{p_l} = a$, $c_{p_r} = b$, $c_{q_l} = c$, $c_{q_r} = b$.

References

1. Alatabbi, A., Iliopoulos, C.S., Langiu, A., Rahman, M.S.: Algorithms for longest common Abelian factors. Int. J. Found. Comput. Sci. **27**(5), 529–544 (2016). http://dx.doi.org/10.1142/S0129054116500143
2. Amir, A., Apostolico, A., Hirst, T., Landau, G.M., Lewenstein, N., Rozenberg, L.: Algorithms for jumbled indexing, jumbled border and jumbled square on run-length encoded strings. In: Moura, E., Crochemore, M. (eds.) SPIRE 2014. LNCS, vol. 8799, pp. 45–51. Springer, Cham (2014). https://doi.org/10.1007/978-3-319-11918-2_5
3. Badkobeh, G., Gagie, T., Grabowski, S., Nakashima, Y., Puglisi, S.J., Sugimoto, S.: Longest common Abelian factors and large alphabets. In: Inenaga, S., Sadakane, K., Sakai, T. (eds.) SPIRE 2016. LNCS, vol. 9954, pp. 254–259. Springer, Cham (2016). https://doi.org/10.1007/978-3-319-46049-9_24
4. Constantinescu, S., Ilie, L.: Fine and Wilf's theorem for Abelian periods. Bull. EATCS **89**, 167–170 (2006)
5. Crochemore, M., Iliopoulos, C.S., Kociumaka, T., Kubica, M., Pachocki, J., Radoszewski, J., Rytter, W., Tyczyński, W., Waleń, T.: A note on efficient computation of all Abelian periods in a string. Inf. Process. Lett. **113**(3), 74–77 (2013)
6. Cummings, L.J., Smyth, W.F.: Weak repetitions in strings. J. Comb. Math. Comb. Comput. **24**, 33–48 (1997)
7. Erdős, P.: Some unsolved problems. Hungarian Academy of Sciences Mat. Kutató Intézet Közl **6**, 221–254 (1961)
8. Fici, G., Lecroq, T., Lefebvre, A., Prieur-Gaston, É.: Algorithms for computing Abelian periods of words. Discrete Appl. Math. **163**, 287–297 (2014)
9. Fici, G., Lecroq, T., Lefebvre, A., Prieur-Gaston, É., Smyth, W.F.: A note on easy and efficient computation of full Abelian periods of a word. Discrete Appl. Math. **212**, 88–95 (2016)
10. Grabowski, S.: A note on the longest common Abelian factor problem. CoRR abs/1503.01093 (2015)
11. Kociumaka, T., Radoszewski, J., Rytter, W.: Fast algorithms for Abelian periods in words and greatest common divisor queries. In: STACS 2013, pp. 245–256 (2013)
12. Kociumaka, T., Radoszewski, J., Wiśniewski, B.: Subquadratic-time algorithms for Abelian stringology problems. In: Kotsireas, I.S., Rump, S.M., Yap, C.K. (eds.) MACIS 2015. LNCS, vol. 9582, pp. 320–334. Springer, Cham (2016). https://doi.org/10.1007/978-3-319-32859-1_27
13. Sugimoto, S., Noda, N., Inenaga, S., Bannai, H., Takeda, M.: Computing Abelian regularities on RLE strings. CoRR abs/1701.02836 (2017). http://arxiv.org/abs/1701.02836

Author Index

Printed in the United States
By Bookmasters